Achieving sustainable cultivation of oil palm

Volume 2: Diseases, pests, quality and sustainability

It is widely recognised that agriculture is a significant contributor to global warming and climate change. Agriculture needs to reduce its environmental impact and adapt to current climate change whilst still feeding a growing population, i.e. become more 'climate-smart'. Burleigh Dodds Science Publishing is playing its part in achieving this by bringing together key research on making the production of the world's most important crops and livestock products more sustainable.

Based on extensive research, our publications specifically target the challenge of climate-smart agriculture. In this way we are using 'smart publishing' to help achieve climate-smart agriculture.

Burleigh Dodds Science Publishing is an independent and innovative publisher delivering high quality customer-focused agricultural science content in both print and online formats for the academic and research communities. Our aim is to build a foundation of knowledge on which researchers can build to meet the challenge of climate-smart agriculture.

For more information about Burleigh Dodds Science Publishing simply call us on +44 (0) 1223 839365, email info@bdspublishing.com or alternatively please visit our website at www.bdspublishing.com.

Related titles:

Achieving sustainable cultivation of oil palm Volume 1: Introduction, breeding and cultivation techniques
Print (ISBN 978-1-78676-104-0); Online (ISBN 978-1-78676-106-4, 978-1-78676-107-1)

Achieving sustainable cultivation of soybeans Volume 1: Breeding and cultivation techniques
Print (ISBN 978-1-78676-112-5); Online (ISBN 978-1-78676-114-9, 978-1-78676-115-6)

Achieving sustainable cultivation of soybeans Volume 2: Diseases, pests, food and other uses
Print (ISBN 978-1-78676-116-3); Online (ISBN 978-1-78676-119-4, 978-1-78676-118-7)

Managing soil health for sustainable agriculture Volume 1: Fundamentals
Print (ISBN 978-1-78676-188-0); Online (ISBN 978-1-78676-191-0, 978-1-78676-190-3)

Managing soil health for sustainable agriculture Volume 2: Monitoring and management
Print (ISBN 978-1-78676-192-7); Online (ISBN 978-1-78676-194-1, 978-1-78676-195-8)

Precision agriculture for sustainability
Print (ISBN 978-1-78676-204-7); Online (ISBN 978-1-78676-206-1, 978-1-78676-207-8)

Chapters are available individually from our online bookshop: https://shop.bdspublishing.com

climate•SMART•publishing
IN AGRICULTURAL SCIENCE

BURLEIGH DODDS SERIES IN AGRICULTURAL SCIENCE

NUMBER 28

Achieving sustainable cultivation of oil palm

Volume 2: Diseases, pests, quality and sustainability

Edited by Professor Alain Rival, Center for International Cooperation in Agricultural Research for Development (CIRAD), France

burleigh dodds
SCIENCE PUBLISHING

Published by Burleigh Dodds Science Publishing Limited
82 High Street, Sawston, Cambridge CB22 3HJ, UK
www.bdspublishing.com

Burleigh Dodds Science Publishing, 1518 Walnut Street, Suite 900, Philadelphia, PA 19102-3406, USA

First published 2018 by Burleigh Dodds Science Publishing Limited

Library of Congress Control Number: 2017956485

British Library Cataloguing in Publication Data
A catalogue record for this book is available from the British Library

ISBN 978-1-78676-108-8 (print)
ISBN 978-1-78676-111-8 (online)
ISBN 978-1-78676-110-1 (online)
ISSN 2059-6936 (print)
ISSN 2059-6944 (online)

Typeset by Deanta Global Publishing Services, Chennai, India

Contents

Part 2 Nutritional and sensory quality

Part 3 Sustainability and supporting smallholders

Series list

Title	Series number
Achieving sustainable cultivation of maize - Vol 1 From improved varieties to local applications Edited by: Dr Dave Watson, CGIAR Maize Research Program Manager, CIMMYT, Mexico	001
Achieving sustainable cultivation of maize - Vol 2 Cultivation techniques, pest and disease control Edited by: Dr Dave Watson, CGIAR Maize Research Program Manager, CIMMYT, Mexico	002
Achieving sustainable cultivation of rice - Vol 1 Breeding for higher yield and quality Edited by: Prof. Takuji Sasaki, Tokyo University of Agriculture, Japan	003
Achieving sustainable cultivation of rice - Vol 2 Cultivation, pest and disease management Edited by: Prof. Takuji Sasaki, Tokyo University of Agriculture, Japan	004
Achieving sustainable cultivation of wheat - Vol 1 Breeding, quality traits, pests and diseases Edited by: Prof. Peter Langridge, The University of Adelaide, Australia	005
Achieving sustainable cultivation of wheat - Vol 2 Cultivation techniques Edited by: Prof. Peter Langridge, The University of Adelaide, Australia	006
Achieving sustainable cultivation of tomatoes Edited by: Dr Autar Mattoo, USDA-ARS, USA & Prof. Avtar Handa, Purdue University, USA	007
Achieving sustainable production of milk - Vol 1 Milk composition, genetics and breeding Edited by: Dr Nico van Belzen, International Dairy Federation (IDF), Belgium	008
Achieving sustainable production of milk - Vol 2 Safety, quality and sustainability Edited by: Dr Nico van Belzen, International Dairy Federation (IDF), Belgium	009
Achieving sustainable production of milk - Vol 3 Dairy herd management and welfare Edited by: Prof. John Webster, University of Bristol, UK	010
Ensuring safety and quality in the production of beef - Vol 1 Safety Edited by: Prof. Gary Acuff, Texas A&M University, USA & Prof.James Dickson, Iowa State University, USA	011
Ensuring safety and quality in the production of beef - Vol 2 Quality Edited by: Prof. Michael Dikeman, Kansas State University, USA	012
Achieving sustainable production of poultry meat - Vol 1 Safety, quality and sustainability Edited by: Prof. Steven C. Ricke, University of Arkansas, USA	013
Achieving sustainable production of poultry meat - Vol 2 Breeding and nutrition Edited by: Prof. Todd Applegate, University of Georgia, USA	014
Achieving sustainable production of poultry meat - Vol 3 Health and welfare Edited by: Prof. Todd Applegate, University of Georgia, USA	015

Introduction

Oil palm is widely cultivated in tropical countries for use in food and feed, personal care products and other applications such as biodiesel. Cultivation faces a range of challenges, such as its environmental impact (deforestation and biodiversity loss), threats from pests and diseases, and the need to better support smallholders. There is thus an urgent need to make oil palm cultivation more efficient as well as environmentally and socially sustainable.

These challenges are addressed in the two volumes of *Achieving sustainable cultivation of oil palm*:

- Volume 1: Introduction, breeding and cultivation techniques
- Volume 2: Diseases, pests, quality and sustainability

Volume 1 covers breeding and cultivation techniques, while Volume 2 (this volume) reviews advances in understanding and managing fungal and other diseases affecting oil palm as well as insect pests. It includes integrated management tools in pest and disease control and the generation of disease-resistant oil palm varieties. It also discusses the latest research on palm oil and human health, addressing debates about the nutritional value and health effects of palm oil. The key issues of sustainability and conservation are also examined. Chapters cover the monitoring of the environmental impacts of oil palm cultivation, including life cycle analysis, sustainability certification, conservation, waste management and recycling. Finally, the volume looks at how best to support smallholders in attaining sustainable oil palm production in practice.

Part 1 Diseases and pests

The first part of the volume reviews the latest research in understanding and managing fungal and other diseases affecting oil palm, such as basal stem rot (BSR), vascular wilt and bud rot, as well as insect pests.

Chapter 1 focuses on fungal disease affecting oil palm. The chapter concentrates on one agent of basal stem rot, *Ganoderma boninense*, but it also addresses *Fusarium* wilt and bud rot disease. It examines the symptoms and sources of *Ganoderma* disease and infection, the latest research on resistant materials, including laccase gene discoveries, and techniques for developing resistance against BSR in plant material. It also reviews future trends in research in this field.

Chapter 2 then looks at the exotic and emerging diseases affecting oil palm production, especially in West Africa, Southeast Asia, and South and Central America. It describes these key diseases, including basal stem rot, dry basal rot disease (*Thielaviopsis paradoxa*), bud rot disease (*Phytophthora palmivora*) and vascular wilt disease (*Fusarium oxysporum* f.sp. *elaeidis*). It also discusses phytoplasma diseases in South and Central America, such as lethal wilt disease (LWD) and leaf stripe disease (LSD). In the case of each disease, the chapter outlines its biology, symptoms and effects on yield, as well as its epidemiology. The chapter also highlights the methods that are currently being applied to control them.

In Chapter 3 attention turns to the insect pests affecting oil palms. Monocultures such as oil palm are known to face stronger parasitic pressure. *Coleoptera* and *Lepidoptera* are the main insect pests affecting the oil palm on all continents where it is cultivated. This chapter offers an overview of the major types of oil palm pests (23 species), which have been specifically selected to illustrate the relationships found between the oil palm and the most damaging groups of herbivorous arthropods. It concentrates on the major oil palm insect pests found in oil palm plantations in Latin America, West Africa, Southeast Asia and the Pacific, and assesses future research trends in the study of oil palm pests.

Continuing on the theme of pests in oil palm, Chapter 4 examines Integrated Pest Management (IPM) protocols which are central to sustainable palm oil production. This chapter explores the rapid expansion of oil palm and its impact on the environment, the range of pest species found in plantations, and the impact of replanting on pest numbers. The chapter introduces the concept and history of IPM in oil palm as well as the diverse range of approaches which form IPM strategies, including plant breeding, targeted chemical applications, management to reduce pest numbers and transmission, and management to increase the numbers of natural enemies and pathogens of pests. The chapter concludes by considering how approaches that are focused towards more diverse oil palm landscapes and more diverse pest control species assemblages can increase the effectiveness of IPM.

Chapter 5 continues to examine integrated management techniques but returns to the subject of diseases in oil palm. It considers the integrated management of bud rot disease and *Phytophthora palmivora* in oil palm. This chapter provides a general overview of the history of research on oil palm bud rot disease, which is believed to constitute the most important limiting factor for palm oil production in Central and South America. It presents the different approaches adopted in identifying the causative agent of the disease and its management. The chapter describes recent research work undertaken by the Colombian Oil Palm Research Centre (Cenipalma) which is aimed at providing evidence that *Phytophthora palmivora* can be considered as the causative agent of bud rot, as well as suggesting strategies for controlling its spread throughout oil palm stands. Such a historical overview is necessary in order to understand the importance of the disease, and to show how knowledge regarding its biological determinants is required for developing a successful integrated pest management programme. This will be a prerequisite go any further development of sustainable palm oil production in Central and Latin America.

In the last chapter in Part 1, attention turns to disease-resistance in oil palm varieties. Chapter 6 describes advances in the breeding of disease-resistant varieties, presenting the key issues associated with oil palm disease-resistance, including the usefulness and sustainability of resistance, simple modelling of disease development and the agricultural practices and screening required to enable it. The chapter focuses on the achievement of improving genetic resistance to three major diseases of oil palm: *Fusarium* wilt, *Ganoderma* basal/upper stem rot, and bud rot. The chapter ends by looking ahead to potential future developments in the field of oil palm disease-resistance.

Part 2 Nutritional and sensory quality

There is a great deal of confusion regarding the nutritional value and health effects of palm oil. The second part of this volume therefore focuses on the health and nutritional impacts of oil palm.

Chapter 7 looks at bioactive compounds in oil palm. There is growing evidence for the protective effect of plant bioactives against diseases, and this has led to interest from the functional food and nutraceutical industries for the development of bioactive-enriched food and supplements. The oil palm fruit is a source of both lipid and water-soluble bioactives. This chapter begins by considering lipid-soluble bioactives and their potential health benefits. It then considers water-soluble bioactives which have only recently received attention. These include phenolic compounds and shikimic acid, which show potential health benefits against a whole spectrum of diseases, including cardiovascular disease, cancer and diabetes. These water-soluble compounds enter the aqueous stream (vegetation liquor) and are usually discarded as palm oil mill effluent. The valorisation of palm oil milling waste for the production of functional food and nutraceuticals offers an opportunity for increased productivity and sustainability of the oil palm industry. The chapter ends with a look at future trends and developments.

Continuing the theme of the previous chapter, Chapter 8 reviews the effects of palm oil consumption on human health, including its effects on cardiovascular risk markers such as fasting lipids and lipoproteins and post-prandial plasma lipids and lipoproteins. The evidence for a connection between cardiovascular problems and the consumption of saturated fatty acids more generally is also considered. While palm oil is often criticised for its supposed effects on human health, due to its high saturated fatty acid content, it is this high saturated fatty acid content (50%) which makes it a solid oil, and therefore a good technological alternative to partially hydrogenated trans fatty acids. Palm oil has therefore enabled a reduction in the intake of trans fatty acids. This chapter argues that there is a place for palm oil in a healthy, balanced diet.

Part 2 closes with a specific case study. Chapter 9 investigates the nutritional value of red palm oil. It looks at the controversy and conflicting views that still persist on whether or not palm oil is atherogenic. The primary focus of this chapter is on the nutritional value of crude (red) palm oil, specifically as a source of vitamin A. The chapter includes a detailed case study on the use of red palm oil in Burkino Faso, specifically the effects on the vitamin A status of women and children. It also looks at the use of red palm oil in food biofortification as well as its effects on women's income and empowerment. It also looks ahead to future trends for research in this area.

Part 3 Sustainability and supporting smallholders

The third part of this volume looks at the environmental impacts of and sustainability of palm oil cultivation, and how best to support smallholders in oil palm production. Chapters 10 and 11 begin by looking at the life cycle assessment (LCA) of palm oil. LCA is a tool for evaluating the environmental impact of a product or process throughout its entire life cycle.

Chapter 10 looks at the methods and applications of LCA of palm oil. Oil palm impacts on the environment both during cultivation and as a result of land use change for new plantations. It is therefore crucial to have adapted models and tools that allow the best oil palm cultivation practices to be identified in order to reduce the environmental impacts of cultivation. Chapter 10 describes the principles and methodology of LCA of oil palm products. It reviews the results from published LCA and greenhouse gas (GHG) assessments on palm oil products, including the environmental impact of oil-based bioenergy and of oil palm fruits and palm oil. Finally, it discusses the information yielded by existing studies on the environmental impacts of palm oil and outlines the remaining challenges regarding LCA development and applications for palm oil products.

Chapter 11 discusses the practical application of LCA, taking Malaysia as an example. The chapter charts the environmental performance of the production of crude palm oil (CPO) and suggests measures to mitigate its environmental impact. Case studies examine the environmental impact of different scenarios, including land use change from logged-over forest to oil palm production, and assesses the impact of biogas emissions from palm oil mills. The chapter argues that biogas capture facilities and methane avoidance are positive steps towards mitigating the environmental impact of the oil palm industry. It closes with a look at future trends.

Chapter 12 moves on to discuss approaches to modelling the environmental impacts of oil palm agriculture. The cultivation of crops affects the environment via flows of energy and materials. Impacts are felt in the atmosphere, hydrosphere, surrounding terrestrial ecosystems and the field itself. Models are useful tools for improving our understanding of the processes and predicting how they might be affected by changes in management. Current models range from simple indicators of risk or impact, based on empirical relationships, to dynamic process-based models. Increasingly complex and comprehensive models with greater spatial and temporal resolution and extent are being developed, mostly by coupling diverse sub-models. This chapter reviews the range of models developed for oil palm systems, and discusses how other existing models might be adapted for oil palm.

Building on the previous chapters, Chapter 13 discusses sustainability certification in oil palm cultivation. In recent years, transnational private regulations, aimed at implementing sustainable development principles, have emerged across several commodity sectors. To address the growing environmental and social concerns raised by oil palm expansion, the sector has adopted voluntary sustainability standards led by industry and civil society, as well as national standards and regulations implemented by the main producing countries. This chapter presents these sustainability initiatives as well as addressing some of their important limitations, due to the increasing complexity of the regulatory framework, the market segmentation associated with a growing demand for non-certified palm oil from emerging countries, and the unresolved issues of smallholders' inclusion in certification. The chapter suggests a way forward for sustainability certification, which includes strengthening global cooperation and including smallholders in the process, and looks ahead to future research in this area.

Chapter 14 looks at the issue of balancing oil palm cultivation with forest and biodiversity conservation. With the formation of the Roundtable for Sustainable Palm Oil, environmentalists and consumers anticipated a decrease in the indiscriminate destruction of tropical rainforests. Ten years later, thousands of hectares of tropical rainforests continue to be cleared for oil palm plantations in the tropical world and endangered species are being lost in the process. This chapter, based on six years of collaboration between Copenhagen Zoo and United Plantations Bhd in Central Kalimantan, Malaysia, describes how measuring, monitoring and managing the environmental impact of plantation operations are key components in sustainable palm oil production. The chapter examines the impacts of oil palm cultivation on biodiversity. The chapter recommends that dedicated companies develop environmental divisions at estate level in order to ensure that biodiversity concerns are integrated into standard operational procedures.

In Chapter 15 the discussion turns to waste management and recycling in oil palm cultivation. In a typical palm plantation, almost 70% of the fresh fruit bunches are turned into wastes in the form of empty fruit bunches, fibres and shells, as well as liquid effluent. Until recently, most of the wastes from palm oil mills were either burnt in the open or thrown away in waste ponds or open areas. This has contributed enormously to global climate change by emitting carbon dioxide and methane. This chapter reviews the technological advancements which now make it possible to convert palm oil waste products into useful energy or to recycle them into value-added products to generate additional profits for the industry.

In the remaining chapters of Part 3, smallholder producers become the focus. Chapter 16 begins this section by defining and categorising smallholders in oil palm cultivation. This chapter provides insights on oil palm smallholders in the Riau Province of Sumatra, Indonesia. The chapter characterises smallholders' production structures in terms of human, social and physical assets. The chapter looks ahead towards forming a typology of independent smallholders and ends with an examination of future trends in this area.

This section then moves on to looking at ways to improve cultivation practices. Chapter 17 looks at ways of closing yield gaps for small- and medium-scale oil palm producers. The proportion of small- and medium-scale growers in the Colombian oil palm sector has increased in recent years, both in terms of the land area planted and the number of growers. However, a yield gap has been found between producers, based on their size, with large-scale producers attaining average yields of 6.6 ton FFB per hectare per year, which is higher than those obtained by small- and medium-scale producers. One reason that could explain this yield gap is the failure to adopt technologies which could improve yields. This chapter makes use of a detailed case study of the strategies that were implemented by Cenipalma in Colombia to narrow the yield gap between large and small oil palm producers. These strategies included the use of organic matter on palm stands to improve soil conservation and nutrition, improved irrigation, developing institutional arrangements and adopting a technology transfer strategy. The chapter proposes future lines of research on further improving the efficiency of oil palm production at both large and small scales.

In the final chapter of Part 3, Chapter 18, the focus returns to artisanal mills and the local production of palm oil by smallholders, this time in Africa. In Africa, there are two oil palm supply chains. The first goes from industrial plantations and surrounding smallholdings to a formal market of commercial oils complying to international quality standards, while the second goes from smallholdings and 'wild groves' to an informal market of artisanal red palm oil. It is the second of these that is discussed in this chapter. The chapter reviews the emergence of artisanal extraction of red palm oil in Africa, considers who is involved in this supply chain and why, and discusses the major operations and equipment involved in artisanal processing, from harvesting, storage, threshing and steam sterilising to crushing, mixing, oil extraction, clarification and oil drying. It then reviews the composition and quality of artisanal palm oil and its various uses (in human health and for human consumption at home and in the food industry). Finally, it considers the sustainable development issues associated with artisanal red palm oil production.

Dedication

The two volumes of *Achieving sustainable cultivation of oil palm* are dedicated to the memory of our very esteemed colleague Hubert de Franqueville. Hubert was a renowned and brilliant scientist with an elegant and precise mind who was highly appreciated by the whole palm community. I would have loved sharing the editing of these two volumes with Hubert. We all miss his sharp understanding and exquisite humor.

Alain Rival
Editor

Part 1

Diseases and pests

Chapter 1

Fungal diseases affecting oil palm

Tan Joon Sheong, Lee Yang Ping and Sharifah Shahrul Rabiah Syed Alwee, FELDA Global Ventures Research and Development, Malaysia; Létizia Camus-Kulandaivelu, Maxime Mercière, Alba Zaremski and Frédéric Breton, CIRAD, France; and Christophe Klopp, INRA, France

1 Introduction

Oil palm has proved to be the most efficient and productive oil crop, giving far higher yields when compared to soybean, rapeseed and sunflower. As of October 2014, the production of palm oil was reported at 53.67 million tonnes, compared to soybean at 41.66 million tonnes, rapeseed at 24.48 million tonnes and sunflower at 14.8 million tonnes (www.oilpalmresearch.org). Based on the statistics obtained from the Malaysian Palm Oil Board in 2015, Malaysia's export of palm oil reached 17.31 million tonnes. The majority of Malaysian palm oil was exported to India (3.25 million tonnes), China (2.84 million tonnes) and the Netherlands (1.60 million tonnes). However, oil palm is also prone to attack by a number of fungal diseases such as basal stem rot (BSR), *Fusarium* wilt (vascular wilt) and bud rot disease. To date, only partial resistance materials have been identified for the BSR, and we will be providing more insights into the BSR that is causing a major problem on plantations particularly in Malaysia and Indonesia. Meanwhile, resistant planting materials were found for the latter two fungal diseases.

In Southeast Asia, one of the most devastating diseases at the moment is BSR. The disease was first described in 1915 in the Republic of Congo, West Africa (Wakefield, 1920). Thompson (1931) detected the disease infecting palms aged over 25 years in

http://dx.doi.org/10.19103/AS.2017.0018.13

Figure 1 The presence of *G. boninense* mature fruiting body (basidiomata) found on an oil palm tree showing basal stem rot disease.

Malaysia on old palms due for replanting. BSR disease, which is caused by the fungal species *Ganoderma boninense* (Fig. 1), is the most serious disease affecting the oil palm in Malaysia and Indonesia (Ariffin et al., 1989; Rao, 1990). The disease progresses slowly and eventually causes the death of infected palms. The disease can infect all stages of the oil palm, and palms infected at the early stage of their life cycle usually show no symptoms until after they are more than 12 years old (Lubis, 1992).

2 BSR caused by *Ganoderma boninense*

Ganodermataceae are cosmopolitan basidiomycetes which cause white rot of hardwoods such as oak, maple, maple, sycamore and ash by decomposing lignin, cellulose and related polysaccharides (Hepting, 1971; Blanchette, 1984; Adaskaveg and Ogawa, 1990; Adaskaveg et al., 1991, 1993). *G. boninense* Pat. is known to be the causal agent of BSR in oil palm (Ho and Nawawi, 1985; Abadi, 1987; Khairuddin, 1993; Utomo, 1997; Soepena et al., 2000).

Figure 2 Plantation with severe BSR disease with vacant planting spots in the field.

Soepena et al. (2000) reported that in the second and third cycles of replanting oil palms in the same place, the symptoms can appear as early as one to two years after planting. A more acute *G. boninense* problem is likely to surface over the next few years as the fungus increases its geographical range and virulence (Murphy, 2007). Singh (1991) conducted a study aimed at quantifying yield losses based on fresh fruit bunch (FFB) production and concluded that FFB production was adversely affected by the incidence of the disease. Losses due to BSR are not only through the direct reduction in oil palm numbers in stands, but also through a decrease in fruit bunch number and weight from diseased standing palms (Fig. 2), as well as from those with sub-clinical infections (Turner, 1981). The disease can result in the death of more than 80% of the plants by the time that they are halfway through their normal economic life and losses, reaching 30% have quite frequently occurred (Turner, 1981).

2.1 Symptoms of *Ganoderma* disease

In young palms, the external symptoms of BSR normally include a one-sided yellowing, or mottling of the lower fronds, followed by necrosis (Singh, 1991a). Ariffin et al. (2000) reported that in cases of infection, the newly unfolded leaves were shorter than normal, chlorotic, and the tips may be necrotic. As the disease progresses, palms may take on an overall pale appearance, with retarded growth and spear leaves remaining unopened.

Similar symptoms are observed in mature palms, with multiple unopened spear leaves and an overall pale leaf canopy. Often, when foliar symptoms are observed, it is found that at least one-half of the basal stem tissue has already been killed by the fungus. Infected young oil palms normally die at 6–24 months after the first appearance of symptoms, although mature palms can take 36 months to die.

2.2 Possible sources of *Ganoderma* infection

There have been numerous theories on the possible source of *Ganoderma* infection in oil palm. Oil palm stands were found to exhibit BSR symptoms at 10–14 years after planting following the conversion of rubber plantations (Flood et al., 2000). Meanwhile, a more widely accepted hypothesis proposes that *Ganoderma* attacks the root system from second and subsequent replanting cycles of oil palm. Through a series of controlled *Ganoderma* nursery trials, Rees et al. (2007, 2009b) clearly indicated that controlled root infection leads to typical symptoms. Rees et al. (2007) showed that root infection in the field originates from multiple, natural infections of different roots in a single palm. This finding, which is consistent with results from Flood et al. (2005), indicated that infection of the seedlings resulted from nearby colonized oil palms, with those seedlings nearer to the colonized trunks becoming diseased more quickly than others.

More recent findings showed that basidiospores were involved in the development of both manifestations (upper and basal) of stem rot (Rees et al., 2012). Authors reported that basidiospores were produced in prolific numbers throughout the sampling period, with a maximum release intensity during the evening, earlier than the midnight maximum release reported by Ho and Nawawi (1986). This suggested a constant potential for inoculum to colonize wounds and palm debris throughout the plantation (Rees et al., 2012; Cooper et al., 2011). The idea that spores from fruiting bodies (basidiomata) were identified is not new, as Lim and Fong (2005) and Utomo et al. (2005) had hypothesized the most likely sources of inocula for further infection via root contact.

2.3 Studies of BSR caused by *Ganoderma* sp. in oil palm

BSR was found to be a significant constraint for the oil palm plantation industry (Miller et al., 1999). In the subsequent year, Flood et al. (2000) stated that BSR has been a serious disease of oil palm for over 80 years, with severe economic loss in both Malaysia and North Sumatra (Indonesia). Bridge and Utomo (2005) suggested that *Ganoderma* infections occur over a period of several years, which is much longer than the timescales for other severe fungal plant pathogens (e.g. *Fusarium oxysporum*).

Ariffin et al. (2000) described how basidiomata develop at either the stem base, leaf base, or infected root, with the location providing a guide to the diseased area inside the palm. Abdullah's (2000) study on *Ganoderma* in coconut hypothesized a spread from independent secondary inocula, and their work did not support root-to-root spread and/or airborne spore spread. In contrast, a review by Paterson (2007) suggested that roots were thought to be the mode of spread. This idea, that infection results from contact between healthy roots and diseased tissue remaining in the soil, appears to be widely accepted. Interestingly, Bridge et al. (2004) mentioned that almost all isolates of *G. boninense* differed from one another. Therefore, the author emphasized the importance of basidiospores, and argued that ignoring this route of infection and the generation of new pathogen variation would be a mistake.

2.4 Effects of BSR on Malaysia's oil palm plantations

Due to the scarcity of available arable land in Malaysia, and the implementation of new land regulations on land use change, the expansion of oil palm plantations is now very limited. Plantation companies have to deal with this limitation by embarking on intensive replanting activities, including planting land that is heavily infected with BSR. Based on the

G. census conducted in Malaysia by Idris et al. (2011), 632 out of 1061 plantations (~60%) have recorded BSR incidence.

It was estimated by the Federal Land Development Authority (FELDA) that 200 million Malaysian Ringgit (US$45 million) had been lost due to the impact of BSR in the first generation of oil palm planting in Malaysia (Sirrul et al., 2016). Astonishingly in the same report, *Ganoderma* infection was found to be on the rise, as 264 plantations (61%) were found to be infected with BSR disease in 2015. Through this census, where all the FELDA's plantation schemes participated, BSR was observed in plantations two years after replanting had been completed.

A field study was conducted by Sirrul et al. (2016) in order to compare the levels of *Ganoderma* incidence between sanitation-windrowing palm debris (*standard practice*), under-planting practices (Fig. 3), and sanitation, against non-sanitation areas. Through these observations, it was found that plots practising under-planting showed a very high incidence of *Ganoderma*, which was to be 110% higher than plots cultivated under standard practice. The results, after eight years of the re-planting programme in the non-sanitation area, recorded *Ganoderma* incidence at about 4.47–4.66%. In comparison, no *Ganoderma* incidence was observed in areas where proper sanitation processes has been practised. Therefore, under-planting is not recommended where oil palm replanting programme takes place, although this method is commonly practised by the non-participating FELDA settlers. The sanitation practice is currently being enforced in all plantations where BSR is recorded.

Figure 3 Scenario where under-planting is adopted in replanting areas, where old palms are not removed to allow proper field sanitation to take place. New planting materials were planted close to the dead palms which will eventually be culled by trunk injection (chemical). Image courtesy of Sirrul Asrar Husain Fatimi (FGV R&D Sdn Bhd).

2.5 SSR genotyping of *G. boninense* strains (Southeast Asia and Africa)

G. boninense, a soil-borne basidiomycete, is the main causative agent of one of the more devastating oil palm diseases, BSR. Several studies focused on detecting *G. boninense*'s infection, assessing isolate aggressiveness, or deciphering oil palm's defence reaction during infection (Ho et al., 2014; Tee et al., 2013; Idris et al., 2004; Utomo et al., 2000). However, only limited information is available on this pathogen's spreading properties as well as its evolvability. Spreading properties encompass the reproduction mode (sexual vs. asexual), the dispersion ability of the propagules, or even the number of isolates effectively contributing to reproduction in a given population. Evolvability is the capacity for a given population to generate adaptive genetic diversity that will help in adapting to new environmental constraints. These biological characteristics are very important in the sense that they govern dynamic changes of the disease, and they should both be accounted for when designing efficient agricultural practice and developing a meaningful strategy for BSR-tolerant oil palm planting material. However, spreading characteristics and evolvability are also difficult to assess directly. A good strategy to evaluate these factors is to estimate population genetic parameters associated with genetic diversity, and population structure by means of an appropriate set of molecular markers.

Among many possible molecular markers, microsatellites, also called simple sequence repeats or SSRs, present very interesting properties. Microsatellites are tracts of repetitive DNA with small motifs (2–6 base pairs) repeated typically 5–50 times. They are characterized by a higher mutation rate than the rest of the genome (Brinkmann et al., 1998) and thus have a higher genetic diversity than other regions of the genome. Other advantages of SSRs include their moderate cost, the technical feasibility of their use on large samples, and their codominance and locus specificity (Jarne and Lagoda, 1996). This provides them with desirable properties for use in population genetics estimations and other demographic inference (Luikart et al., 2003; Cornuet et al., 1999; Michalakis and Excoffier, 1996). In this chapter, we describe the development of a *G. boninense*-specific SSR set from a draft genome assembly and its use in population genetic analysis.

2.6 Developing an SSR set

As shown by Dutech et al. (2007), the development of SSR is generally harder in fungi than in other groups. This difficulty could be due to the scarcity and short length of SSR in fungi genomes. Additionally, and contrary to what can be observed in plants, it is also generally hard to transfer SSR between fungal species of the same genus (Dutech et al., 2007). However, the increasing affordability and availability of sequencing data from next-generation sequencing technologies now enables access to large amount of sequenced data, from which it is possible to isolate SSR motifs.

In *G. boninense*, Mercière et al. (2015) combined 454 sequencing and Illumina data to produce a 63-Mb draft assembly that was used to identify and characterize all SSR loci. They then focused on a subset of these, in order to design primers and define amplification protocols to allow multiplexing. Accounting for SSR with perfect repeats and di- to hexanucleotide motifs only, Mercière et al. (2015) recorded a total of 2487 SSRs from the draft assembly. The SSR content (39 SSRs per Mb) of the *G. boninense* draft genome investigated is thus consistent with those published for other fungal species which ranges from tens to hundreds of SSR per Mb (Labbé et al., 2011; Karaoglu et al., 2005). It is also similar to the SSR content of the *G. lucidum* genome (37 SSRs per Mb) (Qian et al., 2012). While it was not possible to undergo marker development for the complete *G. boninense* SSR set, 145 of them were randomly selected for primer design. A preliminary step of *in silico* polymerase chain reactions allowed 35 of them to be discarded because of non-monolocus targeting. The remaining 110 SSR primers were then evaluated *in vitro* on a small subset of 5 *G. boninense* isolates: 17 primer pairs were finally retained according to their monolocus amplification pattern, polymorphism and the possibility to multiplex their amplification and genotyping. These 17 primers were used to study a collection of 107 *G. boninense* isolates from North Sumatra (Indonesia), Peninsular Malaysia and Sabah (part of Borneo), to validate their desired amplification and sufficient polymorphism level.

2.7 Deciphering *G. boninense* in Malaysia and North Sumatra

Malaysia and North Sumatra are the historical centres for large-scale cultivation of oil palm, and at the same time, for BSR development in Southeast Asia. Therefore, studying *G. boninense* genetic diversity in these particular regions is of key interest. Mercière et al. (2017) collected and genotyped an extensive sample of 311 isolates from Peninsular Malaysia (FELDA plantations) and North Sumatra (Socfindo plantations), and used 11 of the previously mentioned SSRs (Mercière et al., 2015) to investigate the various characteristics of their genetic diversity, in an attempt to decipher the biological processes that may have shaped them. It appeared that *G. boninense* exhibited a high level of genetic diversity (H_e = 0.651) with only a few duplicated genotypes. The five pairs of duplicated genotypes studied, originating from either Peninsular Malaysia or North Sumatra, were all collected from neighbouring trees, at a distance of about nine metres. These results clearly support the predominance of sexual reproduction in shaping global *G. boninense* diversity, confirming the works of Miller et al. (1999) and Pilotti et al. (2003). Additionally, this data enabled the estimation of the effective population size for the pathogen in this region. This was calculated as the number of distinct genotypes involved in breeding (about 50 000 as a rough estimate). The asexual spread from root to root appeared to be marginal, at least within the well-maintained plantations from which the samples were collected. Nevertheless, the importance of sexual reproduction has been highlighted by other authors (Rees et al., 2012; Pilotti et al., 2003; Miller et al., 1999), although those studies were limited by their geographical scope and the genetic marker used. *G. boninense's* sexual reproduction is mediated by basidiospores that are notoriously capable of large-scale dispersal (Rees et al., 2012; Hallenberg et al., 2001). This biological feature is responsible for gene flow over long distances, and for the absence of any genetic structure observed in the area encompassing North Sumatra and Peninsular Malaysia (Mercière et al., 2017). More particularly, these authors showed that *G. boninense* has a two-tier spreading pattern, with both a high capacity for dispersal over short distances (less than 2 km), and over long distances (between 500 and 700 km). Such a spreading pattern seems common to basidiospore-spread fungi as it was also identified in *Heterobasidion irregulare* (Garboletto et al., 2013).

2.8 Deciphering *G. boninense* in Southeast Asia

Besides Peninsular Malaysia and North Sumatra, BSR has been described in other regions of Southeast Asia, and in Africa. Studies were conducted in order to determine whether the development of BSR in these regions was caused by *G. boninense* genotypes closely related to those observed in Peninsular Malaysia and North Sumatra, or, on the contrary, by genotypes belonging to other *G. boninense* sub-populations, (or even due to other *Ganoderma* species). Mercière et al. (2017) genotyped an additional group of isolates using the same set of 11 SSRs previously used for Peninsular Malaysia and North Sumatra (Mercière et al., 2015). These isolates were collected in Thailand (Krabi, four isolates), Borneo (Sabah, 34 isolates), South Sumatra (7 isolates), Papua New Guinea (11 isolates) and Cameroon (32 isolates). In addition to the SSRs used, all the isolates were further confirmed with an internal transcribed spacer (ITS) sequence of *G. boninense* (Latiffah et al., 2002; Rees et al., 2012).

The diversity, as measured by the expected heterozygosity (Nei, 1987) is high, with the order of magnitude of North Sumatra (0.65) and Peninsular Malaysia (0.63), ranging from 0.57 (Papua New Guinea) to 0.62 (South Sumatra). According to El Moussadik and Petit (1996), the estimation of allelic richness buffering uneven sample sizes yielded similarly high values among locations, ranging from 3.4 (North Sumatra and South Sumatra) to 2.9 (Thailand). Based on the Bayesian clustering approach implemented in STRUCTURE software (Pritchard et al., 2000), Southeast Asia's *Ganoderma* diversity is structured in two main differentiated genetic groups (Fst = 0.13). The first group encompasses isolates from Thailand, Peninsular Malaysia, North Sumatra and South Sumatra. The second group mainly gathers isolates from Borneo and Papua New Guinea. A limited number of isolates (6) are classified as intermediate between these two groups. Finally, a significant phylogeographic signal (Hardy et al., 2003) could be detected between the two identified genetic groups, indicating low gene exchange between them. Although the analysis is still at its preliminary stage, this is an interesting discovery as it allows for the possibility of new hypotheses.

The use of SSR markers revealed that the *G. boninense* affecting oil palm in Southeast Asia is highly diverse, and subdivided into two distinct and geographically isolated sub-populations. These results may have important implications for oil palm improvement for tackling *Ganoderma*. Firstly, due to the high genetic diversity of the pathogen, associated with it's large effective population size and active sexual reproduction, one should be careful of the development of multigenic tolerance rather than monogenic total resistance, in order to design sustainable oil palm varieties. On the other hand, the subdividing of *G. boninense* in two Southeast Asian populations calls for more studies in order to better understand their differences, notably in terms of aggressiveness. Here, we have stressed the fact that human-mediated gene flows between the two *G. boninense* groups should be carefully avoided in order to prevent hybridization, possibly leading to the emergence of *Ganoderma* with increased aggressiveness. Finally, and before more knowledge on the characteristics of the *Ganoderma* sub-populations is gained, partially resistant oil palm varieties should be used, taking into account the genetic background of locally present *G. boninense* isolates (e.g. using SSR markers). The use of oil palm which has tolerance against both *Ganoderma* sub-populations will need to be fully ascertained.

3 Cellulolytic degradation as a possible mode of infection

Oil palm is susceptible to attacks from pathogenic fungi such as *Ganoderma* spp. which can cause rotting of the roots and basal stem of the standing palms. *G. boninense* among the *Ganoderma* spp. is capable of degrading the plant's lignocellulosic biomass of the cell wall.

The main components of lignocellulose are cellulose, hemicellulose and lignin (Sjöström, 1993). Studies on lignin-modifying enzymes from *Ganoderma lucidum* indicate that lignin component is the rate-limiting factor in the lignin degradation process (D'Souza et al., 1999). Deposition of lignin in the cell wall can serve as a barrier to microbial attack, because it provides strength and is a water-impermeable seal across cell walls in the xylem tissue. Lignin therefore, may play a major role in disease response mechanisms (Paterson, 2007).

The ability of *Ganoderma* spp. to mineralize lignin to produce carbon dioxide and water is remarkable. It is related to its ability to cleave the monomeric ring structures of the lignin monomers, although the physical properties of lignin are the product of aromatic heteropolymers synthesized from the phenylpropanoid aryl-C_3 units, coniferyl, synapyl and *p*-coumaryl alcohols, with at least 12 different types of linkages such as aryl-ether and carbon-carbon bonds (Boerjan et al., 2003).

Ganoderma spp. produces several types of extracellular oxidative enzymes that are related to the degradation of the lignin component of plant cell walls. For instance, manganese peroxidase and laccase in *G. lucidum* (D'Souza et al., 1999), lignin peroxidase in *G. applanatum* (Morgenstern et al., 2008; Martínez, 2002), and lignin peroxidase, manganese peroxidase and versatile peroxidase of other white rot fungi (Dashtban et al., 2010; Ward et al., 2004), are involved in the degradation of lignin.

Laccases are blue multicopper phenol oxidases, that oxidize various phenolic substrates via the reduction of oxygen to water, and also oxidize non-phenolic lignin (Bourbonnais and Paice, 1990). Laccases work with a wide range of substrates without cofactors, and they have wide applications in biotechnology (Upadhyay et al., 2016; Arora and Sharma, 2010). They are commonly used in feedstock industries (Sharma et al., 2013), textile industries (Couto et al., 2004) and soil bio-remediation (Nyanhongo et al., 2006). Laccases are produced in a wide range of organisms including bacteria, fungi, plants and insects. In fungi, laccases are involved in the degradation of lignin (D'Souza et al., 1999), whilst they are involved in the biosynthesis of lignin in plants (Sato et al., 2001).

4 Laccase gene discoveries

The genome of *G. boninense* was sequenced using Illumina and 454 sequencing technology for the discovery of microsatellite markers (Mercière et al., 2015). A draft genome sequence of 63 Mb was assembled, providing the starting genomic resources for laccase gene discoveries. *Ab initio* gene model prediction on this draft genome was performed with Augustus 2.6.1 (Stanke and Waack, 2003) using *Phanerochaete chrysosporium* as a training organism (unpublished results).

In order to discover laccase genes from the *G. boninense* draft genome *in silico*, Camus-Kulaidevalu et al. (2014) mined for the laccase genes in both the public database and the predicted gene models derived from the draft genome. In order to retrieve known fungi laccases, we searched the NCBI 'nr' protein database using keywords 'laccase' and 'fungi'. This resulted in 841 laccase protein sequences that were longer than 100 amino acids with methionine amino acid start codons, and which contained the protein sequence name 'laccase'. We analysed the essential sequence features of laccases available from the public domain, based on multiple sequence alignments of 841 protein sequences using MAFFT software (Katoh and Standley, 2013) using '-einsi' option to allow for large unaligned regions. We then used HMMER3 software (Mistry et al., 2013) to synthesize sequence profiles common to laccase alignment ('hmmbuild' module), and mined *G. boninense* predicted gene model database for putative laccase ('hmmsearch' module). Based on the full sequence score and *E*-value of 1.1^{e-63} (best fitting domain), 33 *G. boninense* gene models showed significant similarities to the laccase sequence profile based on 841 input fungus laccase alignment using HMMER software.

Among these 33 gene models fitting HMMER laccase pattern, 25 bore the four-domain laccase signature (Kumar et al., 2003). These 25 gene models displayed a predicted coding region length between 1473 and 1926 bp (between 491 and 642 amino acids). These gene models are distributed on seventeen scaffolds. Scaffold 1 bears three gene models with four-domain laccase signature; scaffold 49 bears three gene models with four-domain laccase signature; scaffolds 46, 149 and 197 bear two gene models each with four-domain laccase signature. The predicted intron number varies between 3 and 12.

Twenty-four gene models were predicted using TARGETP 1.1 (http://www.cbs.dtu.dk/services/TargetP) and 23 displayed a predicted peptide signal SIGNALP 3.0 (http://www.cbs.dtu.dk/services/SignalP). At least seven predicted laccase genes are expressed in *G. boninense* isolates, based on the sequence similarity search of 25 predicted gene models with four-domain laccase signature, against the transcriptome assembly (unpublished data) of *G. boninense* isolate cultured in potato dextrose agar plate. Our search for sequence similarities indicated that seven gene models showed strong similarity with assembled transcripts with a sequence similarity between 93 and 99% over more than 1300 bp in length.

In order to investigate the relationship of the 25 predicted *G. boninense* laccase genes with other fungi species, we aligned the putative laccases with other known fungal laccases from *Laccaria bicolor*, *Coprinopsis cinerea*, *Trametes villosa*, *G. lucidum* and *Phanerochaete chrysosporium* (in this latter case Multi-Copper Oxidase – MCO). These were retrieved from NCBI using MAFFT software (Katoh and Standley, 2013). The five species studied belong to basidiomycota, either from the polyporales order (*T. villosa*, *G. lucidum*, *Phanerochaete chrysosporium*) or from the agaricales order (*L. bicolor*, *C. cinerea*). We removed poorly aligned regions with gBlocks software (Castresana, 2000) and produced a phylogenetic tree (Fig. 4) with PhyML (Guindon et al., 2010) using default values for amino acid analysis.

Based on our analysis, predicted laccases from *G. boninense* are positioned in two groups. Gene models g9396 and g4264 are gathered with a *G. lucidum* and two *L. bicolor* laccases belonging to laccase *sensu lato* (Hoegger et al., 2006). More specifically, these two *L. bicolor* laccases are ferroxidases. The other 23 *G. boninense* predicted gene models gather in a large group solely composed of polyporales laccases, including all five *T. versicolor* laccases and 12 *G. lucidum* laccases. Interestingly, the seven laccase predicted gene models showing transcript evidence are located in the polyporales gene cluster.

In summary, Camus-Kulaidevalu et al. (2014) identified 25 gene models exhibiting the four domains with fungal laccase signatures, and the main characteristic of fungal laccases in terms of length, cellular addressing, signalling and intron number. For seven of these 25 laccase predicted gene models, we found transcript evidence supporting gene prediction. Our results suggest the existence of a specific group of polyporales *sensu stricto* laccases (Hoegger et al., 2006) for which ancient duplications occurred (Fig. 4). This group includes laccases from the white rot *G. lucidum*, *G. boninense* and *T. versicolor* but not *Phanerochaete chrysosporium*. Within the polyporales *sensu stricto* laccase group, the phytopathogenic *G. boninense* exhibit a noticeable position, with the specific expansion of some gene sub-family (e.g. gene model g6150, g580, g7800, g11265, g785), leading to a potentially doubled laccase copy number compared to its relative *G.*

lucidum. This important information allowed us to conduct further analysis for verification on lignocellulosic pathway, and the possibility of finding an inhibitor might provide a long-term solution such as systemic fungicide development in the future.

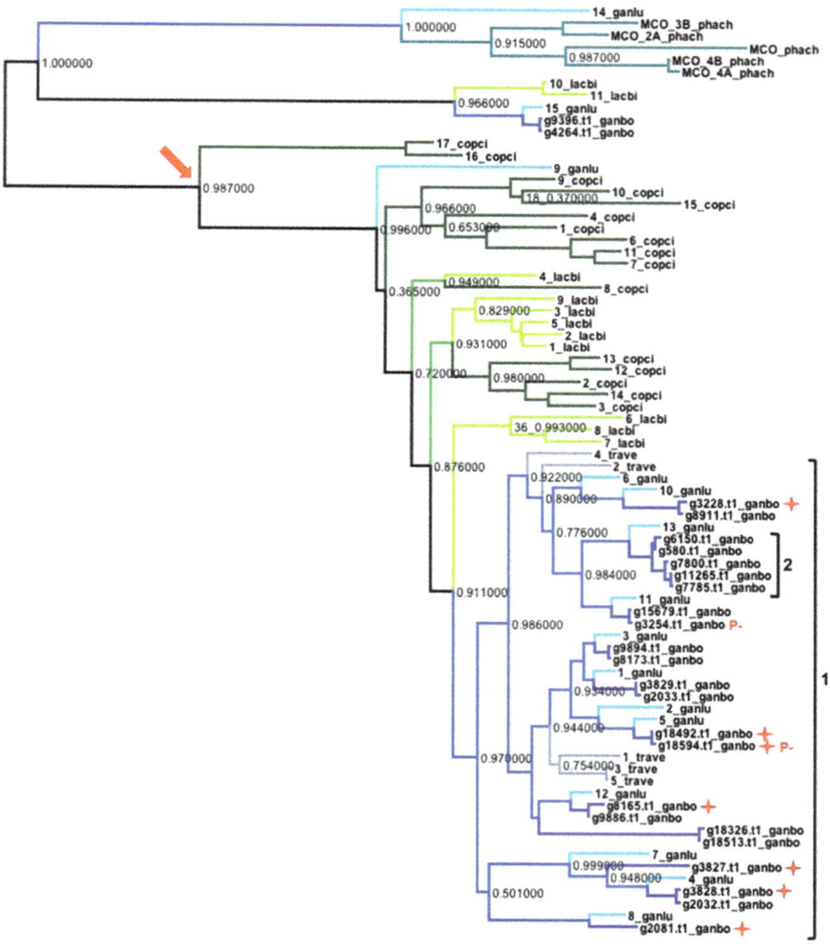

Figure 4 A phylogenetic tree for *G. boninense*-predicted laccases, and laccases of four other basidiomycetes species. The phylogenetic tree was obtained with PhyML (Guindon et al., 2010) using default values for amino acid analysis. It is rooted on the median point. ganbo: *G. boninense* predicted laccase gene models (electric blue), ganlu: *G. lucidum* (cyan), trave: *T. versicolor* (grey), phach: *Phanerochaete chrysosporium* (Turkish green), copci: *C. cinerea* (dark green), lacbi: *L. bicolor* (light green). Plain blue indicates branches specific to polyporales and plain green indicates branches specific to MCO: Multi Copper Oxidase. Branch support values were obtained with LRT SH-like test. A red arrow indicates *sensu stricto* laccases as defined by Hoegger et al. (2006). Bracket one underlines a laccase clade specific to polyporales order, and bracket two pinpoints a possible copy expansion in *G. lucidum*. Red star indicates *G. boninense* gene model with transcript match, and red 'P-' indicates *G. boninense* gene models with no peptide signal.

5 Developing partial resistance planting material against BSR

At present, there is no known oil palm planting material that is resistant to *G. boninense*. However, there are at least two commercial planting materials with partial resistance against BSR that have been released. In 2014, PT Socfindo successfully released 'DxP Socfindo Moderate Gano Resistant (MTG)' in Indonesia. In order to obtain the varieties of *Ganoderma*-partially-resistant seeds, the material is obtained from the genetic seed garden after passing through the census process. This is followed by a series of trials involving subjecting the materials in a *Ganoderma* nursery to a screening test (Idris et al., 2004; Breton et al., 2006), and accompanied by a miniature experiment in the field. Field tests were conducted in parental garden, seed garden and progeny trials.

In 2016, FELDA Global Ventures (FGV) of Malaysia released another partial resistance material against BSR namely 'Yangambi Ganoderma-Tolerant 1 (GT1)'. The approach used by FGV in developing GT1 is very much similar to methodology described before with the exception that planting materials were initially screened with a set of markers (Tan et al., 2011) targeting a disease resistance gene, before being subjected to *Ganoderma* nursery screening phase. The integration of molecular marker (with up to 70% accuracy) enabled FGV to fast track the screening of partially resistant materials.

6 *Fusarium* wilt

Fusarium wilt, also more commonly known as vascular wilt disease, is caused by *F. oxysporum* f.sp. *elaeidis*. It is a soil-borne disease and it was reported to enter its host (oil palm) through the roots and along the stele (Corley and Thinker, 2003). Wardlaw (1950) first reported it during a visit to Belgian Congo in 1946 where he observed a wilt disease in oil palm, and later isolated *F. oxysporum* from the necrosed vascular strands. Subsequently, other plantations in the African and South American continents such as Nigeria, Ivory Coast, Central Africa, Brazil and Ecuador have reported *Fusarium* wilt (van de Lande, 1983; Renard and de Franqueville, 1989; Flood et al., 1989). However, to date no *Fusarium* wilt disease has been reported in Southeast Asia.

In a review by Flood (2006), two disease syndromes were observed. The first syndrome is described as 'acute wilt' wherein the leaves dry out and die rapidly while retaining their original erect positions on the plant until broken off, usually several feet from the base, by wind action. As the disease progresses, the palm will eventually die within two or three months after this syndrome is observed. The second syndrome also known as 'chronic wilt' sees the palms remain alive for many months and even years, but becomes progressively stunted. This eventually leads to flattening of the crown due to the production of the youngest fronds in the crown which are often chlorotic. As in BSR, vascular wilt is also known to be a soil-borne disease and the main culprit of the infection is through contact with dead or infected tissue (Renard and de Franqueville, 1989).

Unlike BSR, it is actually possible to breed for resistance planting material against vascular wilt. It was reported that breeding programmes in Ivory Coast had been successful, with the introduction of resistant varieties that were able to reduce the disease incidence from an area as high as 30% to less than 3%. Rosenquist et al. (1990) reported that pure Dumpy Deli *dura* materials appeared to be immune against vascular wilt. Nursery screening was the initial tool used to conduct rapid screening and identification of potential resistant varieties. Prendergast (1963) was the first to develop a technique to screen for seedlings resistant to the disease at the nursery stage. Subsequently, a few researchers have adopted and modified the methodology (Flood et al., 1989; Renard et al., 1991). However, Corley and Tinker (2003) suggested that the best approach was described by de Franqueville (1984), where a wilt index system (percentage of wilt-infected plants divided by the mean wilt percentage of all progenies in a trial) first introduced by Renard et al. (1972), was used to calculate wilt severity. The modified wilt index system by de Franqueville (1984) involved transformation of percentages for individual plots. Data was statistically analyzed and progenies were only accepted as resistant if they had significantly lower losses than either the mean of the trial or standard crosses with known performance data.

7 Spear rot or bud rot

Spear rot, which was previously known as bud rot, is another important disease predominantly found in Latin America (Mariau, 2001). As the name of the disease suggests, it primarily affects spear leaves and the heart of the palm. Turner (1981) had suggested that the term 'spear rot' should be considered when the primary rotting affects the spear, while 'bud rot' should be considered for the disease primarily destroying unemerged leaves. The latter disease is usually fatal, while the former frequently non-fatal. The first published report of an oil palm plantation destroyed by bud rot in Latin America was from Suriname, where a 4-year-old plantation was completely destroyed in 1920 (Malaguti, 1953). Meanwhile, bud rot was also reported in Colombia and in 1964, complete destruction of an oil palm plantation (~2000 ha) in Turbo near the border of Panama (De Rojas and Ruíz, 1972).

There has been much speculation on the cause of bud rot disease with micro-elements deficiency such as boron (Ferweda, 1954). Observations made by Robertson (1960) in Nigeria suggested that there was an active pathogen involved, as susceptible palms had regular cycles of infection, and prevention by cutting off the spear could prevent the disease from recurring. Duff (1963) later suggested that bacterium of the genus *Erwinia*, similar to *E. lathyri* which was isolated from palm, was found in Congo.

A review by de Franqueville (2003), highlighted that there had been many bacterial and fungal pathogens, such as *Fusarium* and *Thielaviopsis* that were often isolated, but inoculation experiments with these pathogens could not be proven via Koch's postulates in healthy oil palms. It was only recently that *Phytophthora palmivora* had been demonstrated to be the causal agent of bud rot disease in Colombia (Drenth et al., 2013; Torres et al., 2010). Generally, initial visible symptoms such as the formation of small, water-soaked lesions in the leaflets at the base of the spear leaf, are usually followed by an increase in the number and size of the lesions. Sometimes, it is also associated with unusually wet weather, and long periods of continuous precipitation, accompanied by cloudy mornings and wet palm canopies (Martínez, 2009).

Fortunately, one of the most effective ways to overcome bud rot disease is through a genetic pool where resistant planting materials are to be planted in the epidemic area (Avendanõ and Garzon, 2013; Bastidas et al., 2007; Meunier, 1991). Planting materials such as interspecific hybrid crosses of *Elaeis guineensis* × *Elaeis oleifera* have been shown to give rise to desirable agronomic traits and reduced susceptibility towards bud rot. Crucially, it can provide potential options for disease control (Alvarado et al., 2013; Corley and Tinker, 2003). Torres et al. (2016) reported that if the disease is detected in the early stages, remedial action such as 'surgery' can be taken. This involves the removal of all of the affected tissue to just above the meristem, quick flaming of the exposed area to control any surviving spores, followed by the application of a mixture of bactericides, fungicides and insecticides to the exposed area. The same researchers also suggested that an integrated crop management system (cultivar selection, proper drainage, fertilization, monitoring, removal of infected tissues and destruction of tissues or palms), can effectively control bud rot disease to keep it to a minimum.

8 Future trends and conclusion

Although concerted efforts have been made to prevent various fungal infections in oil palm plantations, the ultimate long-term solution is to identify planting materials that are resistant to the respective fungus through a conventional or marker-assisted breeding approach. More importantly, an integrated disease management approach must be incorporated in tandem with the planting materials, as soil sanitation and integration of beneficial microbes that are antagonistic to the fungus are important for their long-term sustainability. In the future, diagnostic kits for both fungus identification and quantification of the virulence level of the fungus, could be used on-site for better plantation management to prevent the spread of fungal diseases

9 Where to look for further information

FGV's world class R&D and Agri-Services Cluster is anchored on five decades of research and development. The Cluster's key objective is to generate cutting-edge agri-business technologies to enhance operational performance and commercial utilization across all facets of FGV. The company's award-winning Yangambi oil palm planting material, which has 40 percent market share in Malaysia, is just one of R&D's innovative products. Besides this, FGV has launched other planting materials such as FGV 3-way and Yangambi GT-1 for niche markets. Our Research & Development centre is led by a team of highly specialized scientists. Its activities support internal growth across all Clusters. These involve improving the yield of selected agri crops through breeding, tissue culture agronomy, crop protection and sustainability. Other activities include the optimization of waste and by-products for use in generating new products with higher-growth, higher-margin industries, and the provision of high quality agro-based products and services. R&D and Agri-Services Cluster's focus on bio-molecular marker research has led to the pioneering of marker-assisted oil palm breeding and selection. For more information, please visit: http://www.feldaglobal.com/our-business/plantation/rd-services/.

CIRAD (French Agricultural Research Centre for International Development) is a public establishment (EPIC) under the joint authority of the Ministry of Higher Education, Research and Innovation and the Ministry for Europe and Foreign Affairs. Its activities concern the life sciences, social sciences and engineering sciences, as applied to agriculture, the environment and territorial management. Its work centres on several main topics: food security, climate change, natural resource management, reduction of inequalities and poverty alleviation. CIRAD works with its partners in southern countries to generate and pass on new knowledge to support agricultural development. It puts its scientific and institutional expertise at the disposal of policymakers in those countries and global debates on the main issues concerning agriculture. It also supports French scientific diplomacy operations. For more information, please visit: http://www.cirad.fr/en/.

10 Acknowledgements

The authors thank the management of FGV for its support and commitment in the joint collaboration work with CIRAD, France, under the 'SeqGano Project'. The authors posthumously thank Dr de Franqueville Hubert for his work on oil palm pathogens, in particular his experiment on BSR disease, without which none of this work would have been possible.

11 References

Abadi, A. L. (1987) Biologi *Ganoderma boninense* Pat. pada kelapa sawit (*Elaeis guineensis* Jacq.) dam pengaruh beberapa mikroba tanah antagonistik terhadap pertumbuhannya. PhD thesis, IPB, Bogor.

Abdullah, F. (2000) Spatial and sequential mapping of the incidence of BSR of oil palms (*Elaeis guineensis*) on a former coconut (*Cocus nucifera*) plantation. In Flood, J., Bridge, P. D. and Holderness, M. (Eds) *Ganoderma Diseases of Perennial Crops*. CABI Publishing, Wallingford, UK, pp. 183–94.

Adaskaveg, J. E. and Ogawa, J. M. (1990) Wood decay pathology of fruit and nut trees in California. *Plant Disease* 74: 341–52.

Adaskaveg, J. E., Blanchette, R. A. and Gilbertson, R. L. (1991) Decay of date palm wood by white-rot and brown-rot fungi. *Canadian Journal of Botany* 69: 615–29.

Adaskaveg, J. E., Miller, R. W. and Gilbertson, R. L. (1993) Wood decay, lignicolous fungi and decline of peach trees in South Carolina. *Plant Disease* 77: 707–11.

Ariffin, D., Idris, A. S. and Singh, G. (2000) Status of Ganoderma in Oil Palm. In Flood, J., Bridge, P. D. and Holderness, M. (Eds) *Ganoderma Diseases of Perennial Crops*. CABI Publishing, Oxon, pp. 49–68.

Ariffin, D., Singh, G. and Lim, T. K. (1989) *Ganoderma* in Malaysia – current status and research strategy. In Jalani, S. and et al. (Eds) *Proceedings of the 1989 PORIM International Palm Oil Development Conference-Module II Agriculture, 5–9 September 1989*. Palm Oil Research Institute of Malaysia, Bangi, Selangor, Malaysia, pp. 249–7.

Arora, D. S. and Sharma, R. K. (2010) Ligninolytic fungal laccases and their biotechnological applications. *Applied Biochemistry and Biotechnology* 160: 1760–88.

Avendanõ, J. C. and Garzon, G. M. (2013) Experiencias del manejo de la pudrición del cogollo (PC) en *Elaeis guineensis* y el híbrido OxG en Astorga S.A. *Palmas* 34: 345–52.

Bastidas, S., Penã, E., Reyes, R., Pãrez, J. and Tolosa, W. (2007). Comportamiento agronómico del cultivar híbrido Rc1 de palma de aceite (*Elaeis oleifera* × *Elaeis guineensis*) v *Elaeis Guineensis*. *Corpoica Ciencia y Tecnología Agropecuaria* 8: 5–11.

Blanchette, R. A. (1984) Screening wood decayed by white rot fungi for preferential lignin degradation. *Applied Environmental Microbiology* 48: 647–53.

Boerjan, W., Ralph, J. and Baucher, M. (2003) Lignin biosynthesis. *Annual Review of Plant Biology* 54: 519–46.

Bourbonnais, R. and Paice, M. G. (1990) Oxidation of non-phenolic substrates. *FEBS Letter* 267: 99–102.

Breton, F., Hasan, Y., Haraidi, Lubis, Z. and Franqueville, H. (2006) Characterization of parameters for the development for an early screening test for basal stem rot tolerance in oil palm progenies. *Journal of Oil Palm Research* (Spec.): Special edition: April 2006, 24–36.

Bridge, P. D. and Utomo, C. (2005) Editorial. *Mycopathologia* 159: 91.

Bridge, P. D., Singh, T. and Arora, D. K. (2004) The application of molecular markers in epidemiology of plant pathogenic fungi. In Arora, D. K. (Ed.) *Fungal Biotechnology in Agriculture, Food and Environmental Applications.* Marcel Dekker, New York, pp. 57–68.

Brinkmann, B., Klintschar, M., Neuhuber, F., Hühne, J. and Rolf, B. (1998) Mutation rate in human microsatellites: Influence of the structure and length of the tandem repeat. *The American Journal of Human Genetics* 62(6): 1408–15.

Camus-Kulandaivelu, L., Mercière, M., Tan, J. S., Lee, Y. P., Klopp C., Durand-Gasselin, T., Sharifah Shahrul Rabiah, S. A. and Breton, F. (2014) Identification of laccase genes in *Ganoderma boninense* draft genome assembly. In *International Oil Palm Conference (IOPC), 17 to 19 June 2014.* Bali Nusa Dua Convention Centre, Indonesia.

Castresana, J. (2000) Selection of conserved blocks from multiple alignments for their use in phylogenetic analysis. *Molecular Biology and Evolution* 17: 540–52.

Cooper, R. M., Flood, J. and Rees, R. W. (2011) *Ganoderma boninense* in oil palm plantations: Current thinking on epidemiology, resistance and pathology. *The Planter* 87(1024): 515–26.

Corley, R. H. V. and Tinker, P. B. (2003) Diseases and pests of the oil palm. In *The oil palm (Forth Edition).* Blackwell Science, Oxford, pp. 391–444.

Cornuet, J. M., Ravigne, V. and Estoup, A. (2010) Inference on population history and model checking using DNA sequence and microsatellite data with the software DIYABC (v1.0). *BMC Bioinformatics* 11: 401.

Couto, S. R., Sanroman, M. A., Hofer, D. and Gübitz, G. M. (2004) Production of laccase by *Trametes hirsuta* grown in an immersion bioreactor and its application in the docolorization of dyes from a leather factory. *Engineering in Life Sciences* 4: 233–8.

D'Souza, T. M., Merritt, C. S. and Reddy, C. A. (1999) Lignin-modifying enzymes of the white rot basidiomycete *Ganoderma lucidum. Applied and Environmental Microbiology* 65: 5307–13.

Dashtban, M., Schraft, H., Syed, T. A. and Qin, W. (2010) Fungal biodegradation and enzymatic modification of lignin. *International Journal of Biochemistry and Molecular Biology* 1: 36–50.

de Franqueville, H. (1984) Vascular wilt of the oil palm: relationship between nursery and field resistance. *Oléagineux* 39: 513–18.

de Franqueville, H. (2003) Oil palm bud rot in Latin America. *Experimental Agriculture* 39: 225–40.

De Rojas, E. and Ruíz, E. (1972) *Investigaciones sobre la pudrición del cogollo y pudrición de la flecha de la palma africana en la plantación la Arenosa de Coldesa S.A. (Turbo, Ant.).* Instituto Geográfico Agustín Codazzi, Medellin, Colombia.

Drenth, A., Torres, G. A. and Martínez, G. (2013) Phytophthora palmivora, la causa de la pudrición del cogollo en la palma de aceite. *Palmas* 34: 87–94.

Duff, A. D. S. (1963) The bud rot little leaf disease of the oil palm. *West African Institute for Oil Palm Research* 4: 176–90.

Dutech, C., Enjalbert, J., Fournier, E., Delmotte, F., Barres, B., Carlier, J., Tharreau, D. and Giraud, T. (2007) Challenges of microsatellite isolation in fungi. *Fungal Genetic Biology* 44(10): 933–49.

El Mousadik, A. and Petit, R. J. (1996) High level of genetic differentiation for allelic richness among populations of the argan tree (*Argania spinosa* (L.) Skeels) endemic to Morocco. *Theoretical and Applied Genetics* 92: 832–83.

Ferwerda, J. D. (1954) Boron deficiency on oil palms in the Kasai region of the Belgian Congo. *Nature* 173: 1097.

Flood, J., Cooper, R. M. and Lees, P. E. (1989) An investigation of pathogenicity of four isolates of Fusarium oxysporum from South America, Africa and Malaysia to clonal oil palm. *Journal of Phytopathology* 124: 80–8.

Flood, J., Hasan, Y., Turner, P. D. and O'Grady, E. B. (2000) The spread of *Ganoderma* from infective sources in the field and its implications for the management of the disease in oil palm. In Flood, J., Bridge, P. D. and Holderness, M. (Eds) *Ganoderma Diseases of Perennial Crops*. CABI Publishing, Oxon, pp. 101–12.

Garbelotto, M., Guglielmo, F., Mascheretti, S., Croucher, P. J. and Gonthier, P. (2013) Population genetic analyses provide insights on the introduction pathway and spread patterns of the North American forest pathogen Heterobasidion irregulare in Italy. *Molecular Ecology* 22(19): 4855–69.

Guindon, S., Dufayard, J.-F., Lefort, V., Anisimova, M., Hordijk, W. and Gascuel, O. (2010) New algorithms and methods to estimate maximum-likelihood phylogenies: Assessing the performance of PhyML 3.0. *Systematic Biology* 59: 307–21.

Hallenberg, N. and Kuffer, N. (2001) Long-distance spore dispersal in wood-inhabiting Basidiomycetes. *Nordic Journal of Botany* 21(4): 431–6.

Hardy O. J., Charbonnel N. N., Fréville, H. and Heuertz, M. (2003) Microsatellite allele sizes: a simple test to assess their significance on genetic differentiation. *Genetics* 163: 1467–82.

Hepting, G. H. (1971) *Diseases of Forest and Shade Trees of the United States*. US Department of Agriculture, Agricultural Handbook 386, USDA Forest Service, Washington, DC: U.S. Government Printing Office. pp. 1–658.

Ho, Y. and Nawawi, A. (1986) Isolation, Growth and Sporophore Development of Ganoderrna boninense from Oil Palm in Malaysia. *Pertanika* 9(1): 69–73.

Ho, Y. W. and Nawawi, A. (1985) *Ganoderma boninese* Pat. from basal stem rot of oil palm in peninsular Malaysia. *Pertanika* 8: 424–8.

Hoegger, P. J., Kilaru, S., James, T. Y., Thacker, J. R. and Kües, U. (2006) Phylogenetic comparison and classification of laccase and related multicopper oxidase protein sequences. *FEBS Journal* 273: 2308–26.

Idris, A, S., Mior, M. H. A. Z., Maizatul, S. M. and Kushairi, A. (2011) Survey on status of *Ganoderma* disease of oil palm. In *Proceedings of the PIPOC 2011 International Palm Oil Congress – Agriculture Conference*. MPOB, Bangi, pp. 235–8.

Idris, A. S., Kushairi, D., Ariffin, D. and Basri, M. W. (2006) Technique for inoculation of oil palm germinated seeds with *Ganoderma*. *MPOB Information Series*. MPOB TT No. 314, 4p.

Jarne, P. and Lagoda, P. J. L. (1996) Microsatellites, from molecules to populations and back. *Trends in Ecology and Evolution* 11: 424–9.

Karaoglu, H., Lee, C. M. Y. and Meyer, W. (2005) Survey of simple sequence repeats in completed fungal genomes. *Molecular Biology and Evolution* 22: 639–49.

Katoh, K. and Standley, D. M. (2013) MAFFT multiple sequence alignment software version 7: Improvements in performance and usability. *Molecular Biology and Evolution* 30: 772–80.

Khairudin, H. (1993) Basal stem rot caused by *Ganoderma boninense*. An update. In *PORIM, International Palm Oil Congress, Update and Vision*. PORIM, Kuala Lumpur, pp. 739–49.

Kumar, S. V. S., Phale, P. S., Durani, S. and Wangikar, P. P. (2003) Combined sequence and structure analysis of the fungal laccase family. *Biotechnology and Bioengineering* 83: 386–94.

Labbé, J., Murrat, C., Morin, E., Le Tacon, F. and Martin, F. (2011) Survey and analysis of simple sequence repeats in *Laccaria bicolor* genome, with development of microsatellites markers. *Currents Genetics* 57: 75–88.

Latiffah, Z., Harikrishna, K., Tan, S. G., Tan, S. H., Abdullah, F. and Ho, Y. W. (2002) Restriction analysis and sequencing of the ITS regions and 5Æ8S,gene of rDNA of *Ganoderma* isolates from infected oil palm and coconut stumps in Malaysia. *Annals of Applied Biology* 141: 133–42.

Lim, H. P. and Fong, Y. K. (2005) Research on basal stem rot (BSR) of ornamental palms caused by basidiospores from *Ganoderma boninense*. *Mycopathologia* 159: 171–9.

Lubis, A. U. (1992) *Kelapa Sawit* (*Elaeis guineensis* Jacq.) *di Indonesia*. Pusat Penelitian Perkebunan Marihat-Bandar Kuala, Pematang Siantar, Sumatera Utara.

Luikart, G., England, P. R., Tallmon, D., Jordan, S. and Taberlet, P. (2003) The power and promise of population genomics: from genotyping to genome typing. *Nature Reviews Genetics* 4: 981–94.

Malaguti, G. (1953) 'Pudrici´on del cogollo' de la palmera africana (*Elaeis guineensis* Jacq.) en Venezuela. *Agronomial Tropical* 3: 13–31.

Mariau, D. (Ed.) (2001) *Diseases of Tropical Tree Crops*. CIRAD-Science Publishers, Montpellier.

Martínez, A. T. (2002) Molecular biology and structure-function of lignin-degrading heme peroxidases. *Enzyme Microbial Technology* 30: 425–44.

Martínez, G. (2009) Identificación temprana y manejo integrado de la pudrición del cogollo. *Palmas* 30: 63–77.

Mercière, M., Boulord, R., Carasco-Lacombe, C., Klopp C., Lee, Y. P., Tan, J. S., Sharifah Shahrul Rabiah, S. A., Zaremski, A., De Franqueville, H., Breton, F. and Camus-Kulandaivelu, L. About *Ganoderma boninense* in oil palm plantations of Sumatera and Peninsular Malaysia: Ancient population expansion, extensive gene flow and large scale dispersion ability. *Fungal Biology*, 121(6–7): 529–40.

Mercière, M., Laybats, A., Carasco-Lacombe, C., Tan, J. S., Klopp, C., Durand-Gasselin, T., Sharifah Shahrul Rabiah, S. A., Camus-Kulandaivelu, L. and Breton, F. (2015) Identification and development of new polymorphic microsatellite markers using genome assembly for *Ganoderma boninense*, causal agent of oil palm basal stem rot disease. *Mycological Progress* 4(11), e103(11p).

Meunier, J. 1991. Una posible solución genética para el control de la pudrición de cogollo en la palma aceitera: Híbrido interespecífico *Elaeis oleifera* × *Elaeis guineensis*. *Palmas* 12: 39–42.

Michalakis, Y. and Excoffier, L. (1996) A generic estimation of population subdivision using distances between alleles with special interest to microsatellite loci. *Genetics* 142: 1061–4.

Miller, R. N. G., Holderness, M., Bridge, P. D. and Chung, C. F. (1999) Genetic diversity of *Ganoderma* in oil palm plantings. *Plant Pathology* 48: 595–603.

Mistry, J., Finn, R. D., Eddy, S. R., Bateman, A. and Punta, M. (2013) Challenges in homology search: HMMER3 and convergent evolution of coiled-coil regions. *Nucleic Acids Research* 41: e121.

Morgenstern, I., Klopman, S. and Hibbett, D. S. (2008) Molecular evolution and diversity of lignin degrading heme peroxidases in the Agaricomycetes. *Journal of Molecular Evolution* 66: 243.

Murphy, D. J. (2007) Future prospects for oil palm in the 21st century: Biological and related challenges. *European Journal of Lipid Science Technology* 109: 296–306.

Nei, M. (1987) *Molecular Evolutionary Genetics*. Columbia University Press, New York, NY.

Nyanhongo, G. S., Couto, S. R. and Guebitz, G. M. (2006) Coupling of 2,4,6-trinitrotoluene (TNT) metabolites onto humic monomers by a new laccase from *Trametes modesta*. *Chemosphere* 64: 359–70.

Paterson, R. R. M. (2007) *Ganoderma* disease of oil palm – A white rot perspective necessary for integrated control. *Crop Protection* 26: 1369–76.

Pilotti, C. A., Sanderson, F. R. and Aitken, E. A. B. (2003) Genetic structure of a population of *Ganoderma boninense* on oil palm. *Plant Pathology* 52(4): 455–63.

Prendergast, A. G. (1963) A method of testing oil palm progenies at the nursery stage for resistance to vascular wilt disease caused by *Fusarium oxysporum*. *Journal of West Africa Institute Oil Palm Research* 4: 156–75.

Pritchard J. K., Stephens, M. and Donnelly, P. (2000) Inference of population structure using multilocus genotype data. *Genetics* 155: 945–59.

Qian, J., Xu, H., Song, J., Xu, J., Zhu, Y. and Chen, S. (2013) Genome-wide analysis of simple sequence repeats in the model medicinal mushroom *Ganoderma lucidum*. *Gene* 512(2): 331–6.

Rao, A. K. (1990) Basal Stem Rot (*Ganoderma*) in oil palm smallholdings – IADP Johore Barat experience. In Ariffin, D. and Jalani, S. (Eds) *Proceedings of the Ganoderma Workshop, 11 September 1990*. Palm Oil Research Institute of Malaysia, Bangi, Selangor, Malaysia, pp. 113–31.

Rees, R. W., Flood, J., Hasan, Y., Wills, M. A. and Cooper, R. M. (2012). *Ganoderma boninense* basidiospores in oil palm plantations: evaluation of their possible role in stem rots of *Elaeis guineensis*. *Plant Pathology* 61(3): 567–78.

Rees, R. W., Flood, J., Hasan, Y. and Cooper, R. M. (2007) Effect of inoculum potential, shading and soil temperature on root infection of oil palm seedlings by the basal stem rot pathogen *Ganoderma boninense*. *Plant Pathology* 56: 862–70.

Rees, R. W., Flood, J., Hasan, Y., Potter, U. and Cooper, R. M. (2009) Basal stem rot of oil palm (*Elaeis guineensis*); mode of root infection and lower stem invasion by *Ganoderma boninense. Plant Pathology* 58: 982–9.

Renard, J. L. and de Franqueville, H. (1991) Effectiveness of crop techniques in the integrated control of oil palm vascular wilt. *Oléagineux*. 46: 255–65.

Renard, J. L., Gascon, J. P. and Bachy, A. (1972) Research on vascular wilt disease of the oil palm. *Oléagineux* 27: 581–91.

Renard, J. L. and de Franqueville, H. (1989) Oil palm vascular wilt. *Oléagineux* 44: 341–9.

Robertson, J. S. (1960) Plant pathology division. In *8th Annual Report West African Institute for Oil Palm Research*. Benin City, pp. 106–12.

Rosenquist, E. A. (1990) An overview of breeding technology and selection in *Elaeis guineensis*. In Jalani, B. S. and et al. (Eds) *Proc. 1989 Int. Palm Oil Dev. Conf. – Agriculture*. Palm Oil Res. Inst. Malaysia, Kuala Lumpur, pp. 5–25.

Sato, Y., Wuli, B., Sederoff, R. and Whetten, R. (2001) Molecular cloning and expression of eight laccase cDNAs in loblolly pine (*Pinus taeda*). *Journal of Plant Research* 114: 147–55.

Sharma, K. K., Shrivastava, B., Sastry, V. R. B., Sehgal, N. and Kuhad, R. C. (2013) Middle-redox potential laccase from *Ganoderma sp.* : Its application in improvement of feed for monogastric animals. *Scientific Reports* 3: 1299.

Singh, G. (1991) *Ganoderma* – the scourge of oil palms in the coastal areas. *The Planter* 67: 421–44.

Sirrul, A. H. F., Mohd Rizuan, Z. A., Noor Hisham, H. and Mohd Na'aim, S. (2016) Kesan kaedah tanaman bawahan dan tanpa sanitasiterhadap kejadian penyakit reput batang sawit akibat *Ganoderma* di kawasan tanaman semula di ladang Felda. *Buletin Kemajuan Penyelidikan* 63: 43–52.

Sjöström, E. (1993) *Wood Chemistry, Fundamentals and Applications*, 2nd Edition. Academic Press, London.

Soepena, H., Purba, R. Y. and Pawirosukarto, S. (2000) A control strategy for basal stem rot (*Ganoderma*) on oil palm. In Flood, J., Bridge, P. D. and Holderness, M. (Eds) *Ganoderma Diseases of Perennial Crops*. CABI Publication, UK, pp. 83–8.

Stanke, M. and Waack, S. (2003) Gene prediction with a hidden Markov model and a new intron submodel. *Bioinformatics* 19: ii215–25.

Stevenson, M., (2017) Untangling the great palm oil debate. http://www.palmoilresearch.org/statistics.html (01 February 2017).

Tan, J. S., Chin, C. F., Massawe, F., Mayes, S. and Sharifah Shahrul, R. S. A. (2011) Markers for Ganoderma Disease Diagnosis and A Use of Thereof. PCT patent filing (31 October 2011).

Tee, S. S., Tan, Y. C., Abdullah, F., Ong-Abdullah, M. and Ho, C. L. (2013) Transcriptome of oil palm (*Elaeis guineensis* Jacq.) roots treated with *Ganoderma boninense. Tree Genetics & Genomes* 9(2): 377–86.

Thompson, A. (1931) Stem-rot of the oil palm in Malaya. *Bulletin, Science Series 6, Department of Agriculture, Straits Settlements and FMS*, Kuala Lumpur.

Torres, G. A., Sarria, G. A., and Martínez, G. (2010) *Identificación temprana y manejo de la pudrición del cogollo de la palma de aceite*, 1st ed. Cenipalma-Fedepalma-Sena SAC, Bogotá.

Torres, G. A., Sarria, G. A., Martinez, G., Varon, F., Drenth, A., and Guest, D. I. (2016) Bud rot caused by *Phytophthora palmivora*: A destructive emerging disease of oil palm. *Phytopathology* 106: 320–9.

Turner, P. D. (1981) *Oil Palm Diseases and Disorders*. Oxford University Press, Oxford, pp. 88–110.

Upadhyay, P., Shrivastava, R. and Agrawal, P. K. (2016) Bioprospecting and biotechnological applications of fungal laccase. *3 Biotech* 6: 15

Utomo, C. and Niepold, F. (2000) Development of diagnostic methods for detecting ganoderma–infected Oil palms. *Journal of Phytopathology* 148(9–10): 507–14.

Utomo, C., Werner, S., Niepold, F. and Deising, H. B. (2005) Identification of *Ganoderma*, the causal agent of basal stem rot disease in oil palm using a molecular method. *Mycopathologia* 159: 159–70.

van de Lande, H. L. (1983) Vasucular wilt disease of oil palm (*Elaeis guineensis* Jacq.) in Para, Brazil. *Oil Palm News* 27: 6–10

Wakefield, E. M. (1920) Diseases of the oil palm in West Africa. *Kew Bulletin* 1920: 306–8.
Ward, G., Hadar, Y. and Dosoretz, C. G. (2004) The biodegradation of lignocellulose by white rot fungi. In Arora, D. K. (Ed.) *Fungal Biotechnology in Agricultural, Food, and Environmental Applications*, Vol 21. Marcel Deckker, New York, pp. 393–407.
Wardlaw, C. W. (1950) Vascular wilt disease of the oil palm caused by *Fusarium oxysporum* Schl. *Tropical Agricultural, Trinidad*, 27: 42–7.

Chapter 2

Diseases affecting oil palm

Elizabeth Alvarez, CIAT, Colombia

1 Introduction

Oil palm production is one of the most important agricultural activities in South American and Central American countries, including Brazil, Colombia, Costa Rica and Ecuador, and constitutes a key alternative for generating employment, income and foreign currency. It is also considered strategic for national economies. In Asia (India) and Africa (Ghana, Cameroon and Nigeria), oil palm is one of the most important oil-producing crops.

Diseases have been some of the most important constraints to oil palm production. The oil palm disease-causing phytoplasmas that remain a major constraint presently were first reported by Turner (1981). An early report of *Phytophthora* disease from Latin America is from Alvarez (1998) and Sanchez et al. (1999). All of the major plant disease-causing agents – fungi and phytoplasmas – cause economically important damage to oil palm through large parts of the regions, countries and continents within which the crop is grown. However, there are important differences in patterns of distribution. Indeed, it is important to highlight here that these differences mean that there continue to be significant risks of the spread of pathogens with a limited current distribution to parts of the world in which they do not currently occur. Although many of the diseases of oil palm are most readily

http://dx.doi.org/10.19103/AS.2017.0018.14

spread through seed, several have important interactions with insect vectors. An important feature of this interaction is that changes in abundance levels or patterns of distribution of vectors can have important consequences for the transmission and geographical spread of the pathogens. Since insects can be particularly sensitive to abiotic factors, such as temperature and rainfall, this also means that present and anticipated climate changes will have a significant impact on transmitted pathogens.

Oil palm diseases affect all parts of the plant: leaves, stem, flowers and roots, although the most common symptoms are chlorosis and blights in leaves and soft or dry rot in the roots. Losses caused by oil palm diseases vary greatly depending on the region, local environmental conditions and oil palm variety. Overall, however, it is likely that more than a third of global palm oil production, equivalent to several billion US dollars, is lost annually. Control efforts have, therefore, become more important although much remains to be done to improve and extend management efforts.

2 Oil palm basal stem rot (BSR) disease, *Ganoderma* spp.

2.1 BSR in Africa

In *Africa*, one of the most devastating diseases, called basal stem rot (BSR) disease, is caused by the genus *Ganoderma*, a soilborne fungus that threatens the oil palm industry, especially in Ghana, Nigeria and Cameroon. *Elaeis guineensis* (African oil palm) palms are the main hosts for BSR disease.

Biology

The genus *Ganoderma* belongs to the class Basidiomicetes. *Ganoderma* spp. are referred to as white rot fungi that degrade lignin as well as other wood components. Mainly, three species of *Ganoderma* (*G. boninense, G. zoanatum* and *G. miniatocinctum*) are the cause of BSR disease (Naher et al. 2015). *Ganoderma* is characterized by basidiocarps, large, perennial and woody brackets that are lignicolous. The fruit bodies grow in a hoof-like form on the trunks of trees. They have double-walled, truncated spores with yellow-to-brown ornamented inner layers (Hushiarian et al. 2013).

Studies of *G. boninense* have found that oil palm plantations have great diversity. The small ribosomal subunit RNA (in prokaryotes 16S and in eukaryotes 18S) gene is one of the most important molecular markers, with a range of applications in biodiversity screening, phylogenetic analyses and evolutionary studies (Meyer et al. 2010).

A phylogenetic study of *Ganoderma* using single-locus mt SSU rDNA led to dividing *Ganoderma* into six distinct monophyletic groups and indicated that complex situations related to the geographic region and the pathogen–host relationship must be considered as well as the phylogenetic relationships (Soon Gyu and Jung 2004).

Symptoms and effects on yield

The symptoms of BSR are decay of the bottom of the stem from where basidiocarps emerge and, sometimes, also decay of the roots. Stem rotting restricts the uptake of water and nutrients to the fronds, causing chlorosis. When the disease is more advanced, the older fronds wilt and hang down to form a skirt around the trunk (Turner and Gillbanks 1974). Other observable symptoms are flattening of the crown and spear leaves that have not opened. In the most severe cases, the stem might even fracture (Rees et al. 2012).

Once the plant dies, colonization of basidiomata can develop rapidly along the entire trunk. In young palm, decay of the soft tissues of the stem occurs rapidly compared with mature palm (Naher et al. 2015).

The disease can decrease the yield of infected palms by killing them or by reducing the weight of the fresh fruit bunch in infected palms (Assis et al. 2016).

Epidemiology

Ganoderma spreads in the soil through roots and the air. Wind, rain and insects all assist in carrying spores to wounds on trees, most commonly those that have been cut. In particular, the *Oryctes* beetle (Turner 1981) and larvae of the *Sufetula* spp. caterpillar play at least a small role in spreading *Ganoderma* spores (Genty et al. 1976). Experiments in which an enormous number of *Ganoderma* spores were released in a field but did not infect most trees (Ho and Nawawi 1986) have indicated that infected tissues in the soil are more likely to spread the disease to healthy roots than airborne spores.

Control

Some control measures are used to minimize economic loss from the disease. Soil mounding, in which soil is heaped around the trunk to a height of 75 cm, is an effective technique to prevent BSR in oil palm. The technique may prolong the yielding life of oil palm affected by BSR. Digging trenches around infected palms prevents mycelium spread by root contact with healthy palms (Ho and Khairudin 1997; Naher et al. 2015). Fungicide treatment is another approach to control BSR in oil palms. It is mainly used to protect seeds from infection and to control BSR. Fungicides such as triadimefon, cycloheximide and benomyl inhibited *Ganoderma* growth. But, they are not recommended. The use of species of *Trichoderma* as biological control agents showed their potential antagonistic effect in controlling other fungi (Soepena et al. 2000). The BSR was lower in a field treated by *Trichoderma harzianum* than in an untreated field (Susanto et al. 2005). The effect of isolates of *Bacillus* was tested on *G. boninense* growth on oil palm seedlings (Suryanto et al. 2012) and resistant cultivars are alternative means to control BSR. The use of resistant-planting material offers the best alternative to control BSR. Removing or destroying infected palms through sanitation remains one of the most important measures for controlling BSR.

2.2 BSR in Asia

In *Asia*, Indonesia and Malaysia produce 87% of the world's total palm oil. Fungi that rot and kill oil palm could cost Southeast Asian countries USD 500 million a year (Ommelna et al. 2012). The BSR, a disease caused by *G. boninense*, is considered a serious threat to the oil palm industry in these countries. The BSR caused by *Ganoderma* spp. causes significant losses up to 80% on oil palm plantations in Southeast Asia (Susanto 2009).

Biology

BSR is caused by numerous species of the fungus *Ganoderma*. Pathogenic species for oil palm have a wide host range. In Southeast Asia, the BSR disease that is caused by the white rot fungi of the genus *Ganoderma* is the major disease affecting the oil palm industry.

Symptoms and effects on yield

Internal symptoms of BSR appear like brown rot marked by darker coloured bands. There are yellow transitional areas between healthy and diseased tissues. Small wide buttons of tissue mark the beginnings of *Ganoderma* sporophores. Typical bracket-shaped sporophores of *Ganoderma* showed a shiny upper surface and whitish border. Approximately 60% of the plantations in Malaysia reported diseased trees. In East Malaysia, the lack of good field training is a contributing factor for production losses (Murphy 2014). It has been reported that economic loss caused by *G. boninense* can reach USD 500 million (Ommelna et al. 2012).

Epidemiology

Survival and development in the soil vary between species. Soil type has been implicated with the pure structure of coastal clays, and the impermeable soil of marine origin is thought to lead to increased *Ganoderma* attack. High incidence has been associated with a high water table limiting rooting or where flooding occurs. Spread and infection can occur through roots (Cooper et al. 2011).

Control

There are sources of natural genetic disease resistance. Genetic resistance to BSR of oil palm is a major component of an integrated control strategy for the disease. Early detection of the level of resistance is of importance for the breeding and sustainability of this crop in Southeast Asia. A set of standard crosses was developed by both Socfindo and Lonsum to provide results from different trials using different *Ganoderma* isolates (Breton et al. 2009). Research efforts have focused on understanding the molecular defence responses in plant–pathogen interactions to find practical solutions to control the disease (Ho and Tan 2015). Lignin structure modification by breeding could result in genetic resistance. Candidate genes for breeding BSR resistance are sought in transcripts and proteins that alter their expression upon infection (Ho and Tan 2015). Phytosanitary practices, biological control with *Trichoderma* spp., and palm endophytes are practical solutions to manage the disease (Mohammed et al. 2014). *Trichoderma* and its efficacy as a biocontrol agent of BSR of oil palm have been reported by Ilias (2000).

Management of *Ganoderma* in peat soil in Indonesia was reported by Lim and Udin (2010).

2.3 BSR in Central America

In Central America and northern regions of South America, *Elaeis oleifera* (Kunt, Cortez) is known as the American oil palm. *Elaeis oleifera* populations grow along riverbanks, tolerating shade and flooding, thus showing environmental adaptability (Corley and Tinker 2003). Colombia, Suriname and northwest Brazil are considered the species' centre of origin (Ooi et al. 1981). Oil palm fruits are the richest plant source of provitamins A and E. The genus *Ganoderma* is a major problem.

In the state of Pará, Brazil, the crop is often considered as an industrial crop, but in many areas, it is a valuable smallholder crop (Feintrenie et al. 2010). Government policies favour smallholder involvement in the oil palm industry.

Biology

Basidiocarps are clear to dark brown in colour with shiny texture. Basidiospores have a truncated apex, in some cases with a 'cap', and free piles (dots in the upper part of the basidiospore), with the width being thin to medium thick.

The morphological characteristics of basidiocarps identified the fungus within the genus *Ganoderma*, subgenera *Elfvingia* and *Lucidum*.

Symptoms and effects on yield

In Colombia, BSR is considered the most important emerging infectious disease threatening oil palm because of its direct effects, with a direct impact on yields. Symptoms include multiple spear leaves, the oldest fronds wilt and hang down to form a skirt around the trunk (Turner and Gillbanks 1974). In the most severe cases, the stem might even fracture (Rees et al. 2012). The proliferation of adventitious roots is visible only when the internal damage is in a very advanced state.

Disease symptoms were described on two plantations in El Copey, Cesar, in northern Colombia.

Epidemiology

Airborne basidiospores are of high importance in the epidemiology of *Ganoderma*.

Control

Ganoderma was found on four plantations of the northern coast of Colombia and eradication was recommended as a disease-control strategy. The number of cases reported was low.

Triazole triadimenol (Bayfidan) 250EC 2.5 g ia/palm by root absorption and PCNB 5 g ia/palm by stem injection were used as chemical controls.

3 Oil palm dry basal rot disease, *Thielaviopsis paradoxa*

3.1 Dry basal rot disease in South America

Ceratocystis paradoxa (anamorph: *Thielaviopsis paradoxa*) is parasitic on a range of economic and food crops and is the cause of dry basal rot, a limiting disease in oil palm.

Biology

The pathogenic and genetic diversity of *Thielaviopsis* isolates from oil palms in Colombia, Ecuador and Brazil was evaluated. A total of 164 strains of *Thielaviopsis paradoxa* were characterized using pathogenicity tests, random amplified polymorphic DNA (RAPD) markers and polymerase chain reaction (PCR) sequencing of the ITS region of 5.8 S ribosomal DNA (Alvarez et al. 2012a).

Population structure analyses of the RAPD data suggested that most of the isolates obtained in this study belonged to a single population. The genetic diversity of the isolates from South America was intermediate and, therefore, *Thielaviopsis paradoxa* is likely to be predominantly clonal compared with *Ceratocystis* species. Sporadic sexual reproduction may occur for *Thielaviopsis paradoxa,* but is secondary to clonal reproduction. Data on pathogen diversity will provide information on breeding strategies and population structures (Alvarez et al. 2012a).

Symptoms and effects on yield

Dry basal rot, caused by the vascular pathogen *Thielaviopsis paradoxa* (teleomorph: *Ceratocystis paradoxa*), can attack the stem (Fig. 1a), leaves and fruits, causing premature fruit drop (Kile, 1993) (Fig. 1b).

The disease is characterized by dry rot of the bud tissue with necrotic lesions of different colours (Buitrago and Nieto 1995).

It can be economically significant, for example, in northern Colombia, where 70% of the plantations are affected (Tovar and Nieto 1998). It is also economically important in the southern region of Bahia, Brazil.

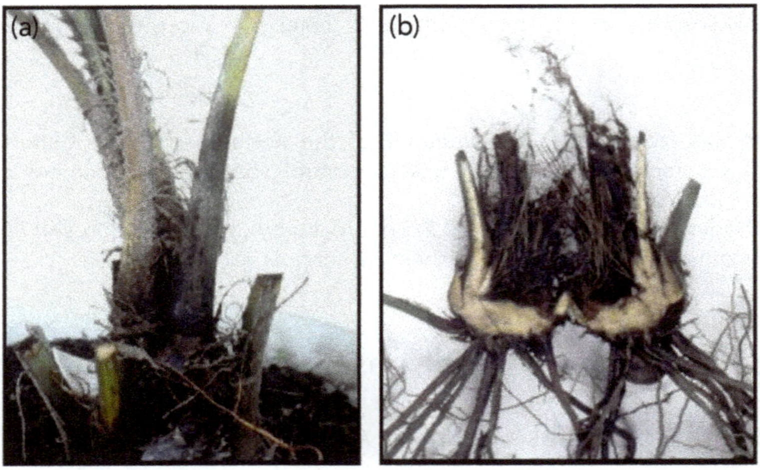

Figure 1 (a) Oil palm inoculated with *Ceratocystis paradoxa.* (b) Transversal cutting of oil palm inoculated with *Ceratocystis paradoxa.*

Epidemiology

This disease affects palms of any age and is prevalent in lowland areas where the average annual rainfall is 2 500 mm/year or above.

Oil palm seeds are mostly free from *Thielaviopsis paradoxa*, so transmission through seedlings is low and some other mechanisms of disease inoculum dispersal must be postulated (Alvarez et al. 2012).

In this sense, the relatively frequent isolation of this pathogen from different plant tissues of oil palm trees indicated to us that dispersal might be by airborne spores or by infected harvesting and pruning equipment. Because old leaves of oil palms are carefully pruned during their growth cycle, this second dispersal mechanism may be common, as it is in other crops according to Kile (1993).

Control

Knowledge of the genetic and pathogenic diversity and population structure of *Ceratocystis paradoxa* may help when designing effective control measures such as breeding for resistance or biological and chemical control (Alvarez et al. 2012).

The association of *Thielaviopsis paradoxa* with different palm genotypes needs to be studied in order to identify resistance sources.

3.2 Dry basal rot disease in Africa and Asia

In *Africa*, dry stem rot caused by *Thielaviopsis* can affect oil palms in commercial cultivars grown in Cameroon and Nigeria (Tovar and Nieto 1998).

Biology

Oil palm *Ceratocystis* isolates from trunks, cut basal ends of leaf fronds and rotting palm nuts were collected in Cameroon. The DNA was isolated with the CTAB protocol described by Moller et al. (1992).

For the identification of *Ceratocystis* isolates, internal transcribed spacers ITS1 and ITS2 and intervening 5.8S rDNA of the ribosomal RNA gene cluster (ITS) were amplified with PCR and sequenced. Additional genetic regions of the β-tubulin gene and TEF-1a were sequenced (Mbenoun et al. 2014).

Multigene DNA phylogenics in combination with mating studies and careful morphological examination revealed that the *Ceratocystis* complex includes more species than the five described previously (Harrington 2009).

Symptoms

Disease symptoms are characterized by bud and trunk rots.

Epidemiology

In the future, a more comprehensive sampling of taxa with larger sample sizes for each taxon may reveal more alternative hosts among, for example, grasses. Other challenges within the field of lethal wilt (LW) epidemiology include the identification of insect vectors (Bila et al. 2015).

Control

In West Africa, biological control of *Ceratocystis paradoxa* causing black seed rot in oil palm sprouted seeds was accomplished by *Trichoderma* species.

Trichoderma viride, Trichoderma polysporum, Trichoderma hamatum and *Trichoderma aureoviride* were used as potential biological agents for seed treatments against *Ceratocystis paradoxa* causing black seed rot in oil palm sprouted seeds (Eziashi et al. 2006).

The emergence (survival) and the growth of oil palm seedlings from *Trichoderma* species–treated oil palm sprouted seeds were significantly higher than the emergence and the growth observed from the control treatment. *Ceratocystis paradoxa*–infested seeds had 63% to 71% survival and average seedling height of 18.4 to 25.1 cm. Seeds without pathogen had 100% survival with average seedling height of 38.2 to 46.5 cm. The effects of the bioagents were similar or higher than the effects obtained from benlate solution.

Occurrence in Asia

Common spear rot (CSR), which is also known as crown disease, was first reported in Indonesia in the 1920s. It has caused considerable losses in young oil palm plantings. Symptomatic spear leaves were collected from oil palm plantations and farm plots in South Sumatra, North Sumatra and Bangka-Belitung, Indonesia (Akino and Kondo, 2012).

In a study conducted by Akino and Kondo (2012), Koch's postulate experiments showed that *Ceratocystis paradoxa* was able to infect wounded oil palm leaves causing a symptom of extensive rotting similar to that found in the field. It was confirmed that *Ceratocystis paradoxa* is one pathogen that is associated with CSR of oil palm in Indonesia.

4 Oil palm bud rot disease, *Phytophthora palmivora*

One of the pseudofungal diseases affecting oil palm is root rot caused by *Phytophthora* species from the genus of plant-damaging oomycetes (Alvarez et al. 1998; de Franqueville 2003; Faparusi 1973). *Phytophthora palmivora* has been reported to attack oil palm (Joseph and Radha 1975; Alvarez et al. 2010; Torres et al. 2016).

4.1 Biology

Oil palm bud rot (OPBR) disease is caused by the pseudofungus *Phytophthora palmivora*. The sporangia of *Phytophthora palmivora* isolates from oil palm are papillate and limoniform. In Colombia, the isolates were characterized by morphology light microscopy, ELISA enzyme immunoassay, PCR and hybridization with a specific probe for *Phytophthora* spp. (Alvarez 2007). A homogeneous and reproducible band of about 900 bp was obtained with PCR (ITS1 and ITS4), which, when digested by restriction enzymes, showed an equal pattern of bands for all isolates. The isolates thus belonged to one species. The RAPD analysis revealed intraspecific genetic diversity, demonstrating a wide range of genetic diversity in the pathogen population. Restriction fragment length polymorphism (RFLP) analyses provided genetic evidence of a well-characterized group among isolates.

Phytophthora palmivora is characterized by oval and ellipsoid sporangia with a short pedicel. The sporangia are obovoid, limoniform or ellipsoid and have tapered bases.

Phytophthora strains were examined for differences in morphology, pathogenicity and DNA polymorphism. The strains were chosen to be diverse according to geographic origin. Characterization of the strains by morphology was difficult because not all strains produce fungal structures. The strains caused typical symptoms in oil palm, but varied in pathogenicity. The DNA polymorphism was determined by both RFLP of the ribosomal ITS and 5.8S DNA amplified by PCR. Restriction digest with *Hin*dI, of the product amplified for the ITS region, showed three different restriction patterns that corresponded to three species, respectively. Twenty random 10-mer primers were tested, and four gave reproducible bands. The results suggested a high genetic diversity among the strains tested (Alvarez et al. 1999).

The PCR amplification of the ITS region followed by digestion with restriction enzymes can be used to separate the species. A collection of *Phytophthora* strains obtained from different regions in Colombia showed high morphological and pathogenic variations (Llano et al. 2006). *Phytophthora palmivora* was identified as the most important species causing bud rot in Colombia (Alvarez 2010; Torres et al. 2016). A *Phytophthora* sp. was isolated from commercial plantations cultivated in Ecuador, Brazil, and Colombia showing typical symptoms of bud rot. Pathogenicity was evaluated in 6- and 24-month-old palms under controlled conditions in the greenhouse and in leaf fragments placed in humidity chambers. Isolates of this *Phytophthora* sp. were found to be pathogenic on both 6 and 24-month-old palms, and *Phytophthora* sp. was re-isolated from infected tissue, therefore fulfilling Koch's postulates. Through PCR, the causal agent was identified using A2/I2 primers specific to the genus *Phytophthora*, which amplified a 788-bp fragment located in the internal transcribed spacer (ITS) region. The PCR product was digested with the restriction enzymes MspI, RsaI, and TaqI, generating fragments that corresponded to *P. palmivora*. With the sequences obtained, molecular beacon probes and primers specific to the species were designed. This new method permitted timely, sensitive, and specific diagnoses of the pathogen (Alvarez et al. 2010). In Colombia, isolate identifications were confirmed using an ITS marker. Host–pathogen interactions were also studied under greenhouse conditions (Alvarez et al. 1998).

Inoculated 2-year-old oil palm variety Gunung Melayu, inoculated with the puncture method at the base of the stem, showed symptoms 20–25 days after inoculation, starting with chlorosis of the spear and flag leaf, followed by a dry necrosis (browning of the tissue) in a progressive form (Sánchez et al. 1999) (Fig. 1).

The pathogen (*Phytophthora* spp.) was re-inoculated from the affected tissue, and its identification was confirmed by morphology in culture medium and light microscopy. *Phytophthora* spp. initially caused a yellowing or chlorosis of leaves, resulting in browning or dry necrosis of the affected tissue (Alvarez et al. 2007).

4.2 Symptoms and effects on yield

OPBR symptoms are encountered where oil palm crops are poorly drained or on farmland recently converted from humid forest and it can be difficult to determine the source of primary infection. Bud rot initiated by primary-infecting pathogens is frequently exacerbated by subsequent colonization by saprophytic species. Disease symptoms on spear and young leaves are characterized initially by chlorosis and finally by necrosis (Fig. 2a) (Alvarez 2007).

In Colombia, up to 80% of the oil palm has been reported to be affected by bud rot, whereas in southern India, production losses of 70% have occurred. Bud rot is a major constraint to oil palm production in areas of the eastern plains and Tumaco in Colombia.

Bud rot has destroyed more than 70 000 ha of oil palm in the western and central oil palm–growing regions of Colombia (Torres et al. 2016).

Figure 2 (a) Necrosis on spear leaf and flag leaf in oil palm inoculated with *Phytophthora palmivora*. (b) Negative control (inoculated with water).

4.3 Epidemiology

Phytophthora species are spread through the use of seeds obtained from an infected plant. Easy dispersion hinders management of the disease. Transmission also occurs through infected soil, which is common in fields that are flooded. The pathogen survives in the soil. Disease incidence is the highest during the rainy season, on humid soils. To prevent introducing the pathogen into disease-free regions, a real-time PCR diagnostic method has been developed to detect *Phytophthora palmivora* in infected tissues (Alvarez et al. 2010).

Phytophthora species attack oil palm plants, especially under favourable conditions such as poor drainage, compacted soil and usage of infected seed.

4.4 Control

Strategies to control *Phytophthora* rot have focused on the use of resistant oil palm hybrids. In a biological control strategy, *Trichoderma harzianum* and *Trichoderma viride* reduced percentage infection and increased production.

Resistance to the disease is reported in the South American species *Elaeis oleifera* (*Elaeis melanococea*) and hybrids between these and *Elaeis guineensis.*

Trials were carried out on oil palm plantations to evaluate the effect of induced resistance to bud rot, applied in two ways: injection to the trunk below the meristem and absorption by the roots. Phosphoric acid was used in concentrations of 25% and 40%, and the product NF (phosphoric acid: 434 g/L P_2O_5, and potassium hydroxide and potassium citrate:

403 g/L K_2O). Sterile water was used as a control. There were no significant differences between injection to the trunk and root absorption. It was concluded that bud rot could be controlled using inducers of resistance based on potassium and phosphorus, as part of integrated management of the disease (Alvarez et al. 2003).

The most effective means of preventing infection and the spread of OPBR is by planting disease-free seed. Strict quarantine and surveillance measures are required to prevent its introduction to Asia or Africa. The use of resistant parental genotypes in breeding programmes combined with phenotypic field selection under high disease pressure should enhance control.

The most economical solution for bud rot control in the medium and long term is the incorporation of American palm resistance into commercial hybrid materials without affecting the performance of commercial palm oil materials and preserving the quality of the oil (Alvarez et al. 2007).

In advanced stages of the disease, the emerged outer part of the spear leaf looks totally dry, while the white unemerged central tissues and the tissues above the meristem are destroyed (Sarria et al. 2013). A final symptom is the destruction of the young tissue and the meristem (Torres et al. 2016).

5 Oil palm vascular wilt disease, *Fusarium oxysporum* f.sp. *elaeidis*

In Africa, vascular wilt is the most destructive disease of oil palm. The disease was first described in the Democratic Republic of Congo (Wardlaw 1946) and also has been reported in Cote d'Ivoire, Nigeria, Ghana, Cameroon and Congo (Flood 2006).

In South America, the disease had been reported in Brazil (Van de Lande 1984) and Ecuador (Renard and de Franqueville 1989). The disease has not yet been reported in Malaysia or in Southeast Asia (Rusli et al. 2013).

5.1 Biology

Fusarium oxysporum Schlechtend.: Fr. f.sp. *elaeidis* Toovey is a soilborne fungus that produces macroconidia, microconidia and chlamydospores. Chlamydospores allow survival in plant debris and soil.

Flood et al. (1992) reported that all pathogenic isolates from Zaire were vegetatively compatible with each other and were designated vegetative compatibility group (VCG) 0140, while another VCG (0141) occurred in Brazil. Later, Mouyna et al. (1996) reported that South American isolates were compatible with isolates from Ivory Coast and allocated to VCG 0141.

5.2 Symptoms

Disease symptoms are acute wilt in which leaves dry out and die rapidly while retaining their original erect positions on the plant until broken off, usually several feet from the base, by wind action. The disease progresses rapidly and palms die within two or three months.

Additional disease symptoms are characterized by chronic wilt in which the palm remains alive for months or years and becomes progressively stunted. The older fronds become desiccated and hang around the stem. These symptoms progress gradually, with younger fronds becoming affected. The youngest fronds produced in the crown remain erect, often chlorotic and reduced in size (stunted). The apex of the trunk may be reduced in diameter (Flood 2006).

5.3 Effects on yield

Renard and de Franqueville reported substantial yield reductions (from 6 to 16%) in 6-year-old replanted palms when only a small per cent (2.5–5.5% of plants) actually showed external symptoms. These authors attributed most of this yield reduction to 20–30% of palms that were actually infected but appeared healthy (Renard and de Franqueville 1989).

5.4 Epidemiology

The pathogen is soilborne and spread is thought to occur through root contact with dead, infected palm tissues (Renard and de Franqueville 1989). Palms killed by the pathogen become sources of nutrients for adjacent palms and these too become infected (Corley and Tinker 2003).

Contamination of oil palm seeds by the pathogen has also been reported, both on the outside of seeds and the kernel surface. Spores can be carried on oil palm seeds on the kernel surface.

Oil palm seeds are important in global breeding programmes, and there is a risk of long-distance transmission on contaminated seeds. The possibility of contaminated seeds giving rise to infected plants has been investigated (Flood 2006).

5.5 Control

The soilborne nature of the pathogen has made management of the disease difficult. The use of *Brachiaria* as a cover crop and a cultural practice such as the removal and burning of infected palms have been promoted. The disease can also be controlled through better field management, e.g., disease incidence can be halved by planting new palms at a distance of more than 2 m from old stumps. The application of potassium is reported to reduce disease incidence. Breeding for resistance remains a practical long-term method of management for this disease. Breeding programmes such as those in Ivory Coast have been successful, with the introduction of resistant varieties reducing losses in some areas from 20 to 30% to less than 3%.

Screening trials in the nursery where palms are deliberately inoculated have been introduced in several West African breeding programmes and generally correlate well with field results (Renard and de Franqueville 1989).

Chemical and heat as seed treatments are measures to control the disease. In addition, biological control with *Trichoderma* has been used (Flood 1994).

6 Oil palm phytoplasma diseases in South and Central America: lethal wilt disease (LWD)

Various groups and subgroups of phytoplasma have been reported to be associated with oil palm diseases in Latin America (Alvarez et al. 2014) and Asia (Jones and Turner, 1979; Mehdi et al. 2012). Also, a phytoplasma has been reported from oil palm in Mozambique (Bila et al. 2015).

Few phytoplasma groups have been reported to cause disease in oil palm. The diversity of phytoplasma is reported to be low from South America (*Candidatus* Phytoplasma asteris) and a similar number of groups are recognized from Southeast Asia (*Candidatus* Phytoplasma asteris) and Africa (*Candidatus* Phytoplasma asteris and *Candidatus* Phytoplasma palmicola). Only two subgroups have been identified as important pathogens of oil palm in South America (Alvarez et al. 2014).

Phytoplasmas are mycoplasma-like organisms that are often mistaken for plant pathogenic viruses. They are specialized bacteria that are obligate parasites of plants and that infect phloem tissue. They are often difficult to diagnose and detect. Discovered in the 1960s, phytoplasmas are pathogens that are becoming increasingly important in Latin America and the Caribbean. The most important oil palm diseases associated with phytoplasma infection are oil palm LWD and oil palm leaf stripe disease (LSD).

In this section, we will focus on LWD in South and Central America. Section 7 will then focus on LSD, while Section 8 discusses the occurrence of phytoplasma diseases in Asia and Africa.

The LWD has been reported in Colombia (Alvarez et al. 2014), Ecuador (Alvarez et al. 2013; Baer et al. 2013) and Brazil (Brioso et al. 2006; Montano et al. 2007). The LWD affects production and can cause severe economic loss. In Colombia, losses of more than USD 85 million have been reported. The LWD is associated with phytoplasma group 16SrI subgroup B (Alvarez et al. 2013).

Colombia is the major producer of oil palm (*Elaeis guineensis* Jacq.) in Latin America, and the fifth-largest grower worldwide, which plays an important role in biodiesel markets (Castiblanco and Mosquera 2011). Oil palm has become a paramount export crop and source of employment for the national economy. However, its production has declined by 7.1% since 2002, with a dramatic drop of 10% in northeastern Colombia (FEDEPALMA 2010). The cause of this decrease has been LWD ('marchitez letal' in Spanish) (Mejia de Los Rios 2014). The LWD was reported for the first time in the Llanos Orientales of Colombia in 1994. In Ecuador, the disease was reported for the first time in 2006 in the region of San Lorenzo, Santo Domingo, Quevedo and Orellana, with total crop losses of 200 hectares (Baer et al. 2013).

LW is present on the oil palm plantations (Pérez and Cayón 2010) of the Upía River area, Palmar del Oriente (1994), Palmas del Casanare (1999), Palmeras Santana (2000) and Palmeras del Upía (2002). The disease had decimated around 690 hectares and a total of 97 619 oil palms by 2010 (FEDEPALMA 2010; Mejia de Los Rios 2014).

The disease has been observed on Palmas Sicarare oil palm plantations in northern Colombia (Elizabeth Alvarez, personal communication, 4 October 2016).

6.1 Biology

A preliminary study detected phytoplasmas in symptomatic plants in commercial crops of the susceptible oil palm hybrid (*Elaeis guineensis* × *Elaeis oleifera*) (Alvarez and Claroz 2003) in Colombia.

Cell characteristics of phytoplasma were detected in sieve tube inflorescence phloem. The phytoplasma structures observed were pleomorphic, comprising round, elongate, dumbbell and ring-shaped elements, mostly 150–250 nm wide and 1000 nm long (Fig. 3). The phytoplasma structures were limited only to phloem tubes and were never seen in large quantities (Alvarez and Mejía 2005).

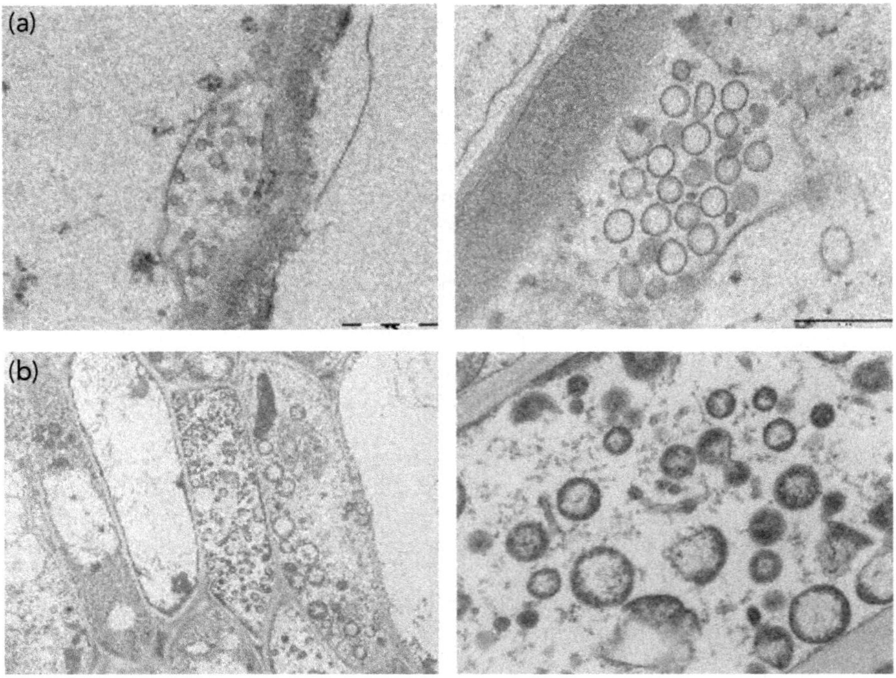

Figure 3 Transmission electron micrographs of phytoplasma cell of LWD. (a) Infected oil palm inflorescences and (b) positive control (affected periwinkle).

Nested PCRs primed by phytoplasma universal primer pair R16F2n/R2 resulted in the amplification of 1.2-kb DNA fragments of the 16S ribosomal DNA, indicating that the symptomatic oil palm plants were infected by phytoplasma (Alvarez 2006).

The RFLP analysis of the 1.2-kb 16S rDNA amplicons indicated that a phytoplasma belonging to subgroup 16SrI-B ('*Ca.* P. asteris') was present in all symptomatic oil palm plants. The RFLP patterns from the positive samples were indistinguishable from each other and from phytoplasma reference strains belonging to subgroup 16SrI-B (Alvarez et al. 2014).

PCR assays with the rpF1/rpR1 primer pair amplified the expected fragment length of about 1200 bp from 18 oil palm samples. RFLP analyses with four restriction enzymes produced RFLP profiles that were identical to each other. These allowed a clear differentiation of the two oil palm phytoplasma strains from those aster yellow (AY) strains, phytoplasma from Colombia (Fig. 4).

More in-depth phytoplasma differentiation can be obtained by studying polymorphisms in other gene sequences (Mitrović et al. 2011).

Phytoplasma strains of the 16SrI group were detected and characterized by multigene typing analysis based on four phytoplasma genes and distinguished from reference strains (Bertaccini et al. 2014).

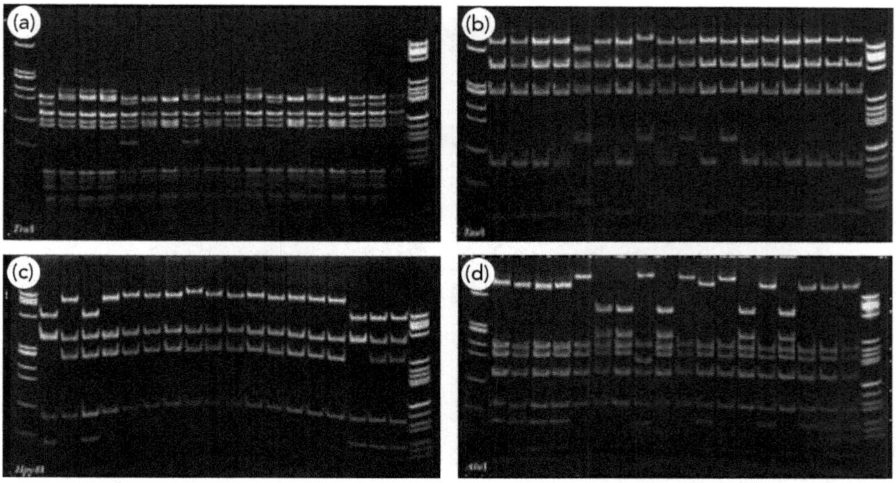

Figure 4 Restriction fragment length polymorphism (RFLP) patterns of oil palm phytoplasma strains OP45 and OP47 compared with several reference strains from periwinkle that were amplified with primers rpF1/rpR1 and digested with restriction enzymes *Tru*I (a), *Taa*I (b), *Hpy8*I (c) and *Alu*I (d). The acronyms of the strains are described in Table 1. Markers: phiX174, phiX174 *Hae*III digested; and pBR322, pBR322 *Hae*I digested.

The RFLP characterization of the *groEL*, *amp* and *rp* genes, together with sequence data, distinguished the AY strain detected in Colombian oil palm samples from other AY phytoplasmas used as reference strains (Alvarez et al. 2014).

The 16S rDNA is a valuable classification tool, but it is not always able to discriminate phytoplasma strains. Multigene sequence analysis based on *amp*, *groEL* and *rp* genes indicated that they are useful molecular markers to follow up on the Colombian oil palm epidemic (Mejía de Los Rios 2014).

The rpF1/rpR1 sequence from OP47 (1168 bp) was deposited in GenBank under accession number KF434318 (Mejía de Los Rios 2014).

A real-time PCR assay was developed to detect and quantify a 16SrI phytoplasma associated with oil palm LWD. A taqMan probe was designed for the microorganism, based on the *rp* gene (16SrI phytoplasma). By qPCR, sensitivity increased from 100- to 1000-fold of that obtained from nested PCR (Alvarez et al. 2012).

In Ecuador, from 16 samples from Palmar del Río in Orellana with LW symptomatology, amplification of the fragment of approximately 1430 base pairs in second nested PCR was obtained, using universal primers reported previously by Alvarez (2006) and Smart et al. (1996), associated with the presence of phytoplasma, whereas in none of the asymptomatic plants was an amplicon obtained (Baer et al. 2013).

6.2 Symptoms and effects on yield

Symptoms of LW commonly first appear as vascular discolouration and leaf yellowing in seven-year-old palms (i.e. flowering and fruiting). These symptoms are followed by leaf

Figure 5 Oil palms in Colombia showing mild (a, severity score 2) and severe (b, severity score 4) lethal wilt symptoms. Lethal wilt symptoms reported in oil palm in northern Colombia in young plants, and leaf drying and leaf discolouration (c and d).

drying, wilt necrosis of infected tissues and eventual plant collapse (Fig. 5). Root necrosis often accompanies leaf discolouration. Internal discolouration of trunk tissue may also occur, but does not represent a distinctive symptom.

LW is potentially destructive because it spreads rapidly and causes plant death within 4 to 6 months after symptoms first appear (Pérez and Cayón, 2010).

In Colombia, the disease causes vascular discolouration and leaf yellowing when the palm is mature. These symptoms are followed by leaf drying, wilt, necrosis of infected tissues and eventual plant collapse (Alvarez et al. 2014).

Oil palm production in Colombia presented a sharp decline in the first six months of 2002, registering a fall of 7.1%. In the regional scenario, the area that most contributed to this drop in production was the eastern region, where production decreased by almost 10%. An approximation of the losses incurred indicates that more than one hectare of oil palm was eliminated per month (150–200 plants in an area of 4000 hectares, which is equivalent to 0.34%) (Alvarez 2006; Alvarez et al. 2014).

LW-like symptoms were also observed in oil palms in Brazil affected by a disease known as 'fatal yellowing' (Brioso et al. 2006; Montano et al. 2007).

In Ecuador, the disease appears in both young plants and adult plants, after growing for three to four years. The first disease symptoms are leaflet yellowing and drying, beginning at the tips and edges, as can be seen in Fig. 6. The last part that reaches necrosis is the spear leaf or meristem (Baer et al. 2013).

6.3 Epidemiology

Phytoplasmas are commonly transmitted by phloem-feeding Hemiptera vectors. These include plant hoppers and leafhoppers of the genera *Macrosteles, Euscelis, Euscelidius, Scaphoideus* and *Cacopsylla* (Weintraub and Beanland 2006). The microorganism first multiplies in the intestinal cells of its insect vectors and subsequently in the hemolymph after passing through the salivary glands. It infects internal organs such as the thoracic ganglion and fatty bodies (lipids) (Kawakita et al. 2000). Transmission can also occur through grafting and the parasitic plant *Cuscuta*. Phytoplasma disease is disseminated through the movement of infected seeds. It has recently been reported that the leafhopper *Myndus crudus* (Cicadellidae) is able to transmit it from infected plants to healthy plants (Arango et al. 2011). In addition, Cortés et al. (2015) concluded that *Haplaxius crudus* is the insect vector of the causal agent of LW. Further studies are required, however, to determine how significant this vector transmission is as a factor in the field- and regional-level epidemiology of oil palm disease (Alvarez 2006).

LW spreads rapidly and causes plant death within 4–6 months after symptom onset (Pérez and Cayón 2010).

6.4 Control

An effective method of controlling phytoplasma diseases is through preventing infection by assuring the health and quality of seed. Improved quarantine procedures are required to prevent long-range disease spread. A thermotherapy treatment has been used in Colombia for cleaning oil palm seed. Use of a resistance inducer (Kendal) produced an increase in yield and reduced disease incidence (Elizabeth Alvarez, pers.

Figure 6 Yellowing and drying of leaflets with lethal wilt. To the left, the initial stage; to the right, the advanced stage.

comm., 3 October 2013). Accurate and rapid diagnosis is a key component of disease management strategies, and nested PCR and real-time PCR amplification assays have been developed to allow specific detection of the phytoplasma causing LWD from field-collected samples (Alvarez et al. 2003; Alvarez and Mejía 2005; Alvarez and Pardo 2012).

Endophytic communities associated with healthy and phytoplasma-diseased plants were described through cultivation-dependent and -independent methods, allowing the identification of putative biocontrol agents. Although the mechanisms of phytoplasma–endophyte interaction are not clear, preliminary experiments in controlled conditions reported a mitigation of phytoplasma-associated symptomatology mediated by endophytes (Bianco et al. 2013).

Tetracycline antibiotics have been found to cause remission of symptoms in a number of diseases of suspected phytoplasma aetiology (Alvarez 2006). Tetracycline therapy produces only a temporary remission of disease before symptoms reappear (McCoy 1972; Agrios 1997).

In the Colombian eastern plains, the control strategy against LW includes establishing leguminous cover crops in order to avoid the presence of grass; scouting fields on a weekly basis in order to guarantee early detection of palms diseased with LW; immediately destructing diseased oil palms and applying insecticide to palms surrounding the diseased palms for preventing LW spread. Additionally, insecticide application was introduced during the season of the year when the population of *Haplaxius crudus* was very high. Results show that after a year of using the control strategy for LW, its incidence and development rate both declined: by 67% on IRHO and by 64% on Golden Hope material (Cortés et al. 2015).

7 Oil palm phytoplasma diseases in South and Central America: leaf stripe disease (LSD)

The LSD has been reported in Colombia (Alvarez et al. 2014). Oil palm samples including leaf tissue, embryo and germinated seeds were obtained at Sicarare and Unipalma plantations.

7.1 Biology

Disease samples including rachis, petiole, basal part of the bulb, leaf half, leaf base, meristem, leaflets and roots from African palm plantations located in the eastern plains of Colombia were evaluated (Table 1).

Table 1 Phytoplasma groups detected in LSD oil palm samples

Tissue	Phytoplasma group
Root	16SrI
Frond base	16SrI
Meristem	16SrI
Apical base	16SrI

Amplification of the 16S rDNA gene with universal primers M1/M2 showed a fragment of approximately 500 bp (Fig. 1). The DNA amplifications correspond to disease samples including roots, leaf base, meristem and apical base (Table 1).

The RFLP patterns were compared with those reported by Contaldo (2012) (Fig. 7).

Additionally, seed embryo tissue samples from African palm plantations located in the eastern Colombian plains were evaluated (Table 2).

Amplification of the 16S rDNA gene with M1/M2 universal primers showed a 500-bp fragment (Fig. 8).

The 16Sr gene of ribosomal DNA was amplified by nested PCR using universal and specific primers: P1/Tint, R16F2n/R2 (Lee et al. 1995) and M1/M2 (Gibb et al. 1995). Positive controls correspond to X disease 16SrIII group. Amplification of the 16S rDNA gene using M1/M2 universal primers showed a 500-bp fragment (Fig. 9). Positive samples were digested with *MseI*.

Sequence analysis of the 16Sr gene was carried out for the identification and classification of phytoplasma.

A high level of homology (99%) was obtained with sequences of the 16SrI ('*Ca.* P. asteris') group. The Blastn-assisted alignment revealed that the nucleotide sequences of the 16SrDNA gene of the phytoplasmas detected in samples 6492, 6497 and 6498 were similar to those of GenBank accessions KC009838, KC243397 and KC808147.

7.2 Symptoms

Symptoms are characterized by stunting, yellow streaking of the leaves and a marked reduction in fruit and stalk size (Fig. 10), which progress to no fruit production in the final

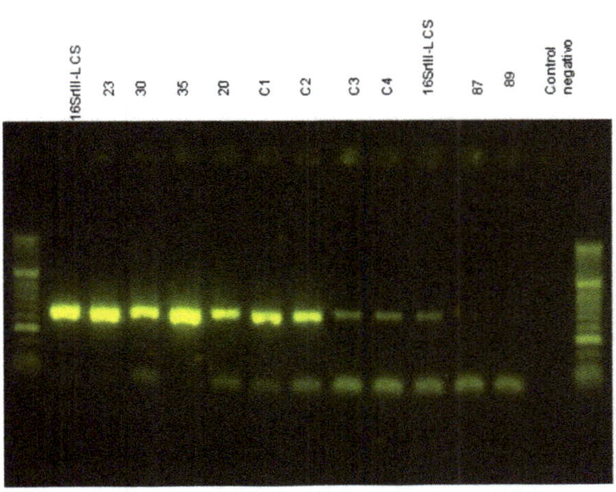

Figure 7 Amplified DNA fragments with M1/M2 primers. Lanes 23, 30, 35 and 20 correspond to leaf base tissue, meristem, apical base and diseased roots, respectively, and lanes 87 and 89 correspond to seed samples 6577 and 6876, respectively. Lanes C1, C2, C3 and C4 correspond to positive controls of 16SrI phytoplasma. The lanes labelled 16SrIII-L CS correspond to a positive control of cassava phytoplasma group III and the lane labelled negative control corresponds to PCR product without DNA (HyperLadder TM II).

Table 2 Phytoplasma groups detected in LSD oil palm embryo seed samples

Sample no.	Source	Phytoplasma group
83	Seed raceme 6874 U = 735 C 5:20	16SrI
87	Raceme 6577 U = 8030 5:20	16SrI
89	Raceme 6876 U = 707 C 5:20	16SrI

stages. Pale or bright yellow streaks on the leaf midrib are also observed (Alvarez, 2013). The described symptoms are similar to the ones reported by Ammar et al. (2005) in date palms in Egypt.

7.3 Epidemiology

The disease is transmitted by an insect vector. The insect vector transmitting the disease has not been identified yet. The disease could be disseminated by an infected seed.

Transmission studies to confirm the vector of the phytoplasma causing LSD in Colombia are required.

Research on phytoplasma–insect vector interaction is needed for the development of strategies to control LSD.

7.4 Control

The effectiveness of resistance inducers was confirmed on oil palm plantations in Sicarare, northern Colombia (Alvarez et al. 2012b, 2013).

The use of resistant germ plasm is the best strategy to manage the disease. Phytoplasma disease-free seed is another strategy to control LSD.

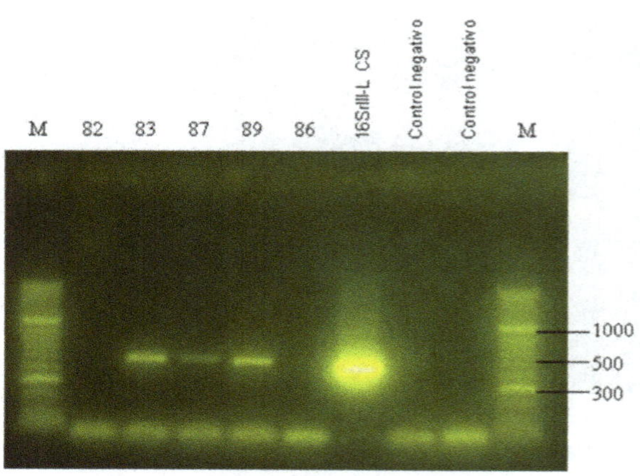

Figure 8 Amplification of DNA with M1/M2 primers. Lanes 82, 83, 87, 89 and 86 correspond to seed samples 6874, 6577 and 6876, respectively. Lane 16SrIII-L CS corresponds to a positive control and lane negative control corresponds to PCR product without DNA. Lanes with the letter M indicate the molecular weight marker 100 bp (HyperLadder TM II).

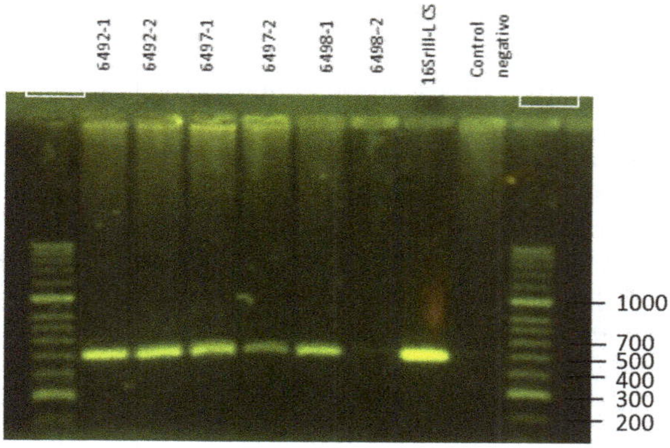

Figure 9 Amplification of DNA from germinated palm seed with primers M1/M. Lanes 6492-1, 6492-2, 6497-1, 6497-2, 6498-1 and 6498-2 correspond to positive samples. Lane 16SrIII-L CS corresponds to a positive control, phytoplasma group III, and lane negative control corresponds to PCR product without DNA (HyperLadder TM II).

8 Oil palm phytoplasma diseases in Asia and Africa

8.1 Phytoplasma diseases in Asia

In Asia, LWD is associated with phytoplasma 16SrI subgroup B. AY phytoplasmas have been detected through RFLP analyses of nested PCR-amplified fragments corresponding to the phytoplasma. Oil palm disease phytoplasmas indicated the presence of strains all related to the 16SrI group. Sequence analyses of partial 16S rDNA fragments showed that phytoplasmas have 99% similarity with *Candidatus* Phytoplasma asteris (Alvarez et al. 2014). Phytoplasma-associated symptoms include leaf yellowing, reduction in fruit and stalk size, stunted growth, wilting and severe or slow decline (Mehdi et al. 2012). The LWD reduces yield, and, in India, the disease causes yield losses in clonally propagated oil palm (*Elaeis guineensis* Jacq.) in the district of West Godavari, Andhra Pradesh (Mehdi et al. 2012). Similarly, oil palm plantations in Malaysia were affected in Banting, Selangor (Nejat et al. 2009).

Figure 10 (a and b) Leaflets showing yellowish lines running parallel to veins. (c) Fruit bunch with low fruit production.

Spear rot disease (SRD) of oil palm was reported on plantations of Kerala (Kochubabu and Nair 1993; Nair and Kochubabu 2000). Oil palm stunting (OPS) disease was reported in oil palms from Andhra Pradesh, India, and was found associated with 16SrI-B subgroup *Ca.* P. asteris-related strain (Azadvar et al. 2012).

Oil palm is one of the most important oil-producing crops in India. The area cultivated each year is 121 128 ha, of which more than 85 000 ha is grown in Andhra Pradesh (Directorate of Oil Palm Research 2010).

In Malaysia, the phytoplasma infecting oil palm belongs to the same 16SrXIV phytoplasma group (Nejat et al. 2009). The same authors also reported another novel phytoplasma species infecting oil palm.

Biology

A single occurrence of what seems to be a different phytoplasma from that occurring in coconut and other palm species has been recorded in an oil palm planting in the West New Britain area of Papua New Guinea, suggesting that oil palm is susceptible to infection by one type of phytoplasma (Turner 1981). Molecular confirmation and the interrelationship of phytoplasmas associated with diseases of palms in southern India were reported by Kochubabu et al. (2014).

Leaf samples of symptomatic oil palms were collected, and the presence of phytoplasma was confirmed by nested PCR using universal phytoplasma-specific primer pairs P1/P7, followed by R16F2n/R16R2 for amplification of the 16S rRNA gene, and semi-nested PCR using universal phytoplasma-specific primer pairs SecAfor1/SecArev3, followed by SecAfor2/SecArev3 for the amplification of a part of the secA gene (Mehdi et al. 2012).

Analysis of 16S rDNA sequence of the phytoplasma associated with stunting of oil palm in Andhra Pradesh, India, revealed its similarity with 16SrI AY group phytoplasma.

The RFLP pattern derived from *in silico* analysis of R16F2n/R16R2-primed sequence of 16S rRNA of OPS phytoplasma using the iPhyClassifier tool assigned the OPS phytoplasma to subgroup B of 16SrI group phytoplasma.

Samples of 28 SRD-affected oil palms from three locations in Kerala and one in Karnataka were collected after assessing disease severity. A phytoplasma presence in symptomatic oil palm was confirmed by nested PCR. The presence of phytoplasma group 16SrXI was determined. Comparison of the ~1.2-kb sequences of two oil palm SRD strains from Wayanad and Kollam (accession numbers KM593237 and KM593241) with other phytoplasma sequences in GenBank revealed that the phytoplasma associated with SRD of oil palm had a 16S rDNA sequence identity of 99–100% with sugarcane grassy shoot phytoplasmas (KJ792743 and KJ435292).

Symptoms and effects on yield

Phytoplasmas are known to induce symptoms that are quite variable and consist of leaf yellowing, reduction in fruit and stalk size, stunting, wilting and severe or slow decline.

Symptoms found in the West New Britain area of Papua New Guinea were some necrosis of the oldest fronds and foliage colour in general was chlorotic, in marked contrast with the dark green appearance of the remainder of the planting (Turner 1981).

In India, severe stunting symptoms were observed in clonally propagated oil palm in the district of West Godavari, Andhra Pradesh. In Kerala and Karnataka, phytoplasma symptoms of yellowing of the inner whorl of leaves and rotting of spears in oil palm were observed. The SRD of oil palm and OPS disease associated with phytoplasma have been described (Kochubabu 1989; Azadvar et al. 2012).

In Malaysia, oil palm samples were collected from spear leaves of six 2-year-old seedlings showing yellowing and necrosis symptoms in a nursery in Banting, Selangor. A phylogenetically distinct oil palm phytoplasma shared 99% similarity with EU371934 (periwinkle) and EU498727 (coconut) (Nejat et al. 2009).

Epidemiology

The LW has no known vector, but spread is known to be mediated by the use of seed obtained from infected fields.

In view of the cross-transmission of phytoplasmas associated with LW by *Proutista moesta* and dodder in Kerala (CPCRI 1995) and the capability of insect vectors (leafhoppers) such as *Aphrodes bicincta* and *Euscelis incisus* in transmitting more than one group of phytoplasmas, the possibility of cross-transmission of other groups (16SrXIV; 16SrI-B) of phytoplasmas by *Proutista moesta* is very much present (Kochubabu et al. 2014).

Control

Identification of alternative hosts is essential to develop sustainable control strategies. The varieties that were shown to be resistant to lethal yellowing in the Caribbean region did not show the same degree of resistance in Africa (Bila et al. 2015).

8.2 Phytoplasma diseases in Africa

The LWD in Africa was associated with phytoplasma.

Biology

Elaeis guineensis is an alternate host of coconut LYD phytoplasma in Mozambique.

Out of 28 plants sampled, one positive PCR for 16S rRNA gene primers was achieved on oil palm samples. A higher ratio of PCR-positive samples with both 16S rRNA and secA gene primers was achieved on *Elaeis guineensis* palm tree samples collected using the stem boring method.

Symptoms and effects on yield

In the district of Nicoadala, African oil palm (*E. guineensis*) exhibiting symptoms of a skirt-shaped brown discolouration (necrosis) of mature and spear leaves (Fig. 11) was associated with a phytoplasma disease.

Epidemiology

Transmission of phytoplasma between different host species was observed in Malaysia, where the causal agent of coconut LY-type diseases was also observed in oil palm (Nejat and Vadamalai 2010).

Challenges within the field of LYD epidemiology include the identification of insect vectors (Bila et al. 2015).

Figure 11 African oil palm (*Elaeis guineensis*) exhibiting skirt-shaped brown discolouration of the older leaves.

Control

One disease management strategy is the removal and burning of symptomatic trees. Removed palms must be replaced by resistant varieties. However, identifying resistance to the palm phytoplasma in Africa has proven challenging (Bila et al. 2015).

9 Conclusion, summary and future trends

9.1 Conclusion

Major oil palm diseases are limited to certain continents: Fusarium in Africa, Bud Rot in Latin America and *Ganoderma* in SE Asia. In various regions of the world, the existence of soils that are in some way inhospitable to Fusarium wilts has prompted exploration of biotic and abiotic mechanisms of disease suppression. The association between Fusarium wilt suppression and environmental factors is typically related, in part, to the effects of these factors on the soil microbiota, members of which may compete directly with or release toxins against the pathogen or incite plant host defence responses (Baker and Cook 1974). A natural biological control system of Fusarium-suppressive soils could explain why vascular wilt disease is not present in Malaysia. Alkaline soils generally are less conducive to Fusarium wilts of many crops (Jones et al. 1989), which is thought to explain why the disease does not limit seed production in Denmark. Another theory regarding the mechanisms of Fusarium wilt suppression achieved with soil limestone application is that Ca provided by the limestone is beneficial for plant defence responses, because Ca fortifies cell wall structure by cross-linking pectins and is involved in cellular defence signalling (Marschner 1995).

Rusli et al. (2013) investigated the possibility of *Fusarium oxysporum*-suppressive soils in Malaysia in order to explain the non-appearance of this vascular disease there and possibly to reveal other potential biocontrol agents. The explanation as to why Malaysia has not yet attained the disease is likely to revolve around the soil properties, in particular the microflora. This review reported that greater disease severity based on visual symptoms occurred in autoclaved soils and compost than in untreated soils when oil palm seedlings were artificially infected with *Fusarium oxysporum*.

Given the limited genetic basis exploited in plantations, natural sources of variation existing in oil palm are low and have led studies to exploit the different sources of variation in order to increase the oil palm genetic diversity. Genetic diversity of oil palm has been narrowed due to the limited number of founder lines used in the breeding population. Broadening the genetic base would help increase the oil palm yield (Mayes et al. 1997).

9.2 Summary

Diseases have been documented as one of the primary constraints to oil palm production. Phytoplasma and fungal pathogens cause important yield losses where oil palm is grown. Some of these pathogens and the diseases that they cause are only partially characterized and understood, and many research gaps still need to be addressed. Progress is limited in the management of oil palm diseases.

In this chapter, we have presented information on oil palm diseases and their causal agents. Several of these are of key importance now. In Latin America, LWD and LSD continue to spread through oil palm-producing areas in Colombia. The significance of this threat has increasingly been recognized by the research, plant protection and extension communities within the country. This and other related strategies from oil palm stakeholder institutions are paving the way for an increased commitment to developing effective, affordable and durable solutions to this problem. In Latin America, great progress has been made in recent years in understanding the pathogen interactions that give rise to the devastating LWD. Moreover, experience gained in researching phytoplasmas associated with this condition is having a powerful effect on new research being conducted on a related pathogen that is causing LSD in the eastern plains (Llanos Orientales) in African oil palm, pertaining to Unipalma and Sicarare. This highlights the importance of one of the most important themes in the management of oil palm diseases, which is the prevention of pathogen spread between countries and continents through the vigorous application of phytosanitary protocols. It is essential that human and technical capacity to enforce phytosanitary controls be greatly strengthened to ensure that all oil palm stakeholders are fully informed of the risk of introducing diseases carried by oil palm seed between territories and continents.

9.3 Future research on diseases discussed in this chapter

- The physical size of the oil palm genome is two-thirds of the maize genome and four times the size of the rice genome. A complete sequencing of the palm genome would help in understanding the key genes affecting disease resistance.

- Multi-parental populations are promising tools for identifying quantitative disease resistance loci. The multi-parental population design allowed us to identify quantitative disease resistance loci among an extended genetic diversity, and to test their effects in various genetic backgrounds, which should enhance the transferability of results and the sustainability of the selected resistances.
- The selection of varieties resistant to a combination of diseases will be essential for multi-criteria improvement, as *Ganoderma* is found throughout the oil palm cultivation area, combined with other devastating diseases like the *Fusarium* wilt in Africa and bud rot in South America. Hence, the global genetic architecture of disease resistance in oil palm needs to be investigated, especially potential trade-off effects related to different pathogens, as shown for other species. The multi-parental population and the implemented quantitative trait loci (QTLs) mapping approach provide powerful tools for investigating such global genetic architecture of disease resistance in oil palm (Tisne et al. 2017).
- The use of metabolomic tools such as liquid chromatography–mass spectrometry/mass spectrometry (LC–MS/MS) profiling technology holds great promises for investigating and understanding metabolism associated with root metabolome (Nurazah et al. 2016). The LC–MS/MS method developed for profiling and identifying oil palm root metabolites can also be used as a model for developing a reliable tool towards distinguishing tolerant palms to oil palm disease such as BSR.
- Identifying metabolites associated with resistance promises development of biomarker tools for screening of oil palm resistant to *G. boninense*. In addition, understanding resistance mechanisms at the metabolome level helps breeders better understand resistant gene function and pyramid suitable resistant gene in elite variety.
- Transcriptome of oil palm roots treated with a causal agent of BSR, *G. boninense* using a cDNA microarray approach. This information contributed to our understanding of the defence mechanisms of oil palm in response to *G. boninense*, the future development of molecular markers for marker-assisted breeding and screening of oil palms that are tolerant to *G. boninense* (Tee et al. 2013).
- Approaches to the phytosanitary management of oil palm health: strict quarantine regulations on the importation of oil palm seeds in order to prevent introduction of these destructive diseases.

10 Where to look for further information

Societies involved in oil palm disease study

American Phytopathological Society (APS). http://www.apsnet.org/Pages/default.aspx
International Society for Plant Pathology (ISPP). http://www.isppweb.org/index.asp

Information sources about plant disease

CABI. http://www.cabi.org/isc/
Plantwise Knowledge Bank. http://www.plantwise.org/KnowledgeBank/Home.aspx

Organizations involved in oil palm disease research

Centro Internacional de Agricultura Tropical (CIAT). www.ciat.org
Oil Palm Research Institute (OPRI) of Ghana. www.csir.org.gh

Council for Scientific and Industrial Research (CSIR). www.csir.org.gh
Colombian Oil Palm Research Center (CENIPALMA). www.cenipalma.org
Indonesian Oil Palm Research Institute (IOPRI). www.iopri.org

11 References

Agrios, G. N. 1997. *Plant Pathology*, 4th edition. Academic Press, USA, 458p.

Akino, S. and Kondo, N. 2012. Common spear rot of oil palm in Indonesia. *Plant Disease* 96(4): 537–43.

Alvarez, E. 2006. DNA sequence analysis of the 16s rRNA region of phytoplasma associated with lethal wilt in oil palm. In: *International Oil Palm Conference. AG25*. Nusa Dua, Bali, Indonesia. 19–23 June 2006.

Alvarez, E. 2013. Control of leaf stripe in oil palm, *Elaeis guineensis*, using resistance inducer. Personal communication, 3 October 2013.

Alvarez, E., and Claroz, J. L. 2003. Characterization and classification of phytoplasmas associated with oil palm (*Elaeis guineensis*). ICPP2003 Memories 2: 284.

Alvarez, E., Marroquin-Guzman, M., Mejia, J. F. and Pardo, J. M. 2010. Early detection of *Phytophthora palmivora* in oil palm, using real-time polymerase chain reaction (QPCR) and molecular beacon probes. *Phytopathology* 101(6): S253.

Alvarez, E. and Mejía, J. F. 2005. DNA sequence analysis of the 16Sr RNA region of phytoplasma associated with lethal wilt in oil palm. *Fitopatología Colombiana* 29(1): 39–44.

Alvarez, E., Mejía, J. F., Contaldo, N., Paltrinieri, S., Duduk, B. and Bertaccini, A. 2014. 'Candidatus phytoplasma asteris' strains associated with oil palm lethal wilt in Colombia. *Plant Disease* 98: 311–18.

Alvarez, E., Llano, G. A., Loke, J. B. and Chacon, M. I. 2012a. Characterization of *Thielaviopsis paradoxa* isolates from oil palm in Colombia, Ecuador and Brazil. *Journal of Phytopathology* 160(11–12): 690–700.

Alvarez, E., Llano, G. A., Feris, M. C., Hernández, M. L. and Rodríguez, S. M. 2003. Control of bud rot in oil palm, *Elaeis guineensis*, using resistance inducers. ICPP2003 Memories 2: 179.

Alvarez, E., Loke, J. B. and Trujillo, R. 2007. Avances en la caracterización patogénica de *Phytophthora* spp. asociado al complejo 'pudrición de cogollo' en palma africana *Elaeis guineensis* Jacq. *Fitopatología Colombiana* 31(1): 15–18.

Alvarez, E. and Pardo, J. M. 2012. Diseño de sonda TaqMan® y cebadores específicos para la detección de fitoplasmas asociados a la marchitez letal en Palma de Aceite (*Elaeis guineensis*). CIAT Annual Report.

Alvarez, E., Sánchez, J., Grajales, O. and Gomez, P. L. 1998. Avances en la caracterización molecular de *Phytophthora* spp. en palma de aceite. II Reunion anual de evaluación. Cenipalma art. 3.

Alvarez, E., Sánchez, J., Grajales, O. and Gomez, P. L. 1999. Pathogenic variation and DNA polimorphims in *Phytophthora* sp., causal agent of bud rot of oil palm. *Phytopathology* 89: S3.

Alvarez, E., Truke, M. J. and Pardo, J. M. 2012b. Alternativas de manejo de la enfermedad del estriamiento foliar en palma de aceite. CIAT Annual Report.

Alvarez, E., Truke, M. J. and Pardo, J. M. 2013. Detección e identificación de phytoplasmas asociados con síntomas de estriamiento foliar en palma de aceite (*Elaeis guineensis*). CIAT Annual Report.

Ammar, M. I., Amer, M. A. and Rashed, M. F. 2005. Detection of Phytoplasma associated with yellow streak disease of date palms (*Phoenix dactylifera* L.) in Egypt. *Egyptian Journal of Virology* 2: 74–86.

Arango, M., Ospina, C., Sierra, J. and Martínez, G. 2011. *Myndus crudus*: vector del agente causante de la Marchitez letal en palma de aceite en Colombia. *Revista Palmas* 32(2): 13–25.

Assis, K., Chong, K. P., Idris, A. S. and Ho, C. M. 2016. Economic loss due to *Ganoderma* disease in oil palm. *World Academy of Science, Engineering and Technology, International Journal of Social, Behavioral, Educational, Economic, Business and Industrial Engineering* 10(2): 631–5.

Azadvar, M., Baranwal, V. K., Kochubabu, M. and Praveena, D. 2012. Sequence analysis of 16S rRNA and secA genes confirms the association of 16SrI-B subgroup phytoplasma with oil palm (*Elaeis guineensis* Jacq.) stunting disease in India. *Journal of Phytopathology* 160: 6–12.

Baker, K. F. and Cook, R. J. 1974. *Biological control of Plant Pathogens.* W. H. Freeman, San Francisco.

Baer, N., Morillo, E. and Bernal, G. 2013. Detección del Agente Causal de la Marchitez Letal en Plantaciones Comerciales de Palma Aceitera en el Ecuador Mediante Técnicas de PCR y vMetagenómica. http://ciencia.espe.edu.ec/wp-content/uploads/2013/05/VID94.pdf.

Bertaccini, A., Duduk, B., Paltrinieri, S. and Contaldo, N. 2014. Phytoplasmas and phytoplasma diseases: a severe threat to agriculture. *American Journal of Plant Sciences* 12(5): 1763–88.

Bianco, P. A., Marzachì, C., Musetti, R. and Naor, V. 2013. Perspectives of endophytes as biocontrol agents in the management of phytoplasma diseases. *Phytopathogenic Mollicutes* 3(1): 56–9.

Bila, J., Högberg, N., Mondjana, A. and Samils, B. 2015. African fan palm (*Borassus aethiopum*) and oil palm (*Elaeis guineensis*) are alternate hosts of coconut lethal yellowing phytoplasma in Mozambique. *African Journal of Biotechnology* 14: 52–67.

Breton, F., Miranti, R., Lubis, Z., Hayun, Z., Umi, S., Flori, A., Nelson, S. P. C., Durand-Gasselin, T., Jacquemard, J. C. and De Franqueville, H. 2009. *Ganoderma* disease of the oil palm: hypothesis on natural infection and implementation of an early artificial inoculation test to screen oil palm progenies for their level of resistance. https://agritrop.cirad.fr/553875/7/ID553875.pdf.

Brioso, P. S. T., Montano, H. G., Figueiredo, D. V., Poltronieri, L. S. and Furlan Junior J. 2006. Amarelecimento fatal do dendezeiro: seqüenciamento parcial do fitoplasma associado. *Summa Phytopathologica*. 32: 50.

Buitrago, V. M. and Nieto, L. E. 1995. Hongos asociados con pudriciones de flecha y cogollo en palma de aceite (*Elaeis guineensis* Jacq.) en los Llanos Orientales. *Revista Palmas* (Colombia) 16: 9–17.

Castiblanco, J. S. and Mosquera, M. 2011. Costos asociados a la marchitez letal en palma de aceite. *Revista Palmas*, 32(3): 39–47.

Central Plantation Crops Research Institute (CPCRI). 1995. Annual Report 1994–95. CPCRI, Kasaragod, India. pp. 1–211.

Contaldo, N., Bertaccini, A., Paltonieri, S., Windsor, H. and Windsor, G. D. 2012. Axenic culture of plant pathogenic phytoplasmas. *Phytopathologia Mediterranea* 51: 607–17.

Corley, R. H. V. and Tinker, P. B. 2003. *The Oil Palm*. John Wiley & Sons, Oxford.

Cortés, N. R., Giraldo, J. A. B. and Uribe, M. A. 2015. Tecnologías para el manejo de la marchitez letal de la palma de aceite validadas en la zona oriental de Colombia. *Revista Palmas* 36(1): 55–62.

Cooper, R. M., Flood, J. and Rees, R. W. 2011. *Ganoderma boninense* in oil palm plantations: current thinking on epidemiology, resistance and pathology. *Planter* 87(1024): 515–26.

de Franqueville, H. 2003. Oil palm bud rot in Latin America. *Experimental Agriculture* 39: 225–40.

Directorate of Oil Palm Research. 2010. Pedwagi, India.

Eziashi, E. I., Uma, N. U., Adekunle, A. A. and Omamor, I. B. 2006. Biological control of *Ceratocystis paradoxa* causing black seed rot in oil palm sprouted seeds by *Trichoderma* species. *Pakistan Journal of Biological Sciences* 9(10): 1987–90.

Faparusi, S. I. 1973. Origin of initial microflora of palm wine from oil palm trees (*Elaeis guineensis*). *Journal of Applied Bacteriology* 36: 559–65.

FEDEPALMA. 2010. Es posible controlar la marchitez letal si se aplican las recomendaciones de Cenipalma. *Palmicultor* 458: 12–15.

Feintrenie, L., Chong, W. K. and Levang, P. 2010. Why do farmers prefer oil palm? Lessons learnt from Bungo district, Indonesia. *Small Scale Forestry* 9: 379–96.

Flood, J. 2006. A review of Fusarium wilt of oil palm caused by *Fusarium oxysporum* f. sp. *elaeidis*. *Phytopathology* 96: 660–2.

Flood, J., Mepsted, R., Turner, S. and Cooper, R. M. 1994. Eradication of Fusarium from oil palm by seed treatments, pp. 201–5. In: *Seed Treatment: Progress and Prospects: Proceedings of a Symposium held at the University of Kent, Canterbury, UK*. BCPC Monogr. No. 57.

Flood, J., Whitehead, D. S. and Cooper, R. M. 1992. Vegetative compatibility and DNA polymorphisms in *Fusarium oxysporum* f. sp. *elaeidis* and their relationship to isolate virulence and origin. *Physiol. Molecular Plant Pathology* 41:201–15.

Genty, P., De Chenon, R. D. and Mariau, D. 1976. Infestation of the aerial roots of oil palms by caterpillars of the genus Sufetula (*Lepidoptera: Pyralidae*). *Oleagineux* 31(8/9): 365–70.

Gibb K. S., A. C. Padovan and B. D. Mogen, 1995. Studies on sweet potato little-leaf phytoplasma detected in sweet potato and other plant species growing in Northern Australia. *Phytopathology* 85: 169–74.

Harrington, T. C. 2009. The genus Ceratocystis. Where does the oak wilt fungust fit? In: Appel, D. N. and Billings, R. F. (Eds) *Proceedings of 2nd National Oak Wilt Symposium*. USDA forest service, forest health protection, Austin, Texas, pp. 27–42.

Ho, C. T. and Khairudin, H. 1997. Usefulness of soil mounding treatments in prolonging productivity of prime-age *Ganoderma* infected palms. *The planter* 73:239–44.

Ho, C. L. and Tan, Y. C. 2015. Molecular defense response of oil palm to *Ganoderma* infection. *Phytochemistry* 114: 168–177.

Ho, Y. W. and Nawawi, A. 1986. Isolation, growth and sporophore development of *Ganoderma boninense* from oil palm in Malaysia. *Pertanika* 9(1): 69–73.

Hushiarian, R., Yusof, N. A. and Dutse, S. W. 2013. Detection and control of *Ganoderma boninense*: strategies and perspectives. *SpringerPlus* 2(1): 555.

Ilias, G. N. M. 2000. *Trichoderma* and its efficacy as a biocontrol agent of basal stem rot of oil palm (*Elaeis guineensis* Jacq.). *Doctoral Dissertation*, University Putra Malaysia, Malaysia.

Jones, J. P., Engelhard, A. W. and Woltz, S. S. 1989. Management of Fusariun wilt of vegetable and ornamentals by macro and microelements, pp. 18–32. In: Engelhard, A. W. (Ed.) *Soilborne Plant Pathogens: Management of Disease with Macroelements*. American Phytopathological Society, St. Paul, MN.

Jones, P. and Turner, P. D. 1979. The occurrence of a mycoplasma-like organism in oil palm (*Elaeis guineensis*). (*Unpublished*).

Joseph, T. and Radha, K. 1975. Role of *Phytophthora palmivora* in bud rot of coconut. *Plant Disease Reporter* 59(12): 1014–17.

Kawakita, H., Saiki, T., Wei, W., Mitsuhashi, W., Watanabe, K., and Sato, M. 2000. Identification of mulberry dwarf phytoplasmas in the genital organs and eggs of leafhopper Hishimonoides sellatiformis. *Phytopathology* 90(8): 909–14.

Kile, G. A. 1993. Plant diseases caused by species of *Ceratocystis* sensu stricto and *Chalara*. In: Wingfield, M. J., Seifert, K. A., and Webber, J. F. (Eds) *Ceratocystis and Ophiostoma Taxonomy, Ecology, and Pathogenicity*. The American Phytopathological Society, APS Press, St. Paul, Minnesota, pp. 173–83.

Kochubabu, M. 1989. Spear rot disease of oil palm (*Elaeis guineensis* Jacq.) in India. *Journal of Plantation Crops* 16(Suppl.): 281–6.

Kochubabu, M. and Nair, K. R. 1993. Distribution of spear rot disease of oil palm (*Elaeis guineensis* Jacq.) and its possible association with MLO diseases of palms in Kerala, India. *The Planter* 69: 59–66.

Kochubabu, M., Shailender, K., Govind, P. R., Krothapalli, R. S. and Sambasiva, R. 2014. Molecular confirmation and interrelationship of phytoplasmas associated with diseases of palms in south India. *Phytopathogenic Mollicutes* 4(2): 41–52.

Lee, I. M., Bertaccini, A., Vibio, M. and Gundersen, D. E. 1995. Detection of multiple phytoplasmas in perennial fruit trees with decline symptoms in Italy. *Phytopathology* 85: 728–35.

Lim, K. H., and Udin, W. 2010. Management of *Ganoderma* in peat soil in Indonesia. Paper presented in 2010 at Second international Seminar on Oil Palm Diseases: Advances in *Ganoderma* research and management. 31 May 2010. Yogyakarta, Indonesia, p. 11.

Llano, G. A., Alvarez, E., Feris, M. C., Hernández, M. L. and Rodríguez, S. M. 2006. Efecto de inductores de resistencia sobre pudrición del cogollo de palma de aceite en los Llanos Orientales de Colombia. *Fitopatología Colombiana* 30(1): 15–18.

Maschner, H. 1995. *Mineral Nutrition of Higher Plant*. Academic press, San Diego, CA.

Mayes, S., Jack, P. L., Corley, R. H. V., and Marshall, D. F. 1997. Construction of a RFLP genetic linkage map for oil palm (*Elaeis guineensis* Jacq.) *Genome* 40 (1): 116–22.

Mbenoun, M., de Beer, Z. W., Wingfield, M. J., Wingfield, B. D. and Roux, J. 2014. Reconsidering species boundaries in the *Ceratocystis paradoxa* complex, including a new species from oil palm and cacao in Cameroon. *Mycologia* 106(4): 757–84.

McCoy, R. E. 1972. Remission of lethal yellowing in coconut palm treated with tetracycline antibiotics. *Plant Disease* 56: 1019–21.

Mehdi, A., Baranwal, V. K., Kochu Babu, M. and Praveena, D. 2012, Sequence analysis of 16S rRNA and *secA* genes confirms the association of 16SrI-B subgroup phytoplasma with oil palm (*Elaeis guineensis* Jacq.) stunting disease in India. *Journal of Phytopathology* 160: 6–12.

Mejia de Los Rios, J. F. 2014. Identification and genetic diversity in phytoplasmas associated with diseases of cassava and other agronomic relevant crops in South-East Asia and Latin America (Doctoral dissertation).

Meyer, A., Todt, C., Mikkelsen, N. T. and Lieb, B. 2010. Fast evolving 18S rRNA sequences from Solenogastres (Mollusca) resist standard PCR amplification and give new insights into mollusk substitution rate heterogeneity. *BMC Evolutionary Biology* 10: 70.

Mitrovic, J., Contaldo, N., Paltrinieri, S., Mejia, J. F., Mori, N., Bertaccini, A., and Duduk, B. 2011. The use of groEL gene for characterisation of aster yellows phytoplasmas in field collected samples. *Bulletin of Insectology* 64(Suppl.): S17–18.

Mohammed, C. L., Rimbawanto, A. and Page, D. E. 2014. Management of basidiomycete root- and stem-rot diseases in oil palm, rubber and tropical hardwood plantation crops. *Forest Pathology* 44: 428–46.

Moller, E. M., Bahnweg, G., Sandermann, H. and Geiger, H. H. 1992. A simple and efficient protocol for isolation of high molecular weight DNA from filamentous fungi, fruit bodies and infected plant tissues. *Nucleic Acids Research* 20(22): 6115–16. doi:10.1093/nar/20.22.6115.

Montano, H. G., Brioso, P. S. T. and Pimentel, J. P. 2007. List of phytoplasma hosts in 432 Brazil. *Bulletin of Insectology* 60: 129–30.

Mouyna, I., Renard, J. L. and Brygoo, Y. 1996. DNA polymorphism among *Fusarium oxysporum* f. sp. *elaeidis* populations from oil palm, using a repeated and dispersed sequence 'Palm'. *Current Genetics* 30:174–80

Murphy, D. J. 2014. The future of oil palm as a major global crop: opportunities and challenges. *Journa of Oil Palm Research* 26: 1–24.

Myrie, W. A., Douglas, L., Harrison, N. A., McLaughlin, W. and James, M. 2012. First report of lethal yellowing disease associated with subgroup 16SrIV, a phytoplasma on St. Kitts in the Lesser Antilles. *New Disease Reports* 26(25): 2044–0588.

Naher, L., Siddiquee, S., Yusuf, U. K. and Mondal, M. M. A. 2015. Issues of *Ganoderma* spp. and basal stem rot disease management in oil palm. *American Journal of Agricultural Science* 2(3): 103–7.

Nair, K. R. and Kochubabu, M. 2000. Oil palm spear rot caused by *Stephanitis typica*, a vector of root (wilt) disease of coconut. *Plant Horticultural Technology* 2(2): 51–2.

Nejat, N. and Vadamalai, G. 2010. Phytoplasmas detection in coconut palms and other tropical crops. *Plant Pathology Journal* 9(3): 110–21.

Nejat, N., Sijam, K., Abdullah, S. N. A., Vadamalai, G. and Dickinson, M. 2009. Phytoplasmas associated with disease of coconut in Malaysia: phylogenetic groups and host plant species. *Plant Pathology* 58: 1152–60.

Nurazah, Z., Idris, S. A., Kushairi, A. and Ranli, U. S. 2016. Metabolome analysis of oil palm root for breeding and engineering new varieties. Malaysian Palm Oil Board TT No. 594.

Ommelna, B. G., Jennifer, A. N. and Chong, K. P. 2012. The potential of chitosan in suppressing *Ganoderma boninense* infection in oil-palm seedlings. *Journal of Sustainable and Scientific Management* 7(2): 186–92.

Ooi, S., Barcelos, E., Muller, A. and Nascimento, J. 1981. Oil palm genetic resources—native *E. oleifera* populations in Brazil offer promising sources. *Pesquisa Agropecuaria Brasileira* 16: 385–95.

Pérez, A. P. and Cayón, G. 2010. Metabolism of carbohydrates in oil palms (*Elaeis guineensis* Jacq.) affected by lethal wilt. *Agronomía Colombiana* 28: 181–7.

Rees, R., Flood, J., Hasan, Y., Wills, M. and Cooper, R. M. 2012. *Ganoderma boninense* basidiospores in oil palm plantations: evaluation of their possible role in stem rots of *Elaeis guineensis. Plant Pathology* 61(3): 567–78.

Renard, J. L., and de Franqueville, H. 1989. Oil palm vascular wilt. *Oleagineux* 44:342–7

Rusli, M. H., Seman, I. A. and Yusof, Z. N. B. 2013. Fusarium vascular infection of oil palm: Epidemiology, molecular diagnostic tools and the potential of fusarium suppressive soil in Malaysia. *Journal of Environmental Science and Engineering. B*, 2: 578–85.

Sánchez, N. J., Ayala, A., and Alvarez, E. 1999. Patogenicidad, identificación y caracterización molecular de *Phytophthora* spp. en palma de aceite. *Boletín Divulgativo Ceniavances* 60: 1–4.

Sarria, G. A., Varón, F. H., Martínez, G., Drenth, A. and Guest, D. I. 2013. Nuevas evidencias del cumplimiento de los postulados de Koch en el estudio de las relaciones *Phytophthora palmivora* y la pudrición de cogollo de palma de aceite en Colombia. *Palmas* 34: 41–5.

Smart, C., Schneider, B., Blomquist, C., Guerra, L., Harrison, N., and Ahrens, U. 1996. Phytoplasm-specific PCR primers based on sequences of 16S- 23S rRNA spacer region. *Applied and Enviromental Microbiology* 62: 2988–93.

Soepena, H., Purba, R. Y. and Pawirosukarto, S. 2000. A control strategy for basal stem rot (*Ganoderma*) on oil palm. In: Flood, J., Bridge, P. D., Holderness, M. (Eds), *Ganoderma Diseases of Perennial Crops*. CABI Publishing, Wallingford, UK, pp. 83–8.

Soon, G. H. and Jung, H. S. 2004. Phylogenetic analysis of *Ganoderma* based on nearly complete mitochondrial small-subunit ribosomal DNA sequences. *Mycologia* 96(4): 742–55.

Suryanto, D., Wibowo, R. H., Siregar, E. B. M. and Munir, E. 2012. A possibility of chitinolytic bacteria utilization to control basal stems disease caused by *Ganoderma boninense* in oil palm seedling. *African Journal of Microbiology Research* 6(9): 2053–9.

Susanto, A. 2009. Basal stem rot in Indonesia: Biology, economic importance, epidemiology, detection and control. In: *Proceedings of International Workshop on Awareness, Detection and Control of Oil Palm Devastating Diseases*. Kuala Lumpur Convention Centre, University Putra Malaysia Press, Malaysia.

Susanto, A., Sudharto, P. S. and Purba, R. Y. 2005. Enhancing biological control of basal stem rot disease (*Ganoderma boninense*) in oil palm plantations. *Mycopathologia* 159(1): 153–7,

Tee, S. S., Tan, Y. C., Abdullah, F., Ong-Abdullah, M. and Ho, C. L. 2013. Transcriptome of oil palm (*Elaeis guineensis* Jacq.) roots treated with *Ganoderma boninense. Tree Genetics & Genomes* 9(2): 377–86.

Tisné, S., Pomiès, V., Riou, V., Syahputra, I., Cochard, B. and Denis, M. 2017. Identification of *Ganoderma* disease resistance loci using natural field infection of an oil palm multiparental population. *G3: Genes, Genomes, Genetics* 7(6): 1683–92.

Torres, G. A., Sarria, G. A., Martinez, G., Varon, F., Drenth, A. and Guest, D. I. 2016. Bud rot caused by *Phytophthora palmivora*: A destructive emerging disease of oil palm. *Phytopathology* 106(4): 320–9.

Tovar, J. P. and Nieto, L. E. 1998. Pudrición de estipe seca *Ceratocystis paradoxa. Ceniavances* 50: 1–3.

Turner, P. D. 1981. *Oil Palm Diseases and Disorders*. Oxford University Press, P.O. Box 262, Kuala Lumpur Malaysia.

Turner, P. D. and Gillbanks, R. 1974. *Oil Palm Cultivation and Management*. Incorporated Society of Planters, Kuala Lumpur, 672p.

Van de Lande, H. L. 1984. Vascular wilt disease of oil palm (*Elaeis guineensis* Jacq.) in Para, Brazil. *Oil Palm News* 28:6–10

Wardlaw, C. W. 1946. *Fusarium oxysporum* on the oil palm. *Nature* 158:712.

Weintraub, P. G. and Beanland, L. 2006. Insect vectors of phytoplasmas. *Annual Review of Entomology* 51: 91–111.

Chapter 3

Insect pests affecting oil palms

Laurence Beaudoin-Ollivier, Université de Montpellier and CIRAD Montpellier, France

1 Introduction

The oil palm is a food source for a large number of actual and potential insect pests. The increasing international movement of material is also introducing new insects. Populations of these invasive insects can grow rapidly in new areas where there are no natural enemies and cause sudden and spectacular outbreaks. However, less than 10% of the thousand arthropods that live on palm plantations have been recognised as serious pests for cultivated palm species. Indeed, Lespesme (1947) reported that insects cause very little damage to wild palms. However, as soon as the oil palm is cultivated on a large scale, it becomes more susceptible to pest attack as pest pressure increases. The rapid and exponential increase in cultivated oil palms over the last 50 years (Rival and Levang, 2013; FAOSTAT, 2015), especially in the form of large-scale monocultures, has favoured outbreaks of several species which have been able to colonise new areas and have acquired pest status (e.g. *Oryctes rhinoceros* in Southeast Asia and *Oryctes monoceros* in Africa). Pest outbreaks also occur when there is an imbalance in the biocenosis (ecological community), mainly due to poor cultivation practices and intensive spraying with contact insecticides.

All types of herbivorous arthropods can be found on oil palms, including polyphagous species that also feed on other plant taxa or specialised species which can only develop on plants in the family Arecaceae. These insects attack all niches offered by the oil palm: fronds, sap, stipe, roots, inflorescences and fruits. Like all palm species, the oil palm has unique anatomical and physiological traits, characterised by rapid growth from a single

http://dx.doi.org/10.19103/AS.2017.0018.15

meristem from which the stem, fronds and inflorescences develop, forming a large amount of tissue that is rich in water and nutrients (Cohen, 2017).

Oil palm insect pests include borers which dig galleries and feed on fresh tissue in the meristem (Coleoptera: Scarabaeidae and Lepidoptera). Pests include weevils (Coleoptera: Curculionidae), leaf-eating caterpillars (Lepidoptera), some mites (Acari), sap-feeders (Hemiptera), long-horned grasshoppers (Orthoptera) and stick insects (Phasmatidae). Species belonging to other Lepidoptera families (Crambidae, Satyriidae, Lymantriidae and Psychidae) are also important oil palm pests. The range of species attacking the oil palm differs between continents (the plant is found in South America, Africa, Asia and the Pacific Islands).

Many species from the Scarabaeidae family are serious oil palm pests in their adult form, whereas in other species (families Curculionidae and Chrysomelidae), damage is caused by their immature stages. Borers are by far the most difficult group to manage but, although the large-scale use of insecticides has been successful, many of these products have been, or are being, banned.

The Lepidoptera are the largest group of pests attacking coconut, date and oil palms, with about 240 species. Among the most damaging caterpillars are several species of Limacodidae, leaf eaters that attack unfolded spears and folded fronds. Other species are miners which bore into fronds, flowers, fruits, stems, nursery seedlings and roots (Mariau, 2001). When the harmful stage of the insect pest can be accessed easily as it is living on the palm (such as defoliators or sap-feeders), any natural or chemical control measure can be effective. The situation is totally different when the caterpillar, larva or adult lives inside the plant (leaf miner or borer). Several control methods are discussed in this chapter, some of which are still being improved.

A selection of the main insect pests found in oil palm plantations in Latin America, West Africa, Southeast Asia and the Pacific is presented in this chapter, which was not designed to update existing reviews, but to serve more as an overview of the main types of oil palm pests (23 species) that have been selected from the most damaging groups to illustrate the relationships between oil palm plants and herbivorous arthropods. This chapter is designed for practical use and intended for growers, agronomists, students, extension officers and any stakeholders interested in sustainable oil palm cultivation. Readers will find descriptions and high-quality photographs which should help them identify species. It also provides useful information compiled during field work and from literature reviews (Lepesme, 1947; Bedford, 1980; Mariau, 2001; Howard et al., 2001; Bedford, 2013). For convenience, the pests we describe are classified according to the part of the palm that they target, fronds, meristem, roots or fruit bunches, and some general features of their habitual behaviour that have practical implications for their control are discussed.

In terms of pest control, Desmier de Chenon (1989) has studied the relationships between host-parasitoid-predator-entomopathogen and alternative hosts to encourage specific natural enemies and prevent unnecessary and ineffective chemical treatments. Careful application and selective use of insecticides can prevent any adverse effect on beneficial insect populations (Kamarudin et al., 1998). The development of integrated pest management (IPM) programmes to control oil palm pests (bagworm, rhinoceros beetle, nettle caterpillar, bunch moth) in Malaysia has been reviewed by Ariffin and Basri (2000). Control measures are only effective when an IPM system is adopted. This includes a combination of several factors: a comprehensive surveillance system; farming practices such as retaining some selected nectariferous plants/planting legumes in the field; the timely use of predators, parasitoids and pathogens; and the appropriate use of specific insecticides (Lay, 1996).

2 Insect pests of oil palm fronds

2.1 Lepidoptera: Limacodidae

Palms (Arecaceae) are the main host for the Limacodidae family, some 68 species of which have been recorded to be found in Asia, particularly in Southeast Asia (Cock et al., 1987), 20 species in tropical South America and 6 in Africa. These are polyphagous species, commonly known as nettle caterpillars. Only six species (four from Southeast Asia, one from South America and one from Africa) of the 32 that cause severe defoliation, as mentioned by Mariau (1999), are illustrated in this section. They have been selected according to the probability of outbreaks and the severity of the damage to the oil palm (Fig. 1). The species *Thosea vetusta* and *Susica malayana* are also illustrated, although significant outbreaks remain rare.

The life cycle of the nettle caterpillar takes from 2 to 3 months (60 days on average for *Setora nitens* and *Darna trima*; 104–129 days for *Latoia viridissima*) according to Mariau et al. (1981, 1991). These caterpillars display some characteristic bright colours and a large number of stinging bristles (hence the name nettle caterpillar or slug caterpillar) (Fig. 2, 3, 5–11).

The use of viruses to control outbreaks has been described by Philippe et al. (1997). Various diseases that affect these caterpillars have been listed by Entwistle (1987), and the importance of natural enemies and their potential use in IPM strategies was reported by Desmier de Chenon (1989). Up-to-date biological controls involving natural enemies

Figure 1 Defoliation due to Limacodidae outbreak in Indonesia © Laurence Ollivier.

are not commonly used during pest outbreaks in plantations, but there is a real potential for managing outbreaks using biological controls, rather than pesticides. Developing biological tools to control Limacodidae would be a good strategy, as successful biological control benefits all, without cost. Potential types of biological control for particular regions are summarised below:

2.1.1 Limacodidae from Southeast Asia (Indonesia) (Mariau, 1999)

- *Thosea vetusta* Walker (Fig. 2): *Thosea vetusta* Walker (Hymenoptera: Ichneumonidae) is the most common parasite to attack caterpillars.
- *Susica malayana* Hering (Fig. 3) caterpillars are affected by pathogens and parasitoids (Fig. 4).
- *Darna diducta* Snellen (Fig. 5) is found in Thailand, Malaysia and Indonesia, and parasitised by Hymenoptera [Braconidae *Apanteles aluella* and the hyperparasitoid *Aphanogmus manila*, *Rogas* sp. (hyperparasitoid)]; Eulophidae *Neoplectrus clavatus*; Ichneumonidae (*Buysmania oxymora*); and Diptera (Tachinidae *Chaetexorista javana*, *Palexorista solennis*).
- *Darna trima* Moore (Fig. 6) has been found in Malaysia and Indonesia and four Hymenoptera families have been recorded as parasitic: Braconidae (*Apanteles aluella* and the hyperparasitoid *Aphanogmus manila*, *Apanteles* sp., *Formicia penang* and *Rogas* sp.); Eulophidae (*Platyplectrus orthocraspedae*); Chalcididae (*Brachymeria lasus* (hyperparasitoid)); Ichneumonidae (*Buysmania oxymora*); and the Diptera Bombyliidae *Systropus roekpei* and Tachinidae *Chaetexorista javana*.
- *Setora nitens* Walker (Fig. 7) is found in Malaysia and Indonesia. Four Hymenoptera families are parasitic: Trichogrammatidae (*Trichogrammatoidea thoseae*); Braconidae {*Fornicia ceylonica*, *Fornicia* sp., *Apanteles parasae*, *Meteorus* sp., *Spinaria spinator*, *Rogas* sp. [*Brachymeria lasus*, n.r. (no record), hyperparasitoid, *triraphis*]}; Ichneumonidae (*Buysmania oxymora*, *Chlorocryptus purpuratus*, *Goryphus mesoxanthus*, *Theronia orientalis*, *Xanthopimpla sp.*); and Eurytomidae (*Eurytoma* sp., hyperparasitoid).

Figure 2 *Thosea vetusta* © Laurence Ollivier.

Figure 3 Healthy *Susica malayana* © Laurence Ollivier.

Figure 4 *Susica malayana* affected by entomopathogens or parasitoids © Laurence Ollivier.

Figure 5 *Darna diducta* © Laurence Ollivier.

Figure 6 *Darna trima* © Laurence Ollivier.

Figure 7 *Setora nitens* © Laurence Ollivier.

- Caterpillars are susceptible to diseases and parasitism (Fig. 8a and b).
- *Setothosea asigna* Van Eecke (Fig. 9a and b) is found in Malaysia, Indonesia and the Philippines. Hymenoptera families have been recorded to attack *Setothosea asigna*: Braconidae (*Rogas* sp., *Spinaria* sp., *Fornicia* sp., *Fornicia ceylonica);* Ceraphronidae (*Aphanogmus manila*, hyperparasitoid); Eulophidae [*Euplectromorpha* nr. (no record) *bicarinata*, *Metaplectrus solitarius*]; Ichneumonidae (*Chlorocryptus purpuratus*, *Goryphus* sp., *Theromia orientalis*); Trichogrammatidae (*Trichogrammatoidea thoseae*), and Diptera: Tachinidae (*Chaetexorista javana*). In 1980–81, a cytoplasmic polyhedrosis virus was occasionally reported to be found on *Setothosea asigna*, along with caterpillars displaying disease symptoms of the *B. nudaurelia* type, although it was impossible to say whether the two diseases were associated (Desmier de Chenon et al., 1988; Sinuraya, 1989).
- Caterpillars can also be affected by some unidentified pathogens (Fig. 10).

Figure 8 (a, b) *Setora nitens* affected by entomopathogens © Laurence Ollivier.

Figure 9 (a, b) Healthy *Setothosea asigna* © Laurence Ollivier.

Figure 10 *Setothosea* affected by unidentified pathogens © Laurence Ollivier.

2.1.2 Limacodidae from South America (Ecuador) (Mariau, 1999; Genty, 1972)

- *Sibine fusca* Stoll (Fig. 11): Various species of Hymenoptera attack *Sibine fusca*, including: Braconidae [*Cotesia* sp. nr. (no record) *glomerata* (*Elasmus maculatus*, *Conura immaculata*, hyperparasitoids]; Ichneumonidae [*Casinaria* sp. (*Palmistichus elaeisis*, *Conura immaculata*, *Conura camescens*, *Conura biannulate*, hyperparasitoids), *Baryceros* sp. Diptera: Tachinidae (*Palpexorista coccyx*] and Bombyliidae (*Systropus nitidus*) also.
- Caterpillars were also found to be predated by entomopathogens (Fig. 12). Epizootic infection by a very aggressive virus has been observed in *Sibine fusca* caterpillars, with high levels of mortality in the early stages (Genty and Mariau, 1975). Cases of attack by densonucleosis virus were also described by Meynadier et al. (1977).

2.1.3 Limacodidae from Africa (Gabon) (Mariau, 1999)

- *Latoia viridissima* Holland (Fig. 13) is found in Western and Central Africa. Four Hymenoptera families attack *Latoia viridissima*: Braconidae (*Apanteles* sp., *Rogas* sp.); Eulophidae (*Tetrastichus* spp. nr. *balteatus*); Chalcididae (*Brachymeria* sp. and *Chrysis spina*); Ichneumonidae (*Coccygodes coccyx*), and two Diptera (Tachinidiae *Palexorista* sp. nr. (no record) *moyneana*, and Bombyliidae *Systropus pelopeus*).
- Figure 13 shows caterpillars affected by a possible entomovirus. *Latoia viridissima* is affected by various other viruses: a picornavirus was identified by Fedière et al. (1990) during an outbreak, a nuclear polyhedrosis baculovirus was detected by Kouassi et al. (1991) and a ribovirus was studied by Zeddam et al. (1990). Larval and pupal stages of *Latoia* viridissima are also attacked by a number of parasitoids (Igbinosa, 1988).
- In most of the species studied, population levels are naturally controlled by parasitism (by Hymenoptera: Braconidae and Eulophidae and Diptera: Tachinidae), but Hemiptera predators also play an active role (Mariau, 1999). Virus pathogens have been reported to play a major role in regulating populations during outbreaks (Mariau, 1999), although this effect remains largely underexploited. Small, free viruses have also been detected in South America and West Africa (Desmier de Chenon et al., 1988).

Figure 11 *Sibine fusca* © Laurence Ollivier.

Figure 12 *Sibine fusca* with pupae of Hymenoptera parasitoids © Laurence Ollivier.

Figure 13 *Latoia* sp. in Gabon © Kevin Fernandez.

- Sex pheromone components have been identified in *Darna trima* (Sasaerila et al. 2000a), *Setothosea asigna* (Sasaerila et al., 1997) and *Setora nitens* (Sasaerila et al., 2000b) and these have enabled effective bait traps to be developed to monitor limacodid pest populations in Asian oil palm plantations. However, these cannot be used for mass trapping.

2.2 Lepidoptera: Erebidae

In this section, which focuses on the Lymantriidae, we only discuss two species:

- *Calliteara horsfieldii* Saunders, 1851: that damages the leaves of adult palms
- *Dasychira inclusa* Walker that damages leaves in young plantations and nurseries

Apart from *Orgyia turbata* Butler (Corbett and Dover, 1927), which causes severe defoliation of young palms, other species such as *Dasychira mendosa* Hubner (Wood, 1968), *Euproctis semisignata* Walker (Abraham and Remamong, 1976), *Euproctis linta* Moore (Mariau et al., 1991) and *Laelia venosa* Moore (Wood, 1968) are considered to be secondary pests of the oil palm. These species are characterised by numerous long hairs.

Figure 14 Healthy caterpillar © Laurence Ollivier.

Calliteara horsfieldii Saunders, 1851 affects oil palms in Southeast Asia, particularly in Indonesia (Mariau et al., 1991). This caterpillar displays the yellow dorsal hairs typical of this genus and tufts at the front (Fig. 14), and its life cycle takes 1.5 months (Mariau et al., 1991). *Calliteara horsfieldii* is a polyphagous lepidopteran (Pholboon, 1965; Browne, 1968; Barlow, 1982; Kuroko and Lewvanich, 1993; Hutacherern and Tubtim, 1995; Chey, 1996) and the caterpillars affect the leaves in adult oil palm plantings where they cause major, though not spectacular, defoliation.

This insect is a typical pest, which can appear in sudden outbreaks that can be prevented from having a severe economic impact through natural control or chemical treatments. It seems that these populations are resident in some plantation blocks and are capable of causing repeated outbreaks. Larval excrement on the ground shows that the foliage is actively being consumed by healthy caterpillars, and early caterpillar detection enables treatment to be planned before the final larval stage starts to transform into a chrysalis. The leaf-eating caterpillars' cocoons can be observed on the undersides of leaves and the chrysalides are covered in white silk. Cocoons remain on the foliage, and it is essential to check the effectiveness of any chemical treatment so that they do not form a population reservoir (diapause) which can often cause an outbreak at a later date.

Two tachinid dipterans were found to keep population levels down (Mariau et al., 1991) and natural control was observed in the field prior to chemical treatment in Indonesia (possibly due to a cytoplasmic polyhedrosis virus). Infected caterpillars are soft and pink in colour (Fig. 15) and immobile, or move very little, when handled. Their cuticle is fragile and breaks easily, allowing a yellowish and particularly foul-smelling liquid to leak from the body. Further investigation is required to identify a natural biological control method using a virus.

Dasychira inclusa Walker (Mariau et al., 1991): The caterpillars affect leaves in nurseries and young oil palm plantings (Fig. 16) and their life cycle takes 51–57 days. They attack the upper epidermis, then right through the lamina. A hymenopteran scelionid parasitises the eggs, while *Apanteles* sp. (a braconid) and Diptera: Tachinidae attack the caterpillars.

Figure 15 Caterpillar affected by virus © Laurence Ollivier.

Figure 16 *Dasychira inclusa* caterpillar © Laurence Ollivier.

2.3 Lepidoptera: Cambridae

Furcivena rhodoneuralis Hampson damages oil palm leaves in West Africa (Fig. 17). The caterpillar is very mobile (Fig. 18) and its life cycle takes approximately seven weeks (Mariau et al., 1981). The limbs of the leaflets are perforated along the main nervure (main vein); they are rectangular in shape, one to five times longer than they are wide. Defoliation due to young caterpillars (<5 mm in size) is negligible compared to that recorded during the last two weeks of the caterpillar stage, which can account for more than 90% of the damage observed. These caterpillars cause similar damage to *Coelaenomenodera lameensis* and these two pests might not be differentiated from each other as both can be found on oil palm leaflets. Ten successive cycles (seven weeks/cycle) during which more than 45 caterpillars per lower leaf are counted are believed to cause at least 10% of the damage. Populations do not usually consist of many individuals for very long, suggesting that a natural biological control could be reducing population levels.

Figure 17 Damage on leaves due to *Furcivena rhodoneuralis* in Ghana © Laurence Ollivier.

Figure 18 *Furcivena rhodoneuralis* caterpillar © Laurence Ollivier.

2.4 Lepidoptera: Psychidae

The two main species are:

- *Mahasena corbetti*
- *Metisa plana*

They are known as bagworms and leaf-eating caterpillars in Southeast Asia. They are serious oil palm pests, but several other species of Psychidae can also cause damage (for example, *Clania* sp. in West Papua).

2.4.1 *Mahasena corbetti* Tams 1928

This is a polyphagous pest found throughout Southeast Asia (Lepesme, 1947; Kalshoven, 1950) and the caterpillar attacks oil palm leaves of all ages. In an early stage, the caterpillar builds a bag made of leaf fragments stuck to a mass of silky threads which it secretes (Fig. 19) and it then develops into this bag. This helps the pest use the wind to disperse over large areas (Syed et al., 1974). The female imago remains vermiform and never escapes from its sheath (Fig. 20). Fertility is very high (up to 3000 eggs/adult) and the life cycle of this species takes about four months (Mariau et al., 1991).

Figure 19 *Mahasena corbetti* eating leaflets © Laurence Ollivier.

Figure 20 *Mahasena corbetti* extracted from the bag © Laurence Ollivier.

Eight parasitoids are common to both of these psychid species, while seven are associated only with *Metisa plana* and three only with *Mahasena corbetti* in Peninsular Malaysia (Kamarudin et al., 1996). The list of parasitoids and predators is quite substantial for all the developmental stages except the eggs (Tiong, 1979; Mariau et al., 2001; Syed et al., 1974). IPM with a granulosis virus (Syed and Saleh, 1998) has been used, and a sexual pheromone emitted by the females has been isolated (Rhainds et al., 1997).

Injecting the trunks of mature palms with systemic chemicals is a viable option, and there are many advantages to trunk injection, the main one being its compatibility with IPM practices. The technique is specific to a target insect species, so natural predators are spared and can still proliferate. The use of appropriate trunk injection techniques, the right choice of chemicals and equipment, and a resistance management strategy all contribute to an effective control (Hean, 2000).

2.4.2 *Metisa plana* Walker

The caterpillar of this species mainly damages mature oil palms, and its morphology and biology are similar to those of *Mahasena corbetti*. The larval sheath is attached to the leaflets by a hooked filament (Fig. 21). Natural enemies of the bagworms that attack oil palms in eastern Malaysia have been described by Sankaran and Syed (1972), and Wahid and Kamaruddin (1993) found that *Metisa plana* was attacked by primary and secondary parasitoids and a predator.

2.5 Lepidoptera: Nymphalidae

The caterpillar of *Amathusia phidippus* Linnaeus, 1763 attacks oil palm leaves in the nursery and on mature palms at any developmental stage. These caterpillars have a gregarious behaviour. The life cycle takes two months (Wood, 1968; Lever, 1979). Early instars have three black spots (Lever, 1979) and caterpillars grow to about 90 mm long at the end of their development. They develop long and numerous hairs (Fig. 22a and b) which are not urticarial. The main natural enemies of this insect are several species of tachinid dipterans,

Figure 21 *Metisa plana* sheath © Laurence Ollivier.

Figure 22 (a) *Amathusia phidippus* caterpillars - young caterpillars © Laurence Ollivier. (b) *Amathusia phidippus* caterpillars – last stage caterpillars © Laurence Ollivier.

including *Pamexorista solennis* and *Exorista sorbillans* (Kalshoven and van der Laan, 1981), and a chalcid hymenopteran (Mariau et al., 1991). Removing the insects by hand as soon as they are seen is the most effective control measure (Mariau et al., 1991).

2.6 Lepidoptera: Saturnidae

Automeris liberia Cramer, 1779 damages oil palm foliage in Colombia, Ecuador, Venezuela, Brazil and Peru (Genty et al., 1978) and the caterpillar is especially harmful to 1- to 3-year-old palms. It is also reported to be a highly polyphagous species (Howard et al., 2001; Cock, 2005). The caterpillar is green in colour with a bold white lateral stripe with purplish-red borders, and it is very urticate (Fig. 23). The length of the life cycle is unknown (Genty et al., 1978). As they are an occasional pest on oil palms, Automeris (*Automeris Liberia*) are usually effectively controlled by natural enemies which include Braconidae, Ichneumonidae and Tachinidae.

Figure 23 *Automeris liberia* © Laurence Ollivier.

3 Coleoptera Curculionidae

Coelaenomenodera lameensis Berti and Mariau (1999) is considered to be the most harmful hispine pest in West Africa, displaying disease symptoms of the *B. nudaurelia* (Nudaurelia B virus) including Cameroon (Jacquemard, 2011; Jolivet et al., 2004) (Fig. 24). Larvae mine the leaflets (Fig. 25). Outbreaks cause significant defoliation to the oil palm, which affects crop yields for 2–3 years. *Coelaenomenodera lameensis* attacks were found to cover thousands or even tens of thousands of hectares of *Elaeis guineensis* in most countries (Cachan, 1957). Research into this species is ongoing due to its economic importance. The life cycle takes about 95 days (Morin and Mariau, 1970). Preliminary observations in the field showed that different levels of damage, based on the type of defoliation, depended on the genetic origin of the oil palm (Philippe, 1977, 2003; Coffi, 2006, 2014; Beaudoin-Ollivier et al., 2015).

Rational methods of chemical control have been proposed (Mariau and Philippe, 1983; Philippe, 1990a,b), but chemical control depends on the size of the infested area and the type of plantation. In areas of sizes up to 500 ha, adult pests were controlled most efficiently by spraying Evisect S (Thiocyclam hydrogen oxalate) (Coffi, 2014). The pesticide

Figure 24 *Coelaenomenodera lameensis* adult and galleries © Laurence Ollivier.

Figure 25 *Coelaenomenodera lameensis* last instar larvae outside of the gallery © Laurence Ollivier.

is very efficient, so only few treatments are required. Parasitoids include three species that infest the eggs and four species that live on the larvae and pupae (Jolivet et al., 2004). Local parasitoids are not effective (Mariau and Morin, 1972).

The survival of *Coelaenomenodera lameensis* has been studied in relation to the physical characteristics of different oil palm populations originating from breeding programmes (Beaudoin-Ollivier et al., 2015). No significant difference in the average number of eggs laid and the average hatching percentage was found between *Elaeis oleifera* and *Elaeis guineensis* populations. A highly significant difference was evidenced at the transition stage from first and second larval instars into third and fourth larval instars: the percentage of individuals passing to third and fourth larval instars reached 87% for *Elaeis guineensis* of La Mé origin, whereas it was only 0–2.8% for two *Elaeis oleifera* breeding populations (Beaudoin-Ollivier et al., 2015).

Breeding populations can be classified according to the positive or negative influence they exert, particularly on insect feeding behaviour. Further research into insect feeding preferences for different oil palm breeding populations is required to identify genetic resistance to insect pests. The effect of certain compounds in the leaflets on larval mortality, which might inhibit *Coelaenomenodera lameensis* feeding behaviour, also requires further investigation.

4 Orthoptera Pyrgomorphidae

Zonocerus variegatus (Linnaeus, 1758), or the stink locust, is widespread in West Africa, but infection has not been reported in Gabon (Chiffaud and Mestre, 1990). The adult has a very distinctive appearance due to its bright colours (red, green and yellow) (Fig. 26). This aposematic colouration warns off possible predators. The complete life cycle takes 10–11 months (Mariau et al., 1981). *Z. variegatus* is a polyphagous species that has been recorded on *Elaeis guineensis* (Brunel and de Grégorio, 1978; Copr, 1982; de Grégorio and Brunel, 1977; Hartley, 1967; Jerath, 1965; Koman, 1983; Mayné, 1914; Page, 1978; Steedman, 1988; Mariau et al., 1981) and *Elaeis* sp. (Appert, 1957; Lepesme, 1947; Mayné, 1917).

Elaeis guineensis is included on the list of the main cultivated plants eaten by *Z. variegatus* (Chiffaud and Mestre, 1990). Damage includes severe defoliation, mainly of young palms, by the imagos and larvae (Fig. 26 and 27). The larvae strip the upper surface of the leaflets, leaving the lower epidermis. We observed a few specimens on young palms under three years of age in both Ghana and Gabon in 2016. The economic importance of

Figure 26 *Zonocerus variegatus* imago in Ghana © Laurence Ollivier.

Figure 27 *Zonocerus variegatus* larva in Ghana © Laurence Ollivier.

Z. variegatus has increased over the last 30 years, and the leaf eater has become a threat in both forest and preforest areas (Chiffaud and Mestre, 1990). Following major changes in their environment, pest populations may increase locally and create new damage foci.

The insect's two main enemies are the dipteran *Blaesoxipha filipjevi* Rohdendorf, 1928 (Sarcophagidae: Miltogramminae) and the fungus *Entomophaga grylli* Fresenius, 1858. *Blaesoxipha filipjevi* parasitises the imago and may be an important cause of reduction in the number of eggs laid by the female. Development is optimised in areas with favourable hygrometric conditions. *Entomophaga grylli* attacks all the developmental stages of the insect pest except the first stage. Other parasites or diseases of *Z. variegatus* that have a low impact are cited by Chiffaud and Mestre (1990).

A push–pull strategy using attractive plants such as *Ageratum conyzoides* can be used, and the attractive plant can also be used to detect and monitor *Z. variegatus* before an outbreak. *Solanum verbascifolium, Chromolaena odorata* and coffee clones IRCC 181 and 202 were found to be attractive to *Z. variegatus* in Ivory Coast (Coulibaly et al., 1988, as published in Chiffaud and Mestre, 1990). Small populations can also be controlled by hand-picking. Controlling *Z. variegatus* is innocuous, environmentally safe and inexpensive.

5 Insect pests of the oil palm meristem

5.1 Coleoptera Scarabaeidae

5.1.1 *Oryctes rhinoceros*

Oryctes rhinoceros (Linnaeus, 1758), or the rhinoceros beetle, is native to India and Indonesia, and it is a major pest of both indigenous and imported palms in Southern and Southeast Asia and the western Pacific Islands (Corley and Tinker, 2016; Howard et al., 2001). The adult attacks soft tissues located inside the growing point (Fig. 28), and this often kills young palms, although rarely mature ones. The tightly folded young fronds inside the bud are damaged, resulting in a characteristic and symmetrical loss of wedge-shaped sections of tissue on both sides of the rachis, which are easy to see later on, when the fronds have expanded. The growing point is also damaged and this strongly affects further growth of the palm (Fig. 29). The beetle has been observed attacking oil palm fruit bunches by boring into the stalk (Ponnamma et al., 2001). Its lifespan ranges from 4 to 10 months, and the larvae feed on and develop in organic woody material, dead palm trunks and softwood logs (Fig. 30). Immature stages are also found in empty fruit bunches which are used in the field as a natural potassium fertiliser (Fig. 31). Ecological studies were undertaken by Cumber (1957) in Western Samoa in 1957, and several other surveys were then published by Peterson (1977), Bedford (1980) and Young (1986).

It is likely that IPM could control *Oryctes rhinoceros* populations if the appropriate combination of methods is used. According to the zero-burning policy guidelines issued by ASEAN in 2003, dead oil palm trunks should not be left standing to rot in the field when replacement palms are planted. Felled palms in replanted areas must be shredded, and the resulting woodchip spread in a single layer to destroy the breeding sites provided by the dead palms (Fig. 32).

Insect control is based on the use of a cover crop that is grown on the palm woodchip and acts as a vegetative barrier. It also relies on the manual collection of larvae from breeding sites before replanting and the removal of adults feeding in galleries burrowed in the

Figure 28 *Oryctes rhinoceros* adult © Laurence Ollivier.

Figure 29 Young oil palm affected by *Oryctes rhinoceros* attacks © Laurence Ollivier.

Figure 30 Abundant last instar larvae of *Oryctes rhinoceros* in a breeding site © Laurence Ollivier.

Figure 31 Empty fruit bunches as source of *Oryctes rhinoceros* breeding sites © Laurence Ollivier.

Figure 32 Pieces of old oil palm stems spread as a layer © Laurence Ollivier.

young palms. These are satisfactory methods of reducing *Oryctes rhinoceros* populations (Fig. 33) and fishing nets are also used to protect patrimonial oil palms used in breeding programmes (Fig. 34) (Moore and Quitugua, 2014). The use of insecticides (cypermethrin and lambda-cyhalothrin) is recommended, with the frequency of application depending on the age of the palm.

There have been many attempts to design biological control strategies, and some success was achieved with a baculovirus strain (*Oryctes rhinoceros* nudivirus OrNV) which was spread by infecting and releasing adults (Huger, 2005; Prasad et al., 2008; Bedford, 2013). However, Jackson et al. (2005) pointed out that population control has diminished over time and suggested that efforts should focus on the genetic selection and dissemination of effective virus strains.

Figure 33 Hand-picking of *Oryctes* adult from young palms © Laurence Ollivier.

Figure 34 Fishing net to protect patrimonial oil palms © Laurence Ollivier.

Figure 35 (a) PVC traps © Laurence Ollivier. (b) Bucket trap © Laurence Ollivier.

The aggregation pheromone ethyl 4-methyloctanoate, produced by *Oryctes rhinoceros* males, is commercially used in the field and captures a large number of insects of both sexes (Hallett et al., 1995; Kalidas, 2014). However, there is no evidence that mass trapping can control pest proliferation satisfactorily (Bessou et al., 2017). The pheromone trap remains a useful tool to detect areas newly affected by *Oryctes rhinoceros* and to monitor the spread of the pest (Fig. 35a and b). The fungus *Metarhizium anisopliae* is an effective biopesticide against immature stages of the insect, and it can be spread by adult stages (Fig. 36) and used in combination with pheromone traps (Cik Mohd Rizuan

Figure 36 *Oryctes rhinoceros* adult infected by *M. anisopliae* © Laurence Ollivier.

et al., 2016). The effect of predators (such as scoliid wasp parasites) on eggs, larvae and pupae is not thought to be significant. Studies into the management of this insect pest in Malaysia (Manjeri et al., 2014) are still ongoing, and recent advances have been published (Bedford, 2013, 2014).

5.1.2 *Scapanes australis*

Scapanes australis (Boisduval, 1832) is economically important and is a serious impediment to the establishment of commercial palm plantations in Papua New Guinea (Bedford, 1976; Waterhouse and Norris, 1987; Beaudoin-Ollivier et al., 1999). The adults cause similar damage in young palms to *Oryctes* spp. (Bedford, 1976) (Fig. 29). The male bears three prothoracic horns (Fig. 37) and emits an aggregation pheromone (Prior et al., 2000; Rochat et al., 2000).

5.2 Coleoptera Dryophthoridae

Palm weevils, *Rhynchophorus* spp. are Coleoptera: Curculionidae syn. Rhynchophoridae. They are characterised by an elongated rostrum ('snout') with small mandibles at the distal end. Most palm-associated weevils are polylectic and have a wide range of hosts within the Palmae/Arecaceae, and sometimes within other monocotyledons such as sugar cane (Gramineae/Poaceae), banana (Musaceae) and pineapple (Bromeliaceae).

Rhynchophorus spp. are considered to be major pests of coconut palms in tropical regions of the world (Wattanapongsiri, 1966), and the main species found on the oil palm are *Rhynchophorus palmarum*, *Rhynchophorus bilineatus* and *Rhynchophorus phoenicis* (Fig. 38). Their entire life cycle lasts 3 to 6 months. Larvae are apodous and move peristaltically, tunnelling into the crown and stem (Fig. 39).

Larval stages are protected from most predators, parasites and external environmental factors while devouring all or some of the inside of the host. The tissues around the growing point begin to decay as the late-instar larvae destroy the apical meristem and the palm may be killed (Fig. 40) (Giblin-Davis and Howard, 1989). The external symptoms

Figure 37 Male of *Scapanes australis* in Papua New Guinea © Laurence Ollivier.

Figure 38 *Rhynchophorus* sp [(a) *R. palmarum*; (b) *R. phoenicis*] © Laurence Ollivier.

Figure 39 Larvae of *Rhynchophorus phoenicis* © Laurence Ollivier.

of the attack have been described as similar to those of Fusarium wilt: the leaves show gradually increasing signs of chlorosis and fracture in strong winds.

The adults live on wounded palms and they are usually cryptic, taking refuge between petiole bases, unopened inflorescences, floral peduncles and sites of damage caused by larval feeding in the crown and/or stem of the palm, where they oviposit. *Rhynchophorus palmarum* is a pest of African oil palms and, in addition to causing direct damage, these palm weevil larvae can be vectors of lethal, nematode-borne red ring disease or chronic little leaf of palms. Pheromone traps are used to control populations in oil palm plantations (Fig. 41).

Rhynchophorus bilineatus infestations are endemic in Papua New Guinea, and *Rhynchophorus bilineatus* is a secondary pest once *Scapanes australis* has damaged the plant. Infestation results in the death of the palm. *Rhynchophorus phoenicis* is known to attack the African oil palm in Africa (Fig. 38b and 40), and it is interesting to note that palm weevil larvae are the insects most widely eaten by humans (Howard et al., 2001). Pheromone trapping is widely used to detect and monitor *Rhynchophorus phoenicis, Rhynchophorus bilineatus* and *Rhynchophorus palmarum* (Rochat et al., 1993a,b).

Figure 40 Palm killed by *Rhynchophorus phoenicis* attack © Laurence Ollivier.

6 Insect pests of the oil palm root system

Sufetula sunidesalis Walker (a root miner) is a lepidopteran of the Crambidae family which attacks the root systems of oil palms in Indonesia. Pest pressure is higher in oil palm plantations on peat soils than on mineral soils. The species has been described by several authors (Desmier de Chenon, 1975; Wahyu et al., 2001) and the characteristics of the caterpillar are detailed in Fig. 42 and 43.

The adult is nocturnal, using ferns surrounding palms as shelter during the day, and the eggs are laid in peat soil near the palms. Palms with young aerial roots are particularly susceptible to *Sufetula sunidesalis* attack (Lavogez, 2012) (Fig. 44). The highly mobile caterpillars attack the apex of the primary roots to feed and develop. They can colonise and attack several roots which, once destroyed, generate a new apex near the point of attack. In a severe infestation, primary roots undergo repeated attacks which force the palm to continually renew its root apices (Fig. 45).

This phenomenon has already been described in coconut palms, where it causes an imbalance in energy distribution to the detriment of the fruit bunches and considerably reduces potential yield (Bonneau et al., 2004, 2007). Observations made at two sites in the Indonesian provinces of Riau and North Sumatra in 2010 have made it possible to estimate the level and dynamics of *Sufetula* spp. attacks on oil palm primary roots. After

Figure 41 Use of pheromone trap to catch *Rhynchophorus palmarum* in Ecuador © Laurence Ollivier.

monitoring for one year, results show the degree of damage caused to the oil palm root system on the plot level. The dynamics of the attacks varied according to the planting date (Beaudoin-Ollivier et al., 2011).

Phytosanitary methods involving the massive use of pesticides on very fragile soils and ecosystems are clearly inadvisable (Bessou et al., 2017). Removing all the vegetation that provides shelter to adult *Sufetula sunidesalis* has had positive results in coconut plantations (Bonneau et al., 2007), but these practices were found to generate high labour costs. An agroecological alternative would be to manage the level of the water table carefully so that an area affected by an *Sufetula sunidesalis* outbreak can be quickly flooded to prevent the caterpillars from completing their life cycles (Bessou et al., 2017). This is one of the Roundtable on Sustainable Palm Oil (RSPO) recommendations for management of oil palm plantations on peat soils.

Figure 42 Last instar caterpillars © Laurence Ollivier.

It is important to understand the behaviour and chemical ecology of *Sufetula sunidesalis* by studying the insect's mating behaviour. Studies on the relationship between the caterpillar/moth and host plant have identified two different molecules which could possibly constitute a kairomone emitted by root apices (Lavogez, 2012; Bessou et al., 2017). Indigenous nematodes which can kill *Sufetula sunidesalis* caterpillars have also been discovered in Indonesia (Lavogez, 2012), and these results now serve as a foundation for different hypotheses for the integrated management of *Sufetula sunidesalis*.

7 Insect pests of oil palm bunches

Castnia daedalus Cramer is a large lepidopteran of the Castniidae family (Fig. 46a). The caterpillars are trunk borers and they are harmful for oil palm fruit bunches and bases of the petioles in tropical South America (Venezuela, Suriname, Guyana, northeast Brazil, Colombia, Ecuador and Peru). The whole life cycle is completed in the stem and takes 410–441 days (Genty et al., 1978). Caterpillars feeding on the fruit bunches can cause considerable production losses and they also dig large cavities under the stem apex that kill the palm. At the end of their development, the huge caterpillars (10 cm long) migrate to the periphery of the stem and make a large fibrous cocoon (Fig. 46b).

This pest can be controlled by following best cultivation and hygiene practices (Schuiling and van Dinther, 1980), and adults can be netted during periods of high moth density and cocoons collected from the leaf axils as additional control measures. *Ooencyrtus*

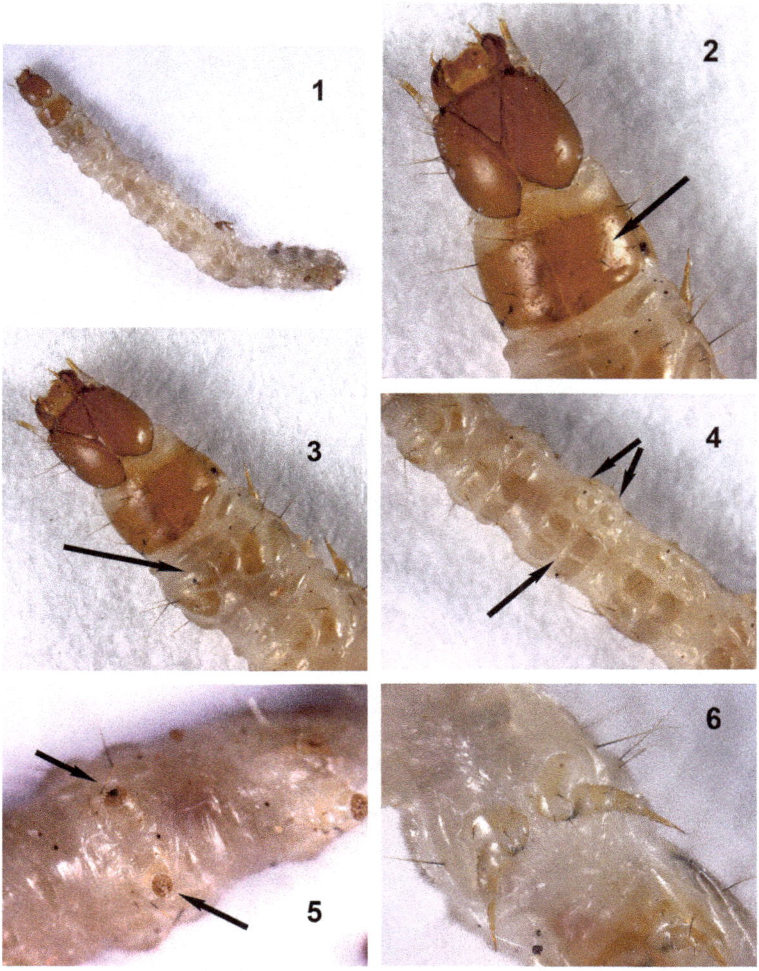

Sufetula larvae characteristics

1. Sufetula larvae final stage (1.7-2 cm)
2. First thoracic segment marked by a large sclerosed transversal plate
3. Second thoracic segment
4. Series of four brown dorsal plates flanked by seven smaller lateral plates on each abdominal segment
5. Abdominal pro-legs crowned by a double row of hooks arranged in a circle
6. Second pair of legs

Figure 43 Details of caterpillars © Laurence Ollivier.

af. *calpodicus* is a hymenopteran in the Encyrtidae family and is a parasitoid for *C. daedalus* eggs (Ruiz and Korytkowski, 1979). Eggs are also attacked by ants of the genera *Odontomachus*, *Pheidole* and *Iridomyrmex*. A nodavirus has also been identified, but its effect is still unknown (Mariau, 2001).

Figure 44 Aerial roots susceptible to *Sufetula sunidesalis* attacks © Laurence Ollivier.

Figure 45 Successive reiterations due to *Sufetula sunidesalis* attacks © Laurence Ollivier.

Figure 46 (a) *Castnia daedalus* imago © Laurence Ollivier. (b) *Castnia daedalus* cocoon © Laurence Ollivier.

8 Future trends and conclusion

This chapter describes only a few representative examples of oil palm pest species which can commonly be found both on young and mature palms in smallholders' plots and agro-industrial plantations. For more detailed information about these insects, the reader should refer to the reviews published by Lepesme (1947), Mariau (2001), Howard et al. (2001) and Bedford (2013).

In April 2011, 26 oil palm companies were granted RSPO certification (14 in Indonesia, 10 in Malaysia, 1 in Papua New Guinea and 1 in Colombia) and the area certified reached 857000 ha (Jacquemard, 2011). RSPO Principle 4 covers the 'use of appropriate best practices by growers and millers' and criterion 4.5 states that 'Growers should apply recognised IPM techniques, incorporating cultural, biological, mechanical and physical methods to minimise the use of chemicals. Native species should be used in biological control where possible'.

According to Sipayung et al. (1989), it is important to understand the various factors involved in the natural biological control of oil palm pests in order to maintain or restore an appropriate environmental balance in the plantation. Several flowering weed species, such as *Euphorbia prunifolia*, *Ageratum conyzoides* and *Turnera subulata*, are maintained and/or actively planted in oil palm fields to encourage parasites and predators (Syed and Shah, 1977) (Fig. 47). A fauna/flora combination which provides food and resting sites for adult parasitoids and predators amongst the palms and along roadsides, where it does not compete with the crop, is always advisable (Delvare and Genty, 1992).

Oil palms should be examined regularly (Morin and Philippe, 1978) and monthly inspections are required for young palms, due to their greater susceptibility to insect damage. Inspections can be less frequent for plantations over four years old. The objectives are to measure the extent and severity of infestation and to determine whether insecticide treatment is required until the pest population returns to normal. There is a natural balance between pests and their enemies; so a small amount of leaf damage might sometimes be seen without any significant effect on palm growth and development. Severe problems only arise when this balance is disturbed.

Figure 47 Nectariferous plants introduced in Indonesia © Laurence Ollivier.

Key measures consist of setting up a network to monitor accidental pest introductions, adopting suitable quarantine and hygiene practices when importing planting material, detecting any changes in the plantation as soon as possible (agronomic practices or products used) and continuously investing in up-to-date staff training. Insects can be trapped using synthetic attractants to detect and monitor new pests, but any trapping strategy should aim to reduce population levels. It is essential to understand the pest's dispersal capacity to effectively combine existing trapping methods and biological controls to achieve a sustainable reduction in pest populations.

Although the use of phytosanitary products is sometimes unavoidable, it must be limited as much as possible because of their environmental impact, and these products must be used in accordance with best practice recommendations. The use of these compounds must only be tolerated in emergency situations where the choice of actions is limited: the selection of a suitable pesticide must take into account its biodegradability rate with respect to local legislation and the RSPO Principles and Criteria. The method of application (trunk injection, or using viruses, entomopathogenic fungi, nematodes, bacteria, predators or parasitoids which are specific to the pest) is also of paramount importance to the agroecological impact of these chemical treatments.

The use of smart IPM methods, adapted to local conditions, is a sensible way to achieve RSPO certification while complying with sustainable development criteria, to the shared benefit of the plantation environment and the people who work in it.

9 Acknowledgements

We are grateful to the management of the different organisations we visited: INRAB in Benin, OLAM in Gabon, SIAT in Ghana, PNG CCRI in Papua New Guinea, PT Bakrie and Socfindo in Indonesia and DANEC in Ecuador, where studies have been undertaken during the last few years and where photographs have been taken.

10 References

Abraham, C. C. and Remanong, T. S. 1976. New record of *Euproctis semisignata* Walker (Lymantriidae-Lepidoptera) as a serious pest of coconut inflorescence. *Current Science India* 45(19), 707–8.

Appert, J. 1957. Les parasites animaux des plantes cultivées au Sénégal et au Soudan. Gouvernement général de l'Afrique Occidentale Française. Centre de Recherches Agronomiques, Bambey, 272p.

Ariffin, D. and Basri, M. W. 2000. Intensive IPM for management of oil palm pests. Oil palm Bulletin 41. Malaysian Palm Oil Board (MPOB), Kuala Lumpur, Malaysia, pp. 1–14.

ASEAN 2003. *Guidelines for the Implementation of the ASEAN Policy on Zero Burning*. The ASEAN secretariat, Jakarta, Indonesia, 40p.

Barlow, H. S. 1982. *An Introduction to the Moths of South East Asia*. the author, Kuala Lumpur, Malaysia.

Beaudoin-Ollivier, L., Morin, J. P., Prior, R. N. B., Kakul, T., Ollivier, J., Rochat, D. and Mariau, D. 1999. Le complexe Scapanes-Rhynchophorus, principal problème entomologique du cocotier en Papouasie Nouvelle Guinée. *Plantations Recherche Developpement* 6, 46–55.

Beaudoin-Ollivier, L., Lazwar, A., Bonneau, X. and Ribeyre, F. 2011. Importance of *Sufetula* sp. attacks on the root system of the oil palm growing on peat soil (*Elaeis guineensis* Jack) in Sumatra. In: *Proceedings of 2011 PIPOC International palm Oil Congress*. MPOB, Kuala Lumpur, Malaysia.

Beaudoin-Ollivier, L., Flori, A., Coffi, A., Cros, D., Glitho, I. and Nodichao, L. 2015. Survival of *Coelaenomenodera lameensis* (Coleoptera: Chrysomelidae) in relation to the physical characteristics of different oil palm (*Elaeis* sp.) breeding populations. *Journal of Insect Science* 15(1), 80. doi:10.1093/jisesa/iev066.
Bedford, G. O. 1976. Observation on the biology and ecology of *Oryctes rhinoceros* and *Scapanes australis*. Pests of coconut palms in Melanesia. *Journal of the Entomological Research Society* 15, 241–51.
Bedford, G. O. 1980. Biology, ecology and control of palm rhinoceros beetles. *Annual Review of Entomology* 25, 309–39.
Bedford, G. O. 2013. Biology and management of palm Dynastid beetles: Recent advances. *Annual Review of Entomology* 58, 353–72.
Bedford, G. O. 2014. Advances in the control of rhinoceros beetle, *Oryctes rhinoceros* in oil palm. *Journal of Oil palm Research* 26(3), 183–94.
Berti, N. and Mariau, D. 1999. *Coelaenomenodera lameensis* n. sp. Ravageur du palmier à huile. *Nouvelle Revue d'Entomologie* 16, 253–67.
Bessou, C., Verwilghen, A., Beaudoin-Ollivier, L., Marichal, R., Ollivier, J., Baron, V., Bonneau, X., Carron, M. P., Snoeck, D., Naim, M., Aryawan, A. A. K., Raoul, F., Giraudoux, P., Surya, E., Sihombing, E. and Caliman, J. P. 2017. Agroecologial practices in oil palm plantations: Examples from the field. *Oilseeds & Fats Crops and Lipids* 24(3), 3–16. doi: 10.1051/ocl/2017024.
Boisduval, J. A. 1832. *Voyage de découvertes de l'astrolabe. Faune entomologique de l'océan Pacifique. Partie Coléoptères*. J. Tastu, Editeur-imprimeur, Paris, France, p. 158, tome 9, Fig. 4.
Bonneau, X., Husni, M., Philippe, R., Somchit, N., Jourdan, C. and Lubis, N. 2004. Discovery of a factor limiting yields in a coconut plantation on peat: The insect pest *Sufetula* spp. *Experimental Agriculture* 40, 53–64.
Bonneau, X., Husni, M., Beaudoin-Ollivier, L. and Susilo, J. 2007. Controlling *Sufetula* spp.: A coconut insect pest on peat soils. *Experimental Agriculture* 43, 289–99.
Browne, F. G. 1968. *Pests and Diseases of Forest Plantation Trees: An Annotated List of the Principal Species Occurring in the British Commonwealth*. Clarendon Press, Oxford, UK.
Brunel, J. F. and de Grégorio, R. 1978. Détermination, à partir de l'étude du contenu du jabot, du régime alimentaire du Criquet puant, *Zonocerus variegatus* (L.), Orthoptera, Pyrgomorphidae: I. Technique et critères de reconnaissance. *Annales de l'université du Bénin, Togo* 4, 103–23.
Cachan, P. 1957. L'Hispidae mineur (*Coelaenomenodera elaeidis* Mlk.), parasite du palmier à huile dans la zone guinéenne. *L'Agronomie Tropicale* 12, 610–32.
Chey, V. K. 1996. Forest Pest Insects in Sabah. Sabah Forest Record 15. Sabah Forest Department, Sandakan, Malaysia.
Chiffaud, J. and Mestre, J. 1990. *Le criquet puant Zonocerus variegatus (Linné, 1758)*. CIRAD- PRIFAS, Département GERDAT, Montpellier, France, 140p.
Cik Mohd Rizuan, Z. A., Noor Hisham, H., Mohd Rozmi, M. and Nazihah Modh, S. 2016. The evaluation of solid substrate formulation of *Metarhizium anisopliae* var. major (M-SS), against *Oryctes rhinoceros* L. in young oil palm plantations. *Planter* 92(1081), 205–18.
Cock, M. J. W. 2005. *Automeris liberia* (Cramer) (Lepidoptera: Saturniidae) in Trinidad. *Living World, Journal of The Trinidad and Tobago Field Naturalists' Club* 2005, 48–9.
Cock, M. J. W., Godfray, H. C. J. and Holloway, J. D. 1987. *Slug and Nettle Caterpillars*. CAB International, Wallingford, UK, 270p.
Coffi, A. 2006. Etude du développement de la mineuse des feuilles *Coelaenomenodera lameensis* Berti (Coleoptera: Chrysomelidae: Hispinae) sur deux backcross de l'hybride interspécifique et quatre origines de palmier à huile. Mémoire de DEA. Faculté des Sciences Agronomiques, Université d'Abomey-Calavi, Bénin, West Africa.
Coffi, A. 2014. Evaluation des facteurs physico-chimiques influençant le développement de la mineuse des feuilles du palmier à huile (*Coelaenomenodera lameensis*) dans les palmeraies de Pobé au Sud-Bénin. Thèse de Docteur en Sciences de la vie. Université de Lomé, Togo, 12 mars 2014.

Cohen, Y. 2017. Morphology and physiology of palm trees as related to *Rhynchophorus ferrugineus* and *Paysandisia archon* infestation and management. In: *Handbook of Major Palm Pests – Biology and Management* (V. Soroker, and S. Colazza, (Eds)). Wiley Blackwell, West Sussex, UK, Volume 2, 39–53.

COPR 1982. *The Locust and Grasshopper Agricultural Manual*. Centre for Overseas Pest Research, London, Volume vii, 690p.

Corbett, G. H. and Dover, C. 1927. The life history and control of some Malayan insects of economic importance I *Orgyia turbata* Butler, a possible pest of rubber. *Malaysian Agricultural Journal* 15(7), 239–95.

Corley, R. H. V. and Tinker, P. B. 2016. *The Oil Palm*, Fifth Edition. Blackwell publishing, 687p.

Coulibaly, N., Decazy, B. and Duverger, B. 1988. *Lutte contre le criquet puant, 1980, avril 1988*. Institut de Recherches du Café et du Cacao, CIRAD, Montpellier, France, Doc multigr, 16p.

Cumber, R. A. 1957. Ecological studies of the rhinoceros beetle, *Oryctes rhinoceros* (L.) in Western Samoa. South Pacific Commission, Nouméa, Nouvelle Calédonie, 31p.

de Grégorio, R. and Brunel, J. F. 1977. Quelques observations nouvelles sur le régime alimentaire du Criquet puant, *Zonocerus variegatus* (L.) (Orthoptera, Pyrgomorphidae). *Bulletin de l'Institut fondamental d'Afrique noire (A)* 39(3), 642–52.

Delvare, G. and Genty, P. 1992. Utility of attractivity plants for beneficial insects in the oil palm plantations of tropical America. *Oléagineux* 47(10), 551–9.

Desmier de Chenon, R. 1975. Présence en Indonésie et Malaisie d'un lépidoptère mineur des racines du palmier à huile, *Sufetula sunidescalis* Walker et relations avec les attaques de Ganoderma. *Oléagineux* 30(1), 449–54.

Desmier de Chenon, R., Mariau, D., Monsarrat, P., Fédière, G. and Sipayung, A. 1988. Recherche sur les agents entomopathogènes d'origine virale chez les défoliateurs du palmier à huile et du cocotier. *Oléagineux* 43(3), 107–17.

Desmier de Chenon, R., Sipayung, A. and Sudharto, P. S. 1989. The importance of natural enemies on leaf-eating caterpillars in oil palm plantations in Sumatra. In: *Indonesia – Uses and Possibilities – Porim International Palm Oil Development Conference*, 5–9 September 1989.

Entwistle, P. F. 1987. Virus diseases of Limacodidae. In: *Slug and Nettle Caterpillar* (M. J. W. Cock,, H. C. J. Godfray, and J. D. Holloway, (Eds)). CAB International, Wallingford, UK, 213–21.

FAOSTAT. 2015. http://www.fao.org/faostat/en/#home

Fedière, G., Philippe, R., Veyrunes, J. C. and Monsarrat, P. 1990. Biological control of the oil palm pests *Latoia viridissima* (Lepidoptera, Limacodidae) in Cote d'Ivoire by a new picornavirus. *Entomophaga* 35(3), 347–54.

Genty, P. 1972. Morphologie et biologie de *Sibine fusca* Stoll., Lépidoptère défoliateur du palmier à huile en Colombie. *Oléagineux* 27(2), 65–71.

Genty, P. and Mariau, D. 1975. Utilisation d'un germe entomopathogène dans la lutte contre *Sibine fusca* (Limacodidae). *Oléagineux* 30(8–9), 349–54.

Genty, P., Desmier de Chenon,, Morin, J. P. and Korytkowski, C. A. 1978. Les ravageurs du palmier à huile en Amérique latine. *Oléagineux* 33(7), 326–419.

Giblin-Davis, R. M. and Howard, F. W. 1989. Vulnerability of stressed palms to attack by *Rhynchophorus cruentatus* (Coleoptera: Curculionidae) and insecticidal control of the pest. *Journal of Economic Entomology* 82, 1185–90.

Hallett, R. H., Perez, A. L., Gries, G., Gries, R., Pierce, H. D., Jr., Yue, J., Oeschlager, C., Gonzalez, L. M. and Borden, J. H. 1995. Aggregation pheromone of the coconut rhinoceros beetle *Oryctes rhinoceros* L. (Coleoptera: Scarabaeidae). *Journal of Chemical Ecology* 21(10), 1549–70.

Hartley, C. W. S. 1967. Diseases and pests of oil palm. In: *The Oil Palm*. Longmans, Green and Co. Ltd., London, UK, 706p.

Hean, Y. T. 2000. The intelligent management of Lepidoptera leaf eaters in mature oil palm by trunk injection (a review of principles). *Planter, Kuala Lumpur* 76(887), 99–107.

Howard, F. W., Moore, D., Giblin-Davis, R. M. and Abad, R. G. 2001. *Insects on Palms*. CABI Publishing, Wallingford, UK, 400p.

Huger, A. M. 2005. The Oryctes virus: Detection, identification and implementation in biological control of the coconut palm rhinoceros beetles, *Oryctes rhinoceros* (Coleoptera: Scarabaeidae). *Journal of Invertebrate Pathology* 89, 78–84.
Hutacherern, C. and Tubtim, N. 1995. *Checklist of Forest Insects in Thailand.* Office of Environmental Policy and Planning, Bangkok, Thailand.
Igbinosa, I. B. 1988. Parasites of the immature stages of *Latoia viridissima* Holland (Lep. Limacodidae), a pest of palms in Nigeria. *Journal of Applied Entomology* 106, 527–30.
Jackson, T. A., Crawford, A. M. and Glare, T. R. 2005. Oryctes virus – Time for a new look at a useful biocontrol agent. *Journal of Invertebrate Pathology* 89, 91–4.
Jacquemard, J. C. 2011. *Le palmier à huile.* Quae, Paris, France; CTA, Wageningen, the Netherlands, Presses agronomiques de Gembloux, Belgium, 240p.
Jerath, M. L. 1965 Notes on the biology of *Zonocerus variegatus* (Linnaeus) from eastern Nigeria. *Revue de Zoologie et de Botanique Africaines* 72(3–4), 243–51.
Jolivet, P., Santiago-Blay, J. A. and Schmitt, M. 2004. *New Developments in the Biology of Chrysomelidae.* SPB Academic Publishing, The Hague, the Netherlands, 803p.
Kalidas, P. 2014. Impact of pheromone baits on the incidence of rhinoceros beetle *Oryctes rhinoceros* (L.) on oil palm. *Pest Management in Horticultural Ecosystems* 20(1), 30–5.
Kalshoven, L. G. E. 1950. *De plagen van de cultuurgewassen in Indonesie.* N.V. Uitgeverij. W. van Hoeve, 's- Gravenhage, Bandoeng.
Kalshoven, L. G. E. and van der Laan, P. A. 1981. *Pest of Crops in Indonesia.* Revised by P. T. van der Laan,. Ichtiar Baru, van Hoeve, Jakarta, 701p.
Kamarudin, N. H., Walker, A. K., Wahid, M. B., LaSalle, J. and Polaszek, A. 1996. Hymenopterous parasitoids associated with the bagworms *Metisa plana* and *Mahasena corbetti* (Lepidoptera: Psychidae) on oil palms in Peninsular Malaysia. *Bulletin of Entomological Research* 86(4), 423–39.
Kamarudin, N. H., Wahid, M. B. and Masijan, Z. 1998. *Handbook of Common Parasitoids and Predators Associated with Bagworms and Nettle Caterpillars in Oil Palm Plantations.* PORIM, Kuala Lumpur, Malaysia, 32p.
Koman, J. 1983. Notes sur le régime alimentaire de *Zonocerus variegatus* (L.) (Orthoptera, Pyrgomorphidae). *Bulletin de l'Institut Fondamental d'Afrique noire (A)* 45(1–2), 118–25.
Kouassi, N. K., Fedière, G., Lery, X., Philippe, R. and Bergoin, M. 1991. Mise en évidence d'un baculovirus de polyédrose nucléarire chez *Latoia viridissima* Lépidoptère Limacodidae, ravageur du palmier à huile et du cocotier en Côte d'Ivoire. *Oléagineux* 46(2), 53–9.
Kuroko, H. and Lewvanich, A. 1993. *Lepidopterous Pests of Tropical Fruit Trees in Thailand.* Japan International Corporation Agency, Tokyo, Japan.
Lavogez, O. 2012. Comportement et écologie chimique de *Sufetula sunidesalis* (Lepidotper: Crambidae: Spilomelinae), ravageur des racines du palmier à huile. Rapport de stage Master Sciences, Technologies, Santé, Spécialité Ecosystèmes, Agrosystèmes et Développement durable. Université de Picardie Jules Verne, Amiens, France, 46p.
Lay, T. C. 1996. Integrated pest management of leaf-eating caterpillars of oil palms in Sabah. *Planter* 72(884), 395–405.
Lepesme, P. 1947. *Les insectes des Palmiers.* Paul Lechevalier (ed), Paris, France, 903p.
Lever, R. J. A. W. 1979. Pests of the coconut Palms. FAO Agricultural studies. N° 77. Food and Agricultural Organization of the United Nations, Rome.
Manjeri, G., Muhamad, R. and Guan Tan, S. 2014. *Oryctes rhinoceros* beetles, an oil palm pest in Malaysia. *Annual Research and Review in Biology* 4(22), 3429–39.
Mariau, D. 1999. The Limacodidae (Lepidoptera) du palmier à huile et du cocotier. Espèces nuisibles et ennemis naturels). *Plantation, Recherche, Développement* 6(3), 149–60.
Mariau, D. 2001. *The Fauna of Oil Palm and Coconut: Insect Mite Pests and Their Natural Enemies.* CIRAD, Montpellier, France, 249p.
Mariau, D. and Morin, J. P. 1972. La biologie de *Coelaenomenodera elaeidis*. La dynamique des populations du ravageur et de ses parasites. *Oléagineux* 27(10), 469–74.

Mariau, D. and Philippe, R. 1983. Avantages et inconvénients des méthodes de lutte chimique contre *Coelaenomenodera minuta* (Coleoptera, Chrysomelidae), Hispine mineur du palmier à huile. *Oléagineux* 38(6), 365–70.

Mariau, D., Desmier de Chenon, R., Julia, J. F. and Philippe, R. 1981. Les ravageurs du palmier à huile et du cocotier en Afrique occidentale. *Oléagineux* 36(4), 170–228.

Mariau, D., Desmiers de Chenon, R. and Sudharto, P. S. 1991. The insect pests of oil palm and their enemies in Southeast Asia. *Oléagineux* 46(11), 400–76.

Mayné, R. 1914. Les ennemis de l'hévéa au Congo Belge. *Bulletin agricole du Congo Belge* 4, 577–600.

Mayné, R. 1917. Insectes et autres animaux attaquant le cacaoyer au Congo Belge. Etudes biol. Agric. Min. Colon. Serv. Agric. Belg. 3.

Meynadier, G., Amargier, A. and Genty, Ph. 1977. Une virose de type densonucléose chez le lépidoptère *Sibine fusca* Stoll. *Oléagineux* 32 (8–9), 357–62.

Moore, A. and Quitugua, R. 2014. Bird netting escape test. Research in support of the Guam coconut rhinoceros beetle eradication project. Cooperative extension service, University of Guam.

Morin, J. P. and Mariau, D. 1970. Etude sur la biologie de *Coelaenomenodera elaeidis*. I. Morphologie et étude du développement. *Oléagineux* 25(1), 11–16.

Morin, J. P. and Philippe, R. 1978. Contrôles sanitaires des plantations adultes de palmiers à huile en Afrique de l'Ouest. I. Contrôles spéciaux. *Oléagineux* 33(6), 277–81: 2 tabl. – Conseils de l'IRHO n. 184.

Page, W. W. 1978. The biology and control of the grasshopper *Zonocerus variegatus*. *PANS* 24(3), 270–7.

Peterson, G. D. 1977. *Research on the Control of the Coconut Palm Rhinoceros Beetle, Fiji, Tonga, Western Samoa. Review of Project Results*. Food and Agriculture Organization of the United Nations. United Nations Development Program, Rome, 130p.

Philippe, R. 1977. Etude du développement de *Coelaenomenodera elaeidis* Mlk (Coleoptera: Chrysomelidae: Hispinae) sur l'hybride E. guineensis × E. melanococca. *Oléagineux* 32, 1–4.

Philippe, R. 1997. Lutte biologique avec des virus entomopathogènes. Application aux ravageurs du palmier à huile et du cocotier. *Plantations, Recherche, Développement, Janvier-février* 1997, 39–45.

Philippe, R. 1990a. Modalités pratiques de lutte chimique contre *Coelaenomenodera minuta* Uhmann (Coleoptera Chrysomelidae). *Oléagineux* 45(4), 165–73.

Philippe, R. 1990b. Etude de l'action de l'Evisect S sur *Coelaenomenodera minuta* (Coleoptera Chrysomelidae Hispinae). *Oléagineux* 45(4),143–63.

Philippe, R. 2003. Visite du centre de recherches agricoles–plantes pérennes (CRA-PP), Bénin, station de Pobé, 13–17 November 2002. 26 p., N ° CP_SIC 1573. CIRAD-CP, Montpellier, France.

Philippe, R., Veyrunes, J. C., Mariau, D. and Bergoin, M. 1997. Biological control using entomopathogenic viruses. Application to oil palm and coconut pests. *Plantations, Recherche, Développement* 4, 39–45.

Pholboon, P. 1965. *A Host List of the Insects of Thailand*. Thai Department of Agriculture & U.S. Operation Mission to Thailand, Bangkok, Thailand.

Ponnamma, K. N., Rethinam, P., Lalitha, N. and Khan, A. S. 2001. Nature and extent of damage by rhinoceros beetle on oil palm bunches. *International Journal of Oil Palm* 2(2), 31–3.

Prasad, G. S., Jayakumar, V., Ranganath, H. R. and Bhagwat, V. R. 2008. Bio-suppression of coconut rhinoceros beetle, *Oryctes rhinoceros* L. (Coleoptera: Scarabaeidae) by Oryctes baculovirus (Kerala isolate) in South Andaman, India. *Crop Protection* 27, 959–64.

Prior, R. N. B., Morin, J. P., Rochat, D., Beaudoin-Ollivier, L., Stathers, T., Kakul, T., Embupa, S. and Nanguai, R. 2000. New aspects of the biology of the Melanesian Rhinoceros Beetle *Scapanes australis* (Col. Dynastidae) and evidence for field attraction to males. *Journal of Applied Entomology* 124, 41–50.

Rhainds, M., Gries, G. and Saleh, A. 1997. Sex specific habitat utilization by bagworms (Lepidoptera: Psychidae). *The Canadian Entomologist* 129(1), 119–200.

Rival, A. and Levang, P. 2013. *La Palme des controverses: Palmier à huile et enjeux de développement*. Quae ed, Versailles, France, 102p.
Rochat, D., Descoins, C., Malosse, C., Nagnan, P., Zagatti, P., Akamou, F. and Mariau, D. 1993a. Chemical ecology of palm weevils, *Rhynchophorus* spp. (Coleoptera Curculionidae). *Oléagineux* 48(5), 225–36.
Rochat, D., Malosse, C., Lettere, M., Ramirez-Lucas, P., Einhorn, J. and Zagatti, P. 1993b. Identification of new pheromone-related compounds from volatiles produced by males of four Rhynchophorinae weevils (Coleoptera, Curculionidae). C. R. Acad. Sci. Paris, t. 316, Serie II, pp. 1737–42.
Rochat, D., Ramirez-Lucas, P., Malosse, C., Rosa Aldana, R., Kakul, T. and Morin, J. P. 2000. Role of solid-phase microextraction in identification of highly volatile pheromones of two rhinoceros beetles *Scapanes australis* and *Strategus aloeus* (Coleoptera, Scarabaeidae, Dynastinae). *Journal of Chromatography A* 885, 433–44.
Ruiz, E. R. A. and Korytkowski, C. A. 1979. *Ooencyrtus* sp. (Hymenoptera: Encyrtidae) parasite de huevos de *Castnia daedalus* (Cramer) (Lepidoptera Castniidae). *Revista Peruana de Entomologia* 22(11), 13–15.
Sankaran, T. and Syed, R. A. 1972. The natural enemies of bagworms on oil palms in Sabah, East Malaysia. *Pacific Insects* 14(1), 57–71.
Sasaerila, Y., Gries, G., Khaskin, G., Gries, R. and Hardi,. 1997. Identification of sex pheromone components of nettle caterpillar, *Setothosea asigna*. *Journal of Chemical Ecology* 23(9), 2187–96.
Sasaerila, Y., Gries, R., Gries, G., Khaskin, G., King, S. and Chor Boo, T. 2000a. Decadienoates: Sex pheromone components of nettle caterpillar *Darna trima* and *D. bradleyi*. *Journal of Chemical Ecology* 26(8), 1969–81.
Sasaerila, Y., Gries, R., Gries, G., Khaskin, G. and Hardi,. 2000b. Sex pheromone components of nettle caterpillar *Setora nitens*. *Journal of Chemical Ecology* 26(8), 1983–90.
Schuiling, M. and van Dinther, J. B. M. 1980. Ecology and control of *Castnia daedalus*, a major pest of oil palm in Brazil. *Zeitschrift für angewandte Entomologie* 2, 161–74.
Sinuraya, L. 1989. Biological control of an oil palm pest in North Sumatera, *Setothosea asigna* Van Eecke (Lepidoptera Limacodidae), using a virus. Biotrop. Spec. Publ. n° 36, pp. 85–94.
Sipayung, A., Desmier de Chenon, R. and Sudharto, P. 1989. Natural enemies of leaf-eating Lepidoptera in the oil palm plantations, North Sumatra. Biotrop. Spec. Publ. n°36, pp. 99–121.
Steedman, A. (Ed.) 1988. *Locust Handbook*, 2nd edition. Overseas Development Natural Resources Institute, London, UK, Volume vii, 180p.
Syed, R. A. and Shah, S. 1977. Some important aspects of insect pest management in oil palm estates in Sabah, Malaysia. In: *International Developments in Oil Palm* (D. A. Erp, and W. Newall, (Eds)). Incorporated Society of Planters, Kuala Lumpur, Malaysia, pp. 570–90.
Syed, R. A. and Saleh, H. A. 1998. Integrated pest management of bagworms in oil palm plantations of PTPP London Sumatra Indonesia TBK (with particular reference to *Mahasena corbetti* Tams) in North Sumatra. In: *International Oil Palm Conference*, Nusa Dua, Bali, 23–25 September.
Syed, R. A., Shah, S. and Liew, K. T. 1974. Occurrence and control of *Mahasena corbetti* Tams in oil palms in Sabah, Malaysia. In: *Advances in Oil Palm Cultivation* (R. L. Wastie, and D. A. Earp, (Eds)). Incorporated Society of Planters, Kuala Lumpur, Malaysia, pp. 402–13.
Tiong, R. H. C. 1979. Some predators and parasites of *Mahasena corbetti* (Tams) and *Thosea asigna* (Moore) in Sarawak. *Planter, Kuala Lumpur* 55, 279–89.
Wahid, M. B. and Kamaruddin, N. 1993. Natural enemies of the bagworm *Metisa plana* Walker (Lepidoptera: Psychidae) and their impact on host population regulation. A 33. In: *PIPOC Porim International palm Oil Congress. Update and vision*. 20–25 September 1993, Kuala Lumpur, Malaysia.
Wahyu, A., Philippe, R., Lasiman, P. and Caliman, J. P. 2001. Pest damage on the root system of oil palms planted on peat soil. In: *Proceedings of 2001 PIPOC International Palm Oil Congress*. MPOB, Kuala Lumpur, Malaysia, pp. 369–81.

Waterhouse, D. F. and Norris, K. R. 1987. *Biological Control. Pacific Prospects*. Inkata Press, Melbourne, Australia, 454p.

Wattanapongsiri, A. 1966. A revision of the genera Rhynchophorus and Dynamis (Coleoptera: Curculionidae). *Department of Agriculture Science Bulletin* 1, 1–328.

Wood, B. J. 1968. *Pests of the Oil Palms in Malaysia and Their Control*. The Incorporated Society of Planters, Kuala Lumpur, Malaysia, 204p.

Young, E. C. 1986. The rhinoceros beetle project: History and review of the research programme. *Agriculture, Ecosystems and Environment* 15, 149–66.

Zeddam, J. L., Philippe, R., Veyrunes, J. C., Fedière, G., Mariau, D. and Bergoin, M. 1990. *Etude du Ribovirus de Latoia viridissima* Holland. ravageur de palmacae en Afrique de l'ouest, *Oléagineux* 45(11), 493–98.

Chapter 4

Integrated pest management in sustainable palm oil production

Edgar Clive Turner and Julie Hinsch, University of Cambridge, UK

1 Introduction

Oil palm is now the number one source of vegetable oil worldwide, forming an important part of the diet for over a billion people (Murphy 2007). It has an annual production of around 65 million tonnes and a crop area of more than 15 million hectares. Production is currently centred in Southeast Asia, with Malaysia and Indonesia alone making up over 85% of global production, with a total combined area of 11.5 million hectares (FAOSTAT 2017). Production has increased rapidly over the last few decades, largely as a result of an increase in the area cultivated. For example, Indonesia, currently the world's largest single producer, increased production by 68% from 2005 to 2011 (Murphy 2014) (Fig. 1). Demand is predicted to more than triple by 2050 (to 240 million tonnes), potentially requiring an additional 12 million hectares of oil palm to be cultivated (Corley 2009).

The expansion of oil palm and its impacts on the environment have received considerable attention and criticism from conservation groups (Anderson 2008). This has perhaps been most marked in Southeast Asia, which is a hotspot of species diversity and endemism (Myers et al. 2000; Sodhi et al. 2010) but, owing to very high levels of logging and forest conversion (Wilcove et al. 2013), has suffered high levels of species loss. Studies have found that conversion of forest to oil palm plantations results in a dramatic reduction in the diversity of most groups studied (Fitzherbert et al. 2008; Danielsen et al. 2009; Foster et al. 2011), with vertebrate taxa showing an overall reduction in richness of 68% compared to forest,

http://dx.doi.org/10.19103/AS.2017.0018.18

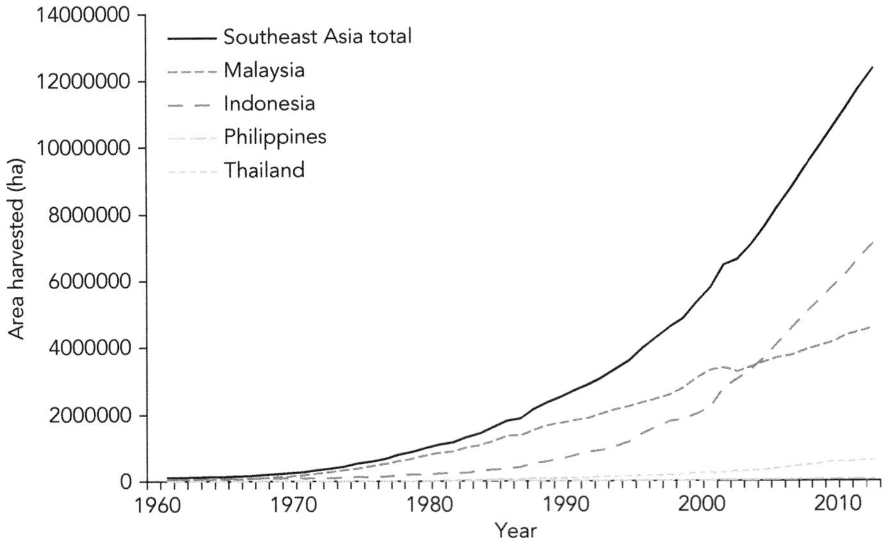

Figure 1 Increase in area of oil palm production in key Southeast Asian countries since 1961. Data from FAOSTAT (2016).

and invertebrates a reduction of 11% (Danielsen et al. 2009). Causes of these losses are generally attributed to a reduction in the physical and biological complexity of plantations compared to forest (Foster et al. 2011; Luskin and Potts 2011), as well as a hotter, drier and less-stable microclimate in plantations (Turner and Foster 2006; Hardwick et al. 2015; Sabajo et al. 2017). Other changes in management in plantations may trigger further losses, for example illegal hunting, road deaths, killing of wildlife by domestic animals (Azhar et al. 2013) and effects of pesticides on non-target organisms (Duckett 1991).

Reductions in species numbers are also likely to have effects on the ecosystem functioning within plantation systems, including pest control. A recent review of the published literature found that 11 out of 14 ecosystem functions assessed showed a reduction in level compared to forest systems (Dislich et al. 2016). The effects of forest conversion to oil palm plantations on biological control are not clear. For example, some studies have found that oil palm plantations contain a lower diversity of insectivorous birds than forest habitats, suggesting that herbivore control by this group may be reduced (Aratrakorn et al. 2006). However, other studies have found no significant difference in the proportion of insectivorous birds between forest and oil palm, suggesting that predation by birds in plantations may remain high (Prabowo et al. 2016). One study in Columbia found lower numbers of large cats in oil palm plantation compared to forest (although higher numbers of medium-sized cats), with possible consequences for predation (Boron (2012), reported in Vargas et al. (2015)). The majority of snake species have also been found to decline in forests versus plantations, again potentially reducing their capacity to control pests (Lynch 2015). Finally, beetle assemblages in plantations have a lower proportion of predators compared to forest habitats (Chung et al. 2000), potentially indicating reduced predation by this group.

2 Pest species in oil palm

Since the rapid expansion of oil palm, there has been a concurrent increase in the number of pest species feeding on palms (Mariau 2001; Duckett 1991; Basri et al. 1995). Pests can sometimes appear very rapidly and quickly reach high and damaging levels. For example, *Norape argyrrhorea,* a lepidopteran pest from Peru, first appeared in the early 2000s, but now displays regular outbreaks, with larvae reaching mean abundances of over 1000 per frond (Zeddam et al. 2003a). Although the most important pests of oil palms are arthropods and mammals (Corley and Tinker 2008), pest species also include a diverse range of other invertebrates and vertebrates (Mariau 2001; Corley and Tinker 2008). Pest species can attack the leaves, trunk, roots and fruits. They commonly act as defoliators (e.g. the caterpillars *Metisa plana* and *Setora nitens*), trunk borers (e.g. the beetles *Oryctes rhinoceros* and *Rhynchophorus ferrugineus*), fruit eaters (e.g. rats *Rattus rattus diardii*) and sap suckers (e.g. the bug *Zophiuma butawengi*) (Dislich et al. 2016). Moth caterpillars of the genus *Sufetula* can also directly attack palm roots (Genty et al. 1976), especially in peatland areas, and have been reported to cause reductions in yield in coconut palms (Bonneau et al. 2004). Palms of different ages and even palm nurseries can be affected by different suites of pests, and different regions of the world can be affected by different taxa (Mariau 2001). Although pests rarely kill the palms themselves (Renard et al. 1998), they can severely damage palms, reduce photosynthetic area, affect water transport and act as vectors for diseases (Gitau et al. 2009; Mariau 2001) (particularly sap suckers; Oehlschlager et al. 2002). These issues are not unique to areas of long-term cultivation, but are also seen in more recently cultivated regions. For example, feeding damage by *N. argyrrhorea* can facilitate a complex of phytopathogenic fungi entering the palms (Zeddam et al. 2003a). For a good summary of insect vectors and their diseases see Gitau et al. (2009).

Damage caused by pests and associated diseases can result in a significant reduction in yield (Renard et al. 1998). Indeed, inadequate control of pests and diseases, leading to crop losses is thought to be one of the main reasons for the large differences in yield seen between different oil palm areas (Murphy 2007), highlighting the global importance of adequate pest control through IPM. Direct damage to palm leaves has been shown to reduce yield. For example, one experimental study on eight-year-old palms found that a 50% defoliation of the palms resulted in a 45% reduction in production (a loss of four tonnes of fresh fruit bunch per hectare), with effects marked for more than a year (Wood et al. 1972). Damage to roots as a result of feeding activity by pests, such as root-feeding cockchafers (*Psilopholis vestita*) or moth caterpillars (*Sufetula* spp.), may also reduce yield (Wood et al. 1972; Genty et al. 1976).

A variety of vertebrate species have been recorded to attack oil palm and particularly to eat oil palm fruits (Corley and Tinker 2008). In most regions, birds are not thought to be a major pest of oil palm, although several species have been reported to eat ripe oil palm fruit (Corley and Tinker 2008; Dhileepan 1989). Various mammals have also been reported to cause damage to palms and fruit, including porcupines, elephants, wild pigs and squirrels (Corley and Tinker 2008). However, rats are the most important mammal pests of oil palm fruits. *Rattus tiomanicus* has been recognised as a pest of oil palm since the 1930s and is often the commonest rodent in oil palm estates (e.g. Buckle et al. 1997), although since the 1980s *Rattus rattus diardii* has replaced it in some areas (Wood and Chung 2003). Estimates of rat densities in plantations from studies carried out in the 1960s in Malaysian estates (before extensive control) recorded very high population levels of between 106 and 578 rats per hectare (Wood and Chung 2003). Rat abundance is commonly associated with the level of fruit damage on palms (e.g. Puan et al. 2011b) and has been reported to cause damage corresponding to losses of 5–10% of plantation yield (Corley and Tinker 2008; Wood and Chung 2003).

A very wide variety of invertebrate taxa have been found to attack oil palms, including butterfly and moth caterpillars (Lepidoptera from at least 11 families), true bugs (Hemiptera), beetles (Coleoptera; particularly Scarabaeidae, Chrysomelidae and Curculionidae), grasshoppers and crickets (Orthoptera), stick insects (Phasmida), termites (Isoptera), thrips (Thysanoptera), and mites (Acari). For an excellent overview of the pests of oil palm see Mariau (2001).The most important lepidopteran defoliators are species of nettle caterpillars (Limacodidae) and bagworm caterpillars (Psychidae). Indeed, 69 species of nettle caterpillars have been recorded to feed on oil and coconut palms in Asia alone. Caterpillar infestations can be common and extensive. For example, a long-term study carried out between 1986 and 2000 across a 63 955 hectare-area of oil palm in Peninsular Malaysia found that over 59% of the area suffered outbreaks of bagworms (*Metisa plana* and *Pteroma pendula*), with 4% of the area being affected each year (Tuck et al. 2011). These infestations can cause significant declines in yield. For example, moderate defoliation by *Metisa plana* of between 10 and 13% can cause yield reductions of 30–40% over a 2-year period (Basri (1993) in Kamarudin et al. (2010)) and crop losses of 44% have been reported due to defoliation by the same species (Basri et al. 1995).

Among the most common beetle pests in plantations are species of rhinoceros beetle (*Oryctes* sp., including *Oryctes rhinoceros* and *Oryctes monoceros*) (Mariau 2001). The larvae develop in decaying plant matter, particularly piles of dead palm material in plantations (Bedford 1980). Adult beetles can then burrow into the base of palm fronds, damaging the developing leaves. This has much more severe effects in younger palms, which in extreme cases can be killed (Bedford 1980). Heavy infestation by *Oryctes* have been reported to reduce plantation yields by 42% and 90% in young palms (Liau and Ahmad (1991) and Chung et al. (1999) (reported in Ooi and Heriansyah (2005)), although the severity of this effect is variable between studies and in some cases has been found to be limited (Wood et al. 1972).

As well as direct effects of individual pests, pest species can also interact to increase the likelihood of other species attacking palms. For example, damage to young palms by *Oryctes rhinoceros*, can act as an entry point for pests that feed on open wounds, such as the weevils *Rhynchophorus palmarum* and *Rhynchophorus ferrugineus* (Bedford 1980), which in turn are vectors for the spread of the red-ring disease-causing nematode *Bursaphelenchus cocophilus* (Oehlschlager et al. 2002; Gitau et al. 2009).

3 The effect of replanting

Several studies report lower incidences of pest outbreaks in older plantations (Mariau 2000), perhaps owing to the build-up of pest control agents, which can reduce pest numbers. For example, one large-scale study in Malaysia found lower rates of bagworm attack in plantations of over 15 years of age (Tuck et al. 2011). However, following the rapid expansion in oil palm area in the 1980s and 1990s (Fig. 1), large areas of plantation are now reaching the end of their productive life and are due for replanting (Snaddon et al. 2013). For example, it is estimated that 365 000 hectares of oil palm in Malaysia are now over 25 years old (Murphy 2014). Replanting dramatically alters and simplifies the habitat structure and can result in pest species increasing in abundance and causing damage to the newly planted palms (Mariau 2000). For example, replanting has been reported to increase the incidence of *Oryctes* beetle attacks (Murphy 2007), probably due to an increase in suitable breeding grounds in the form of dead palm material (Mariau 2000). These piles are also likely to represent breeding sites for rats (Buckle et al. 1997). Predatory groups may also be reduced in number and diversity as a result of replanting. For example, a recent study of frogs found a lower abundance and diversity in replanted compared to mature plantation areas (Kurz et al. 2016), possibly owing to more extreme and variable microclimatic conditions in young plantations.

4 Integrated pest management (IPM) in oil palm

Since as early as the 1960s IPM practices have been applied in oil palm ecosystems (Corley and Tinker 2008) and adopted as oil palm expanded into new areas (e.g. Dewhurst 2011). A key reason for this development was the early realisation that using broad-spectrum, persistent-residue insecticides such as *DDT*, *dieldrin* and *endrin* over large areas could trigger outbreaks of pests (Wood 1971; Basri et al. 1995), by disproportionately reducing the populations of natural enemies that may be more susceptible to pesticide use (Corley and Tinker 2008; Yii et al. 2016; Zeddam et al. 2003a). In addition, widespread use can also lead to selection for resistance to pesticides (Murphy 2014). For example, although rodenticide was initially very successful in controlling rat numbers, rodenticide-resistant populations of the wood rat (*Rattus tiomanicus*) surfaced only 15–20 years after systemic use (Wood and Chung 2003). As a result, applications of pesticides are now generally limited to severe outbreaks (Renard et al. 1998) and more specific chemical applications that are less harmful to natural enemies (Basri et al. 1995).

At the core of IMP is management to maintain and enhance the numbers of natural enemies, where possible without the use of insecticide, which can then keep pest numbers below damaging levels. Under this framework, pesticides are only used when pest numbers reach damaging levels (Dewhurst 2011), generally defined using an economic threshold (Wood et al. 1972; Tuck et al. 2011). There are four key areas of knowledge needed for successful IPM implementation (following Corley and Tinker (2008)):

1 Knowledge of the life history and ecology of pest species.
2 Knowledge of pest numbers through regular monitoring, so control methods can be applied reactively and rapidly.
3 Knowledge of the levels at which pest numbers become damaging, ensuring that potentially damaging and expensive control methods are not implemented until necessary.
4 Identification of key management interventions that can be applied, which target outbreak pests, but, as far as possible, leave the rest of the ecosystem unaffected.

Partly in response to this requirement for information about oil palm pests and pest control agents, research on oil palm has expanded dramatically over the last 50 years (Fig. 2). For example, a search of the scientific search engine, Web of Science (http://wok.mimas.ac.uk/

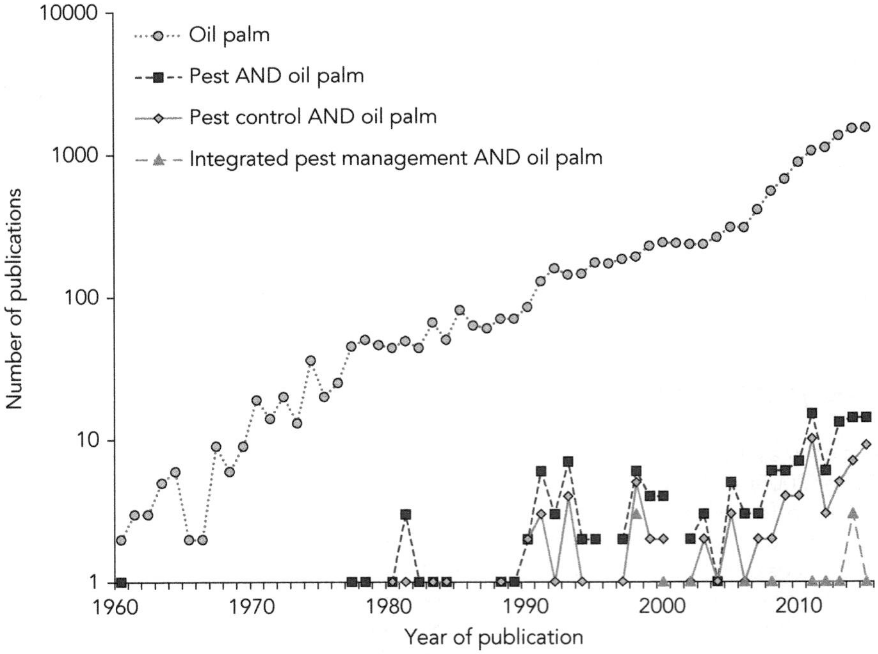

Figure 2 Number of publications returned by the search topic 'oil palm' (14 000), 'pest AND oil palm' (153), 'pest control AND oil palm' (84) and 'integrated pest management AND oil palm' (14) using the scientific search engine, Web of Science (http://wok.mimas.ac.uk/ accessed August 2016).

accessed August 2016), with the search topic 'oil palm' returned 14 000 hits. Research on pest control in plantations has also shown a marked increase over time, with the number of publications returned for 'pest AND oil palm' (153), 'pest control AND oil palm' (84) and 'integrated pest management AND oil palm' (14) increasing, perhaps reflecting a greater research focus on pest control in plantations (Fig. 2).

5 IPM strategies

IPM strategies in oil palm have focused on targeted chemical control, non-chemical management to reduce pest numbers and transmission, targeted management to increase the abundance of key pest control agents (Renard et al. 1998) and breeding of pest-resistant plant stock. So far, much of this management has been at the single species level, although it is clear that suites of control species can interact to control outbreaks more effectively.

5.1 Breeding and genetic approaches

Following the publication of the oil palm genome in 2013 (Singh et al. 2013), there is now a much greater understanding of oil palm genetics, opening up possibilities for targeted breeding and cultivation and even the production of genetically modified palms. So far most of the studies in this area and traditional breeding programmes have focused on the production of high-yielding and disease-resistant varieties (Murphy 2007; Gitau et al. 2009). The value of such strains can be seen in Malaysia, where yields per hectare declined by 9% between 2008 and 2012, partly due to a failure to replant with the highest-yielding new varieties (Murphy 2014). However, relatively little work has focused on pest resistance (Murphy 2007), although interactions between growth and resistance could undermine the suitability of new cultivars (Gitau et al. 2009). The timing of replanting and introduction of resistant strains is also important, although the incentive for this practice needs careful management while palms are still productive, especially as young palms do not yield for the first three to four years (Murphy 2014). Modern molecular approaches may also prove valuable for early detection and characterisation of pests and diseases, allowing early treatment (Panchal and Bridge 2005) and informing IPM.

5.2 Targeted chemical applications

The development of targeted applications of pesticides, such as the use of trunk injection techniques to control pests, such as *Rhynchophorus ferrugineus* (Ferry and Gomez 2014), without the need for widespread spraying has been key to reducing the impacts of chemical control measures on non-target organisms. Other examples include the application of *Carbofuran* granules directly to the axils of young oil palms to reduce attack by *Oryctes* beetles (Bedford 2014). For this to be successful, close monitoring of pest numbers and damage is essential (Nahounou et al. 2010), to ensure outbreaks are detected before yield losses occur (Kathirithamby et al. 1998).

5.3 Management to reduce pest numbers and transmission

Careful management and sanitation in plantations is an important, but often neglected, part of IPM (Gitau et al. 2009). For example, to reduce numbers of the weevil *Rhynchophorus palmarum* and red ring disease, infected trees and secondary host species, such as coconut palms, must be removed quickly from the area (Oehlschlager et al. 2002). Management methods such as chipping and piling dead palms during replanting may also reduce available nest sites for rats (Buckle et al. 1997) and breeding grounds for *Oryctes* larvae (Ooi and Heriansyah 2005).

Trapping of pest species using kairomones or pheromones can also be effective in reducing numbers of individuals. For example, ethyl 4-methyloctanoate has been identified as the main aggregation pheromone for *Oryctes rhinoceros* and *Oryctes monoceros* (Gries et al. 1994) and is widely used in barrier traps to catch these pests (Bedford 2014). Similarly, the pheromone 2-methyl-5(E)-hepten-4-ol can be used to trap the weevil *Rhynchophorus palmarum* (Oehlschlager et al. 1993; Moya-Murillo et al. 2015) and is particularly effective if combined with sugar cane (Hallett et al. 1999). As long as high enough densities of these traps are used, pest numbers can be reduced substantially. For example, trapping in young palms in Southeast Asia has been reported to reduce *Oryctes rhinoceros* damage by 90% (Oehlschlager 2007) and mass trapping of *Rhynchophorus palmarum* in Costa Rica, reduced incidence of red ring disease by 80% (Oehlschlager et al. 2002). Pheromones for both *Oryctes rhinoceros* and *Rhynchophorus ferrugineus* have been used successfully in the same traps, with no negative effect on trapping effectiveness for either species, making this trapping method more efficient (Hallett et al. 1999). Pheromone trapping for bagworm moths (*Metisa plana*) can also be effective, by reducing the populations of males, the mating success of females, and the larval damage, thereby increasing yield (Kamarudin et al. 2010).

5.4 Pest control species

A wide variety of vertebrate and invertebrate control species, as well as pathogens that attack pests, have been identified from oil palm plantations that can contribute to pest control within IPM strategies (Mariau 2001). Along with an increase in the number of pests attacking oil palms, a large number of pest control agents have also switched to this novel ecosystem to attack oil palm pests (Duckett 1991) and there is an active search for new potential biological control agents (e.g. Guerrieri et al. 2011).

5.4.1 Vertebrates

Vertebrate predators highlighted as potential control agents include wild cat species (e.g. Boron (2012) (reported in Vargas et al. (2015)), Duckett 1991), snakes (Lynch 2015, Liat 1999), amphibians (Kurz et al. 2016; Pardo et al. 2013 (in Vargas et al. 2015)) and birds. Given their high occupancy in plantations and preference for rodents as prey items (Jennings et al. 2015) it is possible that leopard cats might be particularly important predator species in oil palm. Early work also investigated the potential of domestic animals, such as domestic cats, as control agents and even the establishment of 'cat farms' (Wood and Chung 2003). Experimental exclusion of birds from baby palms has been shown to increase herbivory levels, demonstrating their potential importance in invertebrate pest control (Koh 2008a). Some of these potential control species may be affected by use of pesticides, such as rodenticides in plantations. For example, predator numbers have been reported to decline when rodenticides such as *warfarin* are widely applied (Wood and Chung 2003), presumably due to secondary poisoning and possibly also low prey numbers. Conversely, if baiting is stopped, other predators, including leopard cats, domestic cats and civets, are reported to increase in number (Duckett 1991). Management to reduce the use of these pesticides may therefore benefit predator groups and pest control.

The most widely studied and used vertebrate predator in oil palm plantations is certainly the Barn owl (*Tyto alba javanica*), used to control rat numbers in plantations and providing a classic example of biological control. In Peninsular Malaysia, this species was only an occasional visitor until the 1960s when the first nesting individuals were reported from an estate house in Johore. Since then they have spread with oil palm cultivation, becoming a common species (Duckett 1991) and being reported from about 15% of areas in Peninsular Malaysia by the 1990s (Wood and Chung 2003). Being an open-habitat species and having high numbers of rats to prey on within plantations, oil palm provides an excellent habitat for this species. As much as 98% of Barn owl diet is rats and small rodents (Duckett 1991) and they do not appear to be selective in relation to prey sex or size (Puan et al. 2011a), making them excellent general rat control agents. However, despite the high potential food supplies in plantations, owls were initially limited by nesting space (Duckett 1991). To aid their spread and colonisation, many estates established artificial nest boxes and some carried out deliberate introductions. Nest boxes appear to have been very effective in increasing owl numbers and resulted in very high population densities in some estates (Duckett 1991). For example, in a survey of 2412 oil palm estates conducted in Malaysia in 1993–94, 188 had established nest boxes for Barn owls. In 78% of these estates Barn owls arrived naturally, while in the rest they were introduced (Basri et al. 1996). In another survey of 68 estates in Peninsular Malaysia in the late 1990s, Barn owls were identified as a major component of rat control in 82% of estates. In 21% of estates surveyed, use of owls were the only control method used, while in 61% of estates owls were used in combination with poison baiting (Hafidzi and Saayon 2001). Despite the widespread use of owls, direct evidence for their effect in reducing rat numbers is mixed, perhaps due to behavioural effects on foraging (Wood and Chung 2003). However, the majority of estates surveyed in Peninsular Malaysia believed that Barn owls were effective in reducing rat numbers and damage (Basri et al. 1996), with 75% of estates no longer considering rats as major pests (Hafidzi and Saayon 2001). The potentially toxic side effects of rodenticides on owls has been of concern for some plantations (Basri et al. 1996; Wood and Chung 2003), with some estates limiting applications to minimise impacts on owls (Hafidzi and Saayon 2001). Generally, first-generation rodenticides, such as *warfarin* seem to be less toxic to owls, but are now less effective owing to resistance, while second-generation anticoagulants are more toxic (Duckett 1991). Use of both first- and second-generation anticoagulant rodenticides, as well as new control agents, such as protozoan-based biorodenticides have also been found to reduce the weight gain of young Barn owls, presumably due to reduced prey access and potentially direct toxic effects (Naim et al. 2010).

5.4.2 Invertebrates

There is a very diverse range of invertebrate predators and parasitoids associated with the control of oil palm pests (e.g. Sipayung et al. 1989, De Chenon et al. 1989, Mariau 2001). This diversity can also act to increase the effectiveness of control, by additively controlling pests. For example, in one study on bagworms in Malaysia, about 37% of bagworm mortality was attributable to predators, 36% to parasitoids and 27% to fungal infection (Cheong et al. 2010). It is, therefore, important that management strategies do not disrupt communities of these natural enemies (Renard et al. 1998), for example through the widespread use of insecticides (e.g. Timti 1991), as this could reduce effective control of pests. Numbers of predatory insects can be increased through targeted management at the habitat level or through mass rearing and releasing of individuals. This has been successfully achieved for several different species including the assassin bug *Sycanus* sp. (Basri et al. 1995).

5.4.3 Predatory insects

A wide range of predatory insect groups have been reported to attack oil palm pests, including several groups of predatory bugs. These can be very effective, although impacts on pest control may vary with the life stage of the oil palm. For example, the assassin bug *Sycanus* sp. can be common in plantations and very effective at pest control, with one individual capable of eating 430 bagworm larvae over its lifespan. However, they seem to be commoner in young plantations, so their importance for pest control over the whole oil palm life cycle is less clear (Basri et al. 1995). Several groups of beetles are also important predators. For example, one species of cleric beetle, *Callimerus arcufer*, is particularly common in plantations in Sumatra and has been found to feed on the eggs and larvae of pests (Basri et al. 1995).

The most abundant group of predatory insects in the tropics are ants, which have been estimated to represent between one-third and one-half of all insect biomass (Way and

Khoo 1992). Ants in the canopy of oil palms can be extremely abundant (Room 1975) and quite diverse (Fayle et al. 2010; Pfeiffer et al. 2008) and are likely to be an important pest control agent (Way and Khoo 1992; Leston 1973). However, some can also act as vectors of plant diseases, increase pest numbers by tending plant-feeding bugs, and attack or irritate plantation workers (Way and Khoo 1992). The Asian weaver ant (*Oecophylla smaragdina)* has been found to reduce the abundance of bag worm *(Pteroma pendula*) caterpillars on oil palms, resulting in lower levels of herbivory and higher crop production. In one study in Malaysia, which compared palms occupied by weaver ants and unoccupied palms, 84% of occupied palms showed no evidence of bagworm infestation and 91% no caterpillar damage, compared to 0% and 2% for unoccupied palms. However, like assassin bugs, this effectiveness seems to be dependent on the life cycle stage of the palm, with ants generally only occupying older palms (Pierre and Idris 2013). Other ant genera have also been reported to control oil palm pests. For example, ants of the genera *Odontomachus*, *Pheidole* and *Iridomyrmex* eat pest insect eggs in oil palm in Peru (Way and Khoo 1992). In Cameroon, the presence of ants in the genus *Crematogaster* was associated with lower abundances of the leaf-mining beetle, *Coelaenomenodera minuta*. Experimental release of *Crematogaster* also demonstrated the ability of this species to attack and reduce the number of leaf miners (Timti 1991).

5.4.4 Parasites and parasitoids

A diverse range of invertebrate parasites and parasitoids, including numerous parasitoid wasps and flies, have been recorded as attacking the eggs, larvae and pupae of oil palm pests (Sankaran and Syed 1972; Basri et al. 1995; Mariau 2001), thus reducing pest numbers. For example, strepsipteran parasites have been used successfully to control long-horn grasshoppers (*Sexava* spp.) in Papua New Guinea (Gitau et al. 2009), by reducing grasshopper feeding and reproduction rates (Solulu et al. 1998; Kathirithamby et al. 1998). Like predators, parasites and parasitoids can be released to control pest outbreaks. For example, inoculum of parasitic nematodes have been deployed to control the pest beetles such as *Strategus aloeus* (Gómez and Sáenz-Aponte 2015). As parasitisation rates can be very variable across seasons and areas (Kalidas and Saravanan 2013), a diverse community is likely to provide more stable pest control (Landis et al. 2000; Letourneau et al. 2011). For a good overview of parasitoid species, see Sipayung et al. (1989) and De Chenon et al. (1989).

5.4.5 Pathogens

Both fungi and viruses have been successfully deployed to control insect pests in oil palm. For example, the fungus *Metarhizium anisopliae* var. *anisopliae* has been used to control the termite *Coptotermes curvignathus* (Yii et al. 2016*)*. In some cases the fungi inoculum can be made more effective by mixing with traditional pesticides such as fipronil. This can weaken the insect and make it more susceptible to attack (Yii et al. 2016). A key advantage of using both viruses and fungi is that they are often specific to certain species and will therefore not harm other species, including pest control agents (Murphy 2014). However, transfer of pathogens to infect pest species can sometimes be a problem. A successful strategy to achieve this for rhinoceros beetles (*Oryctes* sp.) has been direct inoculation of frond heaps, which exposes the larvae to the pathogen and can reduce populations in frond piles by up to 80%. More recent work has investigated the potential of infecting adults and then releasing them to transfer fungal spores more widely through the population. This seems to be particularly effective for the spread of *Oryctes* nudivirus, where infection of breeding sites led to infected adults acting as vectors of the disease, spreading the virus through the population, and reducing beetle numbers and infestation rates (Bedford 1980; Moslim et al. 2011b). Recent work has also showed that pheromone traps can be used effectively to disseminate the pathogenic fungus *Metarhizium anisopliae*, with 67% transmission to adults reported (Moslim et al. 2011a). There is ongoing research to identify new pathogens that can be used to control pest species (e.g. Woruba et al. 2014; Zeddam et al. 2003a,b), as well as more virulent strains of existing pathogens (e.g. Moslim et al. 2011b), which can then be cultivated and released to foster more effective pest control (Ramle et al. 2005). The use of pathogens to control pest numbers can be made more effective by linking this with other IPM strategies. For example, rhinoceros beetle control by nudivirus is more effective if it is combined with removal of beetle breeding sites, the implementation of plantation hygiene measures (Huger 2005), and the use of other pathogens, such bacteria and fungi (Bedford 2014).

5.5 Planting beneficial plants

While targeted management methods and release of key pest control agents can be effective, management for the habitat requirements and food resources of a broader suite of species may represent a more effective and easily implemented IPM strategy. One example of this is the use of cover crops (Baligar and Fageria 2007) in the oil palm understory, particularly nitrogen fixers, which can provide a habitat for predatory species, while also increasing the nitrogen content of the soil, improving soil structure, reducing erosion, increasing soil water-holding capacity, reducing weeds and increasing rates of decomposition (Baligar and Fageria 2007). Cover crops can also help to conceal cut and mulched palms in replanted areas, limiting rhinoceros beetle access and reducing their numbers (Bedford 2014). Examples of species commonly planted in Indonesian plantations include the butterfly bean (*Centrosema pubescens*), white hoary pea (*Tephrosia candida*), and calopo (*Calopogonium mucunoides*) (Baligar and Fageria 2007). Flowering plants in the understory of agricultural systems can also benefit pest control agents, such as adult parasitoids, by providing them with food in the form of nectar that can significantly increase their longevity, fecundity and mobility (Basri et al. 1995; Winkler et al. 2006; Nafziger and Fadamiro 2011). However, not all flowering plants are suitable for all parasitoids. Therefore screening to identify suitable species is important (Gitau et al. 2011; Wäckers 2004; Winkler et al. 2006; Nafziger and Fadamiro 2011; Génaeu et al. 2012). In one study, planting the nectar-producing plant, *Cassia cobanensis*, was found to reduce the abundance of the bagworm *Pteroma pendula* by 50% during an outbreak (Kamarudin and Wahid 2010). A wide variety of different plant species have been used for this purpose in oil palm, including *Turnera subulata, Turnera ulmifolia, Cassia cobanensis, Pueraria phaseoloides ssp. javanica, Stylosanthes humilis, Antigonon leptopus, Vitis japonica and Ageratum gangetica* (Gitau et al. 2011; Nor Sarashimatun et al. 2011).

5.6 Managing for understory plants

Management to reduce herbicide use can benefit a wide suite of naturally occurring understory plants, which could also provide resources for beneficial insects (Gitau et al. 2009), as well as aiding nutrient retention and reducing soil erosion (Anderson 2008). For example, a higher diversity of understory plant species and a higher level of ground cover have been found to correlate with the richness and abundance of understory beetles in a comparative study across habitat types in Malaysia, including oil palm (Chung et al. 2000). A lower understory cover in oil palm plantations is also associated with a lower abundance and diversity of birds and experimental removal of understory reduced bird species richness, indicating that understory cover may also be important for vertebrate predators (Najera and Simonetti 2010). Vegetation cover can also specifically reduce rhinoceros beetle number, by reducing adult access to rotting logs for breeding (Wood et al. 1972). In a study in Malaysia comparing bare ground, sparse vegetation cover and dense vegetation cover, higher incidences of rhinoceros beetle larvae and adults were found in bare ground areas. This resulted in a higher frequency of severely damaged palms in bare ground areas (28.4%), compared to 9.4% in areas of sparse cover and 2% in areas of dense cover (Bedford 1980).

5.7 Forest fragments in plantations

At the landscape scale, maintaining forest in key areas, such as on steep slopes or around rivers can be important for reducing flood risk and soil erosion (Luke et al. 2017) and to aid soil conservation. However, such non-crop areas may also be important for maintaining forest species within the wider landscape (e.g. Maddox et al. 2007; Struebig et al. 2008; Edwards et al. 2010), including predators. These areas can also increase the diversity of functionally important groups, such as dung beetles (Gray et al. 2014). Areas of forest can also impact on the community in adjacent agricultural areas, as forest-dwelling species 'spill-out' into the surrounding habitat. For example, the diversity of birds and butterflies is higher in oil palm landscapes with more forest remaining, presumably as a result of species moving between forest areas (Koh 2008b). The diversity of ants (Lucey et al. 2014), butterflies (Lucey and Hill 2012) and birds (Lees et al. 2015) is also higher at closer distances to forest edges in oil palm landscapes. In some cases, a greater number of predators around river areas can relate to higher levels of predation. For example, one study found evidence for higher rates of invertebrate predator attack on lepidopteran larvae nearby river margins in oil palm, but no effect on vertebrate predator attack (Gray and Lewis 2014). However, it is not clear whether this increased predation links through to higher yield, with a study across 499 plantation parcels in Sabah, Malaysia finding no link between proximity to forest and oil palm yield (Edwards et al. 2014).

6 Achieving IPM

To date, most efforts in understanding pest control have focused on one or a few species, but there is increasing evidence from both within the oil palm sector and throughout the ecological literature that management for a diverse suite of control agents may be more effective in promoting higher and more stable levels of pest control. For example, a long-term study in Malaysia demonstrated the importance of a diverse range of natural enemies, including six different parasitoid species, in controlling bagworm outbreaks (Basri et al. 1995). Each species was found to be more common at different stages of the pest development, while overlapping generations of the parasitoids allowed pest control agents to be consistently present in the plantation system (Basri et al. 1995), rendering pest control more stable. This effect of diversity may be even more marked in areas with seasonality, where different pest control agents are more effective at different times of the year. For example, a study of planthopper pests in Papua New Guinea found that eggs were most commonly parasitised by one species, *Parastethynium maxwelli*, during the wet season, but by another, *Ooencyrtus* sp., during the dry season (Gitau et al. 2011).

Managing for more heterogeneous agricultural systems at both the local and landscape scale may prove a key strategy for supporting higher levels of biodiversity (Tscharntke et al. 2005; Tscharntke et al. 2007; Chaplin-Kramer et al. 2011) and thereby more abundant and diverse pest control agents within oil palm ecosystems (Foster et al. 2011). Because of the growth form of oil palms and their long life, oil palm plantations represent relatively diverse, stable and low-disturbance agricultural habitats (Murphy 2014), meaning that structural complexity and biodiversity could potentially be high compared to other agricultural systems (Anderson 2008). Management methods to increase the heterogeneity of oil palm areas, such as maintaining a diverse understory (Foster et al. 2011) or leaving forest margins around rivers and on steep slopes, and even altering replanting practices (Luskin and Potts 2011), could increase this diversity even further (Anderson 2008). As well as higher average levels of pest control, higher levels of biodiversity can also support higher and more stable ecosystem functions and services (Balvanera et al. 2006; Tylianakis et al. 2008). For example, in a synthesis of 29 studies of crop systems worldwide, Garibaldi et al. (2011) found that non-crop areas left within agricultural landscapes increased the level and stability of pollination in the surrounding crop area, presumably by providing a refuge for pollinators. Several of these management strategies to increase habitat heterogeneity are also in line with recommendations under the principles and criteria for sustainable palm oil production, developed by the Round Table on Sustainable Palm Oil (RSPO) (Round Table on Sustainable Palm Oil 2013), linking these approaches to marketable and accredited sustainable palm oil production.

Taken together, it is clear that IPM within oil palm already represents a sophisticated and diverse suite of approaches that can be effective in suppressing agricultural pests and supporting this highly productive system. New advances in ecological understanding and the interactions between organisms, as well as inputs from new breeding trials and modern genetic approaches, offer an extremely exciting opportunity for the oil palm industry to move towards a more sustainable modern model agricultural system, which combines both very high productivity with a functioning and relatively biodiverse ecosystem.

7 Future trends and conclusion

Although oil palm expansion has come at a significant cost to biodiversity, the very high productivity of the crop itself, as well as its long life and relatively high structural complexity, render it particularly appropriate for IPM strategies. IPM within oil palm has a long and established history, as well as a series of notable successes, particularly the establishment and management of Barn owls to control rat numbers. At the core of the IPM approach is management to maintain and enhance the numbers of natural enemies to keep pest numbers below economically damaging levels. To achieve this, it is crucial to understand the life history and ecology of pest species, to regularly monitor pest numbers (allowing a rapid response), to understand the level at which pest numbers become economically damaging, and to identify a suite of key management interventions that can be applied to target pests but, as far as possible, leave the rest of the ecosystem unaffected. Within oil palm, IPM represents a diverse range of approaches, including plant breeding, targeted chemical applications, management to reduce pest numbers and transmission, and management to increase the numbers of natural enemies and pathogens of pests.

It is clear that IPM represents a valuable tool for controlling pest species within oil palm, with approaches also leading to other benefits for plantation ecosystems, such as improved soil health and reduced erosion. To ensure more stable and productive oil palm ecosystems in the future, especially in the face of a shifting climate, it is important that IPM strategies also include management to increase the resilience and stability of pest control. Managing for more heterogeneous oil palm ecosystems offers a promising approach for achieving this, as habitat diversity is commonly associated with more diverse suites of natural enemies that can operate additively to control pest numbers more effectively across varied environmental conditions. New advances in ecological understanding as well as understanding of the biology of oil palm, offers an exciting opportunity for the oil palm industry to move towards a more sustainable model agricultural system.

Much still needs to be done to fully understand oil palms and oil palm ecosystems to inform more effective IPM strategies. In particular, although well studied, knowledge of the life history and ecology of pest species is still very incomplete, especially given the increasing numbers of species that are recorded as feeding on oil palm each year. This is perhaps most marked in new areas of oil palm expansion, such as in South and Central America and West Africa. For example, from 1990 to 2010 oil palm cultivation in South America increased in area by 3.8% annually (Pacheco 2012). Although this represents an important source of revenue for countries involved, it does mean that oil palm plantations in these areas may experience different problems from pests and diseases. The ecology and effect of expansion in these areas may also differ fundamentally from other regions, compared to plantations in Southeast Asia. For example, in countries like Columbia, where oil palm expansion may be on grassland, including cattle pasture, rather than natural forest areas, there is evidence that biodiversity losses as a result of habitat change may be less severe or non-existent (Vargas et al. 2015).

Another area where active research could be directed is in developing more effective and cutting-edge monitoring methods to assess the point at which pests or diseases become damaging. A recent example of how cutting-edge technology can be applied to this problem is the use of helicopter-mounted, low-altitude remote sensing platforms to assess upper stem rot disease (*Phellinus noxius*). This not only speeds up identification and scope for treatment, but also reduces the costs of processing compared to other remote sensing technologies (Samseemoung et al. 2011).

Finally, more directed research into management interventions that reduce pest outbreaks, but leave the rest of the ecosystem unaffected, would be extremely beneficial. This may include work in understanding the ecology and importance of diversity within oil palm ecosystems in supporting diverse suites of pest control agents, particularly during the replanting process (Mariau 2000), as well as breeding and genetic work to increase the resistance of palms to pests and diseases. For example, the development of new technologies, such as DNA marker-assisted selection and mass propagation methods, offers a novel way of selecting and cultivating plants with favourable characteristics (Murphy 2007). This can occur much more rapidly than using traditional techniques especially given the time lag associated with selective breeding in long-lived crops such as oil palm (Lee et al. 2009).

8 Where to look for further information

For more information on the spread of oil palm and its impact on the environment see: Fitzherbert et al., 2008; Danielsen et al., 2009; Foster et al., 2011.

For more information on the history of IPM in oil palm see: Wood, 2002.

For more information on diversity of diseases and pest species and their enemies in oil palm ecosystems see: Corley and Tinker, 2016a, b; Gitau et al., 2009; Mariau, 2001).

For a review of control methods for rhinoceros beetles, as a case study for invertebrate pest control see: Bedford, 2014.

For a review of control methods for rats, as a case study of vertebrate pest control see: Wood and Chung, 2003.

For online resources associated with oil palm sustainability guidelines and practices see: Round Table on Sustainable Palm Oil, 2013.

9 References

Anderson, J. M. 2008. Eco-friendly approaches to sustainable palm oil production. *J. Oil Palm Res.* 127–42.

Aratrakorn, S., S. Thunhikorn and P. F. Donald. 2006. Changes in bird communities following conversion of lowland forest to oil palm and rubber plantations in southern Thailand. *Bird Conserv. Int.* 16: 71–82.

Azhar, B., D. B. Lindenmayer, J. Wood, J. Fischer, A. Manning, C. McElhinny and M. Zakaria. 2013. Contribution of illegal hunting, culling of pest species, road accidents and feral dogs to biodiversity loss in established oil-palm landscapes. *Wildl. Res.* 40: 1–9.

Baligar, V. C. and N. K. Fageria. 2007. Agronomy and physiology of tropical cover crops. *J. Plant Nutr.* 30: 1287–339.

Balvanera, P., A. B. Pfisterer, N. Buchmann, J. S. He, T. Nakashizuka, D. Raffaelli and B. Schmid. 2006. Quantifying the evidence for biodiversity effects on ecosystem functioning and services. *Ecol. Lett.* 9: 1146–56.

Basri, M., K. Norman and A. Hamdan. 1995. Natural enemies of the bagworm, *Metisa plana* Walker (Lepidoptera: Phychidae) and their impact on host population regulation. *Crop Prot.* 14: 637–45.

Basri, M. W., S. Ismail and N. Kamarudin. 1996. The extent of biological control of rats with barn owls, *Tyto alba javanica* in Malaysian oil palm plantations. *Planter* 72: 5–8.

Bedford, G. O. 1980. Biology, ecology and control of palm rhinoceros beetles. *Annu. Rev. Entomol.* 25: 309–39.

Bedford, G. O. 2014. Advances in the control of rhinoceros beetle, *Oryctes rhinoceros* in oil palm. *J. Oil Palm Res.* 26: 183–94.

Bonneau, X., M. Husni, R. Philippe, N. Somchit, C. Jourdan and N. Lubis. 2004. Discovery of a factor limiting yields in a coconut plantation on peat: The insect pest sufetula spp. *Exp. Agric.* 40: 53–64.

Buckle, A. P., T. H. Chia, M. G. P. Fenn and M. Visvalingam. 1997. Ranging behaviour and habitat utilisation of the Malayan wood rat (*Rattus tiomanicus*) in an oil palm plantation in Johore, Malaysia. *Crop Prot.* 16: 467–73.

Chaplin-Kramer, R., M. E. O'Rourke, E. J. Blitzer and C. Kremen. 2011. A meta-analysis of crop pest and natural enemy response to landscape complexity. *Ecol. Lett.* 14: 922–32.

Cheong, Y., A. Sajap, M. Hafidzi, D. Omar and F. Abood. 2010. Outbreaks of bagworms and their natural enemies in oil palm, *Elaeis Guineensis*, plantation at Hutan Melintang, Perak, Malaysia. *J. Entomol.* 7: 141–51.

Chung, A., P. Eggleton, M. Speight, P. Hammond and V. Chey. 2000. Bulletin of entomological research the diversity of beetle assemblages in different habitat types in Sabah, Malaysia. *Bull. Entomol. Res.* 90: 475–96.

Corley, R. H. V. 2009. How much palm oil do we need? *Environ. Sci. Policy* 12: 134–9.

Corley, R. H. V. and P. B. Tinker. 2008. Diseases and pests of the oil palm. In: *The Oil Palm*, 4th Edition, Wiley-Blackwell, Chichester, UK, pp. 391–444.

Corley, R. H. V. and P. B. Tinker. 2016a. Diseases of the oil palm. In: *The Oil Palm*, 5th Edition, Wiley-Blackwell, Chichester, UK, pp. 399–436.

Corley, R. H. V. and P. B. Tinker. 2016b. Pests of the oil palm. In: *The Oil Palm*, 5th Edition,, Wiley-Blackwell, Chichester, UK, pp. 437–59. Available at: http://onlinelibrary.wiley.com/book/10.1002/9781118953297.

Danielsen, F., H. Beukema, N. D. Burgess, F. Parish, P. F. Donald, D. Murdiyarso, B. E. N. Phalan, L. Reijnders, M. Struebig and E. B. Fitzherbert. 2009. Biofuel plantations on forested lands : Double Jeopardy for biodiversity and climate. *Conserv. Biol.* 23: 348–58.

De Chenon, R., A. Sipayung and P. Sudharto. 1989. The importance of natural enemies on leaf-eating caterpillars in oil palm plantations in Sumatra, Indonesia - Uses and Possibilities. In: *Proc. 1989 Palm Oil Dev. Conf.* Palm Oil Res. Inst., Malaysia Kuala Lumpur, pp. 245–62.

Dewhurst, C. F. 2011. Pests of oil palm in Papua New Guinea, with emphasis on West New Britain. *Phytopathology* 101: S219.

Dhileepan, K. 1989. Investigations on avian pests of oil palm, *Elaeis guineensis* Jacq. in India. *Int. J. Pest Manag.* 35: 273–7.

Dislich, C., A. Keyel, J. Salecker, Y. Kisel, K. Meyer, M. Auliya, A. Barnes, M. Corre, K. Darras, H. Faust, B. Hess, S. Klasen, A. Knohl, H. Kreft, A. Meijide, F. Nurdiansyah, F. Otten, G. Pe'er, S. Steinebach, S. Tarigan, M. Tolle, T. Tscharntke and K. Wiegand. 2016. A review of the ecosystem functions in oil palm plantations, using forests as a reference system. *Biol. Rev.* online: doi:10.1111/brv.12295.

Duckett, J. E. 1991. Management of the barn owl (*Tyto alba javanica*) as a predator of rats in oil palm plantations in Malaysia. *Birds Prey Bull.* 4: 11–24.

Edwards, D. P., J. A. Hodgson, K. C. Hamer, S. L. Mitchell, A. H. Ahmad, S. J. Cornell and D. S. Wilcove. 2010. Wildlife-friendly oil palm plantations fail to protect biodiversity effectively. *Conserv. Lett.* 3: 236–42.

Edwards, F. A., D. P. Edwards, S. Sloan and K. C. Hamer. 2014. Sustainable management in crop monocultures: The impact of retaining forest on oil palm yield. *PLoS One* 9: e91695.

FAOSTAT. 2017. FAO. Food Agric. Organ. United Nations, Stat. Div. http://faostat3fao.org/home/E.

Fayle, T. M., E. C. Turner, J. L. Snaddon, V. K. Chey, A. Y. C. Chung, P. Eggleton and W. A. Foster. 2010. Oil palm expansion into rain forest greatly reduces ant biodiversity in canopy, epiphytes and leaf-litter. *Basic Appl. Ecol.* 11: 337–45.

Ferry, M. and S. Gomez. 2014. Assessment of risks and potential of injection techniques in integrated programs to eradicate the red palm weevil: review and new perspectives. *Fruits* 69: 143–57.

Fitzherbert, E., M. Struebig, A. Morel, F. Danielsen, C. Bruhl, P. Donald and B. Phalan. 2008. How will oil palm expansion affect biodiversity? *Trends Ecol. Evol.* 23: 538–454.

Foster, W. A., J. L. Snaddon, E. C. Turner, T. M. Fayle, T. D. Cockerill, F. Ellwood, G. R. Broad, A. Y. C. Chung, P. Eggleton, V. K. Chey and K. M. Yusah. 2011. Establishing the evidence base for maintaining biodiversity and ecosystem function in the oil palm landscapes of South East Asia. *Philos. Trans. R. Soc. B Biol. Sci.* 366: 3277–91.

Garibaldi, L. A., I. Steffan-Dewenter, C. Kremen, J. M. Morales, R. Bommarco, S. A. Cunningham, L. G. Carvalheiro, N. P. Chacoff, J. H. Dudenhoffer, S. S. Greenleaf, A. Holzschuh, R. Isaacs, K. Krewenka, Y. Mandelik, M. Mayfield, L. A. Morandin, S. G. Potts, T. H. Ricketts, H. Szentgyorgyi, B. F. Viana, C. Westphal, R. Winfree and A. M. Klein 2011. Stability of pollination services decreases with isolation from natural areas despite honey bee visits. *Ecol. Lett.* 14: 1062–72.

Géneau, C. E., F. L. Wäckers, H. K. Luka, C. Daniel and O. Balmer. 2012. Selective flowers to enhance biological control of cabbage pests by parasitoids. *Basic Appl. Ecol.* 13(1): 85–93.

Genty, P., R. Desmier de Chenon and D. Mariau. 1976. Infestation of the aerial roots of oil palms by caterpillars of genus *Sufetula* Walker (Lepidoptera: Pyralidae). *Oléagineux* 31(8–9):365–70.

Gitau, C. W., G. M. Gurr, C. F. Dewhurst, M. J. Fletcher and A. Mitchell. 2009. Insect pests and insect-vectored diseases of palms. *Aust. J. Entomol.* 48: 328–42.

Gitau, C. W., G. M. Gurr, C. F. Dewhurst, H. Nicol and M. Fletcher. 2011. Potential for biological control of *Zophiuma butawengi* (Heller) (Hemiptera: Lophopidae) in coconut and oil palms using the hymenopterans *Ooencyrtus* sp. (Encyrtidae) and *Parastethynium maxwelli* (Girault) (Mymaridae). *Biol. Control* 59: 187–91.

Gómez, A. and A. Sáenz-Aponte. 2015. Susceptibility variation to different entomopathogenic nematodes in *Strategus aloeus* L (Coleoptera: Scarabaeidae). *Springerplus* 4: 620–5.

Gray, C. L. and O. T. Lewis. 2014. Do riparian forest fragments provide ecosystem services or disservices in surrounding oil palm plantations? *Basic Appl. Ecol.* 15: 693–700.

Gray, C. L., E. M. Slade, D. J. Mann and O. T. Lewis. 2014. Do riparian reserves support dung beetle biodiversity and ecosystem services in oil palm-dominated tropical landscapes ? *Ecol. Evol.* 4: 1049–60.

Gries, G., R. Gries, A. Perez, A. Oehlschlager, L. Gonzales, H. J. Pierce, M. Zebeyou and B. Kouame. 1994. Aggregation pheromone of the African rhinoceros beetle *Oryctes monoceros* (Oliver) (Coleoptera: Scarabaeidae). *Z. Naturforsch.* 49: 363–6.

Guerrieri, E., C. W. Gitau, M. J. Fletcher, J. S. Noyes, C. F. Dewhurst and G. M. Gurr. 2011. Description and biological parameters of *Ooencyrtus isabellae* Guerrieri and Noyes sp. nov. (Hymenoptera: Chalcidoidea: Encyrtidae), a potential biocontrol agent of *Zophiuma butawengi* (Heller) (Hemiptera: Fulgoromorpha: Lophopidae) in Papua New Guinea. *J. Nat. Hist.* 45: 2747–55.

Hafidzi, M. and M. Saayon. 2001. Status of rat infestation and recent control strategies in oil palm plantations in Peninsular Malaysia. *Pertanika J. Trop. Agric. Sci.* 24: 109–14.

Hallett, R. H., A. C. Oehlschlager and J. H. Borden. 1999. Pheromone trapping protocols for the Asian palm weevil, *Rhynchophorus ferrugineus* (Coleoptera: Curculionidae). *Int. J. Pest Manag.* 45: 231–7.

Hardwick, S. R., R. Toumi, M. Pfeifer, E. C. Turner, R. Nilus and R. M. Ewers. 2015. The relationship between leaf area index and microclimate in tropical forest and oil palm plantation: Forest disturbance drives changes in microclimate. *Agric. For. Meteorol.* 201: 187–95.

Huger, A. M. 2005. The Oryctes virus: Its detection, identification, and implementation in biological control of the coconut palm rhinoceros beetle, Oryctes rhinoceros (Coleoptera: Scarabaeidae). *J. Invertebr. Pathol.* 89: 78–84.

Jennings, A. P., M. Naim, A. D. Advento, A. A. K. Aryawan, S. Ps, J.-P. Caliman, A. Verwilghen and G. Veron. 2015. Diversity and occupancy of small carnivores within oil palm plantations in central Sumatra, Indonesia. *Mammal Res.* 60: 181–8.

Kalidas, P. and L. Saravanan. 2013. Natural enemies of oil palm defoliators and their impact on pest population. *Pest Manag. Hortic. Ecosyst.* 19: 179–84.

Kamarudin, N. and M. Wahid. 2010. Interactions of the bagworm, Pteroma pendula (Lepidoptera: Psychidae), and its natural enemies in an oil palm plantation in Perak. *J. Oil Palm Res.* 22: 758764.

Kamarudin, N., S. N. Ahmad, O. Arshad and M. B. Wahid. 2010. Pheromone mass trapping of bagworm moths, *Metisa plana* Walker (Lepidoptera: Psychidae), for its control in mature oil palms in Perak, Malaysia. *J. Asia. Pac. Entomol.* 13: 101–6.

Kathirithamby, J., S. Simpson, T. Solulu and R. Caudwell. 1998. Strepsiptera parasites- novel biocontrol tools for oil palm integrated pest management in Papua New Guinea. *Int. J. Pest Manag.* 44: 127–33.

Koh, L. P. 2008a. Birds defend oil palms from herbivorous insects. *Ecol. Appl.* 18: 821–5.

Koh, L. P. 2008b. Can oil palm plantations be made more hospitable for forest butterflies and birds? *J. Appl. Ecol.* 45: 1002–9.

Kurz, D. J., E. C. Turner, A. A. Aryawan, H. C. Barkley, J. P. Caliman, O. Konopik, P. S. Sudharto and W. A. Foster. 2016. Replanting reduces frog diversity in oil palm. *Biotropica* 48: 483–90.

Landis, D. A., S. D. Wratten and G. M. Gurr. 2000. Habitat management to conserve natural enemies of arthropod pests in agriculture. *Annu. Rev. Entomol.* 45: 175–201.

Lee, M., L.-H. Yuen and R. Abdullah. 2009. Expression of *Bacillus thuringiensis* insecticidal protein gene in transgenic oil palm. *Electron. J. Biotechnol.* 9: 117–26.

Lees, A. C., N. G. Moura, A. S. de Almeida and I. C. G. Vieira. 2015. Poor prospects for avian biodiversity in Amazonian oil palm. *PLoS One* 10: e0122432.

Leston, D. 1973. The ant mosaic - tropical tree crops and the limiting of pests and diseases. *Pest Artic. News Summ.* 19: 311–41.

Letourneau, D. K., I. Armbrecht, B. S. Rivera, J. M. Lerma, E. J. Carmona, M. C. Daza, S. Escobar, V. Galindo, C. Gutiérrez, S. D. López, J. L. Mejía, A. M. A. Rangel, J. H. Rangel, L. Rivera, C. A. Saavedra, A. M. Torres and A. R. Trujillo. 2011. Does plant diversity benefit agroecosystems? A synthetic review. *Ecol. Appl.* 21(1): 9–21.

Liat, L. 1999. Reptiles as potential biocontrol agents of pest rodents in plantation areas. In: *Symp. Biol. Control Trop.* Malaysian Agr. Res. Dev. Inst., Train. CTR, Serdang, Malaysia, 18–19 March.

Lucey, J. M. and J. K. Hill. 2012. Spillover of insects from rain forest into adjacent oil Palm plantations. *Biotropica* 44: 368–77.

Lucey, J. M., N. Tawatao, M. J. M. Senior, C. Vun, S. Benedick, K. C. Hamer, P. Woodcock, R. J. Newton, S. H. Bottrell and J. K. Hill. 2014. Tropical forest fragments contribute to species richness in adjacent oil palm plantations. *Biol. Conserv.* 169: 268–76.

Luke, S. H., H. Barclay, K. Bidin, V. K. Chey, R. M. Ewers, W. A. Foster, A. Nainar, M. Pfeifer, G. Reynolds, E. C. Turner, R. P. D. Walsh and D. C. Aldridge. 2017. The effects of catchment and riparian forest quality on stream environmental conditions across a tropical rainforest and oil palm landscape in Malaysian Borneo. *Ecohydrology* 10: 1–14.

Luskin, M. S. and M. D. Potts. 2011. Microclimate and habitat heterogeneity through the oil palm lifecycle. *Basic Appl. Ecol.* 12: 540–51.

Lynch, J. D. 2015. The role of plantations of the African Palm (*Elaeis Guineensis* Jacq.) in the conservation of snakes in Colombia. *Caldasia* 37: 169–82.

Maddox, T., D. Priatna, E. Gemita and A. Salampessy. 2007. The conservation of tigers and other wildlife in oil palm plantations. ZSL Conserv. Rep. No. 7. Zoological Society of London, London, UK.

Mariau, D. 2000. Entomological problems involved in replanting. *Oléagineux* 7: 203–6.

Mariau, D. 2001. *The Fauna of Oil Palm and Coconut. Insect and Mite Pests and their Natural Enemies.* CIRAD, Montpellier, France.

Moslim, R., N. Kamarudin and M. B. Wahid. 2011a. Trap for the auto dissemination of *metarhizium anisopliae* in the management of rhinoceros beetle, *Oryctes rhinoceros*. *J. Oil Palm Res.* 23: 1011–17.

Moslim, R., N. Kamarudin, I. A. Ghani, M. B. Wahid, T. A. Jackson, C. C. Tey and A. M. Ahdly. 2011b. Molecular approaches in the assessment of *Oryctes rhinoceros* virus for the control of rhinoceros beetle in oil palm plantations. *J. Oil Palm Res.* 23: 1096–109.

Moya-Murillo, O. M., R. C. Aldana-de La Torre and A. E. Bustillo-Pardey. 2015. Efficacy of traps to capture *Rhynchophorus palmarum* (Coleoptera: Dryophthoridae) in oil palm plantations. *Rev. Colomb. Entomol.* 41: 18–23.

Murphy, D. J. 2007. Future prospects for oil palm in the 21st century: Biological and related challenges. *Eur. J. Lipid Sci. Technol.* 109: 296–306.

Murphy, D. J. 2014. The future of oil palm as a major global crop: Opportunities and challenges. *J. Oil Palm Res.* 26: 1–24.

Myers, N., R. A. Mittermeier, C. G. Mittermeier, G. A. B. de Fonseca and J. Kent. 2000. Biodiversity hotspots for conservation priorities. *Nature* 403: 853–8.

Nafziger, T. D. and H. Y. Fadamiro. 2011. Suitability of some farmscaping plants as nectar sources for the parasitoid wasp, *Microplitis croceipes* (Hymenoptera: Braconidae): Effects on longevity and body nutrients. *Biol. Control* 56(3): 225–9.

Nahounou, S. C., H. K. Koua, L. R. N. Aboua, G. J. Agnakpa and B. P. Seri-Kouassi. 2010. Comparison of two sanitary control methods of *Coelaenomenodera lameensis* Berti. and Mariau (Coleoptera,

Chrysomelidae: Hispinae), an oil palm (*Elaeis guineensis* Jacq.) leaf miner in West Africa. *Afr. J. Agric. Res.* 5: 114–20.

Naim, M., H. M. Noor, A. Kasim and J. Abu. 2010. Growth performance of nestling barn owls, *Tyto alba javanica* in rat baiting area in Malaysia. *ARPN J. Agric. Biol. Sci.* 5: 1–13.

Najera, A. and J. Simonetti. 2010. Can oil palm plantations become bird friendly ? *Agrofor. Syst.* 80: 203–9.

Nor Sarashimatun, S., C. L. Teh and C.C. Tey. 2011. Evaluation of beneficial plants as hosts for natural enemies of oil palm bagworms. In: *Proceedings of the PIPOC 2011 International Palm Oil Congress*. Malaysia: Malaysian Palm Oil Board, Ministry of Plantation Industries and Commodities, pp. 36–40.

Oehlschlager, A., C. Chinchilla, L. Gonzalez, L. Jiron, R. Mexzon and B. Morgan. 1993. Development of a pheromone-based trapping system for *Rhynchophorus palmarum* (Coleoptera: Curculionidae). *J. Econ. Entomol.* 86: 1381–92.

Oehlschlager, A., C. Chinchilla, G. Castillo and L. Gonzalez. 2002. Control of red ring disease by mass trapping of rhynchophorus palmarum (Coleoptera: Curculionidae). *Florida Entomol.* 85: 507–13.

Oehlschlager, C. 2007. Optimising trapping of palm weevils and beetles. *Acta Hortic.* 736: 347–68.

Ooi, L.-H. and H. Heriansyah. 2005. Palm pulverisation in sustainable oil palm replanting. *Plant Prod. Sci.* 8: 345–8.

Pacheco, P. 2012. Soybean and oil palm expansion in South America. A review of main trends and implications. Cent. Int. For. Res. Work. Pap., 1–28.

Panchal, G. and P. D. Bridge. 2005. Following basal stem rot in young oil palm plantings. *Mycopathologia* 159: 123–7.

Pfeiffer, M., H. C. Tuck and T. C. Lay. 2008. Exploring arboreal ant community composition and co-occurrence patterns in plantations of oil palm *Elaeis guineensis* in Borneo and Peninsular Malaysia. *Ecography (Cop.)* 31: 21–32.

Pierre, E. M. and A. H. Idris. 2013. Studies on the predatory activities of *Oecophylla smaragdina* (Hymenoptera: Formicidae) on *Pteroma pendula* (Lepidoptera: Psychidae) in oil palm plantations in Teluk Intan, Perak (Malaysia). *Asian Myrmecol.* 5: 163–76.

Prabowo, W. E., K. Darras, Y. Clough, M. T. Hernandez, R. Arlettaz and T. Tscharntke. 2016. Bird responses to lowland rainforest transformation in Sumatran smallholder landscapes, Indonesia. *PLoS One* 11: e0154876.

Puan, C. L., A. W. Goldizen, M. Zakaria, M. N. Hafidzi and G. S. Baxter. 2011a. Absence of differential predation on rats by Malaysian Barn Owls in oil palm plantations. *J. Raptor Res.* 45: 71–8.

Puan, C. L., A. W. Goldizen, M. Zakaria, M. N. Hafidzi and G. S. Baxter. 2011b. Relationships among rat numbers, abundance of oil palm fruit and damage levels to fruit in an oil palm plantation. *Integr. Zool.* 6: 130–9.

Ramle, M., M. Wahid, K. Norman, T. Glare and T. Jackson. 2005. The incidence and use of *Oryctes* virus for control of rhinoceros beetle in oil plam plantations in Malaysia. *J. Invertebr. Pathol.* 89: 85–90.

Renard, J., D. Mariau, M. Dollet and H. de Franqueville. 1998. What research into combating oil palm and coconut pests? *Oléagineux* 5: 116–22.

Room, P. 1975. Diversity and organization of the ground foraging ant faunas of forest, grassland and tree crops in Papua New Guinea. *Aust. J. Zool.* 23: 71–89.

Round Table on Sustainable Palm Oil. 2013. Adoption of principles and criteria for the production of sustainable palm oil. Submitt. by RSPO Exec. Board Extraordinary Gen. Assem., 1–70.

Sabajo, C. R., G. le Maire, T. June, A. Meijide, O. Roupsard and A. Knohl. 2017. Expansion of oil palm and other cash crops causes an increase of land surface temperature in Indonesia. *Biogeosci. Discuss* 49: 1–43.

Samseemoung, G., H. P. W. Jayasuriya and P. Soni. 2011. Oil palm pest infestation monitoring and evaluation by helicopter-mounted, low altitude remote sensing platform. *J. Appl. Remote Sens.* 5: 53540.

Sankaran, T. and R. Syed. 1972. The natural enemies of bagworms on oil palms in Sabah, East Malaysia. *Pacific Insects*, 14(1), pp.57–71.

Singh, R., M. Ong-Abdullah, E.-T. L. Low, M. A. A. Manaf, R. Rosli, R. Nookiah, L. C.-L. Ooi, S.-E. Ooi, K.-L. Chan, M. A. Halim, N. Azizi, J. Nagappan, B. Bacher, N. Lakey, S. W. Smith, D. He, M. Hogan, M. A. Budiman, E. K. Lee, R. DeSalle, D. Kudrna, J. L. Goicoechea, R. A. Wing, R. K. Wilson, R. S. Fulton, J. M. Ordway, R. A. Martienssen and R. Sambanthamurthi. 2013. Oil palm genome sequence reveals divergence of interfertile species in Old and New worlds. *Nature* 500: 335–9.

Sipayung, A., R. De Chenon and P. Sudharto. 1989. Natural enemies of leaf-eating Lepidoptera in the oilpalm plantations, North Sumatra. *Biol. Control Pests Biotrop. Spec. Publ.* 36: 99–121.

Snaddon, J. L., K. J. Willis and D. W. Macdonald. 2013. Oil-palm replanting raises ecology issues. *Nature* 502: 170–1.

Sodhi, N. S., L. P. Koh, R. Clements, T. C. Wanger, J. K. Hill, K. C. Hamer, Y. Clough, T. Tscharntke, M. R. C. Posa and T. M. Lee. 2010. Conserving Southeast Asian forest biodiversity in human-modified landscapes. *Biol. Conserv.* 143: 2375–84.

Solulu, T. M., S. J. Simpson and J. Kathirithamby. 1998. The effect of strepsipteran parasitism on a tettigoniid pest of oil palm in Papua New Guinea. *Physiol. Entomol.* 23: 388–98.

Struebig, M. J., T. Kingston, A. Zubaid, A. Mohd-adnan and S. J. Rossiter. 2008. Conservation value of forest fragments to Palaeotropical bats. *Biol. Conserv.* 1: 2112–26.

Timti, I. N. 1991. Control of *Coelaenomenodera minuta* Uhlmann with *Crematogaster* species. *Trop. Pest Manag.* 37: 403–8.

Tscharntke, T., A. Klein, I. Steffan-dewenter and C. Thies. 2005. Landscape perspectives on agricultural intensification and biodiversity – ecosystem service management. *Ecol. Lett.* 8: 857–74.

Tscharntke, T., R. Bommarco, Y. Clough, T. O. Crist, D. Kleijn, T. A. Rand, J. M. Tylianakis, S. Van Nouhuys and S. Vidal. 2007. Conservation biological control and enemy diversity on a landscape scale. *Biol. Control* 43: 294–309.

Tuck, H. C., Y. Ibrahim and K. K. Chong. 2011. Infestations by the bagworms *Metisa plana* and *pteroma pendula* for the period 1986–2000 in major oil palm estates managed by Golden Hope Plantation Berhad in Peninsular Malaysia. *J. Oil Palm Res.* 23: 1040–50.

Turner, E. C. and W. A. Foster. 2006. Assessing the influence of Bird's Nest Ferns (Asplenium spp.) on the local microclimate across a range of habitat disturbances in Sabah, Malaysia. *Selbyana* 27: 195–200.

Tylianakis, J. M., T. A. Rand, A. Kahmen, A. Klein, N. Buchmann and T. Tscharntke. 2008. Resource heterogeneity moderates the biodiversity-function relationship in real world ecosystems. *PLoS Biol.* 6: e122.

Vargas, L. E. P., W. F. Laurance, G. R. Clements and W. Edwards. 2015. The impacts of oil palm agriculture on Colombia's biodiversity: What we know and still need to know. *Trop. Conserv. Sci.* 88: 828–45.

Wäckers, F.L. 2004. Assessing the suitability of flowering herbs as parasitoid food sources: Flower attractiveness and nectar accessibility. *Biol. Control* 29(3): 307–14.

Way, M. J. and K. C. Khoo. 1992. Role of ants in pest management. *Annu. Rev. Entomol.* 37: 479–503.

Wilcove, D. S., X. Giam, D. P. Edwards, B. Fisher and L. P. Koh. 2013. Navjot's nightmare revisited: Logging, agriculture, and biodiversity in Southeast Asia. *Trends Ecol. Evol.* 28: 531–40.

Winkler, K., F. Wäckers, G. Bukovinszkine-Kiss, and J. Lenteren. 2006. Sugar resources are vital for *Diadegma semiclausum* fecundity under field conditions. *Basic Appl. Ecol.* 7(2): 133–40.

Wood, B. 1971. Development of integrated control programs for pests of tropical perennial crops in Malaysia. In: C. Huffaker (Ed.), *Biological Control*, 1st Edition, Plenum Press, New York, NY, pp. 422–57.

Wood, B. 2002. Pest control in Malaysia's perennial crops: A half century perspective tracking the pathway to integrated pest management. *Integr. Pest Manag. Rev.* 7: 173–90.

Wood, B. J. and G. F. Chung. 2003. A critical review of the development of rat control in Malaysian agriculture since the 1960s. *Crop Prot.* 22: 445–61.

Wood, B., R. Corley, and K. Goh. 1972. Studies on the effect of pest damage on oil palm yield. *Planter* 360–79.

Woruba, D. N., M. J. Priest, C. F. Dewhurst, C. W. Gitau, M. J. Fletcher, H. I. Nicol and G. M. Gurr. 2014. Entomopathogenic fungi of the oil palm pest, *Zophiuma butawengi* (Fulgoromorpha: Lophopidae), and potential for use as biological control agents. *Austral Entomol.* 53: 268–74.

Yii, J. E., C. F. J. Bong, J. H. P. King and J. Kadir. 2016. Synergism of entomopathogenic fungus, *Metarhizium anisopliae* incorporated with fipronil against oil palm pest subterranean termite, *Coptotermes curvignathus. Plant Prot. Sci.* 52: 35–44.

Zeddam, J. L., J. A. Cruzado, J. L. Rodriguez, M. Ravallec and E. Candiotti Subilete. 2003a. A cypovirus from the South American oil-palm pest *Norape argyrrhorea* and its potential as a microbial control agent. *BioControl* 48: 101–12.

Zeddam, J. L., J. A. Cruzado, J. L. Rodriguez and M. Ravallec. 2003b. A new nucleopolyhedrovirus from the oil-palm leaf-eater *Euprosterna elaeasa* (Lepidoptera: Limacodidae): preliminary characterization and field assessment in Peruvian plantation. *Agric. Ecosyst. Environ.* 96: 69–75.

Chapter 5

The integrated management of bud rot disease and *Phytophthora palmivora* in oil palm

Gerardo Martinez, José Ignacio Sanz, Gabriel Torres, Greicy Sarria, Diana Velez, Franky Zuñiga, Yuri Mestizo and Francia Varon, Colombian Oil Palm Research Center – Cenipalma, Colombia

1 Introduction

Colombia is the world's fifth largest producer of palm oil. Oil is produced by the fruits of *Elaeis guineensis* Jacq. (African palm), *Elaeis oleifera* Kunth (American oil palm) and the interspecific hybrid between these species (Corley and Tinker, 2003). Oil palm plantations in Colombia covered 476 781 ha by 2014, distributed over four geographical regions according to the Colombian Oil Palm Growers Federation (Fedepalma), namely: the Eastern plains (177 849 ha), the Caribbean (141 099 ha), the Central Magdalena Valley (136 685 ha) and the Pacific Coast (21 149 ha) (Fedepalma, 2014). The annual increase in area of Colombian plantations has been estimated at 7% (Dangond, 2015), thereby making this crop one of Colombia's most important commodities. Palm oil is used in food industries such as bakery, deep-frying, ice cream, and margarines, among others, but also in other industries like soap, resins, fatty acids, animal food, paint and cosmetics (Corley and Tinker, 2003). In Colombia and many South American countries oil palm is the principal source of vegetable oil for human consumption. Recently, it has become a successful source for biodiesel in Colombia and Brazil, which has accelerated the industry's growth all around the world.

http://dx.doi.org/10.19103/AS.2017.0018.30

Bud rot, caused by *Phytophthora palmivora*, is the most widespread disease of the oil palm in Central and South America (Martínez, 2011, 2009a; Torres et al., 2016) and the principal limiting factor for its expansion (Fig. 1). Furthermore, the disease has been reported in all the areas where oil palm is grown (Asia, Africa, Central and South America) and has been known about since the early beginnings of the oil palm business as a large-scale concern. An incident of bud rot destroying a complete oil palm plantation was first reported in Dutch Guiana (Suriname) in 1920, where a four-year-old plantation was devastated by the disease (Herrera-Klindt and Jiménez, 1963). In the same year, Wakefield (1920) reported on the occurrence of the disease in Nigeria, Africa. Bud rot was recorded in 1928 for the first time in Panama, Central America, by Reiking (in de Franqueville, 2003), and an outbreak affecting the nursery stage (27% of losses) forced the United Fruit Company to abandon its oil palm project in the Almirante in Panama and to move it to Siquirres in Costa Rica (Richardson, 1995). The disease was then reported in Malaysia in 1934 (Bunting et al., 1934) and in the Belgian Congo in 1935 (Ghesquiere, 1935). Bud rot was also reported in Venezuela in the mid-1950s (Malaguti, 1953). De Rojas Peña and Ruiz (1972) described an epidemy of the disease in Turbo-Uraba (Colombia, on the border with Panama) between 1965 and 1972, destroying 2,400 ha of oil palm. Dzido et al. (1978) reported on the disease in Ecuador in 1978. Van Den Lande (1999) described epidemies of bud rot in Suriname in 1982. The disease was initially reported in 1974 in the Belem region (Pará) of Brazil, the local name of bud rot being 'Amarelamento fatal' (fatal yellowing) and an outbreak of the disease took place in this region in the western part of the Denpasa plantation in 1983 (Van Slobbe and Rocha de Souza, 1991).

Figure 1 An aerial photograph of a bud rot outbreak in Belen de Bajira, Colombia.

Disease symptoms were found to vary from one region to another and from one oil palm genotype to another and so the disease has received different names (Bull and Robertson, 1959; de Franqueville, 2003; Gómez et al., 1995; Martínez and Torres, 2007; Torres et al., 2016; Turner, 1981). Some of the common names in English for the disease are spear rot, little leaf, bud rot, fatal yellowing, deep bud rot and fatal bud rot (Bull and Robertson, 1959).

The identification of initial symptoms of the disease represents one of the most important challenges in bud rot research. Such symptoms were characterised by very small, light brown lesions (0.5 –1.0 cm wide) having a dark brown border, a chlorotic yellow halo and water-soaked appearance (Fig. 2). Different from the variability of symptoms described for advanced stages of the disease, the initial symptoms are similar among different regions and cultivars affected by bud rot. Internally, the water-soaked brown lesions are darker than external lesions and typically they advance towards the base of leaflets and the spear leaves. As the disease progresses, new tissue becomes affected and the external symptoms become more evident (Fig. 1e Martínez and Torres, 2007). The identification of bud rot symptoms on older palms was corroborated by Ariza et al. (2009) and Aya and Martínez (2011).

Figure 2 The development of bud rot symptoms in oil palm. (a) Healthy spear. (b) Initial stages, small necrotic lesions with aqueous borders. (c) Necrotic lesions reaching infections between leaflets. (d) Lesions are advancing causing bud rot. (e) As bud rot advances, the spear rot is able to infect adjacent neighbour spears. (f) In advance stages the spear collapses. (g) As the tissue degradation continues, more evident is the crater symptom. (h) Internal bud tissue showing a palm with a collapsed spear. Observe the advance of rotten tissue towards the meristematic area.

In addition to symptoms, different hypotheses have been advanced regarding the putative causes of the disease, including abiotic factors, insects and microorganisms (Bull and Robertson, 1959; de Franqueville, 2003; Torres et al., 2016); however, it has not always been possible to reproduce the disease (biotic factors) or correct it using appropriate treatments (abiotic factors) in research conditions.

2 Abiotic factors and insects as putative causative agents

2.1 Abiotic factors

Kovachich (1952) considered spear rot as the initial recovery phase from bud rot disease and not as a different disorder; the author concluded that soil conditions could have been interfering with the plant's physiology. Kovachich associated rainfall patterns with soil characteristics, thereby explaining why the southern Belgian Congo (Democratic Republic of the Congo) showed a 20–25% bud rot incidence, while the northern region experienced only a few cases because of different rainfall pattern and intensity. Kovachich assessed the effect of fertilisation and found a direct relationship with disease incidence; he concluded that nutritional deficiency or imbalance can be a cause for the disease (Kovachich, 1953). Bachy (1954) concluded that the disease observed in Africa was of physiological nature as it was characterised by a blockage of photosynthesis and reduced sap movement due to hormonal disequilibrium. Ferwerda (1954) found a significant difference in the impact of bud rot between plants fertilised with boron and their non-treated counterparts. Incorporating borax into the soil was suggested by Broeshart et al. (1957) in order to promote the recovery of palms showing 'little leaf' symptoms; however, the authors considered the 'little leaf' disorder as different from bud rot disease. Boron deficiency was ruled out as the cause of the disease in Ecuador since soil analyses in the area under study provided homogenous results despite some variability in disease incidence (Renard and Quillec, 1985).

Mineral deficiency has also been proposed as the causative agent of bud rot in South and Central America (Acosta et al., 1996; Gómez et al., 1995). Laing (2009) proposed that strong competition between K^+ and Mg^+ against Ca^{2+} cations reduces Ca^{2+} absorbability, thereby increasing the susceptibility of palms cultivated in soil with low Ca content, low pH (<4.6), high Al^{3+} concentration and high Ca:Mg ratios.

In addition to nutritional imbalance, poor drainage has also been correlated with the occurrence of bud rot disease (Acosta et al., 1996; Cristancho et al., 2012; de Franqueville, 2003; De Rojas and Ruíz, 1972). Indeed, de Rojas and Ruiz (1972) attributed the outbreak in Urabá (Colombia) to an excessive accumulation of nitrites in the soil because of poor drainage. A negative correlation between bud rot occurrence and solar radiation was observed in the Eastern Plains of Colombia (Acosta et al., 1996); the authors also observed a positive correlation between the soil's clay content and bud rot incidence. Cristancho et al. (2012) concluded that soil management practices focusing on reducing nutrient imbalance, and improving plant nutrition and adequate drainage must form an essential part of any integrated bud rot management programme.

2.2 Insects

Several species from various different insect orders have been proposed as port of entry (causing wounds) and carriers (Bull and Robertson, 1959; Gómez et al., 1995; Jiménez, 1991; MacGarvie, 1960; Torres et al., 2008a; Turner and Bull, 1967) or as vectors of the disease (Dollet, 1991; Perthuis, 1991; Renard and Quillec, 1985). MacGarvie (1960) proposed that the 0.5 × 2 cm hole bored by *Temnoschoita quadripustulata* serves as an entry point for the bacteria responsible for causing the bud rot disease; once the bacteria infects the spear leaf tissue, the damage progresses towards the meristem; however, many times its progress is limited by a callus barrier formed by the palm above the meristem (MacGarvie, 1960; Turner and Bull, 1967). The smelling rotting tissue attracts *Rhynchophorus* spp. and *Platygenia* spp. adults, which lay their eggs on infected tissue. Once the larvae hatch, they bore galleries towards the meristem overcoming the callus barrier (MacGarvie, 1960). When insects do not penetrate the callus barrier, palms can recover, but the leaves emitted previously are contorted (Bull and Robertson, 1959; MacGarvie, 1960). When borer insects overcome the callus protective layer, they can destroy the meristem directly by feeding on it or indirectly by facilitating the entry of various microorganisms into the protected area, thereby colonising and killing the meristem (MacGarvie, 1960). In addition to *Platygenia* spp., *Rhynchophorus* spp. and *Temnoschoita* spp., Bull and Robertson (1959) suggested that the beetle *Oryctes* spp. and the pyralid moth *Pimelephila ghesquierei* Tams are also involved in the disease cycle, acting in a similar way. However, MacGarive (1960) concluded that the low frequency and feeding behaviour of *Oryctes* spp. in the affected tissue and the preference of *P. ghesquierei* for lignified tissue would make this hypothesis unlikely.

Urueta (1975) studied more than 30 species of arthropods (including spider mites) in bud rot-affected palms in Colombia. The author concluded that *Cephaloleia vagelineata* was the only species directly involved in bud rot disease, since this species causes similar lesions to bud rot on the base of the spears. Jiménez (1991) identified *Oncometopia rubescens, Herminodes insulsa* and *Tiquadra* sp. in Tumaco (Colombia) as insects associated with bud rot-affected palms. Jimenez (1991) showed that a cricket (Tettigoniidae), some Curculionidae species (*Rhynchophorus palmarum, Metamasius* sp., *Limnobaris* sp. and *Parisoschoenus* sp.), *Oncometopia* sp., and *Persis* sp., together with several species of flies (Diptera) and cockroach, were associated with infected palms (Torres et al., 2008a). Gómez et al. (1995) proposed extending the list of suspected species to include *Imatidium (Demotispa) neiva* and two unidentified beetles (Coleoptera) as possible vectors of the disease in Colombia. In the late 1980s, the French IRHO (Institut de Reherche pour les Huiles et Oléagineux, now Cirad) research group suggested that bud rot was from viroid or viral origin (de Franqueville, 2003). In Brazil, parallel 187 homoptera and 55 heteroptera species were found to be associated with infected palms in Brazil, 12 from such species being considered as good candidates for virus transmission assays (Villanueva-Guerrero, 1988). De Franqueville (2003) reported that research work in Brazil and Ecuador aimed at identifying vectors of bud rot disease included more than 400 homoptera species (56% Membracidae, 15% Derbidae, 13% Delphacidae, 5% Cixiidae, 11% other or not identified species); however, none of the individuals was evaluated.

3 Microorganisms as putative causative agents

Several authors have proposed a microbiological determinism for bud rot in the oil palm. Geostatistical and variogram analysis of the spatial distribution of the disease in Suriname published by Van de Lande and Zadocks (1999) confirmed that bud rot is an infectious disease, which is likely spread by a vector.

Foul smell has frequently been associated with the presence of bacterial populations (Bunting et al., 1934; Duff, 1963; Ghesquiere, 1935; Nieto Páez and Gómez, 1991; Van de Lande, 1991). Ghesquiere (1935) noted that there are two types of bud rot in Africa: one being caused by *P. palmivora* and the other one by *Bacillus coli*. In 1944, Alibert also proposed *B. coli* as a causative agent (Bull and Robertson, 1959). By contrast, Duff (1963) showed that bud rot is reproducible when weakened palms are inoculated with a bacteria similar to *Erwinia lathyri*, but inoculation trials undertaken in 1984 with *Erwinia sp.* by Van de Lande failed to generate symptoms of the disease (Griffith, 1987). Inoculation by MacGarvie (1960) using rotten tissue from affected palms also failed to induce specific symptoms. Bachy (1954) reported the frequent association of bacteria and *Fusarium* spp. with infected plants. The isolation of *Pseudomonas fluorescens, P. putida, P. aeruginosa* and *Erwinia cypripedii* from infected oil palm tissues was also reported by Gómez (1995).

Several fungi and some oomycetes have been studied as possible causative agents of bud rot. In multiple studies *Fusarium* spp. was found with the occurrence of the disease and Reiking isolated *F. moniliforme* from infected palms in Panama in 1928 (in Richardson, 1995). Bachy (1954) stated that studies involving *Fusarium* species which are commonly present on infected palms and some other microorganisms associated with different diseases failed to provide evidence for a role for *Fusarium* as a causative agent of bud rot; the species included *F. oxysporum, F.solani, F. roseum, F. bulbigenum, F. vasinfectum, F. solani* var. *minus, F. camptoceras* and *F. scirpi*. Figueroa reported *F. oxysporum* and *F. roseum* as causative agents of bud rot in Ecuador in 1977 (Martínez et al., 1997). *Botryodiplodia* sp. and *Colletotrichum* sp. were found in Suriname, but to a lesser extent than *Fusarium* spp. (Griffith, 1987). De Franqueville (2003) reported no success in a *Fusarium* inoculation trial in Ecuador involving a mixture with bacteria. Regarding Colombia, Sanchez-Potes isolated some *Fusarium* species on the Eastern Plains in 1967 during preliminary studies on bud rot in this region (Gómez, 1995). Ochoa and Bustamante isolated different species of *Fusarium*, including *F. solani* (Gibson, 1979; Gómez et al., 1995). Ochoa and Bustamante isolated *Arthrobotrys* sp., *Arthrobotryum* sp., *Helminthosporium* sp., *Curvularia* sp. and *Pestalotia* sp. from affected palms (Gibson, 1979; Gómez et al., 1995). Acosta, in 1991, indicated that different *Fusarium* species could be responsible for the disease in the eastern and northern region of Colombia; the author specified that early surgery (mechanical removal of affected tissue above the meristem area) complemented with Carboxin spraying could reduce disease incidence (Acosta, 1991). Nieto Páez (1993) reported *F. solani* as being the most common species isolated from infected tissue from among other species such as *Fusarium, Thielaviopsis, Arthrobotryum, Helminthosporium, Curvularia* and *Pestalotia*. *In vitro* pathogenicity tests by Vargas (1992) on detached spear leaves and leaflets supported the hypothesis of *F. solani* as the causative agent in Colombia.

Thielaviopsis has also been widely associated with bud rot disease. The National Institute for Agronomy in the Belgian Congo (INEAC) proposed *T. basicola* as the causative agent of the disease (Kovachich, 1952). Other authors proposed that the 'little leaf' syndrome was caused by *T. basicola* and bud rot was caused by *P. palmivora* (Bull and Robertson, 1959), respectively. Nieto (1996) evaluated *Thielaviopsis paradoxa* pathogenicity on palms in saturation conditions, observing a dark necrosis of the spear leaves under the experimental conditions at 6–10 days after inoculation. Ayala and Gómez (2000) reported significant differences in pathogenicity among 100 *T. paradoxa* isolates collected in Brazil, Ecuador and Colombia. Similar results with *Ceratocystis paradoxa* were recently reported Suwandi et al. (2012) in Indonesia.

The first association of bud rot with a viral disease was made in 1951. Indeed, Bull and Robertson (1959) mentioned that Wardlaw had described an 'infectious chlorosis' suggesting that it could be caused by a virus. The palm's bud area being affected and the spreading pattern observed in the Belgian Congo also led Kovachich (1953) to propose a virus as the causative agent. Research into a virus as the cause of bud rot was started in Ecuador and Brazil because of failing to reproduce the disease using bacteria and fungi (Perthuis, 1991). Dollet (1991) mentioned that the virological research had two phases: 1984–7 when the problem was studied *in vitro* and then 1987–90 when nucleic acids were studied. The IRHO group in Ecuador and Brazil proposed a viroid or a phytoplasma as causative agent of the disease (Nieto Páez, 1993), but none of the aforementioned hypotheses was finally proven (de Franqueville, 2003) (Table 1).

Oomycetes have also been associated with bud rot. Buitrago and Nieto (1995) assessed *Phytophthora* sp. and *Pythium* sp. but were unable to observe plant infection. Renard (1991) also unsuccessfully tested *Pythium* in Ecuador in 1981. In 1924, Butler had mentioned that according to his own experience the only microorganism, which could develop bud rot in oil palm, was *P. palmivora*. Then Reiking isolated *Phytophthora* sp. in 1928 from infected plants in Panama; because of its similarity with coconut bud rot, the author proposed that it was also responsible for the disease in the oil palm (Richardson, 1995). Ghesquiere distinguished two types of bud rot in 1935, one caused by *Bacillus coli* and another one caused by *P. palmivora*. Alibert reported on the correlation of the disease caused by *Phytophthora* sp. with the topographic conditions being the vicinity with water bodies (Bull and Robertson, 1959). In 1953 Vanderweyen associated *P. palmivora* with bud rot but not with the 'little leaf' disease (Bull and Robertson, 1959). Kovachich (1957) isolated *Phytophthora* sp. from nursery palms and preliminary identification suggested the presence of *P. arecae*, which is now recognised as a synonym for *P. palmivora* (Mchau and Coffey, 1994). Faparusi (1973) was able to isolate *P. palmivora* from tapped oil palms' scar tissue and Joseph and Radha (1975) successfully isolated *P. palmivora* from affected areca nut palm (*Areca catechu*), cardamom (*Elettaria cardamomum*), coconut (*Cocos nucifera*), oil palm, palmyra palm (*Borassus flabellifer*) and rubber (*Hevea brasiliensis*) using glucose nitrate media (5.4 g/L glucose, 1.5 g/L sodium nitrate, 1 g/L potassium dihydrogen phosphate, 0.5 g/L magnesium sulphate, 2 ppm thiamine HCl, 17 g/L agar). Alvarez et al. (1999) described the isolation of *Phytophthora* species from palms affected by bud rot in Colombia; however, this involved neither Koch's postulates nor identification at the species level. Sarria et al. (2008a) isolated *Phytophthora* sp. from infected oil palms in Colombia and were able to complete Koch's postulates both *in vivo* (nursery palms) and *in vitro* (detached immature spear leaves). Morphological and molecular characterisation of the species demonstrated that it belonged to the *P. palmivora* genus (Rodríguez et al., 2009; Sarria et al., 2008b, 2013; Torres et al., 2010).

Table 1 The morphological, cultural and pathogenic characteristics of bud rot in the oil palm

Date	Disease	Location	Symptoms	Diagnosis	References
1917	Bud rot	Nigeria	Fruit rot and rot at youngest leaf leading to bud rot	Unknown	Wakefield (1920)
1921	Bud rot	Suriname	Bud rot	Unknown physiological cause	Herrera and Jiménez (1963)
1922	Bud rot	Malaysia	Base of the youngest leaves being affected, leading to their collapse. Water soak lesions are observed and the meristem is not directly affected.	Only *Phytophthora*-related species can cause this damage. Isolation was not performed.	Sharples (1922)
1928	Bud rot	Almirante, Panama	Bud rot. The disease starts at the tips and sides of the young central group of folded leaves at the heart of the plants and gradually works downward until the bud is killed.	Bacteria, *Fusarium* sp. and *Phytophthora*. Inoculation was not performed.	Richardson (1995)
1929	Bud rot	Africa	Not described in the review for the specific symptomatology	Nitrogen uptake might be related to bud rot.	Bull and Robertson (1959)
1933	Yellow leaf	Nigeria	Yellow leaf	Physiological	West (1938)
1934	Bud rot	Malaysia	Collapse of the unopened leaves (spear)	Bacterial infection associated with damage caused by the beetle *Oryctes rhinocerus* or lightning strike	Bunting (1934)
1935	Bud rot	Coquilhatville province in the Belgian Congo	Bud rot	*Phytophthora palmivora*, *Bacillus coli*. Only isolation was used as diagnosis.	Ghesquière (1935)
1944	Bud rot	Dahomey (present-day Benin)	Bud rot	Bacteria and a *Phytophthora*-like microorganism	Bull and Robertson (1959)
1948	Bud rot	Congo	The disease is similar to bud rot in coconut (caused by *P. palmivora*).	Association of insect damage with bacteria and fungi	Wardlaw (1948)
1950	Bud rot	Venezuela	Reduced development of new tissue, with lysis of the internal tissue above the meristem with or without palm death	Physiological	Malaguti (1953)
1951	Little leaf	Congo	Short and erect leaves. Dark brown rotting of the unopened pinnae, followed by the collapse of the central spear. Great severity may lead to the trunk apical meristem being destroyed.	Deficiency or an unbalanced amount of nutrients	Kovachich (1952, 1953)

1954	Bud rot	Sibiti, Congo (Brazzaville)	Bud rot	Bacteria and *F. oxysporum*, *F. solani* and *F. roseum*	Bachy (1954)
1957	Spear leaf	Yaligimba, Congo	Presence of localised rotten areas on the median leaflets of the spear leaf. Lesions are initially colourless and then take on a water-soaked appearance. Old lesions turn pale brown with a thin marginal orange-brown border.	*Phytophthora arecae* was suggested as the tentative isolated species. Its role as spear leaf causative agent was proven with Koch's postulates.	Kovachich (1957)
1960	Bud rot	Congo	Small lesions in the spear leaves which increase in size and number in new tissue. Spear leaf collapse with increased incidence	Three steps are required: 1) initial damage by *Themnolchoita quadripustulata*, 2) infection by bacteria and 3) damage caused by *Platygenia barbata* and *Rhynchophorus palmarum*.	MacGarvie (1960)
1963	Bud rot/Little leaf	Congo	Bud rot and Little leaf. Wet brown rot on the leaf; the rachis is involved in serious cases and the spear collapses, followed by bud rot. Little leaf may then be observed.	*Erwinia lathyri*	Duff (1963)
1965	Bud rot	Arenosa, Coldesa, Turbo, Antioquia, Colombia	The Spear rot-bud rot complex	Unknown	De Rojas y Ruiz (1972)
					Turner, (1981)
1973	Oil palm wilt	Nigeria	Not described	Tapping palms for palm wine production provides entry for *P. palmivora* in oil palm.	Faparusi (1973)
1975	Bud rot	India	Not described	*Phytophthora palmivora*. The microorganism was isolated, but not re-inoculated	Joseph and Radha (1975)
1976	Bud rot	Arenosa, Coldesa, Turbo, Antioquia, Colombia	The Spear rot-bud rot complex	*F. oxysporum*, *F. solani* and bacteria	Renard (1976)

(Continued)

Table 1 *(Continued)*

Date	Disease	Location	Symptoms	Diagnosis	References
1983	Bud rot	Shushufindi, Ecuador	Spear rot and bud rot	The author proposed *Phytophthora* and *Pythium* as one of the possible causes of bud rot, but only isolated *Pythium*.	de Franqueville (2003)
					Renard and Quillec (1985)
1991	Bud rot	Suriname	Yellowing leaves 1, 2, 3, spear rot and bud rot	Unknown	Van den Lande (1993)
1991	Amareleci-mento fatal – bud rot	Denpasar, Brazil	Yellowing and bud rot	Unknown	Van Slobbe and Rocha de Souza (1991)
1991	Bud rot	Eastern Ecuador	Yellowing and bud rot	Unknown; virus was the hypothesis	Perthius (1991)
1991	Bud rot	Shushufindi, Ecuador	Yellowing, vein banding and bud rot	Unknown	Renard (1991)
1991	Bud rot	Colombian Atlantic coast and eastern region	Yellowing of young leaves, drying of lower leaflets, necrotic spots and spear bud rot	*Fusarium* spp.	Acosta (1991)
1996	Bud rot	Colombian western region	Spear rot and bud rot	*Fusarium* spp., *Pythium* spp. and *Thielaviopsis paradoxa*	Nieto et al. (1996)
2005	Bud rot	Colombian western region	Bud rot	*Phytophthora* sp.; Koch's postulates were not performed.	Alvarez et al. (1999)
2008	Bud rot	Colombian central and south western regions	External symptoms on youngest spear leaf as necrotic small lesions and, in some cases, there is an increase in the number and size of the lesions which, in more severe cases, include the whole spear leaf.	*Phytophthora palmivora*; Koch's postulates were fulfilled.	Sarria et al. (2008) Torres et al. (2010)
2012	Common spear rot	Indonesia	Crown diseases. Necrotic lesions on spear leaf leaflets, followed by extensive rotting of leaflets as the leaf expands	*Thielaviopsis paradoxa*	Suwandi et al. (2012)

Source: Torres et al. 2016.

4 *Phytophthora palmivora*

The discovery of a causative agent opened a wide field of research topics targeting both management and prevention of the disease.

Despite the knowledge on bud rot gathered to date, which now enables the detection of *P. palmivora* in infected tissue by either microscopy, immunological test or molecular analysis, its isolation from diseased oil palm tissue is still challenging. Indeed, our group has obtained at least 30 isolates during recent years, which led to a series of experiments aimed at understanding interactions between the oil palm and *P. palmivora*. Major research work included morphology (Table 2), virulence (Fig. 3), and fungicide-response differences in the oil palm pathogen population (Morales et al., 2008; Sarria et al., 2008a; Torres, 2016; Vélez, 2015) and their relationship with isolates from other crops (Torres, 2016).

Oil palm leaflet infection by *P. palmivora* zoospores has been recently described by Sarria et al. (2015): both light and scanning electron microscopy were used to reveal that zoospores initially 'swim' towards the leaflet trichrome, stomata or intercellular spaces, germinating and forming an appressorium-like structure prior to penetration (Fig. 4) and this phase takes about 30–60 min. Once *P. palmivora* first penetrates, it invades the

Table 2 Morphological, cultural and pathogenic characteristics of *Phytophthora palmivora*

Parameter measured	Mean	Min	Max	Standard error
Number of isolates	150.00	–	–	–
Mating type A1	34.00	–	–	–
Mating type A2	116.00	–	–	–
Sporangia length (μm)	43.24	20.43	75.00	0.14
Sporangia breadth (μm)	25.40	12.72	47.09	0.07
Pedicel length (μm)	3.61	0.83	7.50	0.02
LB ratio	1.71	0.64	3.03	0.01
Sporangial density (x 10^3 spores/ml)	233.00	15.00	1,020.00	0.53
Chlamydospore diameter (μm)	32.72	15.82	58.19	0.15
Chlamydospore/cm^2	2,302.00	768.00	13,312.00	0.86
Oogonium diameter (μm)	28.72	15.17	45.27	0.10
Oospore diameter (μm)	22.6	11.99	38.05	0.08
Antheridium width (μm)	13.15	6.97	24.76	0.07
Oospore cell in wall thickness (μm)	1.77	0.55	4.60	0.02
Colony diameter day 2 (mm)	25.44	9.00	46.00	0.28
Colony diameter day 3 (mm)	32.57	9.00	51.40	0.32
Lesion diameter on apple (mm)	41.55	8.00	67.00	0.46

Torres, 2016.

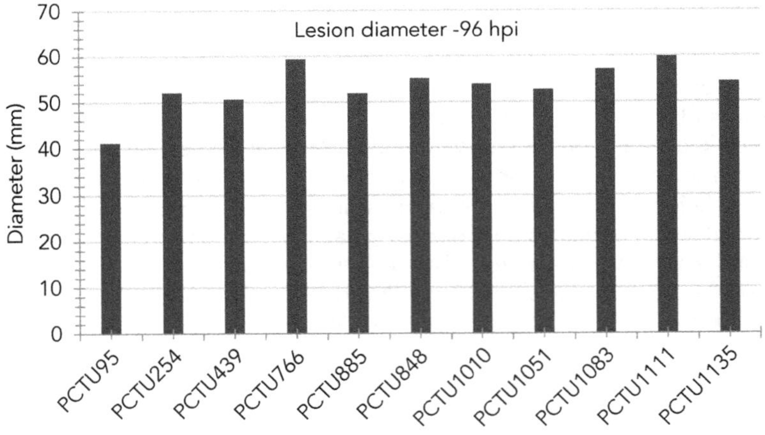

Figure 3 A comparison of virulence among *P. palmivora* isolates from Tumaco (taken from Vélez, 2015).

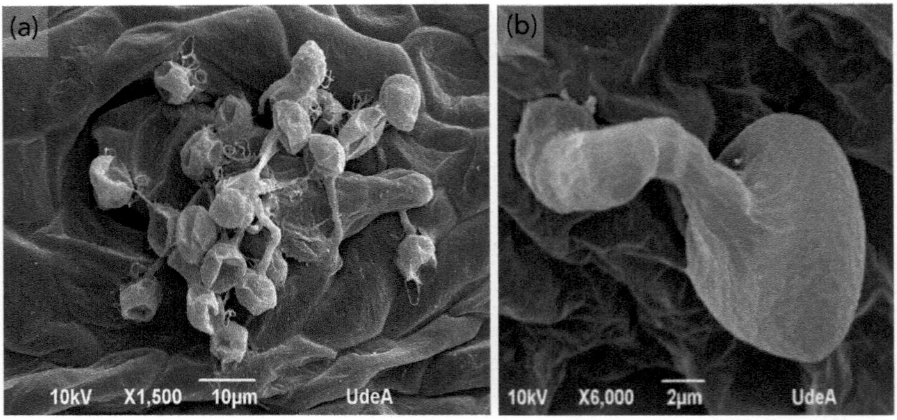

Figure 4 Infection process of *P. palmivora* in immature leaflets of oil palm. (a) Attraction of zoospores to trichomes 0.5 – 1 h occurs after inoculation. (b) Appressorium-like structure formed by *P. palmivora* for penetrating oil palm leaflets (Sarria et al., 2015).

intercellular space and then the intracellular space. The hyphae then emerge onto leaflet surface at 12–24 h after inoculation and new sporangia and chlamydospores are observed 36 h post-inoculation, enabling new cycles of the disease, by direct germination or by formation and release of zoospores from the sporangia (Sarria et al., 2015).

Bud rot lesions on a palm are initially small and can be easily ignored or confused with other disorders; however, if environmental conditions are favourable (>75RH and 20– 35°C), then the pathogen goes into sexual or asexual reproductive phase with more inoculum being produced at every new cycle; this is reflected in larger lesions as time elapses (Sarria et al., 2015).

Advances to understand the biology of *P. palmivora* have led to deciphering the pathogen's life cycle (Fig. 5). During evaluation of inoculum sources of *P. palmivora* in oil palm plantations, it was possible to find structures such as sporangia and chlamydospores in dry necrotic tissues under field conditions (Fig. 6). From these samples, it was possible to recover the pathogen in pure cultures and test its pathogenicity in oil palms. In the same way, pathogen structures were found under soil conditions, which is something expected, as *P. palmivora* is a soil inhabitant. Besides *P. palmivora* was isolated from running water coming from infected oil palm plantations, using fruits (cucumber and pears) as traps, (Sarria et al. 2016).

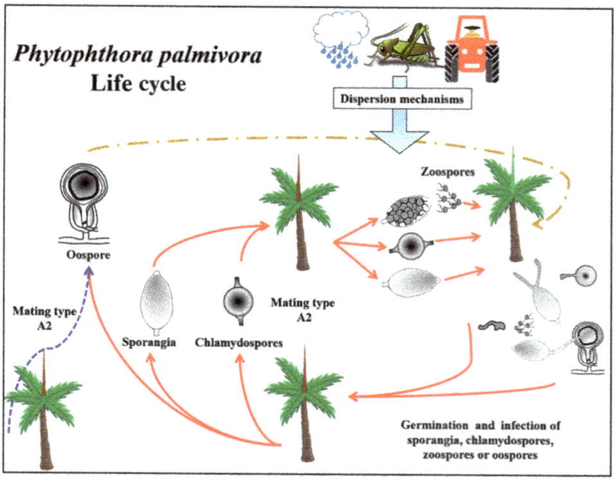

Figure 5 *Phytophthora palmivora* life cycle in oil palm (adapted from Torres, 2016).

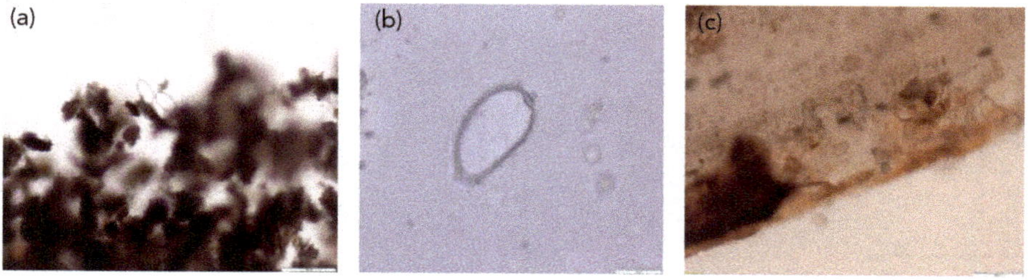

Figure 6 Structures of *Phytophthora palmivora*, coming from necrotic tissue residues infected by pudrición del cogollo – (a) and (b) sporangia. (c) New sporangia formation coming from inoculated sporangia present in necrosed tissue of immature bud foliage of oil palm.

5 Disease incidence and epidemics

Despite the disease being reported as affecting oil palm planting regions worldwide with various levels of severity, Latin America faces more severe detrimental effects. This phenomenon is shown in Fig. 7, which summarises the reported bud rot incidence (number of diseased plants/total plants) in the literature; however, Asia and Africa are not safeguarded as incidence has reached 50–69% in Benin and the Belgian Congo (Bull and Robertson, 1959; Kovachich, 1953). In Malaysia incidence by 1967 did not exceed 5%; however, in some cases it was lethal (Turner and Bull, 1967). Reduction in cumulative yield in Indonesia has been estimated to be 18% for the first six years in the field due to the disease (Suwandi et al., 2012).

In Colombia, four epidemic episodes have been reported to date. The first outbreak started in Urabá near the border with Panama in 1965 (De Rojas and Ruíz, 1972) and it promoted the study of this disease in Colombia. Bud rot had affected around 20 000 palms by mid-1968, entering an exponential phase. The 'Arenosa' plantation started to manage the disease by improving agronomic practices and removing all affected palms. These measures led to the reduction of the incidence of bud rot (Fig. 8), but political instability forced the plantation to abandon all practices. This situation led to the death of more than

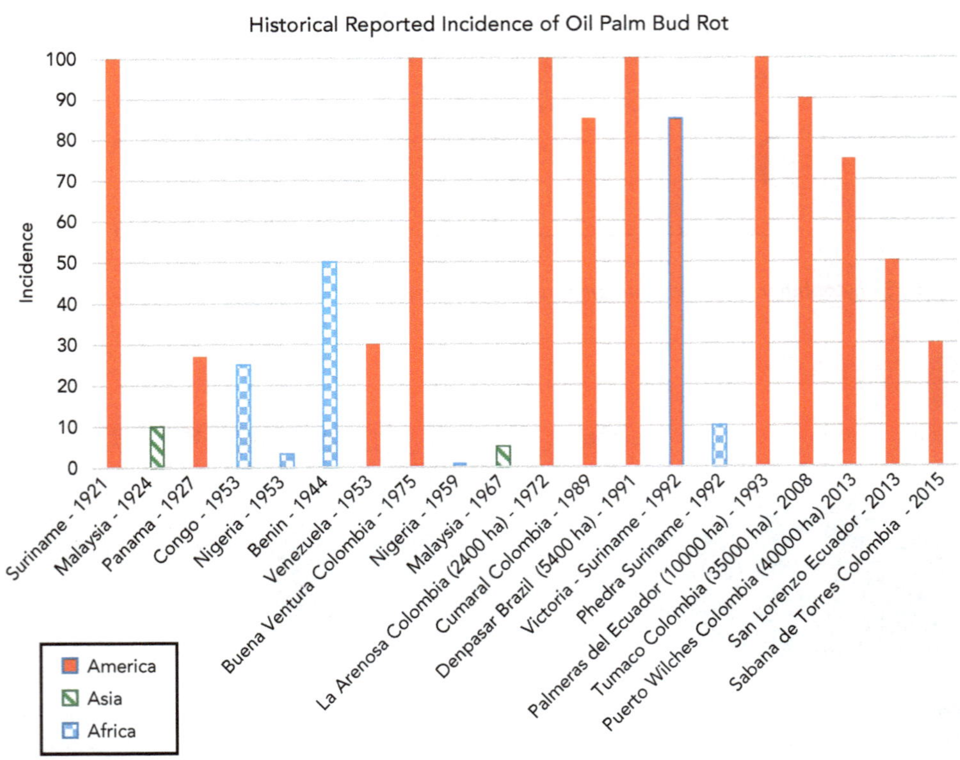

Figure 7 Bar plot representing the reported incidence of bud rot by date and location.

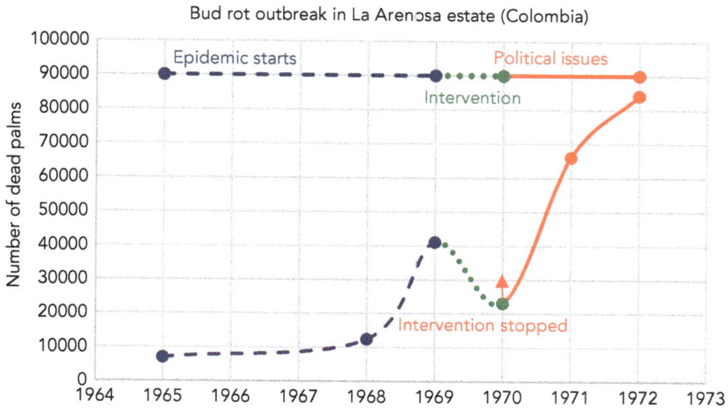

Figure 8 Bud rot progression curve for La Arenosa (Coldesa, Uraba).

60000 palms over a two-year period (1970–2). More than 2400 ha were destroyed during this epidemic episode.

The second large outbreak took place in Colombia's Eastern Plains at the end of the 1980s. In this region, Swinburne et al. (1990) reported an increase of incidence from 11% to 25% in 'La Cabaña' plantation (300 ha affected) from 1988 to 1990 with incidence rising to higher than 85% in some fields of the area. Cenipalma was founded as the National Oil Palm Research Centre due to this outbreak and its initial objective was to study the disease and provide management solutions for oil palm growers (Fedepalma, 1990). The last two outbreaks occurred between 2005 and 2013, in Tumaco, Nariño (western region) and Puerto Wilches, Santander (central Magdalena Valley). The disease was present in Tumaco for many years; however, proper removal of infected palms kept pressure from the disease under control (Gómez, 2010; Jiménez, 1991). However, once this practice was abandoned, the pressure from the inoculum reached untreatable levels and the spread of the disease became more aggressive than in other regions (Gómez, 2010). The bud rot disease progressed rapidly, with incidence rising from 3.9% in January 2007 to 61.3% in February 2008 on a 2,113 ha plantation in Tumaco (Pulido, 2011). One year later, the incidence exceeded 90%, reducing the fresh fruit bunches (FFB) production from 42257 tons in 2007 to 2300 in 2009 (Pulido, 2011). More than 30 000 ha had been destroyed in Tumaco by 2011 (Pulido, 2011). From both economic and social points of view, this outbreak directly or indirectly affected about 73200 people (46% of the population of Tumaco region) (Martínez and Silva, 2008). A similar situation occurred in Puerto Wilches where the incidence of bud rot increased dramatically from 10% in 2009 to 75% in 2013 (Becerra-Encinales et al., 2013).

6 Disease management

Fungicides have been successfully used in Colombia to control bud rot. Indeed, Peña and Jimenez reported lower disease incidence in 1986 when Mancozeb, copper oxychloride and oxycarboxin were used (in Martínez et al., 1997). However, further studies by Jimenez

(1991) suggested that mancozeb and oxycarboxin were ineffective in treating bud rot in its advanced stages in Tumaco. Martínez et al. (1997) evaluated at Casanare department, under field conditions, the effect of carbendazim, captan, triadimefon and chlorothalonil but results showed no significant differences among treatments and control palms.

We developed a strategy for the treatment of infected palms in the field. Briefly, the procedure consisted of a periodical 7–15-day scouting schedule, the identification of early stages of the disease, the removal of infected tissue (surgery), the application of oomyceticide Ridomil Gold or Aliette for *P. palmivora* control, plus insecticide, bactericide and a wide spectrum fungicide for other pests involved in a weekly rotation programme until recovery and setting up of a plastic roof to prevent rain washing off the products (Aya et al., 2011; Martínez and Sarria, 2013; Martínez, 2009, 2009b, 2013a,b; Noreña et al., 2011; Torres et al., 2009). An assessment of this treatment was undertaken by Aya and Martínez (2011) and it revealed a significant 50% reduction in incidence over a 22-month period (Fig. 9). The global incidence of bud rot at Cenipalma La Vizcaina Research Station block 1 (76 ha, with 11091 palms), moved from 20.6% of incidence with 2254 surgery treated palms to a reduction of 4.4%, when the prophylactic measures were implemented to stop the disease.

Despite bud rot continuing to be the most limiting and destructive oil palm disease in Colombia, significant advances have been made regarding disease management during outbreaks. Munevar et al. (2001) demonstrated that soil aeration and fertilisation had a positive impact on crop yield and reduced disease incidence. Martínez et al. (2010), Torres et al. (2010) and Aya and Martínez (2011) evidenced the benefits of early detection and treatment. Optimal disease control was obtained when incidence of PC was lower than 10% at five semi-commercial plots with different soil textures, in three different geographical oil palm regions of Colombia. The plots were properly treated by establishing the recommended agronomical practices, early detection and management programmes (Arias et al., 2014).

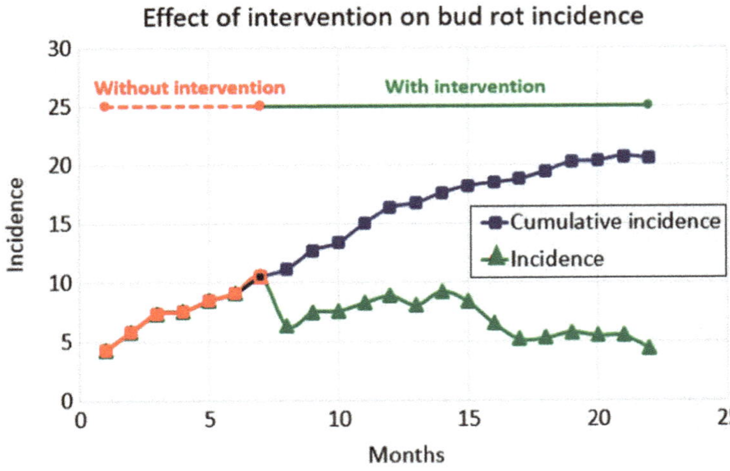

Figure 9 The effect of intervention on bud rot incidence and cumulative incidence. Note that cumulative incidence started to split from incidence as soon as the intervention began (Aya and Martinez, 2011).

Plant breeding is one of the most important tools for the management of plant diseases building on the genetic tolerance of host plant material. Studies have already demonstrated the agronomic potential of *E. oleifera* as a substitute for the African palm (*E. guineensis*) in local conditions (De Rojas and Ruíz, 1972). Other research has demonstrated disease tolerance by the interspecific hybrid *E. oleifera x E. guineensis* (Alegria et al., 2015; Aya et al., 2008; Cochard et al., 2005). All such measures must be incorporated into a specifically designed integrated pest management (IPM) strategy for a successful disease control in the long term (Martínez and Sarria, 2013; Martínez, 2009b; Morales et al., 2008).

Another important aspect in research at Cenipalma has been to find genetic solutions to the bud rot problem by the genetic breeders through cloning-resistant oil palm palms, search of resistant genes at the *E. guineensis* collection; identification of mechanical responses of oil palm to *P. palmivora* and genes implicated, which will produce information to improve oil palm materials to more resistant to PC. In addition, investigation is carried out on selection of palm cultivars resistant to hydric stress and aluminium toxicity, which are implicated in the development of PC. Research on hybrid resistant selection is underway.

7 Future directions for research

Bud rot will remain the most important disease in South and Central America until long-term management strategies are developed. Meanwhile, studies are required regarding fungicide movement within the palm, evaluating new fungicides and a strict analysis of inoculum potential in all the affected regions to improve current management practice. The sensitivity baselines obtained for different fungicides will help in understanding whether a specific isolate is becoming resistant to a specific fungicide; permanent monitoring of the *P. palmivora* population is required. In addition, research on natural enemies has been taken in order to identify potential microorganisms that could be promoted to reduce the detrimental effect of *P. palmivora*. Improved inoculation techniques will provide plant breeders with powerful tools for assessing resistance, not only in *E. guineensis*, but also in *E. oleifera* and the interspecific hybrid species, which in the long term is believed to be the most reliable strategy for managing this fatal disease in oil palm.

Despite breeding programmes being time consuming in general, especially for perennial crops such as oil palm, recent advances in biotechnology, such as whole-genomic analysis and proteomics studies, will help to accelerate the breeding processes by having a better understanding of the plant–pathogen interactions. RNA sequencing will allow us to understand the gene expression at infection and colonisation levels. All these techniques will permit researchers to identify genes that are involved in pathogenicity and virulence from the pathogen standpoint, and in resistant response or susceptibility from the plant genes. Researchers will be able to track the pathogen populations, understand how they are evolving at a molecular base and develop strategies for reducing the detrimental impact of the pathogen evolution.

The understanding of the plant–pathogen interaction and the identification of the gene involved in it will also help researchers to develop markers that assist breeders to identify whether or not a resistant trait is inherited, or how resistant is the bud rot to the progeny of a specific cross of palms. In addition to these tools, upcoming methods in biotechnology such as and gene edition using clustered regularly interspaced short palindromic repeats (CRISPR) will help to easily modify palms genome to recognise the pathogenicity once the resistances genes in oil palm and virulence genes (effectors) in

P. palmivora are identified to block them and reduce the number and the severity of the infections. Other interesting breeding strategies is GWAS (Genome-Wide Association Studies) using GBS Genotyping by Sequencing) to find the SNPs (Single Nucleotide Polimorphism) involved in the plant–pathogen interaction. Another useful tool is the researcher in QTLs (Quantitative Trait Loci) to identify the genome regions involved in the resistance and link it with GWAS results.

Although screening and genome edition cannot be performed for each single palm, once a trait is transferred to a new population by traditional field breeding or by genome edition, mass reproduction of the specific plant will be allowed by the wide recent advances on oil palm cloning techniques (cloning of DxP, and back crosses with bud rot resistant and high oil yielding). These plants would be potentially used in a short period by the breeding programmes as elite progenitors or directly by farmers in the field.

There is no silver bullet to manage bud rot disease. However, a logical integration of the present research with the capabilities of the modern and upcoming tools will help researchers to improve the IPM for *Phytophthora*.

8 Where to look for further information

Diferentes instituciones de investigación científica y universidades en el mundo han concentrado sus esfuerzos en la investigación en Palma de aceite para fortalecer la palmicultura, aquí algunas de las paginas que pueden ser consultadas para mayor información:

- http://www.cenipalma.org/
- http://umr-agap.cirad.fr/equipes-scientifiques/oil-palm-genome-projects-opgp
- http://www.mpob.gov.my/
- http://www.cirad.fr/
- http://www.iopri.org/
- http://sydney.edu.au/agriculture/academic_staff/david.guest.php
- https://qaafi.uq.edu.au/profile/357/andre-drenth
- https://www.researchgate.net/profile/Hernan_Romero3
- http://www.psm.msu.edu/people/mary_k_hausbeck_professor_and_extension_specialist

9 References

Acosta, Á. 1991. Pudrición de cogollo en palma de aceite : observaciones y manejo. Colombia. *Palmas* 12, 49–54.

Acosta, Á., Gómez, P. L. and Vargas, J. R. 1996. Factores físicos de los suelos y su influencia en la predisposición a la pudrición de cogollo de la palma de aceite en Colombia. *Palmas* 17, 71–9.

Alegria, M., Vélez, D. C., Varón, F. H., Martínez, G., Torres, G. and Corredor, J. 2015. Comportamiento frente a la Pudrición del cogollo (PC) de palmas derivadas de cruzamientos entre palmas *Elaeis guineensis* sobrevivientes a la epidemia en Tumaco, Nariño. *Palmas* 36, 69–77.

Alvarez, E., Sánzhez, J., Grajales, O. and Gómez, P. L. 1999. Pathogenic variation and DNA polymorphism in *Phytophthora* spp., causal agent of bud rot of oil palm. *Phytopathology* 89, S3.

Arias, N. A., Beltran, J., Guerrero, J. M. and Sanchez, A. 2014. Technologies for Bud rot treatment in oil palm validated in Colombian palm zones. *Palmas* 35, 39–52.

Ariza, J. G., Torres, G., Sarria, G. A., Varón, F. H. and Martínez, G. 2009. Relación entre los síntomas externos y el avance interno de la lesión causada por la Pudrición del cogollo (PC) en palmas de vivero en Tumaco. *Fitopatologia Colombiana* 32, 35–8.
Avila, D. C., Sarria, G. A., Martínez, G., Torres, G. and Bustillo, A. E. 2015. Observaciones preliminares del papel de *Cocconotus aratifrons* como diseminador de *Phytophthora palmivora* en palma de aceite. In: 42 Congreso de La Sociedad Colombiana de Entomología – SOCOLEN.
Aya, H. A. and Martínez, G. 2011. Manejo de la pudrición del cogollo de la palma de aceite en el Campo Experimental Palmar de la Vizcaína. *Palmas* 35, 32–44.
Aya, H. A., Torres, G., Sarria, G. A., Varón, F. H. and Martínez, G. 2008. Comportamiento de cuatro materiales de palma de aceite frente a la pudrición del cogollo (PC), de la palma de aceite en Tumaco-Nariño. *Fitopatologia Colombiana* 32, 51–6.
Aya, H. A., Varón, F. H., Gerardo, M. and Cayón, G. 2011. Manejo quimico de la pudrición del cogollo en cuatro materiales de palma de aceite en Tumaco, Colombia. *Palmas* 32 (2), 45–52.
Ayala, L. and Gómez, P. L. 2000. Patogenicidad de aislamientos de *Thielaviopsis paradoxa* principal agente causal de la pudrición de cogollo. *Palmas* 21, 121–2.
Bachy, A., 1954. Contribution a l'étude de la porriture de coeur du palmier a huile. *Oleagineux* 9, 619–27.
Becerra-Encinales, J. F., Cooman, A., Roa, S., Penagos, Y., Cortés, C., Guerrero, J. M., Quitero, J. L., Martínez Villorina, R. N., Trujillo, M. and Chinchilla, J. 2013. Avances en la estrategia fitosanitaria de la Zona Central. XXXIII Reunión Tecnica Nacional de Cultivadores de palma de aceite. http://www.cenipalma.org/sites/default/files/files/Cenipalma/22AvancesplanemergenciaZonaCentral_opta.pdf
Broeshart, H., Ferwerda, J. D. and Kovachich, W. G., 1957. Mineral deficency symptoms of the oil palm. *Plant Soil* 8, 289–300.
Buitrago-Alfonso, V. M. and Nieto Páez, L. E. 1995. Hongos asociados con pudriciones de flecha y cogollo en palma de aceite (*Elaeis guineensis* Jacq.) en los Llanos Orientales. *Palmas* 16, 9.
Bull, R. A. and Robertson, J. S. 1959. The problems of little 'leaf' of oil palms – a review. *Journal of the West African Institute for Oil Palm Research* 355–75.
Bunting, B., Georgi, C .D. V. and Milsum, J. N. 1934. The oil palm in Malaya: Malayan planting manual No. 1. Department of Agriculture, Kuala Lumpur.
Cochard, B., Amblard, P. and Durand-gasselin, T. 2005. Oil palm genetic improvement and sustainable development. *Oilseeds and Fats, Crops and Lipids Journal* 12, 141–7.
Corley, R. H. V. and Tinker, P. B. 2003. *The Oil Palm*, fourth edition, 4th edn. Blackwell, Oxford, UK.
Cristancho, J. A., Alfonso, O. A. and Molina, D. L. 2012. Revisión de literatura sobre el papel del suelo y la nutrición de plantas en la Pudrición del cogollo de la palm a de aceite. *Palmas* 33, 9–22.
Dangond, L. F. 2015. Palma de aceite: motor de desarrollo rural. *Palmas* 36, 102–8.
de Franqueville, H. 2003. Review paper Oil palm bud rot in Latin America. *Experimental Agriculture* 39, 225–40.
De Rojas, E. and Ruíz, E. 1972. Investigaciones sobre la pudrición del cogollo y pudrición de la flecha de la palma africana en la plantación la Arenosa de Coldesa S.A. (Turbo, Ant.). Instituto Geográfico Agustín Codazzi, Medellin, Colombia.
Dollet, M. 1991. Etiología de la pudrición de cogollo: investigaciones virológicas conducidas por el IRHO. *Palmas* 12, 33–8.
Duff, A. D. S. 1963. The bud rot little leaf disease of the oil palm. *Journal of the West African Institute for Oil Palm Research* 176–90.
Dzido, J. L., Genty, P. and Ollagnier, M. 1978. Les principales maladies du palmier à huile en Equateur. *Oleagineux* 33, 55–63.
Faparusi, S. I. 1973. Origin of initial microflora of palm wine from oil palm trees (*Elaeis guineensis*). *Journal of Applied Bacteriology* 36, 559–65.
Fedepalma, 1990. Conclusiones del XVIII Congreso Nacional de Cultivadores de Palma Africana realizado en la ciudad de Santa Marta los días 20 y 21 de Septiembre de 1990. *Palmas* 11, 17–19.
Fedepalma, 2014. Anuario estadistico 2014. La agricultura de la palma de aceite en Colombia y el mundo.

Ferwerda, J. D. 1954. Boron deficiency in oil palms in the Kasai region of the Belgian Congo. *Nature* 4414, 1097.

Ghesquiere, J. 1935. Rapport preliminaire sur letat sanitaire de quelques palmeires de la province de Coquilhatville, Serie Scientifique. Publications de l'Institut National pour l'Etude Agronomique du Congo, Brussels.

Gibson, I. A. S. 1979. Two important disorders of oil palm in Latin America. *International Journal of Pest Management* 25, 270–4.

Gómez, J. E. 2010. Aproximación matemática al entendimiento de la dinámica de la enfermedad 'complejo pudrición del cogollo' en palma de aceite, en diferentes regiones productoras de América tropical. *Palmas* 31, 345–52.

Gómez, P. L. 1995. Estado actual de la investigación sobre pudrición de cogollo. *Palmas* 16, 9–23.

Gómez, P. L., Acosta, Á., Guevara, L. A. and Nieto Páez, L. E. 1995. Pudrición de cogollo en Colombia: importancia, investigación y posibilidades de manejo. *Palmas* 16, 198–206.

Griffith, R. 1987. The consultancy in Suriname for diseases in oil palms and coconuts. Inter-American Insittute for Cooperation in Agriculture, Centeno – Trinidad.

Herrera-Klindt, B. and Jiménez-E. C. 1963. Plagas y Enfermedades. In: La Palma Africana de Aceite (*Elaeis guineensis)* Jacq.)Perspectivas Para Su Explotación En Venezuela. Consejo de Bienestar Rural – Oficina de Estudios Especiales. Venezuela, Caracas, pp. 111–21.

Jiménez, O. D. 1991. Pudrición de cogollo en la palma aceitera en la región de Tumaco – Colombia. *Palmas* 12, 45–8.

Joseph, T. and Radha, K. 1975. Role of *Phytophthora palmivora* in bud rot of coconut. *Plant Disease Report* 5, 1014–17.

Kovachich, W. G. 1952. Little leaf disease of the oil palm (*Elaeis guineensis)*, in the Belgian Congo. *Tropical Agriculture* 29, 107–14.

Kovachich, W.G. 1953. Little leaf disease of the oil palm (*Elaeis guineensis)*, in the Belgian Congo. part 2. *Tropical Agriculture* 30, 61–9.

Kovachich, W. G. 1957. Some diseases of the oil palm in the Belgian Congo. *Journal of the West African Institute for Oil Palm Research* 2 (7), 221–9.

Laing, D. 2009. La deficiencia transitoria de calcio como causa primordial de la pudrición de cogollo en palma de aceite. *Information's Agronómicas – IPNI* 3, 26–52.

MacGarvie, Q. 1960. The bud rot and brown germ disease of oil palm in the Belgian Congo, Book.

Malaguti, G. 1953. 'Pudrición del cogollo' de la palmera africana (*Elaeis guineensis* Jacq.) en Venezuela. *Agronomia Tropical* 3, 13–31.

Martínez, G. 2009a. Advances in the research of *Phytophthora palmivora* the causal agent of bud rot of oil palm in Colombia. In: Proceedings Malaysian Palm Oil Board MPOB International Palm Oil Congress PIPOC.

Martínez, G. 2009b. Identificación temprana y manejo integrado de la Pudrición del cogollo. *Palmas* 30, 63–77.

Martínez, G. 2011. Phytophthora bud rot of oil palm in Colombia. In: ACPP APPS 2011 Conference.

Martínez, G., Arango, M., Torres, G., Sarria, G. A., Vélez, D. C., Rodríguez, J., Mestizo, Y., Aya, H. A., Noreña, C., Varón, F. H., Drenth, A. and Guest, D. I. 2013a. Avances en la investigación sobre las dos enfermedades más importantes de la palma de aceite en Colombia: la pudrición del cogollo y la marchitez letal. *Palmas* 34, 39–47.

Martínez, G., Arias, N. A., Sarria, G. A., Torres, G., Noreña, C., Salcedo, S., Aya, H. A., Ariza, J. G., Aldana, R. and Martínez, L. C. 2009. Prácticas de manejo de la pudrición del cogollo (PC) de la palma de aceite. Cenipalma, Bogotá.

Martínez, G. and Sarria, G. A. 2013. Estado del arte de la investigación y control de la Pudrición del cogollo (PC). *Palmas* 34, 47–57.

Martínez, G., Sarria, G. A., Varón, F. H., Romero, H. M. and Sanz, J. I. 2013b. Advances in research on *Phytophthora palmivora* causal agent of Bud rot of oil palm in Colombia. *Palmas* 31, 55–63.

Martínez, G. and Silva, A. 2008. Problemática de la pudrición del cogollo en Tumaco e instrumentos para su manejo y la renovación del cultivo. *Palmas* 29, 11–16.

Martínez, G. and Torres, G. 2007. Presencia de la pudrición del cogollo de palma de aceite (PC) en plantas de vivero. *Palmas* 28, 13–20.
Martínez, J. C., Nieto Páez, L. E. and Hernández, M. L. 1997. Efecto de algunos fungicidas en el control del complejo pudrición de cogollo de la palma de aceite (*Elaeis guieensis* Jacq.). *Palmas* 18, 9.
Mchau, G. R. A. and Coffey, M. D. 1994. Isozyme diversity in *Phytophthora palmivora*: evidence for a Southeast Asian centre of origin. *Mycological Research* 98, 1035–43.
Morales, L. C., Sarria, G. A., Torres, G., Varón, F. H. and Martínez, G. 2008. Experiencias en el Manejo Integrado de la Pudrición del Cogollo (PC) de la Palma de Aceite, en la Zona Central Colombiana. *Fitopatologia Colombiana* 32, 57–61.
Munévar, F., Acosta, A. and Gómez, P. L. 2001. Factores edáficos asociados con la Pudrición de Cogollo. Rev. *Palmas* 22, 9–19.
Nieto Páez, L. E. 1993. El complejo pudrición de cogollo de la palma de aceite (*Elaeis guineensis*). *Palmas* 14, 19–26.
Nieto Páez, L. E. and Gómez, P. L. 1991. Estado actual de la Investigación sobre el Complejo Pudrición de Cogollo de la Palma de Aceite en Colombia. *Palmas* 12, 57–67.
Nieto Páez, L. E., Gómez, P. L. and Lozano-Tovar, J. C. 1996. Identificación y reproducción del complejo pudrición de cogollo de la palma de aceite (*Elaeis guineensis* Jacq.). *Palmas* 17, 63–8.
Noreña, C., Aya, H. A., Lemus, L., Torres, G., Varón, F. H. and Martínez, G. 2011. Evaluación de tratamientos químicos y prácticas culturales para el manejo de la Pudrición de cogollo (PC) de la palma de aceite en plantas de vivero, en Tumaco. *Palmas* 32, 27–32.
Perthuis, B. 1991. Búsqueda de la etiología de la pudrición de cogollo en el Ecuador Oriental. *Palmas* 12, 25–30.
Pulido, G. 2011. Manejo sanitario en Salamanca Oleaginosas. *Palmas* 32, 117–20.
Renard, J. L. 1976. Diseases in Africa and South america. In: Corley, R. H. V., Hardon, J. J. and Wood, B. J. (eds), *Oil Palm Research*. Elsevier Scientific Publishing Company, Amsterdam, pp. 447–66.
Renard, J. L. 1991. Pudrición de cogollo en el Ecuador: aspectos generales - fitopatología. Palmeras del Ecuador, Shushufundi (Ecuador). *Palmas* 12, 31.
Renard, J. L. and Quillec, G. 1985. Enfermedades destructoras de la palma africana en el África y Suramérica. *Palmas* 6, 9.
Richardson, D. L. 1995. The history of oil palm breeding in the United Fruit Company. *ASD Oil Palm Papers* 11, 1–22.
Rodríguez, J., Vélez, D. C., Sarria, G. A., Torres, G., Noreña, C., Navia, M., Romero, H. M., Varón, F. H. and Martínez, G. 2009. Identificación morfologica, molecular y patógenica de los microorganismos asociados a la Pudrición del cogollo de la palma de aceite en Colombia. *Fitopatologia Colombiana* 33, 49–56.
Sarria, G. A., Martínez, G., Varón, F. H., Drenth, A. and Guest, D. I. 2015. Histopathological studies of the process of *Phytophthora palmivora* infection in oil palm. *European Journal of Plant Pathology*, 145 (1): 39-51.
Sarria, G. A., Torres, G., Aya, H. A., Ariza, J. G., Rodríguez, J., Vélez, D. C., Varón, F. H. and Martínez, G. 2008a. *Phytophthora* sp. es el responsable de las lesiones iniciales de la pudrición del cogollo (PC) de la palma de aceite en Colombia. *Palmas* 29, 31–41.
Sarria, G. A., Torres, G., Vélez, D. C., Rodríguez, J. J., Noreña, C. C., Varón, F. H., Coffey, M. D., Elliott, M. L., Martínez, G. and Castro, B. L. 2008b. Caracterización morfológica y molecular de *Phytophthora palmivora* agente causal de las lesiones iniciales de la pudrición del cogollo (PC) de la palma de aceite en Colombia. *Fitopatologia Colombiana* 32, 39–44.
Sarria, G. A., Varón, F. H., Martínez, G., Drenth, A. and Guest, D. I. 2013. Nuevas evidencias del cumplimiento de los postulados de Koch en el estudio de las relaciones *Phytophthora palmivora* y la pudrición del cogollo de la palma de aceite en Colombia. *Palmas* 34, 41–5.
Sarria, G. A., Mestizo, Y., Betancourt, W., Garcia, A., Varón, F., Becerra, J. 2016. Pudrición del cogollo: avances, retos y oportunidades en el manejo integrado de esta enfermedad. *Palmas* 37 (4): 91–107.
Sharples, A. 1922. Observations in Malaya on bud-rot of coco-nuts. *Annals of Botany* XXXVI, 55–70.

Suwandi, A. S. and Kondo, N. 2012. Common spear rot of oil palm in Indonesia. *Plant Disease* 96, 537–43.

Swinburne, T. R. 1990. Amarillamiento fatal, pudrición del cogollo y pudrición de flecha de la palma africana: Comparación de los síntomas de estas enfermedades en Brasil, Ecuador y Colombia. *Palmas* 11, 61–8.

Torres, G. 2016. Morphological characterization, virulence, and fungicide sensitivity evaluation of *Phytophthora palmivora*. Ph.D. Thesis. Michigan State Univeristy.

Torres, G., Sarria, G. A., Martínez, G., Varón, F. H., Drenth, A. and Guest, D. 2016. Bud rot caused by *Phytophthora palmivora*: a destructive emerging disease of oil palm. *Phytopathology* 106, 320–9.

Torres, G., Sarria, G. A., Salcedo, S., Varón, F. H., Aya, H. A., Ariza, J. G., Morales, L. C. and Martínez, G. 2009. Opciones de manejo de la pudrición del cogollo (PC) de la palma de aceite en áreas de baja incidencia de la enfermedad. *Palmas* 29, 63–72.

Torres, G., Sarria, G. A., Varón, F. H., Coffey, M. D., Elliott, M. L. and Martínez, G. 2010. First Report of Bud Rot Caused by *Phytophthora palmivora* on African Oil Palm in Colombia. *Plant Disease* 94, 1163.

Torres, G., Sarria, G. A., Varón, F. H. and Martínez, G. 2008a. Evidencias Circustanciales de la Asociación de Especies de la Familia *Tettigoniidae* con el Desarrollo de las Lesiones Iniciales de la Pudrición del Cogollo de la Palma de Aceite. *Palmas* 29, 53–62.

Torres, G., Sarria, G. A., Varón, F. H. and Martínez, G. 2008b. Evidencias circustanciales de la asociación de especies de la familia *Tettigoniidae* con el desarrollo de las lesiones iniciales de la pudrición del cogollo de la palma de aceite. *Palmas* 29, 53–62.

Turner, P. D. 1981. Oil palm diseases and disorders. Published for the Incorporated Society of Planters [by] Oxford University Press, Kuala Lumpur.

Turner, P. D. and Bull, R. A. 1967. Diseases and disorders of the oil palm in Malaysia. Inc. Society of Planters, Kuala Lumpur.

Urueta, E. J. 1975. Insectos asociados con el cultivo de palma africana en Urabá (Antioquia) y estudio de su relación con la pudrición de la flecha - pudrición del cogollo. *Revista Colombiana de Entomologia* 1, 15–31.

Van de Lande, H. L. 1991. Pudrición de Cogollo en Palma Aceitera en Suriname. *Rev. Palmas* 12, 11–16.

Van de Lande, H. L., Zadoks, J. C. 1999. Spatial patterns of spear rot in oil palm plantations in Surinam. *Plant Pathology* 48, 189–201.

Van Slobbe, W. G. and Rocha de Souza, R. L. 1991. Amarillamiento fatal o pudrición de cogollo en Denpasa – Brasil. *Palmas* 12, 17–24.

Vélez, D. C. 2015. Morfología y patogenicidad de aislamientos de *Phytophthora palmivora* Bult., provenientes de palma de aceite en Tumaco. Tesis de Especialización en Cultivos Perennes. Univseridad Nacional.

Villanueva-Guerrero, A. 1988. Apreciaciones acerca del seminario internacional sobre identificaciones y control de organismos y/o factores causantes del síndrome conocido como spear rot en palma de aceite. *Palmas* 9, 37–46.

Wakefield, E. M. 1920. Diseases of the oil palm in West Africa. Bull. Misc. Inf. (Royal Gard. Kew), 306–8.

Wardlaw, C. W. 1948. Notes on a visit to the Cameroons, Nigeria and the Belgian Congo, The United Africa Company. London.

West, J. 1938. A preliminary list of plant diseases in Nigeria. Bull. Misc. Inf. (Royal Gard. Kew), 17–23.

Chapter 6

Advances in disease-resistant oil palm varieties

Tristan Durand-Gasselin, Benoît Cochard and Hubert de Franqueville, PalmElit–CIRAD, France

1 Introduction

There are many known pests that can damage the oil palm (Corley and Tinker, 2016). Although some cases of limited difference between varieties may occur, such discrepancies have been poorly documented and no specific breeding work has been undertaken to generate pest-resistant varieties. However, it is remarkable that interspecific hybrids (*E. oeifera* × *E. guineensis*) are generally less appealing to pests as compared to *E. guineensis* varieties. We will only consider resistance to diseases in this chapter.

2 Key issues

2.1 Usefulness of genetic resistance for disease control

An almost exhaustive list of diseases affecting the oil palm was given by Turner (1981); this book is a still a reference for oil palm phytopathologists, although some newer considerations or findings are missing (Corley and Tinker, 2016). Although some bacteria, virus and phytoplasmas can cause serious damage, fungus is generally the most frequent cause of disease.

Genetic resistance may exist for most diseases, but has been observed only for a limited number of pathogens (Table 1). In addition, before embarking on a disease-resistant breeding programme, one has to take into consideration the severity of

http://dx.doi.org/10.19103/AS.2017.0018.19

Table 1 Oil palm diseases for which genetic resistance has been observed

Disease	Stage	Genetic resistance (Reference)	Control
Cercospora (Fungus)	Mainly pre-nursery and nursery	Durand-Gasselin personal communication	Agricultural practices and chemical control
Curvularia (Fungus)	Mainly pre-nursery and nursery	Nouy and Durand-Gasselin personal communication (Fig. 1, Table 2)	Agricultural practices and chemical control
Blast (*Phytoplasma* is suspected, transmitted by *Recilia mica*)	Mainly pre-nursery and nursery	de Franqueville (1999, 2001)	Agricultural practices (mainly) and chemical control of the vector
Leaf mottle/Mancha anular. (Very likely a virus transmitted by an insect (*Sogatella*?))	Nursery and young age	Claude Louise personal communication (Table 3)	Agricultural practices (mainly) and chemical control of the vector
Fusarium wilt (fusariosis) (*Fungus*)	Field adult age (> 5 year), rarely in young age	Bachy and Fehling, 1957; Franqueville and de Greef, 1988) (Fig. 2)	Agricultural practices (limited effect). Breeding for resistance
Ganoderma basal or upper stem rot *(Fungus)*	Field adult age (> 5 year), rarely in young age	Akbar (1971); Franqueville et al. (2001) (Fig. 2)	Agricultural practices. Breeding for resistance
Bud rot (Very likely *Phytophthora palmivora* as primary agent)	Field any age	For *E. oleifera* x *E. guineensis:* Turner (1981) For *E. guineensis*: Franqueville, (2003); Amblard et al. (2009) (Fig. 2)	Agricultural practices (limited effect). Breeding for resistance

Figure 1 Susceptible and highly resistant crosses to *Curvularia* (Thailand, © T. Durand-Gasselin).

the damage as well as the availability of simple and economically viable control methods based on field practices or the limited use of chemicals. Indeed, breeding programmes aimed at selecting for genetic resistance have been undertaken only for

Table 2 Levels of resistance to *Curvularia* among Deli progenies at eight-month-old nursery stage (from Nouy, pers. com.)

Deli origin	Cross	Intensity of *Curvularia* damage
A	A_1	Obvious within cross disjunction between affected and healthy palms
A	A_1	Obvious within cross disjunction between affected and healthy palms
B	B_1	High incidence
B	B_2	High incidence
B	B_3	No damage
C	C_1	No damage
D	D_1	High incidence
D	D_2	High incidence
D	D_3	Extremely high damage

Table 3 Effect of Pisifera 1 in the protection of crosses from leaf mottle (*mancha anular*) (two-year-old palms, Ecuador). Pisifera 1 is compared to other pisiferas while using the same Deli mother palm. Results are given as index mean (from six genetic trials which have more than 14% incidence of the disease, Louise pers. com.)

	Father palm		
Mother palm	Other pisifera	Pisifera 1	Comment
Deli 1	126 (2 cross)	5 (2 cross)	25 times less
Deli 2	67 (2 cross)	8 (1 cross)	8 times less
Deli 3	119 (2 cross)	21 (1 cross)	6 times less
Deli 4	130 (3 cross)	22 (1 cross)	6 times less
Deli 5	157 (2 cross)	27 (1 cross)	6 times less
Deli 6	112 (3 cross)	50 (3 cross)	2 times less
Deli 7	138 (2 cross)	97 (1 cross)	1.5 times less

Fusarium wilt, upper or basal stem rot caused by *Ganoderma*, and for various forms of bud rot.

Fusarium oxysporum f. sp *elaeidis* is a soilborne and biotrophic fungus. It is an imperfect fungus, that is, its sexual reproductive stages have never been observed and it produces conidia and chlamidospores (Flood, 2006). It is probably one of the reasons why the genetic diversity of *F. oxysporum* f. sp *elaeidis* is so limited (Mouyna et al., 1996).

The *Ganodermataceae family* are basidiomycota fungi. Within the *Ganoderma* genus, the predominant species, *G. boninense*, appears to be the most aggressive (Cooper et al., 2011). In Africa, *G. ryvardense* has been reported to be the cause of basal stem rot (Kinge and Mih, 2011). *G. boninense* is a soilborne fungus and a polypore which bears double-walled basidiospores. Spores give rise to monokaryonic mycelium which eventually mate to generate dikaryons if they are sexually compatible (Pilotti et al., 2002). Pilotti

Figure 2 Field evidence of genetic resistance differences (a) *Fusarium* wilt, (b) *Ganoderma* Basal Stem Rot and (c) Bud Rot, respectively, in Côte d'Ivoire, Indonesia and Ecuador, © Google Earth.

has conducted studies on the genetic variability of Ganoderma within plantations, using RAPD, concluding that genetic variation was very high even within small plots (Pilotti et al., 2003). Mercière et al. (2015) developed SSR markers from *G. boninense* which will allow interesting diversity studies of *G. b.* in Asia in comparison with Africa and probably with America. *G. boninense* is a necrotrophic fungus.

Bud rot seems to be associated with *Phytophthora palmivora* (Sarria et al., 2008, 2016; Torres et al., 2010). In its early stages, many of the symptoms caused by *P. palmivora* in oil palm have been reported to be typical of *Phytophthora* diseases in general (Torres et al., 2016). In later stages, saprophytic microorganisms might be able to develop, leading to a variety of symptoms (Franqueville, 2003; Louise et al., 2015; Torres et al., 2016). *Phytophthora palmivora* is an oomycete which is often classified as a fungus-like microorganism, but this class is now included in stramenopiles. It produces abundant sporangia which generate zoospores. Sexual reproduction between compatible mating types leads to oospores and chlamydospores which provide the main way *P. palmivora* survives naturally. Even if *P. palmivora* has a soilborne phase, it cannot be classified as a soilborne disease but rather as a biotrophic fungus. Genetic diversity of *Phytophthora* was found to be generally high, although the various pathogen strains might show only limited diversity (Maroa et al., 2017).

2.2 Resistance sustainability

The perennial nature of the oil palm makes disease-resistance the focal point of breeding strategies. Van der Plank (1963, 1968) first developed the concept of horizontal resistance, which is expected to be multigenic, and of vertical resistance, which is expected to be single-gene based. The former – being non-race/strain-specific resistance – is expected to be sustainable, the latter being specific and non-sustainable. This concept has evolved with the understanding of gene–gene interactions between host and pathogen and the description of some mechanisms of partial resistance. Indeed, Robinson (2006) refined that concept although it is of great interest to understand how specific resistance can evolve towards non-specific resistance (McDonald and Linde, 2002). Roux et al. (2014) proposed that plant disease resistance is intrinsically dual, thus combining both a qualitative-like and a quantitative-like behaviour. These authors also consider that our knowledge on the determinants of qualitative resistance represents only the 'tip of the iceberg'. Breeders will have to deal with this lack of knowledge and must bear in mind that things are not as 'black and white' as stated in the 1960s. Nevertheless, it provides an interesting framework for breeders, and breeding for partial resistance is certainly the way to go.

It is probably useful to share here the meaning of some useful words, following the recommendations from the International Seed Federation (ISF) (*https://www.euroseeds.eu/system/files/publications/files/esa_12.0605.pdf*). Indeed, ISF encourages plant pathologists and breeders to make a clear distinction between immunity, resistance and susceptibility. Immunity is when a plant is not subjected to any attack and is generally race/strain-specific. Resistance is the ability of a plant variety to restrict the growth and development of a disease. Resistances are generally non-specific. Immunity usually (though not always) results from a gene–gene interaction between the plant and the pathogen and it can be overpassed. For oil palm, the breeder's strategy will aim at selecting for multiple defence genes involved in (partial) resistance, also called quantitative disease resistance (Roux et al., 2014). This selection strategy will favour sustainable non-specific resistance to a larger diversity of isolates of the pathogens, rather than specific resistance. It is important to note that consequently selecting plants for multiple partial disease resistance will result in varieties that exhibit some disease symptoms but will be more efficient in limiting their number. In line with these recommendations the authors ranked resistances into different levels: highly resistant, intermediate resistant or susceptible varieties. The term 'tolerance' should be avoided for this purpose as it should be used only to describe the ability of a variety to endure abiotic stresses.

2.3 Simple modelling of disease development

Understanding how a disease develops is of great importance in order to predict its evolution and consequently its possible damage. To this end, the development of mathematical models can be of great help. It is not a simple question as each single parameter has to be part of the equation. A non-exhaustive list of parameters could be as follows: % of infected host, % of healthy palms, % of remission, gain of immunity after the first infection or not (could be partial), transmission mode (air, contact or soil), etc. As an example, remission does exist in *Fusarium* wilt (Franqueville and Diabate, 1995) or Bud rot, but it has never been described for basal stem rot (*Ganoderma*). Bud Rot outbreaks generally occur during or just after a heavy rainy season and so some climatic parameters should be incorporated into the models.

One advantage of the mathematical models is that they enable the calculation of the threshold of epidemic take-off, and thus can encourage planters to adopt strict agronomical practices to keep the disease incidence below a given threshold (Jacquez, 1996).

One of the questions which must be addressed by both phytopathologists and breeders is how to describe a disease in the field or after artificial inoculation. Should it be a binary notation (infected/non-infected, alive/dead)? Should the intensity or severity of the damage be taken into account? Should we observe external symptoms only or both internal and external ones?

Two reflections may help address these questions. The first will be to ensure that the observation method does not bypass a potential defence barrier of the plant. For example, if you inoculate a petiole with *Fusarium*, you may bypass barriers that prevent the fungus from entering the roots. The second is the cost/benefit ratio of the experiment. An observation that is difficult to operate and does not significantly improve the discrimination power of the screening should not be upheld.

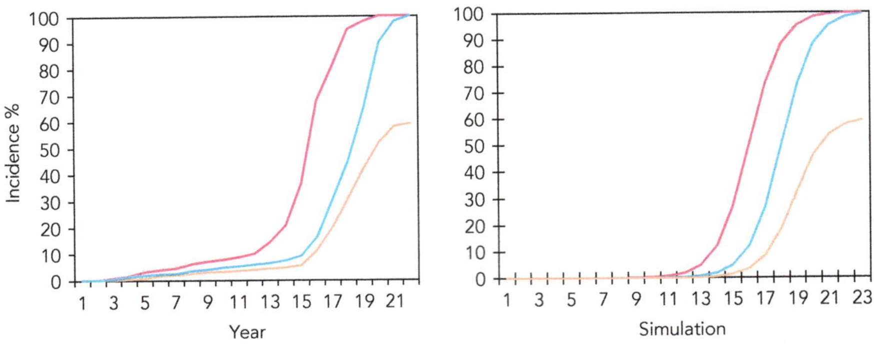

Figure 3 Changes in Bud Rot incidences as monitored for 23 years in trial SHGP 1, Ecuador (planting 1983) (left) and simulation of the data using a logit adjustment (right).

For breeding purposes, simple binary models which are close enough to real-life observations may be sufficient. As an example, Fig. 3 shows some data from a trial planted in 1983 in Ecuador for Bud Rot monitoring together with simulated data using the classic logit function. The figure shows that the function does not describe precisely the early stage of the disease, although it is very difficult to clearly differentiate the different types of materials at this stage. During the epidemic stage it is important to analyse the data when the maximum differences between varieties do occur (after 17 years in that case). At a younger age, only susceptible materials can be distinguished and conversely at a later age, only the most resistant are evidenced. Another very important point, at least for field observations, is to know the maximum percentage of infected palms that the disease will induce. This can be done only at the end of the trial: 100% for susceptible varieties and around 50% for the moderately resistant ones in the present case (Fig. 3).

In most cases, our experience shows that the maximum level of discrimination is reached when the percentage of infected palms in a trial averages 30% or more.

Three different varieties are analysed, namely very susceptible (pink); susceptible (blue); and modertaely resistant (orange).

2.4 Agricultural practices

Agricultural practices and genetic resistances are building blocks of Integrated Pest Management (IPM). The economic efficiency of agricultural practices is also part of the picture as it will impact dramatically the sustainability of oil palm cultivation.

The very first point is to take into account what happened during the previous generation. For *Fusarium* wilt, the success of a replantation is linked to the phytosanitary situation of the previous generation, including the severity of the disease (Franqueville and Renard, 1988). Singh (1991) collected data from different fields infected by *Ganoderma* and underlined the importance of the previous crop on the development of *Ganoderma* infection, coconut being one of the worst.

In the case of replantation, the distance to the previous palm is of tremendous importance for *Fusarium* wilt (Renard and Franqueville, 1991) as well as for basal stem rot caused by *Ganoderma* (Flood, 2000). However, this is not the only infection gate as *Ganoderma* dispersion occurs through root to root infection (Tuner, 1965); a result which

Figure 4 Infected adult palm and bait palms could be infected by the same Ganoderma strain (from Durand-Gasselin et al., 2015).

was confirmed by Flood et al. (2000) using bait palms. These authors confirmed that bait palms were infected by the same *Ganoderma* strain as the source palm (Fig. 4).

From an experiment conducted in Costa Rica, Albertazzi et al. (2009) hypothesised that Bud Rot development can be related to some deterioration of the whole plant due to poor root development which led to the possible development of the disease. The authors recommend improvement of drainage, avoidance of soil compaction and taking into account the fertilisation regime.

2.5 Early screening test

Most of the time, it is preferable to control diseases by agricultural practice or reasonable chemical treatment when possible. The development of early screening tests will obviously help. Prendergast was the first to propose an early screening test for *Fusarium* wilt resistance in oil palm in 1963 and Arrifin for *Ganoderma* (Arrifin et al., 1995). The two methods were nursery tests that were later refined as pre-nursery tests (Renard et al., 1972; Breton et al., 2006) which allowed implementation of much more extensive work.

In oil palm, early screening tests have been developed for two patho-systems, namely *Fusarium* and *Ganoderma*, and they are now used as routine tools. The recent finding that *Phytophthora* is the major causing agent of Bud Rot may lead to the development of another early screening test in the coming years; however, this work might be complicated by interaction with secondary opportunist pathogens. To be useful, an early screening test must comply with a set of characteristics described in the following paragraphs.

2.5.1 Index scores and trial comparison

When screening for resistance to disease, one of the difficulties is to compare the results from one trial to another or to compare field results with those generated by the early screening method. It is difficult to compare percentage of incidence *per se* as environmental conditions are always different. This has led phytopathologists to work on an index score. This has been defined by Renard et al. (1972) and refined by Franqueville (1984). The idea is to give a score of 100 to the mean of the trials and to compare all progenies to that mean. A score of 120 means that 20% more plants were infected as compared to the mean, and logically a score of 80 means that 20% fewer plants were infected. Such index scoring is not a perfect method, although it can be used to compare data from one trial to another as long as the trials (and therefore the mean of the trial) represent a subset of material with a comparable range of resistance. Otherwise, one has to connect the trials using a set of standard crosses which should be included in all the tests. The number of standard crosses to be used is linked to the required level of precision for comparison.

2.5.2 Repeatability

An early screening test must be repeatable: indeed, from one test to another, one must be able to find the same result. Using index scores, Franqueville was able to provide very early results for *Fusarium* nursery tests (Franqueville, 1984) and Durand-Gasselin (2015) provided an example (Table 4) of *Ganoderma* early screening tests.

2.5.3 Host–pathogen interaction

As we are looking for non-specific (partial) resistances, it is important to check if significant interaction between isolates and oil palm progenies exists. When using two different strains of *Fusarium* and 52 progenies to study inheritance of *Fusarium* wilt resistance, Meunier et al. (1979) concluded there was no interaction between strains and progenies. From preliminary data, Durand-Gasselin et al. (2015) reported similar results for *Ganoderma* (Table 5).

2.5.4 Precision

Unfortunately, even if a sizeable number of plants are used (e.g. 100 to 200), the precision of early screening tests will be weak (Renard et al., 1972). This is linked to the binary nature

Table 4 Performance of two crosses which have been repeatedly included in various tests. The resistant cross is consistently classified as resistant when the susceptible one is classified as susceptible or very susceptible

Trial	Susceptible cross	Resistant cross
Test 1	143	89
Test 2	98	73
Test 3	101	75
Test 4	102	72
Test 5	102	77
Test 6	112	78
Test 7	143	82
Test 8	144	85
Test 9	152	89
Test 10	121	89

Table 5 Interaction between Ganoderma isolates (strain) of various levels of aggressiveness and 6 progenies of various degrees of resistance: A and B being the most resistant progenies, F being the most susceptible followed by E, D and C

	Progenies					
Isolate (aggressiveness)	A	B	C	D	E	F
V (+)	1	2	5	4	3	6
W (++)	1	2	3	5	6	4
X (+)	2	1	3	5	4	5
Y (+)	1	2	3	6	4	5
Z (−)	2	1	4	3	6	5

of the observation which participates in making such tests a burden (Renard et al., 1972). Idris et al. (2004) showed data from nursery tests, using 40 palms per progeny, in which 40% infected palms were not significantly different from 22.5% (Idris et al., 2004). Field tests are even weaker in terms of accuracy. Flood et al. (1989) were able to distinguish differences in resistance between clones through a refined protocol and very precise monitoring (Flood et al., 1989), but such a protocol has not yet been generalised and correlation with field data still remains to be assessed.

2.5.5 Additivity

For breeders, it is important to know the nature of the observed variation which can be linked to either genetic or non-genetic effects (environmental factors, field practices, etc.) or to the interaction between the two. Genetic effects arise from either additive, dominance or epistasis variation, the former being the most helpful for breeding.

Meunier et al. (1979) highlighted the predominant role of additive effects in *Fusarium* wilt resistance. Later, de Franqueville and de Greef (1988) also used an additive model to explain field resistance to *Fusarium* at the Binga plantation. Durand-Gasselin et al. (2015) reported the same observation for *Ganoderma* in field experiments as in pre-nursery early screening tests.

2.5.6 Correlation between field and early screening tests

The concern of phytopathologists and breeders is that any early screening test may not be well correlated to the behaviour of the palms in the field. As an example, using a simple test on the leaf petiole may bypass some resistance mechanisms that may arise in the roots or around the roots (Diabaté et al., 1990). It is probably one of the reasons why early screening tests have been developed using whole living plants. Such a system is not easy to implement and further systems such as the one proposed by Mepsted et al. (1995) were based on rachis inoculation. Further efforts are required to make it fully operational, to ensure that the test is correlated with field results for all genotypes. This has been a great concern for all the teams working on early screening tests.

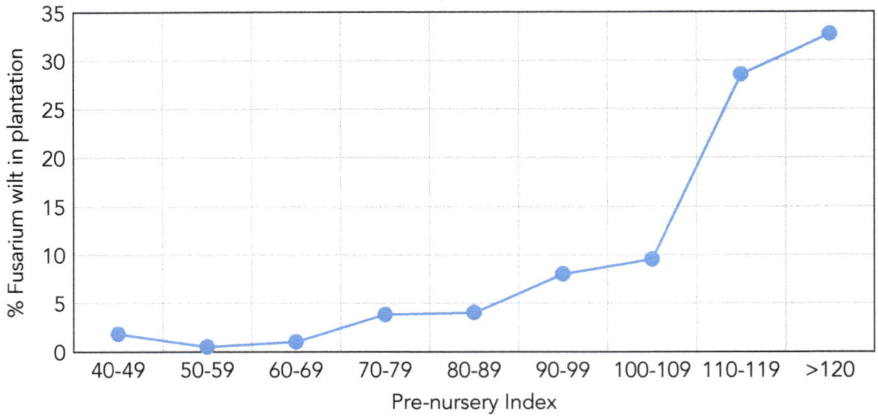

Figure 5 Relationship between pre-nursery index and the degree of resistance in the field (from Franqueville et al., 2011).

For *Fusarium* wilt, Renard et al. (1972) from the IRHO group published the first attempt to establish a correlation between pre-nursery tests and field results which was later confirmed by Franqueville and Diabate (1995), whereas Franqueville undertook similar observations for the Unilever group (Franqueville, 1984). An illustration of this relationship was published by Franqueville et al. (2011) and provided in Fig. 5.

For *Ganoderma*, the very first report on the relationship between field results and pre-nursery tests was presented by Purba et al. (2012), and additional data were later published by Durand-Gasselin et al. (2015). More experiments are needed to reach the same level of confidence as obtained for *Fusarium* wilt.

3 Achievements in improving genetic resistance to oil palm diseases

3.1 *Fusarium* wilt resistance

3.1.1 Sources of *Fusarium* wilt resistance

The Nigerian Institute for Oil Palm Research ranked its material for resistance to *Fusarium* wilt. Indeed, more than 300 progenies were assessed and results show that two palms of Calabar origin bear significant resistance to *Fusarium*, namely 32.364T and 32.3005T; the former being well known for its outstanding general combining ability (GCA) for yield, despite the fact that it shows an extremely rapid vertical growth. Conversely, the palm 4.1811T from Aba population was identified as very susceptible to the disease (Rajagopalan et al., 1978).

At the research station of Unilever and Sipef in Binga (Zaire now Congo RDC), Franqueville and de Greef (1988) described the 69 MAB palm as being very resistant. This palm is from pure Djongo origin, and it is worth highlighting that the palm1212A of the same origin displays high susceptibility. This palm originates from the self-pollination of the palm 16R which is moderately susceptible. Two relations of palm1212A, namely palms 1245A and 1342A, were crossed and the *Fusarium* wilt resistance of two palms resulting from that cross was assessed: one was significantly resistant and the other significantly susceptible. Again, it is clear that within-origin variability is high.

When the French research institute IRHO assessed a large part of its collection, it found sources of resistance in most of the tested population (Renard et al., 1980). As an example, parents from Deli origin were ranked from highly resistant (mean index of 85 for DA300D, from 16 tests) to very susceptible (mean index of 141 for LM412D from 18 tests). The same result was found within the La Mé population and the Yangambi/Sibiti population where the index range was 58 to 120 and 86 to 131, respectively. In the same paper, the authors reported the only known attempt to assess Fusarium wilt resistance of the interspecific hybrid *Elaeis oleifera* x *Elaeis guineensis*. Also *E. o.* populations from Brazil appeared to transmit susceptibility to the hybrids. Several populations from Colombia or Central America have a genetic factor that transmits extremely high resistance (at least during early screening pre-nursery tests, as the connection with field data has not been reported yet). Further information was provided by Durand-Gasselin et al. (2000) and again high variability within the population was highlighted.

Therefore, it is not surprising that the combination of parents from two origins will generate highly variable resistances to Fusarium wilt depending on the resistance of the parents. Among Deli x La Mé crosses reproduced through self-pollination of the parents (Jacquemard et al., 1981), the DA115D x LM2T cross showed an average index of 79 (excellent resistance), whereas LM407D x LM451T index was 132 (poor resistance). Such a huge genetic variation for resistance to *Fusarium* was also found among palms belonging to the same progeny, even when this progeny results from a selfing of a unique palm (Fig. 6).

Such results were used by the breeders to develop commercial material carrying moderate-to-high resistance to Fusarium wilt. In Cameroon, Pamol has developed Fusarium wilt-resistant material based on Binga results, and Cerapah did the same based on IRHO results (Ngando et al., 2012).

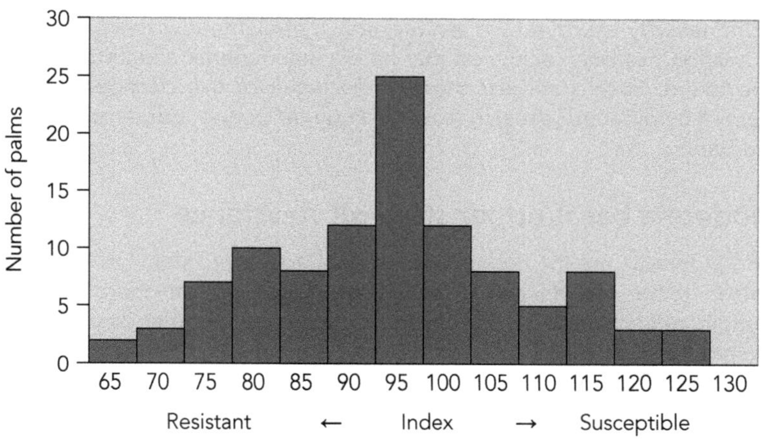

Figure 6 Fusarium wilt resistance assessed among 94 relations within the family LM2T self. Palms with an index over 110 are clearly susceptible; each parent has been tested more than 20 times.

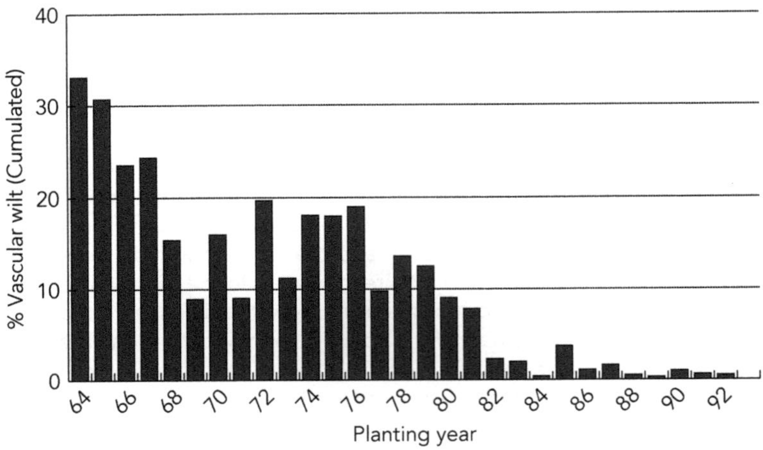

Figure 7 Decrease in vascular wilt (*Fusarium o.*) incidence over the years in Robert Michaud plantation (4000 ha, Dabou Côte d'Ivoire).

A very convincing outcome of breeding for *Fusarium* wilt resistance has been presented by Franqueville and Diabaté (1995). The authors described the evolution of the cumulative incidence of *Fusarium* wilt in a semi-commercial plantation (4000 ha). The incidence is the mean for the entire plantation, all ages gathered. This plantation, called Dabou, was one of the most impacted plantations in Africa. Indeed in 1964, 33% of the palms were sick. The policy of the company was to thoroughly replant palms using progenies which had been assessed to be resistant through early screening tests. In 1992, the incidence was found to be negligible and in addition most of the remaining cases were furtive as palms recovered rapidly from the disease (Franqueville, pers. com.) (Fig. 7).

Phytopathologists and breeders have been able to select highly resistant material which has dramatically improved the sanitary status of plantations in Africa. Such a high level of protection has been achieved that hardly any material affected by the disease can now be found. Such a success story is the result of five decades of continuous work conducted by different research teams in Nigeria, Congo, Cameroon, Ghana, Côte d'Ivoire and Benin.

3.2 *Ganoderma* basal/upper stem rot resistance

Malaysia and Indonesia are the countries in which *Ganoderma* stem rot causes the most damage. Most of the reports and achievements regarding selection for *Ganoderma* resistance originated from research conducted in these two countries despite the fact that *Ganoderma* has been detected and described as pathogenic in some parts of Africa for quite a long period (Congo (RDC) and Cameroon).

Akbar et al. (1971) first reported that pure Deli material was more susceptible to *Ganoderma* stem rot than Deli x Yangambi material. The susceptibility of Deli material as a whole was later confirmed by Durand-Gasselin et al. (2005).

Idris et al. (2004) published initial results from an early screening test conducted in the nursery. Table 6 displays a summary of these results for which we have gathered performance by origin. Here again, the pure Deli material is the most susceptible, whereas African materials are more resistant. Among them, Avros material seems to transmit a higher susceptibility as compared to pure Nigeria or Cameroon material. It is worth noting that the appellations 'Cameroon' or 'Nigeria' are far from being representative of the entire countries. This work is based on a few palms from those origins only. In the same work, three crosses of pure *E. oleifera* (originating from Colombia and Honduras) were found to be susceptible as well as one interspecific cross between *E. g.* and *E. o.*

Setiawati et al. (2010) described the very interesting work carried out by Sumatra Bioscience (also well known as Lonsum) to screen their own genetic material in the field and in pre-nursery. From early results in the field, showing that the average percentage of palms infected by *Ganoderma* was around 10%, the authors were able to identify clear sources of susceptibility among some material from Congo, while one origin from Lobe research station (Cameroon) was found to be highly resistant. In addition, from early results from a multilocal field trial, the authors underlined, as already pointed out by different papers, the susceptibility of Deli material, whereas Binga x Pobè crosses (Congo, Benin) proved to be the most resistant material.

In the same paper, an important set of trials conducted in pre-nursery is presented. The first work was to define a set of standard crosses ranking from susceptible, moderately resistant to highly resistant crosses. This set of crosses was used to connect 35 pre-nursery trials that led to the evaluation of 217 progenies from 69 different breeding origins. Four of those origins were represented by more than 5 crosses: it is worth noting that Deli x Avros crosses showed an intermediate level of resistance between pure Avros crosses and pure Deli crosses (Table 7).

Table 6 Results from nursery test screenings of 19 progenies, grouped by origin (adapted from Idris et al., 2004)

Origin	No. of tested progenies	% of dead palms
Pure Deli	3	36.7
(Nigeria or Cameroon) x Avros	6	25.4
Deli x Avros	4	24.4
Pure Nigeria	2	21.3
Pure Cameroon	4	15.6

Table 7 Average % of infected seedlings in four origins represented by more than five crosses in pre-nursery tests (Adapted from Setiawati et al., 2010)

Type of crosses	% of infected seedlings
Avros x Avros	14
Deli x Avros	18
Deli x Deli	23
'Binga' (Congo)	24

In addition, 65 from the 217 studied crosses were also evaluated in the field. As expected, the correlation was poor, because of i) the low accuracy of field monitoring, especially in fields where the number of planted palms is low; and ii) the fact that the disease in the field has yet to express its maximum. Nevertheless, results clearly showed that susceptible crosses in pre-nursery tend to be susceptible or moderately resistant in the field and rarely resistant (1 out of 16). Furthermore, the resistant palms in nursery were either resistant or moderately resistant in the field, and were found to be rarely susceptible (1 out of 16). The results for moderately-resistant crosses were of course less clear (Table 8).

In 2015, Durand-Gasselin et al. presented outputs of research work in North Sumatra with the Socfindo planting company. A range of 368 commercial crosses were evaluated in pre-nursery. To this end, Deli x La Mé or Deli x Yangambi crosses were planted in an almost complete factorial design (Table 9). In such assessment, it is normal that Deli origins and La Mé or Yangambi origins share the same mean, despite the fact that pure La Mé or pure Yangambi material is generally more resistant than pure Deli.

Table 8 Distribution of 65 families according to their degree of resistance expressed in the field and in the nursery (Setiawati et al., 2010)

		Field		
Number of families		Susceptible	Medium	Tolerant
Nursery	Susceptible	6	9	1
	Medium	9	14	10
	Tolerant	1	10	5

Table 9 Observed index in pre-nursery tests of a set of crosses organised in a factorial design. Five Deli (A group) origins were used as female parents and five La Mé or Yangambi (B group) origins as male parents (adapted from Durand-Gasselin et al. 2015)

Origin	la Mé A	la Mé B	la Mé C	Yangambi A and A'	Yangambi B	Average
Deli A	**82** (36 tests)	**88** (23 tests)	99 (22 tests)	109 (58 tests)	114 (13 tests)	98 Moderately resistant
Deli B	**90** (5 tests)	**86** (16 tests)	122 (12 tests)	107 (8 tests)	107 (8 tests)	102 Moderately resistant
Deli C	95 (13 tests)	107 (8 tests)	–	105 (10 tests)	129 (5 tests)	109 Susceptible
Deli D	86 (20 tests)	103 (13 tests)	102 (9 tests)	*139* (11 tests)	*122* (10 tests)	110 Susceptible
Deli E	97 (11 tests)	118 (12 tests)	115 (15 tests)	*125* (22 tests)	*117* (8 tests)	114 Susceptible
Average	90 Resistant	102 Moderately resistant	109 Susceptible	117 Susceptible	118 Susceptible	General mean: 107

The range of partial resistance was greater for La Mé and Yangambi (from 90 to 118) as compared to Deli (from 98 to 114); this result is in line with the genetic diversity of Deli origin which is smaller than African origins (Cochard et al. 2009). Among the three studied La Mé origins, one was clearly more resistant than the others (Origin la Mé 'A'). The best results were obtained when crossing the two best La Mé origins (A and B) with the most resistant Deli origins (A and B). The average index of such crosses was 86 (in bold in Table 9), indicating that this material has 20% fewer affected palms than the mean of the trials (Index of 107). On the other hand, when crossing the two worst Deli origins with the two Yangambi origins, we observed 20% more infected palms (Index of 127) (in italics in Table 9). This result is in line with the predominant additive inheritance of partial resistance to *Ganoderma*.

Because of the predominant additivity inheritance, the authors have undertaken some work to assess the female origins (most of them being of pure Deli origin, A group) as well as male origins (most of them being of pure La Mé or Yangambi origin, B group) using testers.

This work allowed the authors to rank 16 different female origins (16 pure Deli and 1 Deli x Angola) and 12 different male origins (9 pure La Mé and 3 pure Yangambi). Each origin is represented by different crosses (Fig. 8).

In addition, the authors illustrated some within-origin variation. As an example, the Deli I origin is represented by 14 crosses, each of them having been tested 15 times on average (from 5 to 34) (Fig. 9). It is worth observing that the range of resistant palms within this origin is almost as wide as those obtained within Deli origin. This kind of observation is very useful for breeders as they can expect to find resistant material in a majority of origins. In this example, the best family, which was assessed 11 times, originating from a moderate resistant Deli origin (Deli I) appeared as good as the best-performing Deli origin (Deli A). Of course this is just an example as one can find an even better family within the Deli A origin.

In the same paper, field results were presented. Results obtained from a dedicated factorial design were found to provide similar results as those performed in pre-nursery (results not shown here) that also illustrate the additive inheritance of *Ganoderma* partial resistance.

It has been possible to calculate the GCAs for *Ganoderma* partial resistance of many parents of crosses involved in genetic trials planted in North Sumatra in an area affected by *Ganoderma* (480 crosses with at least 72 palms planted per cross). The analysis by parent

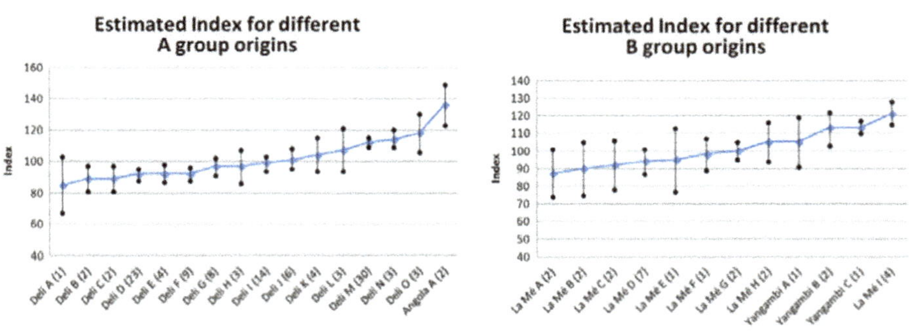

Figure 8 Estimated index for different A (female parent) or B (male parent) group origins. Number of studied progenies within each origin is given in brackets (from Durand-Gasselin et al., 2015).

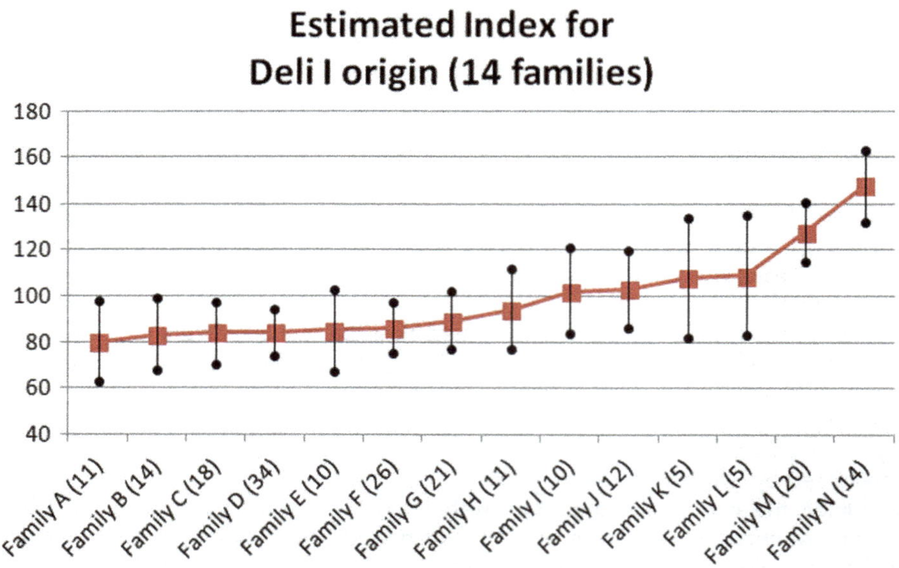

Figure 9 Estimated index for 14 families of Deli I origin. The number of tests performed for each family is given in brackets.

did not give significant results because of the limited number of palms per progeny and relatively low *Ganoderma* incidence among the trials (around 10%). Nevertheless, regrouping the parents in five classes for female (A group) from −2 (susceptible) to +2 (resistant) and for male (B group) according to their partial resistance to the disease gives the clearest picture (Table 10). Crosses between resistant parents showed fewer than 5% *Ganoderma* infection in the field (in bold in Table 10) as compared to 10.7% for the whole experiment.

The authors have identified commercial planting material which showed some intermediate resistance to *Ganoderma* and originates from the combination of the best Deli dura parents, with Pisifera palms originating from the best La Mé family.

This planting material has been tested in nursery in its original form (without use of test crosses). To this end, parents taken from selected families coming from three of the best Deli origins (Deli E, F and I) were crossed to parents from the La Mé A family. The resulting progenies showed very low indexes, thus confirming their estimated levels of resistance (Table 11).

It is important to underline that genetic resistance, while of major influence, is only one facet of any IPM strategy aimed at controlling *Ganoderma* disease.

Impressive research work carried out in Malaysia and Indonesia, the main achievements of which are presented here, has led to the generation of commercial planting material (varieties) which express moderate resistance to basal/upper stem rot caused by *Ganoderma*. There is no doubt that within a few years and step by step, this resistance will be improved to a highly resistant level. Partial resistances are expected to be sustainable and the genetic progress achieved here is of great value for many stakeholders in the plantation sector. However, planters must continue to use any combination of agronomic practices (Flood et al., 2000) together with resistant varieties through integrated management of the disease.

Table 10 Average percentage of *Ganoderma* infection for 480 crosses planted in Aek Loba Timur genetic block (North Sumatra) as grouped by origin and class of resistance. Parents are ranked from −2 to +2 according to the GCA values for *Ganoderma* resistance trait. In bold are the crosses with good resistance behaviour

		B Group					
	Class	−2	−1	0	+ 1	+2	Average
	−2	17.1	8.3	18.9	15.2	5.8	15.3
	−1	12.9	–	13.4	8.6	7.1	10.9
A Group	0	12.4	20.1	11.1	10.7	6.4	11.0
	+1	14.7	8.2	8.9	8.7	**2.0**	9.8
	+2	10.1	9.2	4.6	**7.3**	**4.0**	6.1
	Average	13.3	15.9	10.9	10.4	5.5	10.7

Table 11 Nursery testing of the intermediate resistant commercial varieties for Ganoderma resistance. Results are given as mean index and number of tests

Group A origin	Group B origin Origin La Mé A
Deli I x La Mé A	**85** (11 tests)
Deli F x La Mé A	**88** (7 tests)
Deli E x La Mé A	**71** (34 tests)

3.3 Bud rot resistance

For long the aetiology of Bud rot has remained unclear. It is only recently that Bud rot has been associated with *P. palmivora* (Torres et al., 2010), although it remains very difficult to inoculate the pathogen and trigger the disease. Nevertheless, based on field results, some achievements were published a few decades ago for interspecific hybrids *E. oleifera* x *E. guineensis* and more recently for pure *E. guineensis* material.

3.3.1 Bud rot resistance of interspecific hybrids *E. oleifera* x *E. guineensis*

Bud rot caused massive losses at the La Arenosa estate (Turbo region) in Colombia during the 1960s. In 1981, Turner reported that *Elaeis oleifera* x *Elaeis guineensis* interspecific crosses planted in 1968 were spared by the disease (Renard and Quillec, 1984). The same resistance was observed in Suriname; indeed, Van de Lande described a small plot of interspecific hybrids that had survived the disease. This plot was planted in 1978 at Victoria estate, Suriname (Van de Lande, 1986) and was still intact by 2007 (Corredor pers. com.).

In the Tumaco region (Colombia) many plots of old O x G hybrids (Sinu x Yangambi origins) planted in the 1970s developed resistance to the disease (Durand-Gasselin et al., 2009).

One can probably qualify the interspecific varieties as being highly resistant. Franqueville (2003) found Bud Rot symptoms affecting sporadically interspecific hybrids in Ecuador. It is only when the disease is very active and highly aggressive that interspecific material can be significantly affected and some palms may not survive (Franqueville, pers. com.). The Cenipalma team confirmed this result for the Coari x La Mé interspecific hybrids for which the degree of severity of the disease hardly reached the severity degree of two out of five (Avila-Diazgranados et al., 2016), including one cross that seemed to be free of the disease.

Interspecific hybrids may not yield as much as *E. guineensis* by up to 30% (Franqueville, 2003), although much better results have been obtained in the Llanos region of Colombia (Zambrano, 2004). The drawback of this type of material is that it has to be hand-pollinated, a disadvantage that is somewhat compensated for less frequent harvesting rounds (Corley, 2016) and simpler pest management (Herrera, pers. com.). Finally, yield is no longer an issue (Durand-Gasselin, 2010; Amblard, 2010), and planters' investments are strongly protected against Bud Rot when using the interspecific hybrid.

3.3.2 Bud rot resistance of *E. guineensis* crosses

For a long time there were thought to be no differences in resistance to Bud Rot between pure *E. guineensis* materials planted in America (Chinchilla-López, 2010). However, two teams have shown since then that *E. guineensis* may express partial resistance to Bud Rot.

Unipalma (Unilever) reported that some material, planted in 1991 near Puerto Wilches in Colombia, showed promising partial resistance in regions which were devastated by the disease (Cadavid, 2013) (Fig. 10). The author undertook further research in order to confirm this result in different plantations in Colombia with

Figure 10 Unipalma material partially resistant to Bud Rot (left) and susceptible material (right) (from Cadavid, 2013).

genetically-related material and very early results showed a similar trend (Cadavid, 2013).

The CIRAD/PalmElit team reported results from a set of trials. At first, material showing partial resistance in the field was identified and an example is given in Fig. 2. An attempt was made to confirm that result by comparing some promising material against six other materials (trial SHGP 8) (Louise, 2016). A subset of this trial is presented in Fig. 11, which shows the evolution of the incidence of Bud Rot from the date of planting (2/2000) until the stabilisation of the disease evolution (end of 2013) for three types of material. Origin 1 was the most resistant and around 50% of the palms were decimated by the disease. Because of the presence of a majority of susceptible material, we have to keep in mind that the pressure of the disease was very important throughout the monitoring period. Without such an intense inoculum pressure, the results could have been much better. Nevertheless, we found that such partial resistance did not provide sufficient protection. There are also some differences between the two most susceptible materials, indicating that some partial resistance of small effect does exist and it is worth cumulating it with the strongest one.

In a similar trial (SHGP 9) the same origins were represented by a set of 7 or 8 crosses. Significant but variable resistance to Bud Rot was evidenced in all origins (Louise, 2016). Figure 12 illustrates the observed variability for 3 crosses of each origin as presented in Fig. 11.

It is the hope of breeders that selecting that best progenies within a given origin will lead to significant genetic progress. This seems to be the case within the best origin (Origin 1) for which the best cross (C1) showed only 32% of infected palms as compared to 52% for the mean of the origin (Fig. 13).

Early results confirmed that the most resistant material to Bud Rot in a given location is also the best in other locations. Louise (2016) concluded that such results are promising

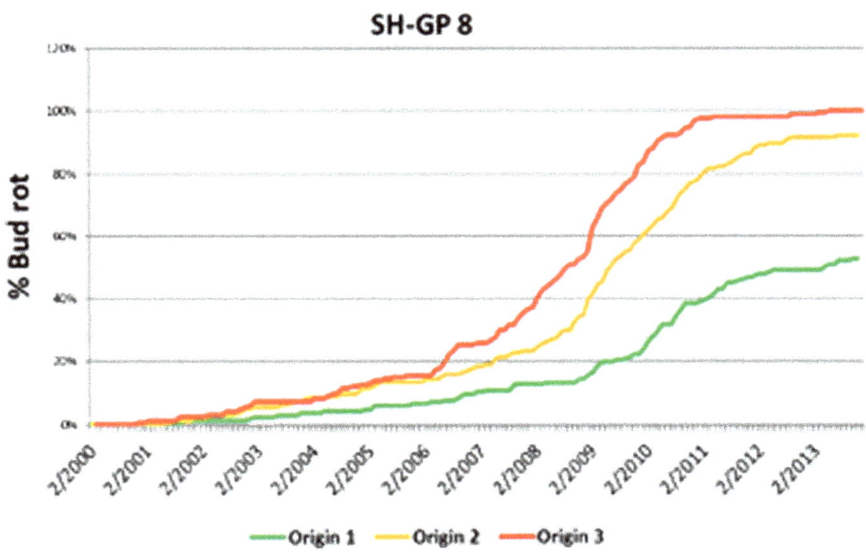

Figure 11 Comparison of the partial resistance of three different types of material in Shushufindi, Ecuador. Adapted from Louise (2016).

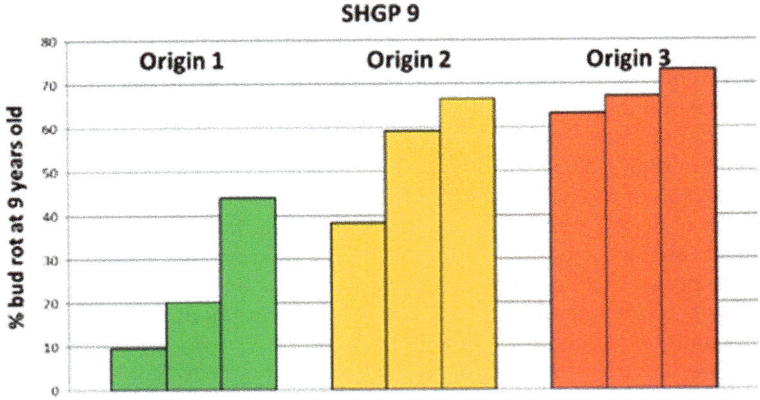

Figure 12 Comparison of the partial resistance of three crosses per type of material at nine years in Shushufindi, Ecuador. Adapted from Louise (2016).

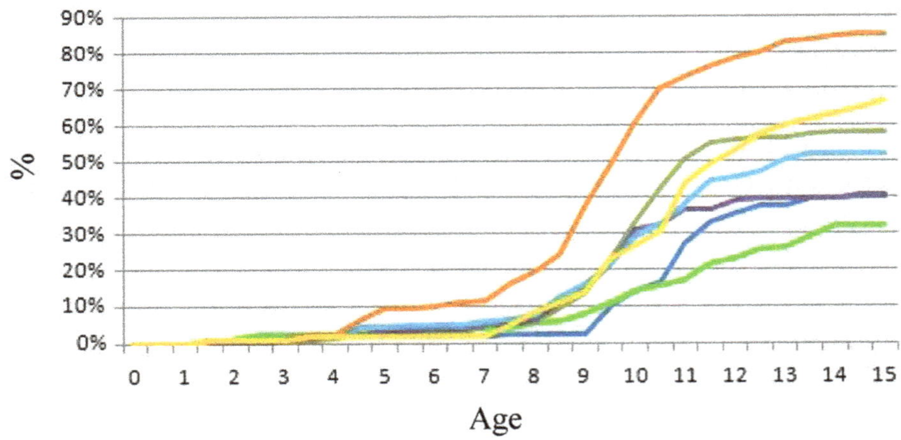

Figure 13 Comparison of the partial resistance of seven crosses within origin 1 at nine years old in Shushufindi, Ecuador (Adapted from Louise, 2016).

and he estimated that within a few years, resistant *E. guineensis* materials with intermediate resistance to Bud Rot will be commercially available in Ecuador.

4 Conclusion

During the past decade breeding for resistance to diseases obviously became a main objective for breeders. Breeding for yield remains of course a long-term target, although both account for the global sustainability of the oil palm industry but in a slightly different way. Indeed, planting material, which expresses resistance to a disease, protects the planter's investment, whereas yield enhances his resources.

Oil palm is a perennial crop and, as a consequence, the number of generations in a given period is limited. For breeders the crop is not as agile as an annual crop. Three to four decades are spent in order to obtain some durable protection against *F. oxysporum*. After two decades of hard work, phytopathologists and breeders are able to generate planting material with intermediate resistance to basal/upper/stem rot caused by *Ganoderma*. For Bud Rot, the most devastating disease, we have been lucky to find a pre-existing strong resistance in the interspecific hybrid. Resistances in pure *E. guineensis* material were rare, but at least two teams have generated some solid hopes that at least intermediate resistances will become exploitable soon.

Such findings have been possible thanks to long-running and steady work undertaken by research institutions and R&D centres from private plantation companies. One of the main challenging issues will be to combine resistances in order to face more than one disease at the same time. In Africa, especially in the Congo basin and in Cameroon, some plantations are facing damage caused by *Fusarium* as well as *Ganoderma*. Spear rot-like diseases do exist in Asia, and they may become epidemic in this region one day (Corredor, pers. com.).

Such a target will be very likely attained because of the diversity evidenced within most oil palm origins, as long as the breeders accept working with a broad genetic base. Two examples can be given to illustrate that point. There is a famous palm named Deli E206 (from Elmina Estate in Malaysia) which is also known as the dumpy palm because of its short stipe and larger girth. This palm is known to be resistant to *Fusarium* wilt (Rosenquist, 1985), but it is highly susceptible to *Ganoderma* (Franqueville, pers. com.). On the contrary, the La Mé palm LM2T, whose resistance to wilt is well documented (Durand-Gasselin, 2000), has also been found to be a promising source of resistance to *Ganoderma* (Durand-Gasselin, 2005). It is for this reason that breeders need a broad genetic base for their breeding programme.

Should planters face compromise on yield when using materials that are resistant to diseases? This question does not have an easy answer. Of course, the yield of interspecific hybrids used to be lower than that of normal *E. guineensis* commercial materials, but this is no longer the case. A possible answer to that question is that most of the planting material that was identified to be resistant to diseases during the first step of selection in the field was commercial material: as a consequence, disease-resistant materials do originate from excellent material.

The sustainability of partial resistance is very likely to be confirmed. Such resistances are the bottom line of breeders' work. The fact is that no example of total immunity has yet been reported. Even the highly resistant interspecific hybrids or the highly resistant *Fusarium* wilt materials can be affected by Bud Rot or fusariosis when planted under high pressure of an aggressive strain of the pathogen.

The development of early screening tools for resistance to *Fusarium* and to *Ganoderma* has enhanced the efficiency of our research work. Nonetheless, before using it, it has proven pivotal to assess some of their main properties: repeatability, precision, low or no host–pathogen interaction, and connection with field data. Breeders will also assess the inheritance of the resistant traits, which have been found to be mainly an additive transmission, at least for resistance to *Fusarium* and *Ganoderma*.

Last but not least, planters must not solely rely on genetic resistance to disease. Partial resistance is part of IPM strategies and any field practice that helps to limit the presence or the development of the inoculum will also limit the number of affected palms.

Today, planters benefit from oil palm varieties that effectively protect their plantation from *Fusarium* wilt and Bud Rot and can provide some moderate protection against *Ganoderma*. The highest level of protection against Bud Rot relies on interspecific hybrids – which need assisted pollination – and on pure *E. guineensis* varieties with moderate resistance, which have been made available recently. As researchers pursue their efforts, there is no doubt that new incremental progress will be steadily achieved.

5 Future trends

At first, efforts must be made to simplify and improve the existing early screening tests and an evaluation of the current resistance screening methods is needed (Kull et al., 2003). Nevertheless, one will have to comply with the specifications described earlier in this chapter, particularly avoiding bypassing possible defence mechanisms. For Bud Rot, it is worth underlining the tremendous efforts made by Cenipalma (Colombia) to artificially duplicate the disease caused by *P. palmivora* in the greenhouse. Without any doubt this work will help in developing a very sought-after early screening test.

Because of its perennial character, the oil palm is unlikely to become a model plant. Nonetheless, any tool that will help to save time will have a huge impact because of this perennial character.

For a long time the work of phytopathologists and breeders has focused on total resistances which were expected to provide plants with a sort of immunity (Dodds and Rathjen, 2010). Despite the fact that immunity has been of huge help to farmers, the risk that pathogens will overcome such resistance remains high, and thus, partial nonspecific resistances are needed as they have been proved to be very durable. Some in-depth understanding of the genetic patterns of such resistance is needed, and despite the diversity of situations and the complexity of research questions, new knowledge is being generated (McDonald and Linde, 2002; Roux et al., 2014).

Breeders have been using partial resistance for decades, but their work will certainly be more relevant if they can integrate such knowledge in their practice. An intermediate situation will be the use of molecular markers as the detection of QTLs is particularly suitable for partial resistance. Historically, the need for specific experiments has slowed down such research, and in addition the inaccurate data available to assess genetic resistance led to the detection of weak QTLs (Cochard, pers. com.). Recent work has generated new hopes; identifying QTLs using the information given by a broad set of relatives. The fact is that oil palm pedigree is relatively well documented and this knowledge can help to detect QTLs in nonspecific genetic trials (Tisné et al., 2015).

It might be the case that some strong resistance factors exist, for example, resistance of interspecific hybrid to Bud Rot, and in this case it might be relatively easy to trace those using molecular markers. This will help the introgression of such traits from *E. oleifera* – which is the source of that strong resistance – into *E. guineensis* through backcrossing.

Now that annotated oil palm genome sequences have been made available, we will be able to discover gene sequences of great interest in durable disease resistance, and new genome editing tools can also be used (CRISPR/Cas9).

6 Tribute to Hubert de Franqueville

This chapter is dedicated to the memory of Hubert de Franqueville, our oil palm phytopathologist colleague, who dramatically passed away in November 2015. He certainly should have been the co-leading author of this chapter. Our remembrance is not only because Hubert was a great colleague, a friend with whom we shared good and bad times, but also because we have been continuously building bridges between pathologists and breeders' knowledge. The work of our team has gained in relevance because of these bridges: taking into account the constraints of the oil palm, the constraints of the phytopathologists and those of the breeders who have made our research much more meaningful. Hubert and I built a strategy to breed for disease resistance which is summarised hereafter.

To breed for disease resistance, we will have to face many challenges, each in interaction with the other. In addition to numbers, just taking into account the first-order interactions already deliver a very complicated picture.

The factors correlated to disease resistance may be related to the following.

The disease itself:

- the type of pathogen: fungi, phytoplasmas, bacteria, viruses, etc;
- the genetic diversity of the pathogen which is generally linked to its sexual cycle and reproduction mode;
- its propagation: in the soil (for soilborne pathogens), in the air, spores quantities and viability, through vectors (for phytoplasmas, nematodes, etc); and
- the pathogen can be either biotrophic or necrotrophic (sometimes hemibiotrophic), which will obviously lead to different responses from the host.

The host:

- annual or perennial crop, and obviously, the perennial character of oil palm is a key factor;
- the genetic diversity of the host; and
- the age of the host is also a key factor. Diseases occurring at the seed, pre-nursery or nursery stage could be easier to control as the acreage involved is much smaller than for field diseases.

The environmental factors:

- climatic conditions, such as the length of dry season, rainfall and temperature;
- soil receptivity and soil infectivity;
- soil microflora or micro fauna; and
- agricultural practices, which are human-dependent, can affect drainage, cover plants or spore transmission.

Accurate knowledge or observations about the pathogen will obviously help breeders to develop resistant varieties. Of course, the identification of the disease is highly desirable even if breeding can in principle (but with great difficulties) be initiated without it. As soon as they have identified a pathogen which could be potentially linked to a disease, phytopathologists will make best efforts to comply with Koch's Postulates.

Evidence of field genetic resistance is of great help, at first to consolidate the idea that breeding could be efficient, and later to confirm in the field that selected resistant varieties are behaving as expected.

Resistance to diseases can be either total or partial and this point has been discussed. Whatever the type of resistance, the challenge is to develop – together with phytopathologists – an early screening test. Ideally, an early screening test should be repeatable, manageable as a routine tool for breeders (high throughput) and as fast as possible. It is also important to check if significant interaction between isolates inducing a given disease and oil palm progenies exists. Last but not least, the early screening test has to be straightforward and highly correlated to field resistance.

The inheritance of resistance to a given disease also needs to be assessed. First, one has to understand that while moving from one generation to another, the additive effects are passed to the offspring, whereas interactions (non-additive effects) have vanished. Under some conditions, non-additive effects can be retained for seed production. Oil palm breeding is still in its early stages; therefore we can expect that additive effect will be largely predominant.

Having in hand most of this knowledge and tools, breeders and phytopathologists will be able to develop oil-palm-resistant varieties to several diseases. The impracticality of obtaining within a short period of time a sufficient number of generations of oil palms is a major constraint. This leads to devising long-term strategies that will have to consider the durability of the resistance as the main objective. For the same reason, the number of traits that can be addressed is very limited: breeding for sustainable resistance to only one disease is such a burden that to date has not been reasonable to undertake a breeding programme for resistance to a disease that can be controlled by any agricultural practice.

7 References

Akbar, U., Kusnadi, M. and Ollagnier, M. (1971). Influence de la nature du matériel végétal et de la nutrition minérale sur la pourriture sèche du tronc du palmier à huile due à *Ganoderma*. *Oléagineux*, 26, 527.

Albertazzi, H., Chinchilla, C. M. and Ramírez, F. (2009). Soil characteristics and root development in young oil palms (*Elaeis guineensis* Jacq.) planted in sites affected by Bud Rots (pudrición del cogollo) *ASD Oil Palm Papers* (Costa Rica), 33: 1–14.

Amblard, P. (2010). Programa de mejoramiento de la palma aceitera de PalmElit y sus socios en Ecuador y Colombia. *Revista Palmas*, 31(especial): 144–54.

Avila-Diazgranados, R. A., Edison, S. D., Navia E.; Romero H. M. (2016). Response of various oil palm materials (*Elaeis guineensis* and *Elaeis oleifera* × *Elaeis guineensis* interspecific hybrids) to Bud Rot disease in the southwestern oil palm-growing area of Colombia. *Agronomía Colombiana*, 34(1), Bogota.

Bachy, A. and Fehling, C. (1957). La fusariose du palmier à huile en Côte d'Ivoire. *Journal d'agriculture Tropicale et de Botanique Appliquée*, 4: 228–40 .

Breton, F., Hasan, Y., Hariadi, Lubis, Z. and de Franqueville, H. (2006). Characterization of parameters for the development of an early screening test for basal stem rot tolerance in oil palm progenies. *Journal of Oil Palm Research*, Special Issue, 24–36.

Cadavid, I. O. (2013). Avances en el desarrollo de materiales DxP con resistencia de campo al complejo de la Pudrición del cogollo (PC) en el programa de mejoramiento genético de Unipalma SA. *Revista Palmas*, 34(2): 121–5.

Chinchilla-López, C. M. (2010). There is no known evidence of a relationship between incidence and severity of spear rot (PC= pudrición de cogollo) in oil palm and the commercial origin of oil palm varieties planted in the past in tropical America. *ASD Oil Palm Papers*, (35): 1–12.

Cochard, B., Adon, B., Rekima, S., Billotte, N., Desmier De Chenon, R., Koutou, A., Nouy, B., Omoré, A., Razak, P., Glaszmann, J. C. and Noyer, J. L. (2009). Geographic and genetic structure of African oil palm diversity suggests new approaches to breeding. *Tree Genetics and Genomes*, 5(3): 493–504.

Cooper, R. M., Flood, J., and Rees, R. W. (2011). *Ganoderma boninense* in oil palm plantations. Current Thinking on Epidemiology, Resistance and Pathology., *The Planter*, Kuala Lumpur 87(1024): 515–26.

Corley, R. H. V. and Tinker, P. B. (2016). *The Oil Palm* (World Agriculture Series) 5th Edition, Wiley Blackwell. ISBN 9781405189392.

Diabaté, S., Taquet, B., Renard, J. L., de Franqueville, H. and Reiser, P. (1990). HPLC analysis of substances produced by oil palm and implicated in resistance to *Fusarium oxysporum* f. sp. *elaeidis*. Prospect for selection. *Oléagineux*, 45(2): 49–55.

Dodds, P. N. and Rathjen, J. P. (2010). Plant immunity: towards an integrated view of plant–pathogen interactions. *Review, Nature Reviews Genetics*, 11: 539–48.

Durand-Gasselin, T. (2010). Visión futura del mejoramiento genético de la palma de aceite en Latinoamérica: varias resoluciones sobre cooperación en Colombia para mejoramiento del material de siembra. *Revista Palmas*, 31(especial): 155–70.

Durand-Gasselin, T., de Franqueville, H., Diabate, S., Cochard, B. and Adon, B. (2000). Assessing and utilising sources of resistance to *Fusarium* wilt in oil palm (*Elaeis guineensis* Jacq.) genetic resources. *In* 'International Symposium on oil Palm Genetic Resources and Utilization.' Rajanaidu, N. and Ariffin, D. (eds), Malaysian Palm Oil Board (MPOB), Kuala Lumpur, pp. U1–24.

Durand-Gasselin, T., Asmady, H., Flori, A., Jacquemard, J. C., Hayun Z., Breton, F., and de Franqueville, H. (2005). Possible sources of genetic resistance in oil palm (*Elaeis guineensis* Jacq.) to basal stem rot caused by *Ganoderma boninense* - Prospects for future breeding. *Mycopathologia*, 159(1): 90–100.

Durand-Gasselin, T., Corredor, J., Sanz, J. I., de Franqueville, H., and Amblard, P. (2009). Future vision of oil palm genetic improvement in Latin America. Several resolutions on cooperation in Colombia for planting material improvement. 'Challenges in sustainable oil palm development'. In *XVI Palm Oil International Conference*, 23.

Durand-Gasselin, T., Turnbull, N., de Franqueville, H., Breton, F., Syahputra, I. and Cochard, B. (2015). Findings and advances on *Ganoderma* in oil palm. XVIII Conferencia International Sobre Palma de Aceite, 22–25 de septiembre de 2015, Cartagena de Indias, Colombia.

Flood, J., Cooper, R. M. and Lees, P. E. (1989). An investigation of pathogenicity of four isolates of *Fusarium oxysporum* from South America, Africa and Malaysia to clonal oil palm. *Journal of Phytopathology*, 124: 80–8.

Flood, J., Hasan, Y., Turner, P. D. and O'Grady, E. B. (2000b). The spread of *Ganoderma* from infective sources in the field and its implication for management of the disease in oil palm. In: *Ganoderma Diseases of Perennial Crops*, Flood, J., Bridge, P. D. and Holderness, M. (ed.), CABI Publishing, Wallingford, U. K., pp. 101–12.

Flood, J. (2006). A Review of Fusarium Wilt of Oil Palm Caused by Fusarium oxysporum f. sp. Elaeidis. *Phytopathology*, 96: 660–2.

de Franqueville, H. (1984). Vascular wilt of the oil palm: relationship between nursery and field resistance. *Oléagineux*, 39: 513–18.

de Franqueville, H. and de Greef, W. (1988). Hereditary transmission of resistance to vascular wilt of the Oil Palm: facts and hypothesis. In: *Proc. of 1987 Int. O. P./P. O. Conference*. PORIM - Kuala Lumpur, pp. 118–29.

de Franqueville, H. and Renard, J. L. (1988). Oil palm wilt in replantings: study methods and determination of certain environmental factors on the expression of this disease. *Oléagineux*, 43: 149–57.

de Franqueville, H. and Diabaté, S. (1995). Oil palm vascular wilt in West Africa. Plantations, recherche, développement PRD Juillet - Août, 5–13.

de Franqueville, H. (1999). Résistance variétale. In « Les maladies des cultures pérennes tropicales », Dominique Mariau ed. CIRAD, collection repères, ISBN 2876143402.

de Franqueville, H. (2001). – Varietal resistance. In: *'Diseases of Tropical Trees Crops'*, Mariau Dominique (ed.), Thanikomoni, K. (trad.), Majithia, M. (trad.). CIRAD-Science Publishers, pp. 87–130. ISBN 1578081750; 2876144689.

de Franqueville, H., Asmady, H., Jacquemard, J. C., Hayun, Z. and Durand-Gasselin, T. (2001). Indications on sources of oil palm (*Elaeis guineensis* Jacq.) genetic resistance and susceptibility to *Ganoderma* sp., the cause of basal stem rot. In: *Proc. 2001 Int. Palm Oil Congr. - Agriculture*, Malaysian Palm Oil Board, Kuala Lumpur, p. 420.

de Franqueville, H. (2003). Review paper: Oil palm Bud Rot in Latin America. *Experimental Agriculture*, 39 : 225–40.

de Franqueville, H., Durand-Gasselin, T., Cochard, B., Nouy, B., Breton, F. and Dossa, J. (2011). *Fusarium* Disease of the Oil Palm (*Elaeis guineensis* Jacq.). Proceedings of the third MPOB-IOPRI international seminar integrated oil palm pests and diseases management Malaysia. Malaysian Palm Oil Board and Indonesian Oil Palm Research Institute. 268 p.

Idris, A. S., Khushairi, A., Ismail, S., and Ariffin, D. (2004). Selection for partial resistance in oil palm to *Ganoderma* basal stem rot. *Journal of Oil Palm Research*, 16: 12–18.

Jacquemard, J. C., Meunier, J. and Bonnot, F. (1981). Genetic study of the reproduction of an *Elaeis guineensis* oil palm cross. *Oléagineux*, 36(7): 343–52.

Jacquez, J. A., Simon, C. P. and Koopman, J. (1996). Core groups and the ROs for subgroups in heterogeneous sis and si models, in Epidemics models: their structure and relation to data, D. ed., Cambridge University Press, pp. 279–301.

Kinge, T. R. and Mih, A. M. (2011). *Ganoderma ryvardense* sp. nov. associated with basal stem rot (BSR) disease of oil palm in Cameroon. *Mycosphere*, 2(2): 179–88.

Kull, L. S., Vuong, T. D., Powers, K. S., Eskridge, K. M., Steadman, J. R., and Hartman, G. L. (2003). Evaluation of resistance screening methods for Sclerotinia stem rot of soybean and dry bean. *Plant Disease*, 87: 1471–6.

Louise, C., Poveda, R., Flori, A. and Amblard, P. (2013). Tolerancia al complejo de Pudrición del cogollo en el material Elaeis guineensis: caso del material C07 en el oriente ecuatoriano. *Palmas*, 34(1), spec: 126–34.

Louise, C., Poveda, R. and Amblard, P. (2015). Avances en el desarrollo de genotipos de palma de aceite resistantes al complejo pudricion del cogollo. XVIII Conferencia International Sobre Palma de Aceite, 22–25 de septiembre de 2015, Cartagena de Indias, Colombia.

Louise, C. (2016). Resistencia a la Pudrición del cogollo en el material *Guineensis*. *Palmas*, 37(Especial Tomo I): 183–9.

Maora, S. J., Liew, E. C. Y. and Guest, D. I. (2017). Limited morphological physiological and genetic diversity of *Phytophthora palmivora* from cocoa in Papua New Guinea., *Plant Pathology*, 66(1): 124–30.

McDonald, B. A. and Linde, C. (2002). Pathogen population genetics, evolutionary potential and durable resistance. *Annual Revue Phytopathology*, 40: 349–79.

Mepsted, R., Flood, J. and Cooper, R. M. (1995). *Fusarium* wilt of oil palm II. Stunting as a mechanism to reduce water stress. *Physiological and Molecular Plant Pathology*, 46: 373–87.

Mercière, M., Laybats, A., Carasco-Lacombe, C., Tan, J. S., Klopp, C., Durand-Gasselin, T., Alwee, S. S. R. S., Camus-KuLandeaivelu, L. and Breton, F. (2015). Identification and development of new polymorphic microsatellite markers using genome assembly for *Ganoderma boninense* causal agent of oil palm basal stem rot disease., *Mycological Progress*, 14(11): e103 (11 pp.)

Meunier, J., Renard, J. L. and Quillec (1979). Heredity of resistance to *Fusarium* wilt in the oil palm *E. guineensis* Jacq. *Oléagineux*, 34: 555–61.

Mouyna, I., Renard, J. L. and Brygoo, Y. (1996). DNA polymorphism among *Fusarium oxysporum* f. sp. *elaeidis* populations from oil palm, using a repeated and dispersed sequence 'Palm'. *Current Genetics*, 30: 174–80.

Ngando, E. G., Etia, C. and Koona P. (2012). Breeding for *Fusarium wilt* tolerance: an overview of research programs and seed production potentialities in Cameroon. International seminar on breeding for oil palm resistance, 21–24 September 2012 Hotel AR Salitre, Bogota Colombia.

Pilotti, C. A., Sanderson, F. R. and Aitken, E. A. B. (2003). Genetic structure of a population of *Ganoderma boninense* on oil palm. *Plant Pathology*, 52(4): 455–63.

Prendergast, A. G. (1963). A method of testing oil palm progenies at the nursery stage for resistance to vascular wilt disease caused by *Fusarium oxysporum*. *Journal of West African Institute for Oil Palm Research*, 4: 156–75.

Purba, R., Setiawatï, U., Susanto, A., Rahmaningsih, M., Yenni, Y., Rahmadi, H. Y. and Stephen, P. C. Nelson (2012). Indonesia's experience of developing *Ganoderma* tolerant/resistant oil palm planting material. International seminar on breeding for oil palm resistance, 21–24 September Hotel AR Salitre, Bogota Colombia.

Rahmaningsih, M., Setiawati, U., Breton, F., Sore, R., Nelson, S. and Caligari, P. D. S. (2010). Genetic Combine Ability (GCA) for *Ganoderma* Tolerance via Nursery Screening Test. International oil palm conference IOPC 2010 1–3 June Yogyakarta Indonesia.

Rajagopalan, K., Aderungboye, F. O., Obasola, C. O. and Eme, A. (1978). Evaluation of oil palm progenies for reaction to the vascular wilt disease. *Journal of the Nigerian Institute for Oil Palm Research*, 5(20): 87–95.

Renard, J. L., Gascon, J. P. and Bachy, A. (1972). Research on vascular wilt disease of the oil palm. *Oléagineux*, 27 : 581–91.

Renard, J. L., Noiret, J. M. and Meunier, J. (1980). Sources et gammes de résistance a la fusariose chez les palmiers à huile *Elaeis guineensis* et *Elaeis melanococca*. *Oléagineux*, 35 : 387–93.

Renard, J. L. and Quillec, G. (1984). Les maladies graves du palmier à huile en Afrique et en Amérique du Sud. *Oléagineux*, 39(2): 57–67.

Renard, J. L. and de Franqueville, H. (1991). Effectiveness of crop techniques in the integrated control of oil palm vascular wilt. *Oléagineux*, 46: 255–65.

Robinson, R. A. (2006). Return to resistance [electronic resource]: breeding crops to reduce pesticide dependence/by Raoul A. Robinson. - 3rd. ed., rev. 2006. ISBN 978-0-9783634-2-0.

Rosenquist, E. A. (1985). The genetic base of oil palm breeding populations. In 'Workshop on oil palm germplasm and utilisation', PORIM, (ed.), 10, Palm Oil Research Institute of Malaysia, Selangor, Malaysia, pp. 27–56.

Roux, F., Voisin, D., Badet, T., Bagalé, C., Barlet, X., Huard-Chauveau, C., Roby, D. and Raffaele, S. (2014). *Molecular Plant Pathology*, 15(5): 427–32.

Sarria, G. A., Torres, G. A., Aya, H. A., Ariza, J. G., Rodríguez, J., Vélez, D. C., Varón F., and Martínez, G. (2008). *Phytophthora* sp. es el responsable de las lesiones iniciales de la pudrición del cogollo (PC) en la palma de aceite en Colombia. *Palmas*, 29 (Special Issue): 31–41.

Sarria, G. A., Martinez, G., Varon, F. et al. (2016). Histopathological studies of the process of *Phytophthora palmivora* infection in oil palm. *European Journal of Plant Pathology*, 145: 39.

Setiawati, U., Rahmaningsih, M., Breton, F., Sore, R., Nelson, S. and Caligari, P. D. S. (2010). Screening of Sumatra Bioscience Progenies for *Ganoderma* Partial Resistance in Field and Nursery Trials. Proceeding of the International oil palm conference IOPC 2010 1–3 JuneYogyakarta Indonesia.

Tisné, S., Denis, M., Cros, D., Pomiès, V., Riou, V., Syahputra, I., Omoré, A., Durand-Gasselin, T., Bouvet, J.-M. and Cochard, B. (2015). Mixed model approach for IBD-based QTL mapping in a complex oil palm pedigree. *BMC Genomics*, 16(1): 1.

Torres, G. A., Sarria, G. A., Varon, F., Coffey, M. D., Elliott, M. L. and Martinez, G. (2010). First report of Bud Rot caused by *Phytophthora palmivora* on African oil palm in Colombia. *Plant Disease*, 94: 1163.

Torres, G. A., Sarria, G. A., Martinez, G., Varon, F., Drenth, A. and Guest, D. I. (2016). Bud Rot caused by Phytophthora palmivora: A destructive emerging disease of oil palm. Phytopathology, 106: 320–9.

Turner, P. D. (1965a). The incidence of *Ganoderma* disease of oil palms in Malaya and its relation to previous crop. *Annals of Applied Biology*, 55: 417–23.

Turner, P. D. (1981). *Oil palm diseases and disorders*. Oxford University Press, Kuala Lumpur.

Van de Lande, H. L. (1986). Diseases of fatal character to oil palm in Suriname and North-Brazil *Surinaamse Landebouw.*, 34(1–3): 15–32.

Van der Plank, J. E. (1963). *Plant Diseases: Epidemics and Control.* Academic Press, New York and London, 349.

Van der Plank, J. E. (1968). *Disease resistance in plants.* New York, USA, Academic Press, 223.

Zambrano, R. J. E. (2004). Híbridos interespecíficos *Elaeis oleifera* HBK. x *Elaeis guineenses* Jacq.: una alternativa de renovación para la zona oriental de Colombia. *Palmas*, 25(Special, tome II) : 339–49.

Part 2

Nutritional and sensory quality

Chapter 7

Bioactive compounds in oil palm

Ravigadevi Sambanthamurthi, Ng Mei Han and Choo Yuen May, Malaysian Palm Oil Board, Malaysia

1 Introduction

Bioactive compounds in plants are defined as substances that elicit pharmacological or toxicological effects in humans and animals. They are typically secondary metabolites produced by the plant in addition to the primary metabolites such as amino acids, proteins, carbohydrates and lipids required for growth and development. Unlike primary metabolites which participate in nutrition and essential metabolic processes within the plant, secondary metabolites are considered non-essential for sustaining life but crucial for the survival of the producing organism (Hadacek, 2002). They influence ecological interactions between plants and their environment and are often involved in defence mechanisms to protect the plant against biotic and abiotic stresses. They are ubiquitous in plants on account of their important protective biological functions. The oil palm being a tropical crop is exposed to high temperature and intense sunlight that result in the generation of reactive oxygen species that elicit oxidative stress responses. The palm has thus evolved cellular and metabolic processes and produces bioactive compounds as protective measures against the ravages of the weather as well as the surrounding ecosystem. While plants are the prime source of bioactive compounds for drug discovery and development of functional foods and nutraceuticals, supply of these compounds is often curtailed by the low content and limitation in supply. The tremendous productivity of the oil palm and the abundance of cultivated palms on a global scale, however, provide an abundant source of lipid- and water-soluble bioactives. A major environmental advantage

http://dx.doi.org/10.19103/AS.2017.0018.20

of palm bioactives is that they are readily available products of palm oil milling and refining processes. This can play a key role in value addition and improved sustainability of the oil palm. The palm fruit which represents one of only a few oleaginous fruits is poised to play an increasingly substantiated role in nutrition and health, not only through the delivery of important dietary fats but also its distinct bioactive compounds.

2 Lipid-soluble bioactives

The oil palm mesocarp yields an orangish red oil commonly known as crude palm oil (CPO). CPO is bestowed with a bouquet of minor components constituting 1% of its total content (Goh et al., 1985; Choo et al. 2000, 2004a,b,c; Choo, 2000). Although small in amount compared to the major lipid components, these minor components are of nutritional value and confer important health benefits.

The major fat-soluble phytonutrients and/or bioactives in palm oil include carotenes, tocols (tocopherols and tocotrienols), squalene, sterols, phospholipids (better known as lecithin) and coenzyme Q_{10} (also known as ubiquinone) (Goh et al. 1985; Hamid et al. 1995; Choo et al. 2000, 2004a,b,c; Choo 2000; Lau et al. 2006a). The compositions of these phytonutrients, together with other minor components of CPO, are shown in Table 1.

2.1 Carotenes

Palm oil gets its orangish red colour from carotenoids. It is the richest source of natural plant carotenoids in retinol (provitamin A) equivalent among all the edible oils and fats (Hamid et al., 1995; Choo, 2000; Latip et al., 2000) – contains 15–300 times more retinol (provitamin A) equivalent than carrot, green leafy vegetables and tomato. The carotenoids have a large content of β-carotene, which also occurs in other vegetable oils, such as olive, corn, sunflower and soybean oil (Choo et al., 1999; Cabrini et al., 2001; Ranalli and Demattia, 1997; Ranalli et al., 1997).

Table 1 Concentrations of minor components in CPO

Component	Concentration (ppm)
Carotenoids	500–700
Tocopherols and tocotrienols	600–1000
Sterols	326–527
Phospholipids	5–130
Triterpene alcohols	40–80#
Methyl sterols	40–80#
Squalene	200–500
Aliphatic alcohols	100–200
Ubiquinones	10–80
Aliphatic hydrocarbons	50

Estimated.
Source: Goh et al. (1985).

The carotenoids in palm oil comprise carotenes and xanthophylls (Ranalli et al., 1999). While the carotenes are hydrocarbons, the xanthophylls contain at least one oxygen moiety. The main carotenes α- and β-carotene and their contents, as well as the contents of other carotenes, differ in CPO and the residual oil obtained from palm pressed fibre (palm pressed fibre oil, PFO), with the largest difference in lycopene (Goh et al., 1985; Choo et al., 1996; Tay et al., 2001). The content of lycopene is much higher in PFO than CPO. Table 2 shows the composition of the individual carotenes in CPO and PFO as reported by various researchers, while their structures are depicted in Fig. 1 (Tan et al., 1986; Yap et al., 1991; Sundram et al., 1989; Tay and Choo, 2000a).

Carotene content increases with fruit ripening and is highest in ripe palm mesocarp oil (Sambanthamurthi et al., 2000; Tay and Choo 2000b; Sundram et al., 2003). Much less is known about the chemistry of xanthophylls in palm oil, although a few compounds such as dehydroretinal, ξ-carotene-dione and β-carotene-5,6-epoxide isomers have been reported (Tan, 1987).

It has been documented that palm carotenes exhibit beneficial properties, such as anti-cancer and antioxidative effects (Yap et al., 1991; Ashfaq et al., 2001; Murakoshi). Carotenes are known to be potent antioxidants and anti-cancer agents (Britton, 1995; Spiller, 2004; Nesaretnam et al., 2002; Sundram et al., 1989; Tay and Choo, 2000a). Continual assessment of palm carotenes on breast cancer cells revealed their ability to inhibit the growth of new cancer cells (Nesaretnam, 2000, 2007a). Studies have shown that mice fed with palm carotenes were protected against DNA damage in bone marrow and had better survival following X-ray irradiation (Umegaki et al., 1997). α- and β-carotenes have been proven to be provitamin A, and protect the cells against free radical damage as well as xerophthalmia, or night blindness, in humans (Umegaki et al., 1997; Dimascio et al., 1989; Ziegler, 1989; Ziegler et al., 1996a,b; Sundram et al., 1989; Tay and Choo, 2000a).

Table 2 Individual carotenes as percentage of total carotenes in CPO and PFO

Carotene	CPO (%)	PFO (%)
Phytoene	1.27	11.87
Phytofluene	0.06	0.40
β-Carotene	56.02	30.95
α-Carotene	35.06	19.45
cis-α-Carotene	2.49	1.17
ξ-Carotene	0.69	7.56
γ-Carotene	0.33	2.70
δ-Carotene	0.83	6.94
Neurosporene	0.29	3.38
β-Zeacarotene	0.74	0.37
α-Zeacarotene	0.23	Trace
Lycopene	1.30	14.13

Source: Choo et al. (1996).

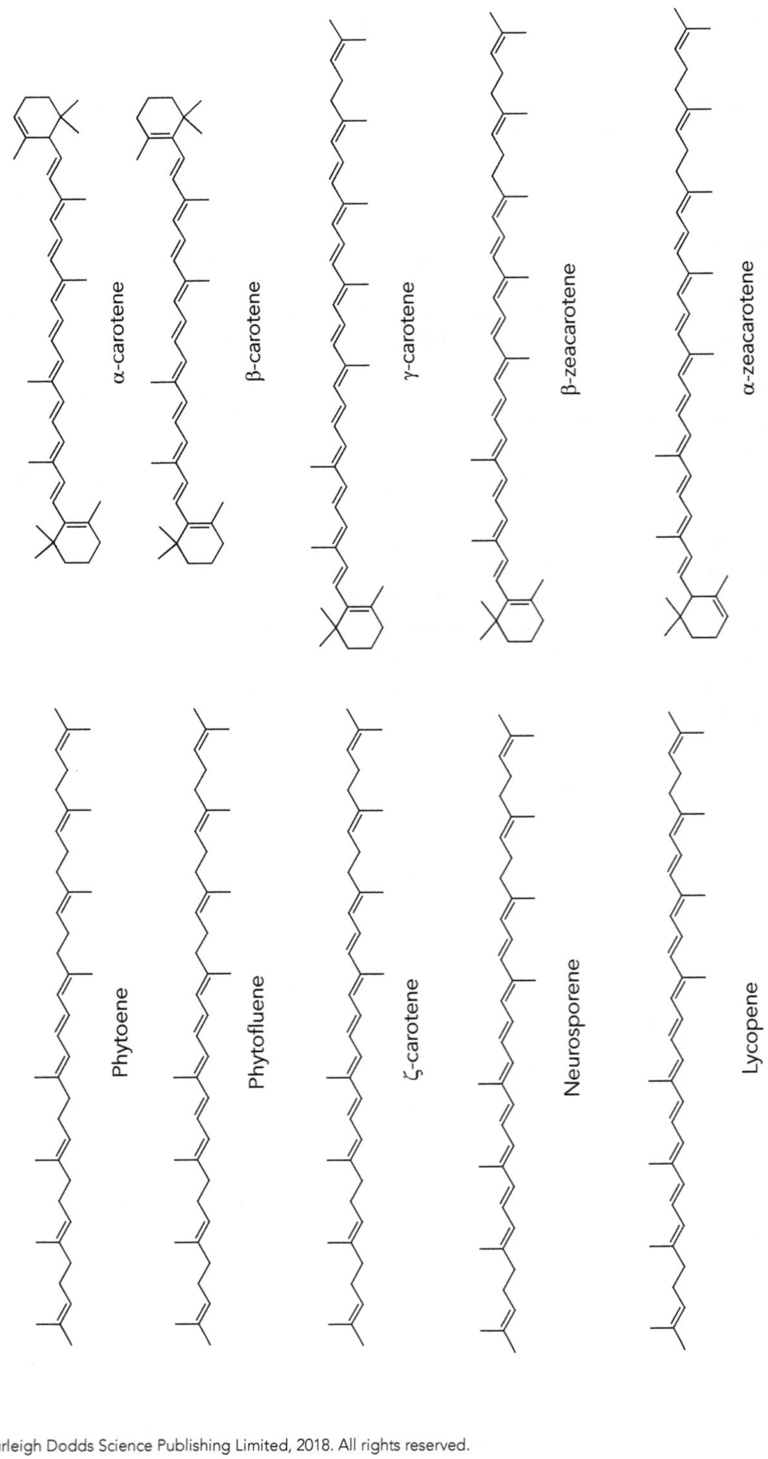

Figure 1 Molecular structures of carotenes in palm oil.

Rabbits fed diets enriched in palm carotenes were found to have different carotenes in varying amounts in their plasma and organs, mainly the liver and adrenal glands. Most of the supplemented α- and β-carotenes were metabolized to retinal and retinyl esters, and mainly stored in the liver and, to a lesser extent, pancreas. Small amounts of other carotenes – phytoene, phytofluene and ξ-carotene – were present in the rabbit plasma and organs. The contents of retinal and retinyl esters in the liver of rabbits fed a diet supplemented with palm carotenes were 3× and 30× times higher, respectively, than in the rabbit groups not fed palm carotenes, suggesting that palm carotenes are converted to retinal and retinyl esters *in vivo*.

2.2 Tocopherols and tocotrienols

The generic name 'tocopherol' was derived from the Greek words, 'tokos', which means *offspring* and 'pherein' meaning to bear, with 'ol' appended to indicate the phenolic hydroxy group (Tay and Choo, 1999). Tocopherol was first isolated and named by Evans et al. (1936). The trivial name 'tocotrienol' was introduced by Bunyan et al. (1961) to represent vitamin E homologues with an unsaturated isoprenoid side chain.

Natural tocols comprise eight individual components which are classified into two homologous series: tocopherols and tocotrienols. Tocotrienols differ from tocopherols in having three unsaturated bonds at the side chain.

The tocols in palm oil consist of both tocopherols and tocotrienols. Oil palm is unique as it contains natural tocotrienols, which are rarely found in other vegetable oils, the exception being rice bran oil (Table 3) (Tay and Choo, 2000b; Bunyan et al., 1961). There are 600–1000 ppm and 2000–4000 ppm tocols, in CPO and PFO, respectively (Goh et al., 1985; Tay et al., 2001; Sundram et al., 1989; Tay and Choo, 2000a; Qureshi et al., 2000; Choo et al., 2003a, 2005a; Lau et al., 2006a,b, 2008; Ng et al., 2004a).

Table 3 Tocol contents in vegetable oils (ppm)

	Tocopherol				Tocotrienol				
Vegetable oil	α-T	β-T	γ-T	δ-T	α-T_3	β-T_3	γ-T_3	δ-T_3	Total
Coconut	5	–	–	6	5	1	19	–	36
Corn	112	50	602	18	–	–	–	–	782
Cottonseed	389	–	387	–	–	–	–	–	776
Olive	51	–	–	–	–	–	–	–	51
Palm	152	–	–	–	205	–	439	94	890
Peanut	130	–	216	21	–	–	–	–	367
Rice bran	324	18	53	–	236	–	349	–	980
Safflower	387	–	174	240	–	–	–	–	801
Sesame	12	6	244	32	–	–	–	–	294
Soyabean	101	–	593	264	–	–	–	–	958
Sunflower	487	–	51	8	–	–	–	–	546

Source: Ng et al. (2004a).

Table 4 Individual tocols as percentage of total tocols in CPO and PFO

Tocol	CPO (%)	FPO (%)
α-Tocopherol	22	61.1
α-Tocotrienol	20	15.4
γ-Tocotrienol	46	18.0
δ-Tocotrienol	12	5.5

Source: Choo et al. (1996).

All tocols have the same chroman structure and a side chain. However, as mentioned earlier, the tocopherol side chain is saturated whereas tocotrienols have three unsaturated bonds in the side chain. The difference between the individual tocopherols and tocotrienols in palm oil is in the positions they are methylated. Of the eight tocol isomers, four are reportedly present in palm oil – α-tocopherol (α-T), α-tocotrienol (α-T_3), γ-tocotrienol (γ-T_3) and δ-tocotrienol (δ-T_3) (Tan et al., 1986; Ng et al., 2004a,b), in varying amounts with γ-T_3 being the dominant (Sundram et al., 1989; Tay and Choo, 2000a; Ng et al., 2004a; Choo et al., 2005b; Gapor et al., 1981), while α-T is more abundant in residual oil obtained from palm pressed fibre (PFO) (Tay et al., 2001). Table 4 shows the composition of tocols in CPO and PFO with their molecular structures in Fig. 2. Choo et al. (2004a) detected tocopherols in the developing fruit mesocarp as early as four weeks after anthesis.

The vitamin E activity in palm oil is largely mediated through γ- and δ-tocotrienols. Many studies have found that palm vitamin E possesses antioxidant properties in scavenging free radicals which have been linked to cancer risk (Ong and Goh, 2002; Spiller, 2004; Tomeo et al., 1995; Yurttas and Addis, 2001; Yu et al., 2008; Maniam et al., 2007; Nesaretnam et al., 2007a; Das et al., 2005). It was also reported that tocols in palm oil inhibit human platelet coalescence and improve glycaemic control *in vivo* (Holub et al., 1989; Hornstra, 1988; Wan Nazaimoon and Khalid, 2002). In addition, palm tocols were also found to exhibit anti-cancer properties (Ashfaq et al., 2001; Nesaretnam, 2000; Nesaretnam et al., 1992, 1995, 1998, 2007b, 2008; Chan and Chan, 2003; Goh et al., 1994; Guthrie et al., 1993; Sylvester and Sumit, 2003; Wong et al., 2009; Li et al., 2010).

α-Tocotrienol was found to inhibit glutamate-induced activation of phospholipase A_2 and conferred neuroprotection (Khanna et al., 2005, 2010). Another isoform, γ-tocotrienol, has also been found to exhibit neuroprotection (Mishima et al., 2003). δ-Tocotrienol was able to protect mouse and human hematopoietic progenitors from gamma irradiation (Li et al., 2010). γ-Tocotrienol, on the other hand, inhibited pancreatic tumours and sensitized them to gemcitabine treatment by modulating the inflammatory microenvironment (Kunnumakkara et al., 2010). Palm tocotrienols have been found to exhibit anti-angiogenic potential *in vitro* (Miyazawa et al., 2004). δ-Tocotrienol is also required for normal vitamin D metabolism in female rats (Norazlina et al., 2005).

2.3 Sterols

Sterols are found in most plants and are easily recognizable by their common steroid structure. The phytosterols in palm oil differ only in the side chains from Carbon 12 of the main steroid structure. There are ~1050 ppm sterols in olive oil with the campesterol: stigmasterol ratio being 2.2:5.7 (Cabrini et al., 2001; Ranalli and Demattia, 1997; Ranalli

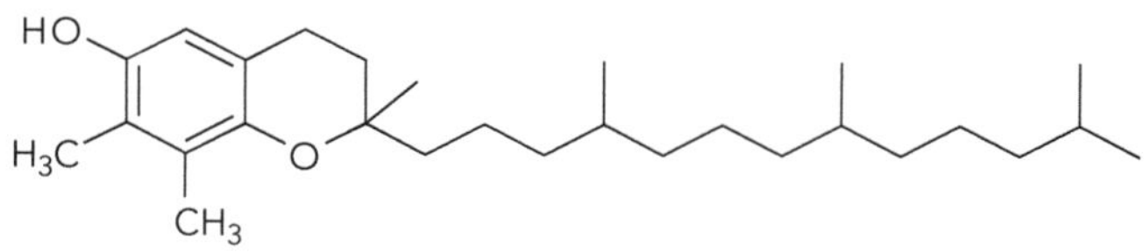

γ-tocopherol

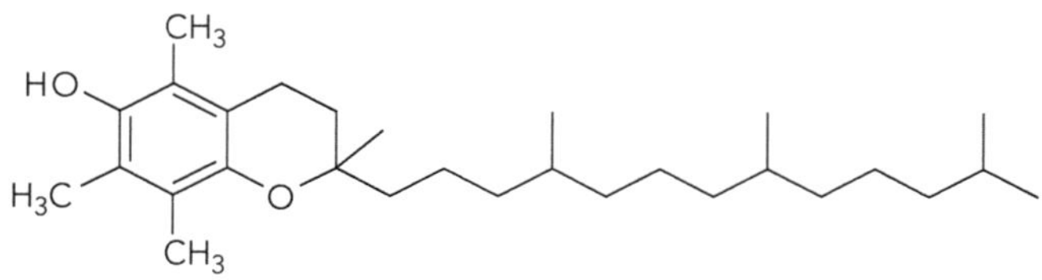

α-tocopherol

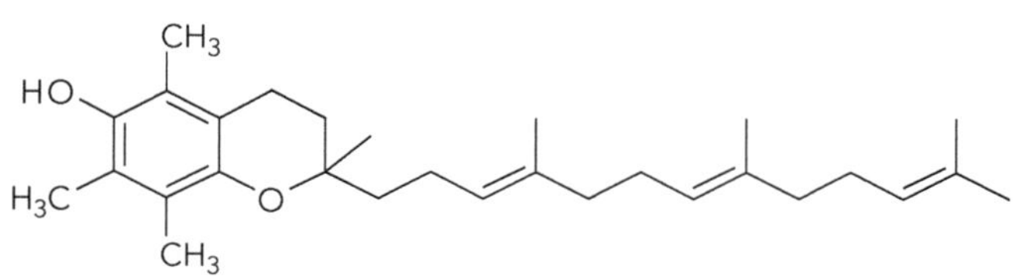

α-tocotrienol

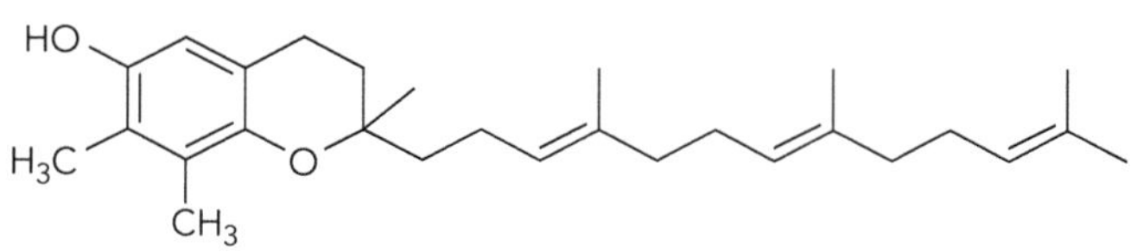

γ-tocotrienol

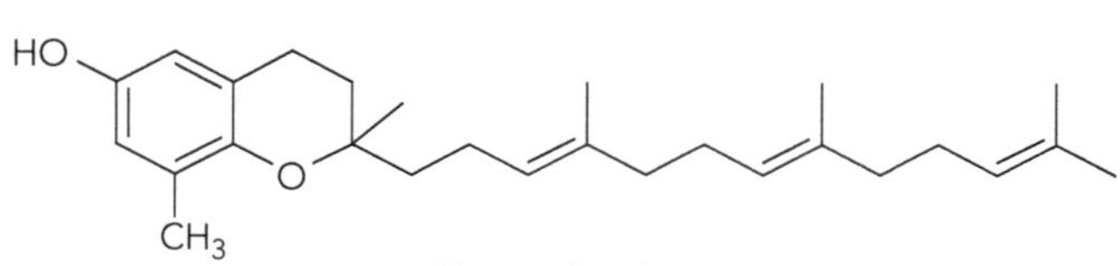

δ-tocotrienol

Figure 2 Molecular structures of tocols in palm oil.

et al., 1997). In CPO, sterols are present at 250–620 ppm, but in higher content (4000–6500 ppm) in PFO (Goh et al., 1985; Choo et al., 2003b, 2004b; Tay et al., 2001; Ng et al., 2004b).

Several investigations have been carried out on the occurrence of sterols in palm oil. The main sterols in CPO and PFO are β-sitosterol, campesterol, stigmasterol and cholesterol, together with traces of avenasterol and brassicasterol (Goh et al., 1985; Choo et al., 2003a,b; Lau et al., 2005). As the amount of cholesterol in palm oil is very low, it can be considered 'cholesterol-free'. The composition of the individual sterols in palm oil is depicted in Table 5 and their molecular structures are given in Fig. 3.

Sterols are converted to steroid derivatives such as β-sitosterol, which have reportedly beneficial hypocholesterolemic effect (Sundram et al., 1989; Tay and Choo, 2000a; Farquhar, 1996; Hedtmann et al., 1988; Yankah and Jones, 2011). However, most of the sterols in CPO are removed during the refining process.

Table 5 Individual sterols as percentage of total sterols in CPO and PFO

	Content (%)	
Sterol	CPO	PFO
β-Sitosterol	60	59
Campesterol	13	22
Stigmasterol	24	18
Cholesterol	3	1

Source: Choo et al. (1996).

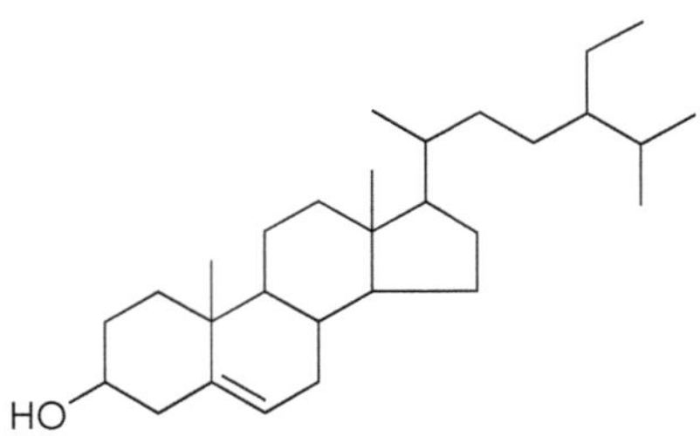

β-sitosterol

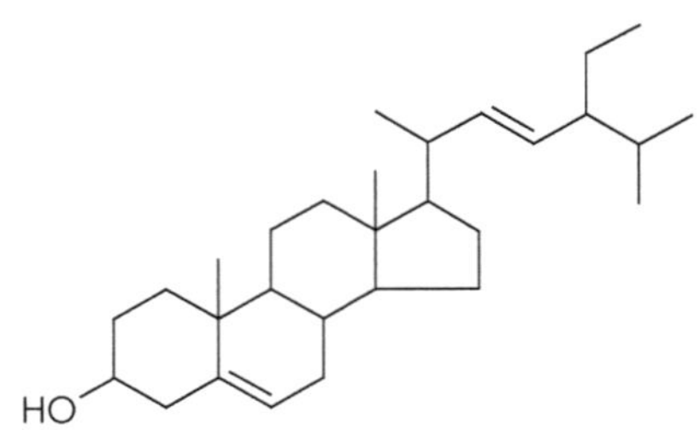

Stigmasterol

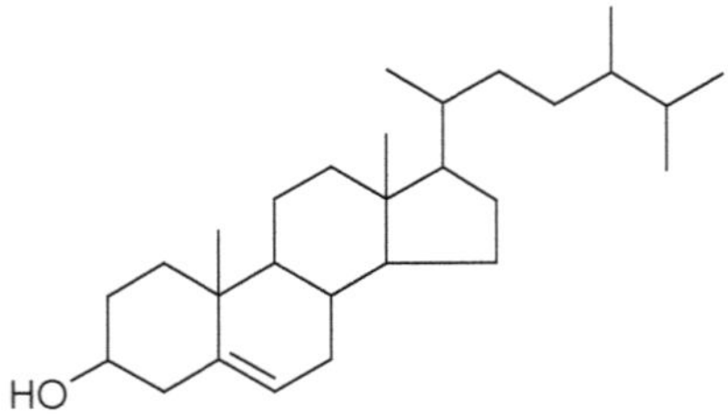

Campesterol

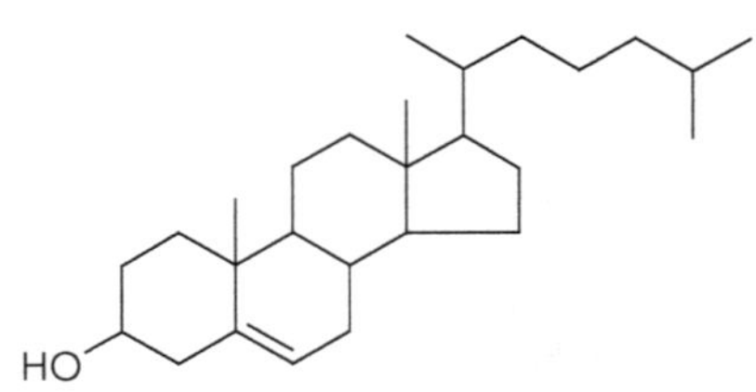

Cholesterol

Figure 3 Molecular structures of major sterols in palm oil.

Plant sterols are similar in structure to dietary cholesterol but are not absorbed by the human gastrointestinal tract. They can thus inhibit the absorption of dietary cholesterol and enable its removal from the body. Plant sterols exert a positive effect on blood lipid profiles by significantly reducing total and LDL cholesterol.

2.4 Ubiquinones

Coenzyme Q compounds have a quinone structure with a side chain of varying isoprene units as shown in Fig. 4. The difference in chain length distinguishes one coenzyme Q from another. Coenzyme Q_{10}, for instance, has the basic quinone ring with 10 isoprene units in its side chain. Coenzyme Q_{10} is a natural compound in the human body where it is a potent antioxidant combating the ageing process (Sundram et al., 1989; Tay and Choo, 2000a; Beyer, 1989; Kagan and Quinn, 2001; Lenaz, 1985b; Lipshutz et al., 2002; Mellons and Tappel, 1996; Kueper et al., 2008; Ooe and Fujii, 2008; Takada et al., 2009; Blatt et al., 1999). With age, the biogeneration of coenzyme Q_{10} becomes less active. It is, therefore, increasingly taken as a health supplement. About 10–80 ppm coenzyme Q_{10} and 5 ppm coenzyme Q_9 were detected in CPO (Lau et al., 2006a; Sundram et al., 1989; Tay and Choo, 2000a; Miettinen and Vanhanen, 1994; Tan and Kuntom, 1994; Hazura et al., 1995; Ng et al., 2009). However, only coenzyme Q_{10} has been reported in red palm oil (18–25 ppm) (Tan, 1987). Ng et al. (2006) reported the presence of coenzyme Q_9 in the developing palm mesocarp as early as four weeks after anthesis. However, its concentration slowly diminishes as the fruit matures. As with other antioxidants, coenzyme Q compounds are light and heat sensitive. It was reported that coenzyme Q_{10} helps in angina relief (Tran et al., 2001). Angina is the pain or discomfort experienced when blood flow to the heart is restricted. Coenzyme Q_{10} has also been attributed for boosting the immune system as well as lowering high blood pressure and, thus, reducing the risk of heart attack (Sundram et al., 1989; Tay and Choo, 2000a; Kagan and Quinn, 2001; Lipshutz et al., 2002; Lenaz, 1985b; Wyman et al., 2010; Sinatra, 1998; Yamamura, 1985; Nafeeza and Kang, 2005).

Recent studies found that coenzyme Q is able to slow the functional decline in neurodegenerative diseases such as Parkinson's and Huntington's disease (Ferrante et al., 2002; Shults et al., 2002; Beal, 1999). Supplementation of coenzyme Q with vitamin E has been shown to reduce the circulating markers of inflammation (Wang et al., 2004).

Coenzyme Q can be extracted from natural sources as well as produced synthetically (Lipshutz et al., 2002). The difference in activity, if any, between natural and synthetic coenzyme Q is yet to be studied.

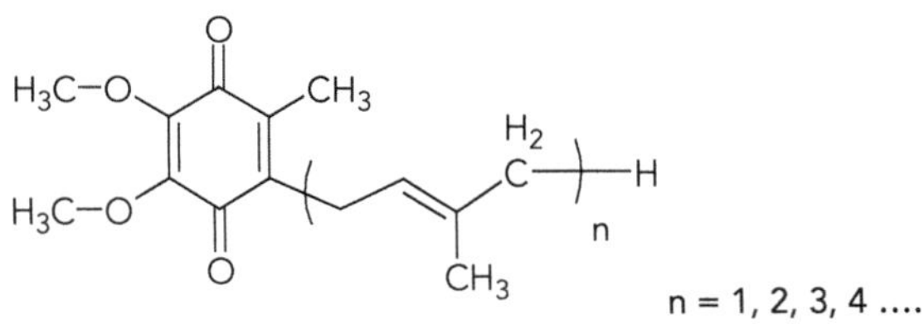

Figure 4 Molecular structure of coenzyme Q.

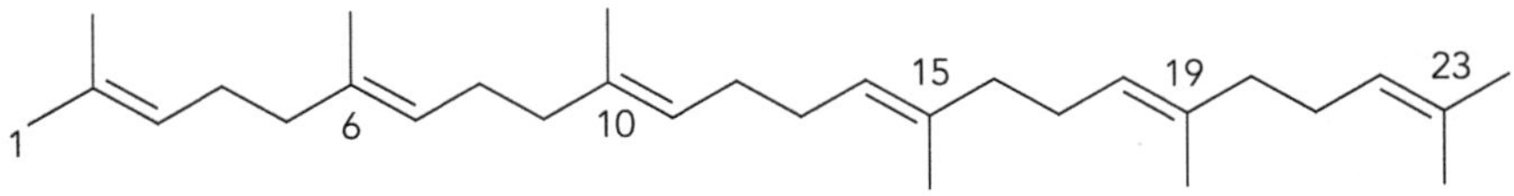

Figure 5 Molecular structure of squalene.

2.5 Squalene

Squalene was first discovered in shark liver oil and is also found in many marine organisms (Lau et al., 2003; Ackman et al., 2000; Deprez et al., 2000). It is known to be an intermediate in the cholesterol biosynthesis (Tay and Choo, 2000b; Kayama and Mankura, 1998; Barrowman et al., 1989). It is also found in CPO (Tan et al., 1986; Lau et al., 2005; Miettinen and Vanhanen, 1994; Berger, 2005; Tan and Kuntom, 1993). The molecular structure of squalene is shown in Fig. 5.

There are 200–500 ppm squalene in CPO, and even higher in PFO (500–1000 ppm) (Lau et al., 2005; Gapor and Hazrina, 2000; Gapor et al., 2000b, 2009; Gapor, 2002; Chua et al., 2007). Squalene is also extracted from palm fatty acid distillate (PFAD) (Gapor and Hazrina, 2000; Gapor et al., 2000, 2009; Gapor, 2002; Chua et al., 2007). Squalene possesses slight antioxidant activities and is able to retard the degradation of unsaturated fatty acids at high temperature (Rao and Achaya, 1970). It is also beneficial to health in helping body cells renew themselves and in boosting the immune system (Carlson et al., 2000).

Squalene is also an excellent emollient, or conditioning agent, in cosmetics (Gasparaoli et al., 1998; Kaiya, 1990). For pharmaceutical applications, squalene is converted to squalane as the saturated derivative has greater oxidative stability, and is better as a carrier for lipid-soluble drugs. It is also used as bactericide, intermediate for the manufacture of organic chemicals, rubber chemicals, aromatics and surfactants.

Studies have shown that squalene has some effect in chemoprevention of cancer (Rao et al., 1998; Smith, 2000). In recent years, squalene has been used in the preparation of vaccine against flu (Beck et al., 2010).

Besides squalene, there are numerous other shorter aliphatic and aromatic hydrocarbons in palm oil. Some of them are the natural constituents of palm oil, but some may be oxidation, or degradation, products.

2.6 Phospholipids and glycolipids

The chemistry of palm oil is complex as it contains many compounds with different behaviours. The more commonly known minor components are divided into non-polar, for example, carotenes, tocols and sterols, and polar types, for example, phospholipids and glycolipids. PFO has relatively high amounts of phospholipids (Choo et al., 2002a,b, 2004b).

Phospholipids in palm oil have gained considerable attention due to the suspected deleterious effect of phosphorus on oil quality. Most of the phosphorus in palm oil is present as inorganic rather than as organic phospholipids. There are 20–80 ppm phospholipids in CPO. Choo et al. (2004b) isolated and identified several of them – phosphatidylcholine (better known as lecithin), phosphatidylinositol, phosphatidylamine and phosphatidylethanolamine. Table 6 shows the phospholipids present in CPO. Phospholipids in palm oil were reported earlier by Goh (1984) and Goh et al. (1985).

Table 6 Individual phospholipids as percentage of total phospholipids in CPO

Phospholipid	Percentage (%)
Phosphatidylcholine	36
Phosphatidylethanolamine	24
Phosphatidylinositol	22
Phosphatidylglycerol	9
Diphosphatidylglycerol	4
Phosphatidic acid	3
Lysophosphatidylethanolamine	2
Phosphatidylserine	Trace
Lysophosphatidylcholine	Trace

Source: Goh (1984).

Choo et al. (2004b) also reported ~46 800 ppm phospholipids in the ethanol extract of PFO with the major ones being phosphatidylcholine, phosphatidylethanolamine, phosphatidylglycerol and phosphatidic acids.

Preliminary studies indicate 1000–3000 ppm glycolipids in CPO (Kulkarni, 1991a,b). They include monogalactosyl diacylglycerols, digalactosyl diglycerides and esterified steryl glycerides (Lau et al., 2006a, 2008; Kulkarni, 1991a,b). Both phospholipids and glycolipids are removed during the refining of CPO through degumming, washing, phosphoric acid treatment and absorption by clays or earths.

Both phospholipids and glycolipids constitute important components of cellular membranes, which possess unique structures with both lipophilic and hydrophilic functionalities. Palm phospholipids have been found to act synergistically with tocols to exhibit powerful antioxidative effects (Choo et al., 2009). In addition, phospholipids are essential for the biosynthesis of acylglycerols in plants (Dybing and Craig, 1969; Tremolieres and Lepage, 1971).

3 Water-soluble bioactives

3.1 Phenolics

Phenolic compounds are a large class of plant secondary metabolites, showing a diversity of structures. All phenolic substances have in common an aromatic ring with one or more hydroxyl substituents. Several thousands of phenolic compounds have been classified based on their basic chemical configuration, for example, number of phenol rings, and into various subclasses based on specific substitutions or functional derivatives in the basic structure including glycosylation, esterification and polymerization. Phenolic compounds thus range from those with relatively simple structures, for example phenolic acids, through polyphenols such as flavonoids that have the characteristic C6–C3–C6 backbone structure (structural variations in the three rings of flavonoids characterize different types of flavonoids) to polymeric compounds based on these different classes (Harborne, 1998). Phenolic acids can be further divided into two main types, benzoic acid and cinnamic acid derivatives with C1–C6 and C3–C6 backbones respectively. The majority of plant phenolics exist as glycosides with different sugars and acylated sugars at various positions of the phenolic structure. The degree of hydroxylation, the methylation pattern of the aromatic rings and the glycosylation pattern all influence the physical and physiological properties of phenolics.

Phenolic compounds are important for the quality of plant-based foods: they are responsible for the colour of fruits, juices and wines; are substrates for enzymatic browning; and are also involved in flavour properties. Phenolic-rich diets have been linked to numerous health benefits. The phenolic ring and hydroxyl substituents of these compounds confer antioxidative properties, as they are able to capture free radicals by donating hydrogen atoms or electrons. However, the bioactivity of phenolics extends well beyond the modulation of oxidative stress and may involve their interaction with cellular signalling pathways and associated machinery that regulate cell activities under both normal and pathological conditions. Their biological activities may also be associated with their capacity to chelate metals, inhibit lipoxygenases, inhibit nitrosation and modulate certain cellular enzyme activities as well as endocrine function. For example, soya isoflavones affect endocrine properties by interacting with oestrogen receptors (Kuiper et al., 1998). Nitric oxide (NO) bioavailability influences insulin-stimulated glucose uptake and vascular tone (Grassi et al., 2005). There is scientific evidence that phenolics exert significant vascular protection because of their antioxidant properties and increased NO bioavailability.

Whilst much attention has been paid to lipid-soluble bioactives of oil palm, the description of water-soluble bioactives, namely palm phenolics, has been relatively recent. Sambanthamurthi et al. (2008,2011a) described the presence of phenolics with potent antioxidant and bioactive function in the vegetation liquor of oil palm milling. Being water-soluble, these phenolic compounds enter the aqueous waste stream and are discharged as palm oil mill effluent (POME). Sambanthamurthi et al. (2008, 2011a) developed a process for the recovery of phenolic-enriched extract, which they called oil palm phenolics (OPP) by diverting the aqueous stream before it is discarded as waste. MS-coupled HPLC and NMR confirmed the presence of several phenolic acids. OPP contains *p*-hydroxybenzoic acid and three isomers of caffeoylshikimic acid as major phenolic compounds that accord a unique signature profile to the extract (Sambanthamurthi et al. 2011a). Figure 6 shows the HPLC profile of OPP. It also contains protocatechuic (PCA), caffeic, vanillic, ferulic and other phenolic acids in addition to soluble fibres, sugars and minerals. Table 7 shows the major phenolic components in OPP. An important non-phenolic component of OPP is shikimic acid.

Caffeoylshikimic acid

Caffeoylshikimic acid is an ester of caffeic acid and shikimic acid. It is an intermediate of the lignin biosynthesis pathway and not commonly found in nature. The date palm, *Phoenix dactylifera*, is one of the few plants reported to contain this phenolic compound, the 3-*O*-caffeoylshikimic acid isomer being one of the key enzymatic browning substrates (Maier et al, 1964). 3-, 4- and 5-*O*-caffeoylshikimic acids are also known as dactilyfric, isodactylifric and neodactylifric acid, respectively. They are among the resistance factors of date palm roots against *Fusarium oxysporum* (Modaffar et al., 2000; Hassni et al., 2004). It would be interesting to investigate if they act as resistance factors in the oil palm.

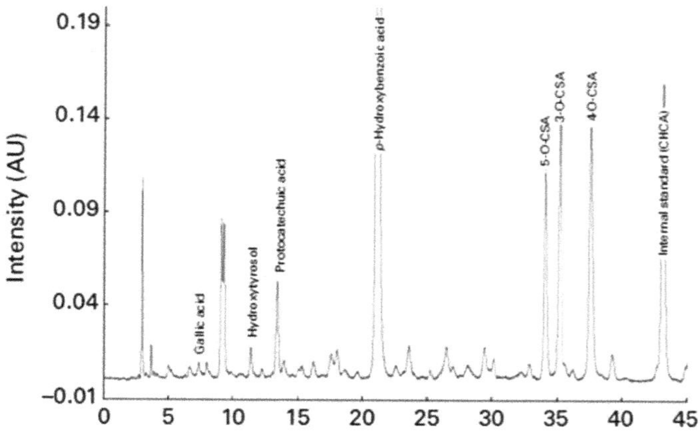

Figure 6 HPLC profile of OPP.

Table 7 Major phenolic components in OPP

Phenolic compounds	Concentration (mg/kg)		
	Mean	SD	Range
Protocatechuic acid	600	100	400–800
p-Hydroxybenzoic acid	7000	1000	5300–8600
Caffeoylshikimic acid (total of 3 isomers)	10 800	2400	7700–15 100
Total major phenolics	18 400	2900	13 800–24 300
Gallic acid equivalents	18 200	1700	15 700–21 300

Source: Sambanthamurthi et al. (2011a).

Caffeoylshikimic acid is a minor constituent of the leaves of *Vaccinium* species – bilberry (*Vaccinium myrtillus* L.), hybrid bilberry (*Vaccinium x intermedium* Ruthe L.) and lingonberry (*Vaccinium vitis-idaea* L.) (Hokkanen et al., 2009); Yerba mate (*Ilex paraguariensis* L.); and green tea (*Camellia sinensis*) (Bastos et al. 2007). More recently, Chideha et al. (2014) reported the presence of high levels of 5-*O*-caffeoylshikimic acid in *Solanum somalense* leaf which are used in Djibouti for its medicinal properties. Caffeoylshikimic acid makes up more than half of the total phenolic content of OPP. Palm vegetation liquor, from which caffeoylshikimic acid (as a constituent of OPP) is extracted, is thus the largest known source of this bioactive compound. Caffeoylshikimic acid is relatively similar in structure to caffeoylquinic acid (also known as chlorogenic acid), which also exists in three isomer forms, namely 3-, 4- and 5-*O*-caffeoylquinic acids. Both are esters of caffeic acid. The only difference between shikimic acid and quinic acid is a double bond in the ring structure of shikimic acid, which is removed by the presence of an additional hydroxyl group in quinic acid. Figure 7 shows the molecular structures of caffeoylshikimic and caffeoylquinic acids. Several health benefits associated with the consumption of chlorogenic acid have been documented including reduced incidence of type 2 diabetes (Salazar-Martinez et al., 2004; Ranheim et al., 2005), cardiovascular disease (Ranheim et al., 2005) and anti-cancer

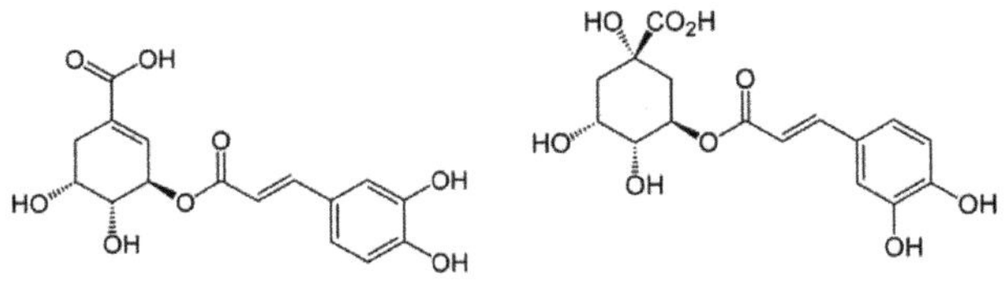

Figure 7 Molecular structures of caffeoylshikimic acid and caffeoylquinic acid.

properties (Tanaka et al., 1990). They also have anti-inflammatory (Santos et al., 2006) and antibacterial properties (Lou et al., 2011; Karunanidhi et al., 2013). Green coffee is a major source of chlorogenic acid (3-, 4-, and 5-caffeoylquinic acids). OPP extracts exhibit several health benefits (see Section 2.3) similar to those of chlorogenic acid. The three isomers of caffeoylshikimic acid may be responsible in part at least for these benefits.

Para-hydroxybenzoic acid

After caffeoylshikimic acid, *para*-hydroxybenzoic acid is the next most abundant phenolic acid in OPP. Also known as 4-hydroxybenzoic acid, it is an isomer of 2-hydroxybenzoic acid commonly known as salicylic acid, which is used for the production of aspirin. *Para*-hydroxybenzoic acid is an important substrate for the preparation of its esters parabens, which are used as preservatives in cosmetics and ophthalmic solutions. The coconut, *Cocos nucifera*, and the acai palm, *Euterpe oleracea*, are rich sources of 4-hydroxybenzoic acid. It is also present in cloudy (unfiltered) olive oil and blueberries.

Protocatechuic acid

Another important phenolic compound in oil palm is PCA (3,4-dihydroxybenzoic acid). PCA has been shown to have strong *in vitro* and *in vivo* antioxidant activity. It is an active component of traditional Chinese herbal medicines (Li et al., 2011) and present in most edible plants used in folk medicine (Liu et al., 2004). A wide range of potential pharmacological benefits has been attributed to PCA. These include anti-inflammatory, anti-diabetic, anti-cancer, cardioprotective, hepatoprotective, antibacterial and antiviral activities (Kakkar and Bais, 2014). PCA is found in a multitude of berries including blueberry, strawberry, mulberry, cranberry and gooseberry. It also occurs in *Olea europaea* (olives), *Hibiscus sabdariffa* (roselle), *Vitis vinifera* (white wine grapes), wine, honey and soybean. It is also a breakdown product of anthocyanin and may be considered the bioavailable form of anthocyanins.

Based on a human feeding trial, Vitaglione et al. (2007) suggested that the phenolic acid degradation product, PCA, was a major metabolite of anthocyanins.

Caffeic acid

Caffeic acid (3,4-dihydroxycinnamic acid), identified as a component of OPP, has been established to have multiple health benefits. Coffee is the primary source of this phenolic compound. It is an active component of propolis extract and widely distributed in several plants. There is extensive evidence from *in vitro* and *in vivo* studies to confirm that it has antioxidant, anti-inflammatory, anti-carcinogenic, immuno-modulatory and antimicrobial activities (Kang et al., 2009; Touaibia et al., 2011; Rodríguez-Vaquero et al., 2015).

3.2 Shikimic acid

An interesting metabolite found in OPP is shikimic acid (3,4,5-trihydroxy-1-cyclohexene-1-carboxylicacid) (Sambanthamurthi et al., 2010). Shikimic acid itself is not a phenolic compound, as it does not contain an aromatic ring. It is a cyclohexene but is a precursor in the biosynthesis of aromatic compounds including phenolics. The benzene ring, which is the basic element of all aromatic compounds, is synthesized in plants and microorganisms through the shikimic acid pathway. Shikimic acid derivatives have wide use in agriculture as herbicides and antimicrobial agents as they can inhibit the shikimic acid pathway in plants and bacteria without negative consequences in animals and humans, as the pathway does not exist in animals. Important bioactivities of shikimic acid have been documented. These include anti-inflammatory, anticoagulant, antithrombotic, antibacterial and antiviral properties (Estevez and Estevez, 2012). Besides its bioactive properties, shikimic acid is also essential in the synthesis of related bioactive compounds in the pharmacological industry (Estevez and Estevez, 2012). It is an important substrate for the industrial synthesis of the antiviral drug oseltamivir, which is used for the treatment H1N1 influenza and against all strains of influenza virus. Shikimic and its derivative, triacetylshikimic acid, have anti-inflammatory properties through inhibition of COX-1 and COX-2 activities, and also reduction of platelet aggregation and blood clot formation (Huang et al., 2002; El-Seedi et al., 2003). Shikimic acid in combination with quercetin was reported to have immunomodulatory effects by regulating the secretion of interleukin-6 and interleukin-8, which mediate lymphocyte recruitment and activation (Bertelli et al., 2008). In addition, shikimic acid derivatives are important in agriculture as herbicides and as antibacterial agents as they can block the shikimate pathway in plants and bacteria without adverse effects towards mammals. Currently, the main source of shikimic acid is the Chinese star anise (*Illicium verum*) but the abundance of vegetation liquor from palm oil mills is an attractive reliable source of shikimic acid.

4 Bioactive properties of OPP

4.1 Antioxidant potential of OPP

It has been established that free radicals and related reactive oxygen and nitrogen species, ROS/RNS, play a key role in the pathogenesis of numerous diseases and disorders. OPP has been demonstrated to have potent antioxidant activity based on its high efficacy in scavenging 2,2-diphenyl-1-picrylhydrazyl (DPPH), a stable free radical (Sambanthamurthi et al., 2011a). The antioxidant potential of OPP results from the ability to donate hydrogen atoms and scavenge free radicals. Antioxidant and radical scavenging activities increase with the level of hydroxylation of the phenolic substance (Rice-Evans et al., 1996). Caffeoylshikimic acid, the major phenolic acid of OPP, has four hydroxyl groups, thus accounting for the powerful antioxidant activity of OPP. The other phenolic compounds in OPP such as hydroxybenzoic acid and PCA would also be expected to contribute to the antioxidant activity observed and in fact may have a synergistic effect.

4.2 Cardioprotective effect of OPP

Atherosclerosis is the main cause of morbidity and mortality worldwide. Oxidation of LDL is a key initial event in the development of atherosclerosis (Steinberg et al., 1989). Oxidized LDL induces complex inflammatory and immunologic reactions in the endothelial cells of the vascular wall. Uptake of oxidized LDL by macrophages and smooth muscle cells leads to the formation of cholesterol-rich foam cells and development of fatty streaks, a principle event in early atherosclerosis. Phenolics inhibit LDL oxidation by scavenging free radical species or sequestering metal ions (Salah et al., 1995; Bernatova et al., 2002), thus protecting against cardiovascular disease. The resistance of LDL to oxidation is positively correlated to the severity of coronary atherosclerosis in human subjects (Regnstrom et al., 1992). Sambanthamurthi et al. (2011a) showed that OPP inhibited Cu-mediated oxidation of human LDL in a dose-dependent manner *in vitro*, suggesting potential efficacy against atherogenesis. Wine, cocoa and green tea have also been shown to inhibit LDL oxidation (Frankel et al., 1995; Osakabe et al., 2002; Yang and Koo, 2000).

Rabbits on a high-fat, cholesterol-rich atherogenic diet were reported to have extensive fibrous and fatty plaques and fatty streaks in their aortas. However, rabbits on the same diet supplemented with OPP had significantly decreased fatty streaks and fatty plaques and more lesion-free areas. This observation indicated bioactivity and bioavailability of OPP in providing nett protection against the development and progression of atherosclerosis (Sambanthamurthi et al., 2011b). Gene expression studies (Leow et al., 2013a) showed that the atherogenic diet resulted in oxidative stress, inflammation and increased turnover of metabolites and cells in the liver, heart and spleen. OPP, however, attenuated/reversed these effects and increased unfolded protein response in the liver, upregulated antioxidant genes in the heart and reduced antigen presentation and processing in the spleen. OPP also returned antioxidant activity in the serum to normal levels. Che Idris et al. (2014) studied the protective effects of OPP and tocotrienol-rich vitamin E singly or in combination against atherosclerosis in a rabbit model fed an atherogenic diet. The results confirmed that while OPP and tocotrienol-rich vitamin E individually reduced the development of atherosclerotic lesions, OPP in combination with tocotrienol-rich vitamin E conferred the highest protective effect. The effect was not additive but synergistic.

Numerous plant compounds such as wine (Bernatova et al., 2002), tea (Grassi et al., 2008), coffee (Suzuki et al., 2006) and soy isoflavones (Chin-Dusting et al., 2004) improve vascular function in whole animals and human subjects by producing vasodilation. OPP induces vascular relaxation via endothelial NO, a key endogenous vasodilator in maintaining cardiovascular homeostasis as evidenced in an *ex vivo* study with isolated vascular preparations (Sambanthamurthi et al., 2011a). OPP also decreased blood pressure in an NO-deficient rat model of hypertension confirming the *in vitro* observations and indicating its direct effect on the endothelial NO synthase system (Sambanthamurthi et al., 2011b).

4.3 Anti-diabetic effects of OPP

Polyphenols have been established as powerful antioxidants that can ameliorate type 2 diabetes through multiple mechanisms. These include anti-inflammatory and immunomodulatory effects, modulation of glucose-induced oxidative stress, inhibition of carbohydrate hydrolysing enzymes such as α-amylase and α-glucosidase and inhibition of glucose absorption as well as preserving vascular functions in diabetic complications.

The potency of phenolic compounds on carbohydrate metabolism and glucose homeostasis has been studied *in vitro*, in animal models and human intervention trials (Bahadoran et al., 2013). Anderson (2008) reported that cinnamon phenolics improved insulin sensitivity *in vitro* and in human subjects with type 2 diabetes. Individuals with the metabolic syndrome who consumed an aqueous extract of cinnamon had improved parameters including fasting blood glucose, systolic BP, percentage body fat and lean body mass compared with the control group. Mulvihill et al. (2009) provided evidence that naringenin attenuated hyperinsulinaemia and dyslipaemia in LDL receptor null mice with diet-induced insulin resistance.

OPP has been shown to protect against early type 2 diabetes characterized by hyperglycaemia and hyperlipaemia in a rat model (Nile rat, *Arvicanthis niloticus*) that spontaneously develops diabetes on a high carbohydrate diet. As in humans, the model used typically exhibits elevated glucose, total cholesterol, triacylglycerol, liver and kidney weights, fatty liver and depressed HDL as markers for type 2 diabetes and the metabolic syndrome. The animals on OPP showed significant decrease in glucose levels and all the criteria for diabetes (Sambanthamurthi et al., 2011b). Bolsinger et al. (2014) further demonstrated that OPP mitigated several aspects related to hyperglycaemia in the Nile rat without any toxic effects.

4.4 Anti-tumour effects of OPP

OPP was shown to have anti-cancer properties in both *in vitro* and *in vivo* studies. It effected dose-dependent growth arrest and cell death of various cell lines including human lung carcinoma, oestrogen receptor-positive human breast adenocarcinoma and mouse IgA-secreting myeloma cell lines (Sekaran et al., 2010). OPP was shown to reduce tumour growth in BALB/c mice injected with myeloma cells. Gene expression studies indicated that OPP inhibited tumour growth in the mice by inducing a G1/S phase arrest in the cell cycle (Sambanthamurthi et al., 2011b). Other phenolics including hydroxytyrosol (Fabiani et al., 2002), resveratrol (Joe et al., 2002), genistein (Raffoul et al., 2006) and luteolin (Lim et al., 2007) have also been reported to inhibit tumour growth of several cancer cell lines by inducing cell cycle arrest. A point of interest is that several of the genes that were significantly changed in expression by OPP were also found to be differentially expressed, but in the opposite direction, in colorectal tumour cells that were resistant to a combined chemotherapy of 5-fluorouracil, folinic acid and irinotecan (Sambanthamurthi et al., 2011b). This suggests that OPP may sensitize cancer cells towards chemotherapy and may enhance cytotoxicity in cancer cells when used in combination with chemotherapeutic agents. Several cytokines including the chemokine CXCL12 and interleukin IL-6 are known to be important mediators of advanced metastasis and significantly contribute to the lethal phenotype (Loberg et al., 2007). CXCL12 was significantly downregulated in tumours of mice treated with OPP suggesting that OPP played a role in reducing the aggressiveness and metastatic potential of these tumours (Sambanthamurthi et al., 2011b). Time course microarray analysis on spleens following injection of J558 myeloma cells in mice indicated that the immune response of tumour-bearing mice given OPP was lower than that in control mice, thus implying delayed inflammation in response to OPP (Leow et al., 2013b). Cholesterol biosynthesis genes were upregulated and inflammatory genes were downregulated in the livers, further suggesting attenuation of systemic inflammation and cachexia. These effects correlated with the delayed *in vivo* development of syngeneic tumours in mice given OPP (Leow et al., 2013b).

OPP suppressed proliferation of cells in two pancreatic cancer cell lines (PANC-1 and BxPC-3) in a dose-dependent manner (Ji et al., 2015). The anti-proliferative, apoptotic (programmed cell death) and anti-metastatic properties of OPP were confirmed based on evidence of cell cycle arrest in the S phase, induction of apoptosis associated with decrease in survivin and Bcl-XL expressions and increased expression of cleaved caspase-3, caspase-9 and PARP. Its anti-metastatic effects were confirmed based on decreased expressions of vascular endothelial growth factor and MMP-9.

Tannic acid, a water-soluble phenolic abundant in Chinese herbal medicines and tea, inhibits CXCL12 and this may explain its anti-inflammatory, anti-tumour properties (Chen et al., 2003).

4.5 Role of OPP in neuroprotection and improved cognitive function

Mice supplemented with OPP showed superior cognitive and spatial learning skills based on their performance in a water maze. They also exhibited better motor function and coordination based on their performance on a rotarod (rotating rod where the rotating speed is gradually increased). Gene expression analysis indicated that OPP upregulated genes involved in brain development and activity, such as the signalling genes under the control of the brain-derived neurotrophic factor. OPP also downregulated genes involved in inflammation. The cognitive and motor function enhancement is thus attributed to the neuroprotective and anti-inflammatory properties of the extract (Leow et al., 2013c). Similar results were reported by van Praag et al. (2007), where supplementation of C57BL/6 mice with catechins resulted in upregulation of genes involved in learning and downregulation of genes associated with inflammation. Furthermore, genes involved in focal adhesion, notably the beta-actin gene that encodes a cytoskeletal protein that is elevated in the brains of patients with Alzheimer's disease and in reactive glia were downregulated by OPP (Leow et al., 2013c). Similar observations were made by Deshane et al. (2004) in the brains of mice that were supplemented with a grape seed extract enriched with proanthocyanidins. Wang et al. (2006) also showed that moderate consumption of Cabernet Sauvignon attenuated Abeta (*amyloid beta*-protein) neuropathology in a mouse model of Alzheimer's disease while Hartman et al. (2006) reported that pomegranate juice decreased amyloid load and improved behaviour in a mouse model of Alzheimer's disease. Haque et al. (2006) also reported that long-term supplementation of green tea catechins in rats significantly improved spatial cognition in rats and also lowered reactive oxygen species in the hippocampus of the brain.

4.6 Protective effect of OPP against mitochondrial dysfunction

It is increasingly evident that several chronic diseases including ageing, diabetes, cardiovascular diseases and neurodegenerative disorders such as Alzheimer's, Parkinson's and Huntington's disease are related to mitochondrial dysfunction (Swerdlow, 2011; Reddy, 2008; Victor et al., 2009). Mitochondrial dysfunction is characterized by diminished efficiency in the electron transport chain and entry of metabolites into the mitochondria and decreased synthesis of high-energy molecules such as adenosine-5′-triphosphate. The nucleotide reverse transcriptase inhibitor 3′-Azido-3′-deoxythymidine

(AZT) is a common component of combination anti-HIV regimens to treat HIV/AIDS. However, prolonged use of this antiviral results in accumulation of mitochondrial mutations and mitochondrial dysfunction. Oxidative stress and damage has been postulated to be one of the reasons for this toxicity. Osborne et al. (2014a) reported that OPP mitigated AZT-induced mitochondrial genotoxicity (mutagenesis) as well as dose-dependent cytotoxicity in cultured HepG2 cells. Three regions of the mitochondrial genome studied (HV2, CO2 and ND1) had 35% of the number of mutations observed in cultures treated with AZT alone and co-treatment of AZT with OPP increased cell survival by up to 350%. It was confirmed that the effects were not because of degradation or inactivation of AZT by OPP. The discovery offers a potential method of mitigating AZT-induced mutations and related contraindications of prolonged AZT use with OPP supplementation. Osborne et al. (2015) also went on to show that oxidative damage was only a minor contributor to AZT-induced mutations. Instead, most of the mutations (80%) arose from AZT-induced mitochondrial DNA polymerase errors suggesting that OPP preserved DNA polymerase fidelity or prevented possible AZT-induced alterations in cellular nucleotide pools that could lead to errors in mitochondrial DNA polymerase. Natural supplements have been used in the treatment of chronic disease resulting from mitochondrial dysfunction (Nicolson, 2014). Osborne et al. (2014b) described the use of OPP for the prevention and treatment of mitochondrial dysfunction and DNA damage, especially in vulnerable populations, such as the aged and individuals being treated with AZT and other drugs.

4.7 Multiple effects of OPP and safety

While the positive outcomes of OPP observed in the multiple studies described are mainly attributed to phenolic compounds, it is most likely that a combination of bioactives such as shikimic acid and soluble fibres worked in synergy with the phenolic compounds and played a key role in the effects observed. It is pertinent to bear in mind that the OPP extract in its entirety conferred the outcomes. The potent effects conferred by this combination of water-soluble bioactives present an opportunity for the formulation of functional foods and nutraceuticals. As OPP is a newcomer to the bioactive scene, it is imperative that its safety is ensured. Toxicological and teratological studies confirmed that it is safe for consumption (Sambanthamurthi et al., 2011a). A Phase 1 human intervention trial (Syed Fairus, pers. comm.) also showed that consumption of OPP had no serious adverse effects.

5 Bioactives in different palm sources

Bioactives such as carotenes, tocols, sterols, coenzyme Q, squalene and phospholipids are present in the oil palm fruits as early as 4 weeks after anthesis. On pressing the fruits for palm oil, these phytonutrients, being oil-soluble, are extracted with the oil (CPO). However, CPO is not the only source of palm phytonutrients. These compounds also end up in various products/by-products in the different stages of oil palm milling and refining processes. CPO, when further processed to downstream products, transfers some of these phytonutrients to the end products, which then also become good sources of palm phytonutrients.

5.1 Oil palm milling

Crude palm oil

The main source of lipid-soluble palm bioactives is derived from CPO, following the milling of fresh fruit bunches. Apart from the *Elaeis guineensis*, *E. oleifera* (the American oil palm) and the hybrids of *E. oleifera* and *E. guineensis* are a rich source of phytonutrients (Latip et al., 2000; Tan et al., 1986; Choo et al., 2002b, 2005a; Kulkarni, 1991; George and Arumughan, 2006; Tan et al., 2007). *E. oleifera* oil is especially rich in carotenes, tocols and sterols, more so than *E. guineensis*. Tables 8–10 depict the concentration of carotenes, tocols and sterols present in different species and varieties of oil palm.

Palm pressed fibre

Palm pressed fibre is the residue of the oil palm fruit mesocarp after pressing for oil. The fibre is normally burnt as fuel to generate electricity for mill operation. These fibres still contain about 5–6% residual oil as the single pressing in the milling process does not completely remove the oil from the fibre. The residual oil can be extracted using solvent or other extraction technology such as supercritical fluid extraction. The oil obtained from the fibre contains much higher concentration of phytonutrients than CPO (Goh et al., 1985; Tan et al., 1986; Choo et al., 2005a; Lau et al., 2006b, 2008). Phospholipids are the major phytonutrients present in PFO, and their concentrations are more than 300-fold that of CPO. The detailed phytonutrient profile in PFO is depicted in Table 11.

Vegetation liquor

The palm oil milling process generates an aqueous stream known as vegetation liquor as a by-product. Vegetation liquor is derived primarily from the sterilizer condensate and centrifuge sludge and usually discarded as POME. About 2.5–3.5 m^3 of vegetation liquor is generated for every tonne of CPO produced. Malaysia generates about 45 million tonnes of vegetation liquor annually while global output exceeds 100 million tonnes. Fresh fruit bunches are typically sterilized using high pressure steam (120 to 140°C at 40 psi) to soften the fruits, enable the separation of the fruit from bunches and inactivate enzymes that hydrolyse the oil and cause fruit deterioration. The steam extracts the water-soluble components of the fresh fruit bunches which then enter the sterilizer condensate. A substantial amount of water is added during the clarification step of palm oil milling to separate the oil from oil-water-sludge emulsion. The oil is skimmed off while the sedimented fraction (centrifuge sludge) enters the aqueous stream and discarded as POME. Like the sterilizer condensate, the centrifuge sludge is a rich source of water-soluble phytonutrients. Sambanthamurthi et al. (2008) developed a process whereby the fresh vegetation liquor from the sterilizer condensate and centrifuge sludge are centrifuged in a three-phase high-speed decanter to remove debris and residual oil and the middle water phase passed through a series of filters to produce a phenolic-enriched filtrate termed OPP, whcih contains among others, soluble fibres, sugars and shikimic acid in addition to phenolics. The sterilization process ensures that polyphenol oxidases are inactivated and the phenolics are bioactive. While the heat-sensitive phenolics may have been inactivated, the extraction process selectively enriches for highly stable phenolics. The phenolics in OPP are thus highly stable.

Table 8 Individual carotenes in mesocarp oils from *E. guineesis* and *E. oleifera* and their hybrids as percentage of total carotenes

	E. guineesis (E.G), %				*E.G x E.O*				
	Tenera (T)	*Pisifera (P)*	*Dura (D)*	*E. Oleifera (E.O)*	*E.O x P*	*E.O x D*	*E.O x P*	Albescens	Fibre
Phytoene	1.27	1.68	2.49	1.12	1.83	2.45	1.3	1.1	10.2
Cis-β-Carotene	0.68	0.1	0.15	0.48	0.38	.55	tr	0.9	0.6
Phytofluene	0.06	0.9	1.24	tr	tr	0.15	0.42	0.3	tr
β-Carotene	56.02	54.39	56.02	54.08	60.53	56.42	51.64	61.1	33.8
α-Carotene	35.06	33.11	24.35	40.38	32.78	36.4	36.4	29.8	21.3
Cis-α-Carotene	2.49	1.64	0.86	2.3	1.37	1.38	2.29	3.1	1.6
ξ-Carotene	0.69	1.12	2.31	0.36	1.13	0.7	0.39	0.7	7.3
γ-Carotene	0.33	0.48	1.16	0.08	0.23	0.26	0.14	0.3	2.2
δ-Carotene	0.83	0.27	2	0.09	0.24	0.22	0.19	0.2	7.2
Neurosporene	0.29	0.63	0.77	0.04	0.23	0.08	0.08	0.3	2.9
β-Zeacarotene	0.74	0.97	0.56	0.57	1.03	0.96	1.53	1	0.4
α-Zeacarotene	0.23	0.21	0.3	0.43	0.35	0.4	0.52	0.2	tr
Lycopene	1.3	4.5	7.81	0.07	0.05	0.04	0.02	1	12.5
Total (ppm)	500–700	300–500	900–1000	4300–1800	1250–1800	1200–2400	800–900	90–110	4000–6000

Source: Choo et al. (1999).

Table 9 Concentrations of tocopherols and tocotrienols in mesocarp oils from different palm varieties and species and their hybrids as percentage of total tocols

	E. guineesis (E.G)				*E.G x E.O*				
	Tenera (T)	*Pisifera (P)*	*Dura (D)*	*E. oleifera (E.O)*	*E.O x P*	*E.O x D*	*E.O x P*	Albescens	Fibre oil
α-Tocopherol	21	24	31	15	19	24	11	11	53.6
α-Tocotrienol	23	38	21	27	28	22	31	31	16.2
γ-Tocotrienol	45	32	40	54	42	49	51	51	24.0
δ-Tocotrienol	11	6	8	4	15	5	7	7	6.2
Total (ppm)	600–1000	600–800	700–1500	700–1500	600–1600	800–1700	700–900	700–900	2500–4000

Source: Choo et al. (1999).

Table 10 Composition of sterols in mesocarp oils from different oil palms as percentage of total sterols

	E. guinensis (E.G)				*E.G x E.O*				
	Tenera (T)	*Pisifera (P)*	*Dura (D)*	*E. oleifera (E.O)*	*E.O x P*	*E.O x D*	*E.O.Dx*	Albescens	Fibre oil
β-Sitosterol	60	54	55	64	59	61	58	70	58.5
Campesterol	13	17	25	19	20	22	20	20	22.3
Stigmasterol	24	22	14	15	16	13	19	8	18.0
Cholesterol	3	7	6	2	5	4	3	2	1.2
Total (ppm)	250–620	1500–2000	2000–2500	3500–4000	1100–1250	1200–1400	700–800	500–600	5000–6300

Source: Choo et al. (1999).

Table 11 Concentrations of phytonutrients in CPO and PFO

Phytonutrient	CPO (ppm)	PFO (ppm)
Carotenes	500–700	4000–6000
Tocols (tocopherols and tocotrienols)	600–1000	2000–4000
Sterols	326–527	4000–6000
Squalene	200–500	500–1000
Phospholipids	5–130	37 000–40 000

Source: Choo et al. (1996).

5.2 Palm oil refining

Palm fatty acid distillate

CPO is refined either physically or chemically for use. Chemical refining is carried out by neutralizing the fatty acids using an alkali. The process yields refined, bleached and deodorized palm oil with PFAD produced as a by-product. PFAD is also rich in phytonutrients, especially tocols, squalene and sterols (Gapor, 1993, 2002; Gapor et al., 1993, 2000, 2009; Ahmad et al., 2009; Gapor and Hazrina, 2000).

Gapor et al. (2000) found 2128–13 505 mg/kg squalene in PFAD, more than 10x the amount in CPO, and these sterols have since been successfully extracted as pure crystals).

Red palm oil

Palm phytonutrients – carotenes, tocols, coenzyme Q and sterols, are largely preserved in red palm oil, a premium oil. In normal refining of CPO, the carotenes and tocols are largely removed/destroyed. A new refining process that retains the carotenes and tocols in the end product (red palm oil/olein) has since been introduced. In commercial red palm olein (CRPo), the average carotene content is ~665 ppm, tocols 717–863 ppm, sterols 325–365 ppm and coenzyme Q 18–25 ppm (Sundram et al., 1989). Commercial red palm oil products, marketed under the trade name 'CAROTINO', are produced using PORIM/MPOB technology (Ong and Goh, 2002).

Today, the products are sold in more than 15 countries, providing the much needed nutrients to consumers. The nutritional information on red palm oil is shown in Table 12.

Red palm oil like standard palm oil is free from harmful *trans*-fatty acids. The carotenes, tocols and coenzyme Q in red palm oil act as antioxidants scavenging oxygen free radicals, and are hypothesized to play a protective role in cellular ageing, atherosclerosis and cancer prevention. The natural carotenes in red palm oil provide provitamin A that is readily converted to vitamin A in the body in accordance with its requirements. Synthetic vitamin A can be toxic in excess.

In a storage stability study of CRPo over 12 months under the typical conditions in Malaysia, the carotene level remained stable with only a minor drop in tocols at the end of the period (Sundram et al., 1989; Tan, 1987).

Due to its nutritional benefits, red palm oil has wide applications in the food, healthcare and skincare industries. The oil can be incorporated in biscuits, breads, cakes, noodles,

Table 12 Nutritional contents per serving of 15 mL and 100 mL commercial red palm oil

	15 mL	100 mL
Energy	510 kJ	3400 kJ
Protein	0 g	0 g
Total fat	14 g	92 g
• Monounsaturated		52.5 g
• Polyunsaturated		25.3 g
• Saturated		14.3 g
Cholesterol, carbohydrate, dietary fibre		Nil
Sodium		
Carotenes		
• β-carotene (provitamin A)	0.80 mg	
• α-carotene (provitamin A)	0.61 mg	100%
• Other carotenes	0.34 mg	40%
Natural vitamin E	10 mg	100%
(Tocopherols and tocotrienols)	0.50 mg	
Natural vitamin K	1 mg	
Coenzyme Q_{10}		

* 1 kJ = 0.24 kcal.
Source: http://www.carotino.com.

margarine (as colourant), frying oil, mayonnaise, salad dressings, health foods and skincare products (Benade, 2001). It not only provides natural ingredients, but also adds value to the products by improving their health and nutritional attributes.

The bioavailability of beta-carotenes from red palm oil and synthetic beta-carotene is similar (Van Het Hof, 1999). However, red palm oil is a better choice as it contains a bouquet of carotenoids rather than just a single carotene. In a study, mothers and children in Burkina Faso experienced a 40% fall in vitamin A deficiency after one year of taking red palm oil (Zagre et al., 2002). The taste and appearance of red palm oil prepared dishes are readily acceptable. Primary school children in South Africa who consumed four CAROTINO red palm oil biscuits a day had significantly higher serum retinol (Stuijvenberg et al., 2000). In studies, pregnant women supplemented with red palm oil had significantly higher alpha- and beta-carotenes in their plasma and breast milk (Canfield et al., 2001). A separate study by the National Institute of Nutrition in Hanoi, Vietnam, showed that CAROTINO red palm oil supplementation significantly improved serum retinal and Hb levels in rural children under five years of age. The haemoglobin level of the red palm oil supplemented group was as high as in those given vitamin A capsules (Nguyen et al., 2001). A vitamin A intervention programme was carried out in Gansu Province, China, where biscuits made from red palm oil were given to school children. Through this programme, the incidence of vitamin A deficiency was reduced from 21.6% to 6.1% (MPOB, unpublished data).

5.3 Other downstream products: palm oil methyl esters (palm biodiesel)

Palm oil methyl esters (palm biodiesel) are derived from esterification/transesterification of CPO with methanol (Choo et al., 1993; Ong et al., 1992). The mild conditions (temperature <70°C and under atmospheric pressure) of the reaction largely conserve the phytonutrients, which are retained in the palm biodiesel from where they can be recovered before it is burnt as fuel. The phytonutrients are recovered as a dense, thick semi-liquid termed as phytonutrient concentrate.

5.4 Oil palm leaf

The oil palm leaf is a rich source of polyphenols. Runnie et al. (2003) showed that oil palm leaf extracts contain 8% higher polyphenol content on a dry weight basis than green tea. Its main phenolics mirror several catechins contained in green tea, namely catechin (0.30%), epigallocatechin gallate (0.28%), epigallocatechin (0.08%), epicatechingallate (0.05%), epicatechin (0.01%) and their glycosides (Jaffri et al., 2011a), and may thus be a potential inexpensive source of such catechins (Mohamed, 2014). Alcoholic extract of oil palm leaves was reported to contain 24.3 mg gallic acid equivalent (GAE)/g dry weight of total phenolic content (Tan et al., 2011). The antioxidant capacity of oil palm leaf extract was shown to be higher than those of papaya shoot, green chilli and lemon grass *in vitro* (Abeywardena et al., 2002). Jaffri et al. (2011b) demonstrated anti-hypertensive and cardioprotective effects of oil palm leaf extract in NO-deficient rats. Abeywardena et al. (2002) reported that polyphenol-enriched extract of oil palm (*E. guineensis*) fronds promotes vascular relaxation via endothelium-dependent mechanisms and Tan et al. (2011) reported that the leaf extract reduced hyperglycaemia and lipid oxidation in streptozotocin-induced diabetic rats. Palm leaf extract also inhibited Cu^{2+}-mediated oxidation of low-density lipoprotein (Salleh et al., 2002). Supplementation of the oil palm leaf extract to healthy volunteers over a 2-month period resulted in significant cognitive enhancement based on computer-assisted cognitive tests that assessed spatial visualization ability and processing speed. Supplementation of the oil palm leaf extract conferred neuroprotection to NO-deficient rats via systemic and cellular modulations that eventually enhanced neuron survival (Mohamed et al., 2013).

6 Future trends

While the oil palm offers a whole array of promising bioactives with potential health benefits, mechanistic studies are required to establish their long-term benefits. As is the case for most plants, much of the scientific evidence on the beneficial effects of palm bioactives is drawn from *in vitro* or animal studies. While human intervention studies have been carried out on various aspects of palm carotenes, tocotrienols and OPP, more intervention studies are needed to confirm the whole spectrum of effects of the bioactives *in vivo*. Also, the doses applied in most animal experiments are high and translations of doses in humans are necessary. Efficacy and toxicity tests will also be intensified. Future research will also involve identification and improvement of the bioactive compounds individually and in combination. Allopathic medicine generally employs purified single compounds while alternative or traditional medicine uses multiple combinations of compounds working in synergy. There may be an increasing trend in this direction as society moves towards natural bioactives as these may have fewer side effects compared to single pure compounds. Advanced delivery systems are important for optimal efficiency of bioactives. Liposomes, hydrogels, virosomes, microspheres and miniature implants are examples of superior delivery systems that are increasingly being used and improved upon to deliver active molecules to target sites.

7 Conclusion

In view of the importance of palm oil phytonutrients, extensive research has been carried out to discover bioactives and authenticate their biological activities. Extracting palm phytonutrients from various products of oil palm, such as palm oil, PFO, PFAD and palm oil mill waste has been carried out with success. Today, palm phytonutrients are available for various applications, such as in the food and non-food industries, cosmetics, nutraceuticals and pharmaceuticals. Valorization of by-products of the oil palm industry to harness value-added bioactives decreases the environmental burden and fulfils the triangle of sustainability – economy, society and environment. The harnessing of valuable phenolic compounds from the waste stream of oil palm milling is one such example.

8 References

Abeywardena, M., Runnie, I., Nizar, M., Suhaila, M., Head, R. and Momamed, S. (2002). Polyphenol-enriched extract of oil palm fronds (*Elaeis guineensis*) promotes vascular relaxation via endothelium-dependent mechanisms. *Asia Pac. J. Clin. Nutr.* 11 Suppl 7: S467–72.

Ackman, R. G., Macpherson, E. J. and Timmins, A. (2000). Squalene in oils determined as Squalane by gas-liquid chromatography. *J. Am. Oil Chem. Soc.* 77 (8): 831–5.

Ahmad, A. L., Chan, C. Y., Abd Shukor, S. R. and Mashitah, D. (2009). Separation of oil and carotenes from palm oil mill effluente by adsorption chromatography with silica based adsorbent. *Asia Pac. J. Chem. Eng.* 4 (5): 717–22.

Anderson, R. A. (2008). Chromium and polyphenols from cinnamon improve insulin sensitivity. *Proc. Nutr. Soc.* 67 (1): 48–53.

Ashfaq, M. K., Zuberi, H. S. and Waqar, M. A. (2001). Vitamin E and beta carotene affect anti cancer immunity: *In vitro* and *in vivo* studies. In *2001 PORIM Int. Conf.*, Kuala Lumpur, Malaysia.

Bahadoran, Z., Mirmiran, P. and Azizi, F. (2013). Dietary polyphenols as potential nutraceuticals in management of diabetes: a review. *J. Diab. Metab. Disord.* 12, article 43.

Barrowman, J. A., Rahman, A., Lindstrom, M. B. and Borgstorm, B. (1989). Intestinal absorption and metabolism of hydrocarbons. *Am. J. Clin. Nutr.* 59: 356–63.

Bastos, D. H., Saldanha, L. A., Catharino, R. R., et al. (2007). Phenolic antioxidants identified by ESI-MS from Yerba mate (*Ilex paraguariensis*) and green tea (Camelia sinensis) extracts. *Molecules* 12: 423–32.

Beal, M. F. (1999). Coenzyme Q10 administration and its potential for treatment of neurodegenerative diseases. *BioFactors* 9 (2–4): 261–6.

Beck, F., Butters, D., Matsumoto, G. And Whitehouse, M. (2010). Queries about vaccines containing squalene. *Immunol. Cell Biol.* 88: 497–9.

Benade, A. J. S. (2001). The potential of red palm oil based shortening as a food fortificant for vitamin A in the baking industry. *Food Nutr. Bull.* 22 (4): 416–18.

Berger, L. G. (2005). *The Use of Palm Oil in Frying*. Malaysian Palm Oil Promotion Council, Kuala Lumpur.

Bernatova, I., Pechanova, O., Babal, P., et al. (2002). Wine poly-phenols improve cardiovascular remodeling and vascular function in NO-deficient hypertension. *Am. J. Physiol. Heart Circ. Physiol.* 28: H942–8.

Bertelli, A. A., Mannari, C. et al. (2008). Immunomodulatory activity of shikimic acid and quercitin in comparison with oseltamivir (Tamiflu) in an *in vitro* model. *J. Med. Virol.* 80 (4): 741–5.

Beyer, R. E. (1989). The role of coenzyme Q in free radical formation and antioxidation. In CRC *Handbook of Free Radicals and Antioxidants in Biomedicine* (Minguel, J., Quinanilha, A. T. and Weber, H. eds). Vol. II, CRC Press, Inc., Boca Raton, FL, 45–62.

Blatt, T., Mundt, C., Mummert, C., Maksiuk, T., Wolber, R., Keyhani, R., Schreiner, V., Hoppe, U., Schachtschabel, D. O. and Stäb, F. (1999). Modulation of oxidative stresses in human aging skin. *Paul Gerson Unna Forschungszentrum Beiersdorf AG* 32 (2): 83–8.

Britton, G. (1995). UV/Visible spectroscopy. In *Carotenoids*, Vol. 1B (Britton, G., Liaaen-Jensen, S. and Pfander, H.-P.). Birkhauser-Verlag, Basel, Switzerland, 13–64.

Bolsinger, J., Pronczuk, A., Sambanthamurthi, R. and Hayes, K. C. (2014). Anti-diabetic effects of palm fruit juice in the Nile rat (*Arvicanthis niloticus*). *J. Nutr. Sci.* 2014; 3: e5.

Bunyan, J., Michale, D., Green, J. and Marcinkiewicz, S. (1961). Biological potenices of ξ- and ζ1-tocopherol and 5-methyltocol. *Brit. J. Nutr.* 15 (2): 253–7.

Cabrini, L., Barzanti, V., Cipollone, M., Florentini, D., Grossi, G., Tolomelli, B., Zambonin, L. and Landi, L. (2001). Antioxidant and total Peroxyl radical trapping ability of olive and seed oils. *J. Agric. Food Chem.* 49: 6026–32.

Canfield, L. M., Kaminsky, R. G., Taren, D. L., Shaw, E. and Sander, J. K. (2001). Red palm oil in the maternal diet increases provitamin A carotenoids in breastmilk and serum of the mother-infant dyad. *Eur. J. Nutr.* 40: 30–8.

Carlson, B. C., Janssen, A. M., Larsson, A., Bucht, A. and Lorentzen, J. C. (2000). The Endogeneous adjuvant squalene can induce chronic T-cell-mediated arthritis in rats. *Am. J. Pathol.* 15 (6): 2057–65.

Chan, H. H. and Chan, K. K. (2003). Effects of tocotrienols on cell viability and apoptosis in normal murine liver cells (BNL CL.2) and liver cancer cells (BNL 1ME A.76.1) *in vitro*. In *PORIM Int. Conc.*, Putrajaya, Malaysia.

Che Idris, C. A. Karupaiah, T., Sundram, K., Tan, Y. A., Balasundram, N., Soon-Leow, S. S, Nasruddin, N. S. and Sambanthamurthi, R. (2014). Oil palm phenolics and vitamin E reduce atherosclerosis in rabbits. *J. Funct. Foods.* 7: 541–50.

Chen, X., Beutler, J. A., McCloud, T. G., et al. (2003). Tannic acid is an inhibitor of CXCL12 (SDF-1alpha)/CXCR4 with antiangiogenic activity. *Clin. Cancer Res.* 9: 3115–23.

Chideh, S., Serge Pilard, S., Attoumbre, J., Saguez, R., Hassan-Abdallah, A., Cailleu, D., Wadouachi, A., Sylvie Baltora-Rosset, S. (2014). 5-*O*-Caffeoylshikimic acid from *Solanum somalense* leaves: Advantage of centrifugal partition chromatography over conventional column chromatography. *J. Sep. Sci.* 37: 2331–9.

Chin-Dusting, J. P., Boak, L., Husband, A., et al. (2004). The iso-flavone metabolite dehydroequol produces vasodilatation in human resistance arteries via a nitric oxide-dependent mechanism. *Atherosclerosis* 176: 45–8.

Choo, Y. M., Ma, A. N. and Yusof, B. (1993). Palm Oil Methyl Esters as Fuel: Palm Diesel. PORIM TT No. 12.

Choo, Y. M. Yap, S. C., Ooi, C. K., Ma, A. N., Goh, S. H. and Ong, A. S. H. (1996). Recovered oil from palm-pressed fiber: A good source of natural carotenoids, vitamin E and sterols. *J. Am. Oil Chem. Soc.* 73 (5): 599–602.

Choo, Y. M., Ma, A. N. and Yap, S. C. (1999). Corotenes, vaitamin E and sterols in oils from Elaeis guineensis, Elaies oleifera and their hybrids. *Palm Oil Techn. Bull.* 5 (1).

Choo, Y. M. (2000). Specialty products: Carotenoids. Book Chapter. In: *Advances in Oil Palm Research Vol:* II (Basiron, Y., Jalani, B. S. and Chan, K. W.) Pub: Malaysian Palm Oil Board, 1036–60.

Choo, Y. M., Ma, A. N. and Basiron, Y. (2000). A Method of Chromatographic Isolation for Vitamin E Isomers. European Patent No. EP1097985.

Choo, Y. M., Lau, H. L. N., Puah, C. W., Ng, M. H., Bong, S. C., Ma, A. N. and Yusof, B. (2002a). Production of Phytonutrients (Carotenes, Vitamin E, Sterols, Squalene, Co-enzyme Q and Phospholipids) from Palm Methyl Esters. MPOB TT No. 151.

Choo, Y. M., Ng, M. H., Ma, A. N. and Yusof, B. (2002b). Production of Individual Carotene, Tocols and Cholesterol-free Sterol from Crude Palm Oil. MPOB TT No. 150.

Choo, Y. M., Lau, L. L. N., Ng, M. H., Ma, A. N. and Yusof, B. (2003a). Value Added Products from Palm Pressed Fibre. MPOB TT. No. 193.

Choo, Y. M., Ng, M. H., Ma, A. N. and Yusof, B. (2003b). A Method of Extracting and Isolating Palm Oil Minor Components. European Patent Application No. 03257486.5

Choo, Y. M., Ma, A. N., Chuah, C. H., Khor, H. T. and Bong, S. C. (2004a). A developmental study on the appearance of tocopherols and tocotrienols in developing palm mesocarp (Elaeis guineensis). *Lipids* 39 (6): 561–4.

Choo, Y. M., Bong, S. C., Ma, A. N. and Chuah, C. H. (2004b). Phospholipids from palm pressed fiber. *J. Am. Oil Chem. Soc.* 81 (5): 471–5.

Choo, Y. M., Ng, M. H., Ma, A. N. and Yusof, B. (2004c), Production of Co-enzyme Q10 from Crude Palm Oil. MPOB TT No. 232.

Choo, Y. M., Lau, H. L. N., Ma, A. N. and Yusof, B. (2005a). Residual Oil and Phytonutrients Concentrate from Palm-Pressed Fibre. Malaysian Patent Application No. PI 20053390.

Choo, Y. M., Ng, M. H., Ma, A. N., Chuah, C. H. and Mohd Ali, H. (2005b). Application of supercritical fluid chromatography in the determination of palm squalene, carotenes, vitamin E and sterols in palm oil. *Lipids* 40 (4): 429–32.

Choo, Y. M., Lau, H. L. N., Yung, C. L., Ng, M. H., Puah, C. W., Rusnani, A. M., Ma, A. N., Yahaya, H. and Andrew, Y. K. C. (2009). Value Addition from Crude Palm Oil – Integrated Production of Palm Biodiesel, Phytonutrients and Other Value-Added Products. MPOB TT No. 428.

Chua, C. S. L., Baharin, B. S., Che Man, Y. and Tan, C. P. (2007). Separation of squalene from palm fatty acid distillate using adsorption chromatography. *Eur. J. Lipid Sci. Tech.* 109 (11): 1083–7.

Das, S., Powell, S. R., Wang, P., Divald, A., Nesaretnam, K., Tosaki, A., Cordis, G. A., Maulik, N. and Das, D. K. (2005). Cardioprotection with palm tocotrienol: antioxidant activity of tocotrienol is linked with its ability to stabilize proteasomes. *Am. J. Physiol. Heart Circ. Physiol.* 289: H361–7.

Deshane, J., Chaves, L., Sarikonda, K. V., Isbell, S., Wilson, L., Kirk, M., et al. (2004). Proteomics analysis of rat brain protein modulations by grape seed extract. *J. Agric. Food Chem.* 52:7872–83.

Deprez, P. P., Volkman, J. K. and Daverport, S. R. (1990). Squalene content and neutral lipid composition of livers from deep-sea sharks caught in Tasmanian waters. *Austr. J. Mar. Freshwater Res.* 41: 375–87.

Dimascio, P., Kaiser, S. and Sies, H. (1989). Lycopene as the most efficient biological carotenoid singlet Oxygen Quencher. *Arch. Biochem. Biophys.* 274: 532–8.

dos Santos, M. D., Almeida, M. C., Lopes, N. P. and de Souza, G. E. (2006). Evaluation of the anti-inflammatory, analgesic and antipyretic activities of the natural polyphenol chlorogenic acid. *Biol. Pharm. Bull.* 11: 2236–40 .

Dybing, C. D. and Craig, B. M. (1969). Fatty acid biosynthesis and incorporation into lipid classes in seeds and seed tissues of flax. *Lipids* 5: 422–9.

El-Seedi, H. R., Ringbom, T. et al. (2003). Constituents of Hypericum laricifolium and their cyclooxygenase (COX) enzyme activities. *Chem. Pharmaceut. Bull.* 51(12): 1439–40.

Estevez, A. M. and Estevez, R. J. (2012). A short overview on the medicinal chemistry of (-)-Shikimic acid. *Mini Rev. Med. Chem.* 12 (14): 1443–64.

Evans, H. M., Emerson, O. H. and Emerson, G. A. (1936). The isolation of an alcohol, α-tocopherol, having properties of vitamin E. *J. Biol. Chem.* 113: 319–32.

Fabiani, R., De Bartolomeo, A., Rosignoli, P., et al. (2002). Cancer chemoprevention by hydroxytyrosol isolated from virgin olive oil through G1 cell cycle arrest and apoptosis. *Eur. J. Cancer Prev.* 11: 351–8.

Farquhar, J. W. (1996). Plant sterols: Their biological effects in humans. In *Handbook of Lipids in Human Nutrition*. CRC Press, Boston, 101–5.

Ferrante, R. J., Andreassen, O. A., Dedeoglu, A., Ferrante, K. L., Jenkins, B. G., Hersch, S. M. and Beal, M. F. (2002). Therapeutic effects of Coenzyme Q_{10} and Remacemide in transgenic mouse models of Huntington's disease. *J. Neurosci.* 1 March 2002, 22 (5): 1592–9.

Frankel, E., Waterhouwe, A. and Teissedre, P. (1995). Principal phenolic phytochemicals in selected Californian wines and their antioxidant activity in inhibiting oxidation of human low-density lipoprotein. *J. Agric. Food Chem.* 43: 890–4.

Gapor, A., Berger, K. G., Hashimoto, T. A., Kato, A., Tanabe, K., Manuro, H. and Yamaoka, M. (1981). In *Proc. of The Int. Conf. On Palm Oil Product Technology in The Eighties*. PORIM Press, Kuala Lumpur, Malaysia, 145.

Gapor, A. M. T. (1993a). Palm Based Vitamin E. MPOB TT. No. 8.

Gapor, A. M. T., Leong, W. L., Ong. A. S. H., Kawada, T., Watanabe, H. and Tsuchiya, N. (1993b). Production of high concentaration tocopherols and tocotrienols from palm oil by products. US patent No. 5190618.

Gapor, A. and Hazrina, A. R. (2000). Squalene in oils and fats. *Palm Oil Dev.* 32: 36–40.

Gapor, A., Wan Hasamudin, W. H. and Sulong, M. (2000). Phytochemicals for nutraceuticals from the by-product of palm oil refining. *Palm Oil Dev.* 36: 13–19.

Gapor, M. T. (2002). Production of Phytosterols from Palm Fatty Acid Distillate. MPOB Information Series, 185, MPOB TT No. 183.

Gapor, A. M. T., Sulong, M., Rosnah, M. S., Noorshamsiana, A. W. and Astimar, A. A. (2009). Developments in utilization of palm fatty acids distillate. In Proceedings of 2009 PIPOC, Kuala Lumpur.

Gasparaoli, A., Tagliabue, S. and Mariani, C. (1998). Phytosqualene: Genuineness of a raw material used in the cosmetic field (in Italian). *Riv. Ital. Sostanze Grasse* 75: 241–5.

George, S. and Arumughan, C. (2006). Distribution of lipids in the exocarp and mesocarp of three varieties of oil palm fruit (*Elaeis guineensis*). *J. Sci. Food Agric.* 56: 219–22 .

Goh, S. H. (1984). Total phospholipids in crude palm oil: Quantitative analysis and corrrelations with oil quality parameters. *Proc. JAOCS.* 61 (10): 1597–1600 .

Goh, S. H., Choo, Y. M. and Ong, A. S. H. (1985). Minor constituents of palm oil. *J. Am. Oil Chem. Soc.* 62 (2): 237–40.

Goh, S. H., Hew, N. F., Norhanom, A. E. and Yadar, M. C. (1994). Inhibition of tumour growth promotion by various palm oil Tocotrienols. *Int. J. Cancer.* 57: 529–31.

Grassi, D., Lippi, C., Necozione, S., et al. (2005). Short-term administration of dark chocolate is followed by a significant increase in insulin sensitivity and a decrease in blood pressure in healthy persons. *Am. J. Clin. Nutr.* 81: 611–14.

Grassi, D., Aggio, A., Onori, L., et al. (2008). Tea, flavonoids, and nitric oxide-mediated vascular reactivity. *J. Nutr.* 138: 1554S–60S.

Guthrie, N., Nesaretnam, K., Chambers, A. F. and Carroll, K. K. (1993). Inhibition of breast cancer cell growth by Tocotrienols. *FASEB J.* 7: A70.

Hadacek, F. (2002). Secondary metabolites as plant traits: Current assessment and future perspectives. *Crit. Rev. Plant Sci.* 21 (4): 273–322.

Hamid, H. A., Choo, Y. M., Goh, S. H. and Khor, H. T. (1995). *The Ubiquinones of Palm Oil in (*Ong, Niki *and* Packer *Eds) Nutrition, Lipids, Health and Disease*. AOCS Press, USA, 122–8.

Haque, A. M., Hashimoto, M., Katakura, M., Tanabe, Y., Hara, Y. and Shido, O. (2006). Long-term administration of green tea catechins improves spatial cognition learning ability in rats. *J. Nutr.* 136: 1043–7.

Harborne, J. B. (1998). Phenolic compounds. In: *Phytochemical Methods*. 3rd edn. Chapman and Hall, London, 40–106.

Hartman, R. E., Shah, A., Fagan, A. M., Schwetye, K. E., Parsadanian, M., Schulman, R. N., et al. (2006). Pomegranate juice decreases amyloid load and improves behavior in a mouse model of Alzheimer's disease. *Neurobiology* 24: 506–15 .

Hassni, M., Hadrami, A., Daayf, G., et al. (2004). Chitosan, antifungal product against Fusarium oxysporum f. sp. Albedinis and elicitor of defence reactions in date palm roots. *Phytopathol. Mediterr* 43: 195–204.

Hazura, Abd. Hamid, Choo, Y. M., Goh, S. H. and Khor, H. T. (1995). The Ubiquinones of palm oil. In *Book Chapter in Nutrition, Lipids, Health and Diesease* (Ong, Nike and Parker, M.). AOCS Press, USA, 122–8.

Hedtmann, U., Hobert, K., Mikova, T. and Weizel, P. (1988). Synthesis of potential Ecdysteroid (moulting horomone) precursors from DELTA sterols. *Tetrahedron Letts.* 44: 1941–52.

Hokkanen, J., Mattila, S., Jaakola, L., et al. (2009). Identification of phenolic compounds from lingonberry (*Vaccinium vitis-idaea* L.), bilberry (*Vaccinium myrtillus* L.) and hybrid bilberry (*Vaccinium £ intermedium Ruthe* L.) leaves. *J. Agric. Food Chem.* 57, 9437–47.

Holub, B. J., Sicilia, F. and Mahadevappa, V. G. (1989). Effect of tocotrienol derivatives in collagen and ADP-induced human platelet aggregation. *PORIM Int. Palm Oil Dev. Conf.*

Hornstra, G. (1988). Dietary lipids and cardiovascular disease: Effect of palm oil. *Oleagineux* 43: 75–81.

Huang, F. Y., Xiu, Q. P. et al. (2002). Anti-platelet and anti-thrombotic effects of triacetylshikimic acid in rats. *J. Cardiovasc Pharmacol.* 39 (2): 262–70.

Jaffri, J. M., Suhaila, M., Ahmed, I. N., Mustapha, N. M., Manap, Y. A. and Rohimi, N. (2011a). Effects of catechin-rich oil palm leaf extract on normal and hypertensive rats' kidney and liver. *Food Chem.* 128: 433–41.

Jaffri, J. M., Mohamed, S., Rohimi, N., Ahmad, I. N., Noorodin, M. M. and Manap, Y. A. (2011b). Antihypertensive and cardiovascular effects of Catechin-rich oil palm (*Elaeis guineensis*) leaf extract in NO deficient rats. *J. Med. Food.* 14: 775–83.

Ji, X., Usman, A., Razalli, N. H., Sambanthamurthi, R. and Gupta, S. V. (2015). Oil palm phenolics (OPP) inhibit pancreatic cancer cell proliferation via suppression of NF-κB pathway. *Anticancer Res.* 35 (1): 97–106.

Joe, A. K., Liu, H., Suzui, M., et al. (2002). Resveratrol induces growth inhibition, S-phase arrest, apoptosis, and changes in biomarker expression in several human cancer cell lines. *Clin. Cancer Res.* 8: 893–903.

Kagan, V. E. and Quinn, P. J. eds. (2001). *Coenzyme Q: Molecular Mechanism in Health and Disease.* CRC Press. Boca Raton, FL.

Kaiya, A. (1990). The use of natural squalene and squalane, and the latest situation of the raw materials (in Japanese). *Yukagaku* 39: 525–9 .

Kakkar, S. and Bais, S. (2014). A Review on Protocatechuic Acid and Its Pharmacological Potential. Hindawi Publishing Corporation. ISRN Pharmacology Volume 2014, Article ID 952943, 9 pages http://dx.doi.org/10.1155/2014/952943.

Kang, N. J., Lee, K. W., Shin, B. J., Jung, S. K., Hwang, M. K., Bode, A. M., et al. (2009). Caffeic acid, a phenolic phytochemical in coffee, directly inhibits Fyn kinase activity and UVB-induced COX-2 expression. *Carcinogenesis* 30: 321–30, .

Karunanidhi, A., Thomas, R., Avan Belkum, A. and Vasanthakumari Neela, V. (2013). *In vitro* Antibacterial and Antibiofilm Activities of Chlorogenic Acid against Clinical Isolates of Stenotrophomonas maltophilia including the Trimethoprim/Sulfamethoxazole Resistant Strain. BioMed Research International Volume 2013, Article ID 392058, 7 pages.

Kayama, M. and Mankura, M. (1998). Natural Oleochemicals in marine fishes. *INFORM* 9: 794–9.

Khanna, S., Roy, S., Slivka, A., Craft, T. K. S., Chaki, S., Rink, C., Notestine, M. A., DeVries, C., Parinandi, N. L. and Sen, C. K. (2005). Neuroprotective properties of the natural vitamin E -Tocotrienol. *Stroke* 36: e144–52.

Khanna, S., Parinandi, S. L., Kotha, S. R. and Roy, S. (2010). Nanomolar vitamin E α-tocotrienol inhibits glutamate-induced activation of phospholipase A_2 and causes neuroprotection. *J. Neurochem.* 112 (5): 1249–60.

Kuenzig, W., Chau, J., Norkus, E., Holowaschenko, H. and Newmark (1984). Caffeic and Ferulic acid as blockers of nitrosamine formation. *Carcinogenesis* 5: 309–14.

Kueper, P. S., Biemoth, T., Wöhrmann, Y., Münster, A. and Beiersdorf. (2008). Aging skin is functionally anaerobic: importance of coenzyme Q10 for anti aging skin care. *Biofactors* 32 (1–4): 245–55.

Kuiper, G. G., Lemmen, J. G., Carlsson, B. et al. (1998). Interaction of estrogenic chemicals and phytoestrogens with estrogen receptor beta. *Endocrinology* 139: 4252–63.

Kulkarni, A. S. (1991a). Phospholipids and glycolipids in the oil from tenera variety of oil palm. *Indian Oil Palm J.* 1 (4): 7–11.

Kulkarni, A. S. (1991b). Phospholipis and glycolipis in the oil fom some varienties of *Elaeis guineesis* in India. *Elaeis* 3 (2): 363–8.

Kunnumakkara, A. B., Sung, B., Ravindran, J., Diagaradjane, P., Deorukhkar, A., Dey, S., Koca, C., Yadav, V. R., Tong, Z., Gelovani, J. G., Guha, S., Krishan, S. and Aggarwal, B. (2010). γ-Tocotrienol inhibits pancreatic tumors and sensitizes them to gemcitabine treatment by modulating the inflammatory microenvironment. *Cancer Res.* 70: 8695.

Latip, R. A., Baharin, B. S., Che Man, Y. B. and Abdul Rahman, R. (2000). Evaluation of different types of synthetic adsorbents for carotene extraction from crude palm oil. *J. Am. Oil Chem. Soc.* 77 (12): 1277–81.

Lau, H. L. N., Choo, Y. M., Ma, A. N. and Chuah, C. H. (2003). An improved and efficient method for analysing palm oil and other vegetable oils. In *PORIM Int. Conf.*, Putrajaya, Malaysia.

Lau, H .L .N., Puah, C. W., Choo, Y. M., Ma, A. N. and Chuah, C. H. (2005). Simultaneous quantifications of free fatty acids, free sterols, squalene and acyl glycerols molecular species in palm oil by high – temperature gas chromatography – flame ionizatio detection. *Lipis* 40 (5): 523–8.

Lau, H. L. N., Choo, Y. M., Ma, A. N. and Chuah, C. H. (2006a). Characterization and supercritical carbon dioxide extraction of palm oil *(Elaeis guineensis)*. *J. Food Lipids* 13: 210–21.

Lau, H. L. N., Choo, Y. M., Ma, A. N. and Chuah, C. H. (2006b). Quality of residual oil from palm-pressed mesocarp fiber (*Elaeis guineensis*) using supercritical CO_2 with and without ethanol. *J. Am. Oil Chem. Soc.* 83 (10): 893–8.

Lau, H. L. N., Choo, Y. M., Ma, A. N. and Chuah, C. H. (2008). Selective extraction of palm carotene and vitamin E from fresh palm-pressed fiber *(Elaeis guineensis)* using supercritical CO_2. *J. Food Eng.* 84: 289–96.

Lenaz, G. (Ed.) (1985a). *A Biochemical Rationale for the Therapeutic Effects of Coenzyme Q. Coenzyme Q.* John Wiley & Sons Ltd, New York, 474–5.

Lenaz, G. (Ed.) (1985b). *Coenzyme Q: Biochemistry, Bioenergetics and Clinical Applications of Ubiquinone.* Wiley, New York.

Leow, S. S., Sekaran, S. D., Sundram, K., Tan, Y. A. and Sambanthamurthi, R. (2013a). Oil palm phenolics attenuate changes caused by an atherogenic diet in mice. *Eur. J. Nutr.* 52: 443–56.

Leow, S. S., Sekaran, S. D., Sundram, K., Tan, Y. and Sambanthamurthi, R. (2013b). Gene expression changes in spleens and livers of tumour-bearing mice suggest delayed inflammation and attenuated cachexia in response to oil palm phenolics. *J. Nutrigenet. Nutrigenomics* 6 (6): 305–26.

Leow, S. S., Sekaran, S., Sundram, K., Tan, Y. A. and Sambanthamurthi, R. (2013c). Oil palm phenolics confer neuroprotective effects involving cognitive and motor functions in mice.

Li, X. H., Fu, D., Latif, N. H., Mullaney, C. P., Ney, P. H., Mog, S. R., Whitnall, M. H., Srinivasan, V. and Xiao, M. (2010). Delta-tocotrienol protects mouse and human hematopoietic progenitors from gamma-irradiation through Erk/mTOR signaling. *Haematologica* 2010, doi: 10.3324/haematol.2010.026492.

Li, X., Wang, X., Chen, D. and Chen S. (2011), Antioxidant activity and mechanism of protocatechuic acid *in vitro*. *Funct. Foods Health Dis.* 7: 232–44.

Lim, D. Y., Jeong, Y., Tyner, A. L., et al. (2007). Induction of cell cycle arrest and apoptosis in HT-29 human colon cancer cells by the dietary compound luteolin. *Am. J. Physiol. Gastro-intest Liver Physiol.* 292: G66–75.

Lipshutz, B. H., Molland, P., Pheiffer, S. S. and Chrisman, W. (2002). A short highly efficient synthesis of coenzyme Q_{10}. *J. Am. Chem. Soc.* 124: 14282–3.

Liu, R. H. (2004). Potential synergy of phytochemicals in cancer prevention: mechanism of action. *J. Nutr.* 134 (12): 3479S–85S.

Loberg, R. D., Bradley, D. A., Tomlins, S. A., Chinnaiyan, A. M. and Pienta, K. J. (2007). The lethal phenotype of cancer: the molecular basis of death due to malignancy. *CA Cancer J. Clin.* 57: 225–41.

Lou, Z., Wang, H., Zhu, S., Ma, C. and Wang, Z. (2011). Antibacterial activity and mechanism of action of chlorogenic acid. *J. Food Sci.* 76 (6): M398–403.

Maier, V. P., Metzler, D. M. and Huber, A. F. (1964). 3-O-Caffeoylshikimic acid (dactylifric acid) and its isomers, a new class of enzymic browning substrates. *Biochem. Biophys. Res. Commun.* 14: 124–8.

Maniam, S., Ima Nirwana, S., Ahmad Nazrun, S. and Norazlina, M (2007). Palm tocotrienols exerted better antioidant activities than a- tocopherols in bone. *In Proceeding 2007 PIPOC International Palm Oil Congress: Empowering Change*, August 26–30.

Mellons, A. and Tappel, A. L. (1996). The inhibition of mitochondrial peroxidation by ubiquinone and ubiquinol. *J. Biol. Chem.* 241: 4353–6.

Miettinen, T. A. and Vanhanen, H. (1994). Serum concentration and metabolism of cholesterol during rapeseed oil and squalene feeding. *Am. J. Clin. Nutr.* 59: 356–63.

Mishima, K., Tanaka, T., Pu, F., Egashira, N., Iwasaki, K., Hidaka, R., Matsunaga, K., Takata, J., Karube, Y. and Fujiwara, M. (2003). Vitamin E isoforms α-tocotrienol and γ-tocopherol prevent cerebral infarction in mice. *Neurosci. Lett.* 337 (1): 56–60.

Miyazawa, T., Inokuchi, H., Hirokane, H., Tsuzuki, T., Nakagawa, K. and Igarashi, M. (2004). Anti-angiogenic potential of Tocotrienol *in vitro*. *Biochem. (Moscow)* 69 (1): 67–9.

Modafar, C., Tantaoui, A. and Boustani, E. (2000). Effect of caffeoylshikimic acid of date palm roots on activity and production of Fusarium f. sp. albedinis cell wall-degrading enzymes. *J. Phytopathol.* 148: 101–8.

Mohamed, S., Lee Ming, T. and Jaffri, J. M. (2013). Cognitive enhancement and neuroprotection by catechin-rich oil palm leaf extract supplement. *J. Sci. Food Agric.* 93 (4): 819–27.

Mohamed, S. (2014). Oil palm leaf: A new functional food ingredient for health and disease prevention. *J. Food Process Technol.* 5: 300.

Mulvihill, E. E., Allister, E. M., Sutherland, B. G., Telford, D. E., Sawyez, C. G., Edwards, J. Y., Markle, J. M., Hegele, R. A. and Huff, M. W. (2009). Naringenin prevents dyslipidemia, apolipoprotein B overproduction, and hyperinsulinemia in LDL receptor-null mice with diet-induced insulin resistance. *Diabetes* 58 (10): 2198–210.

Murakoshi, M. and Nishino, H. (2001). Potent cancer chemopreventive action of palm fruit carotene and various kinds of dietary carotenoids. In *2001 PORIM Int. Conf.*, Kuala Lumpur, Malaysia.

Nafeeza, M. I. and Kang, T. T. (2005). Synergistic effects of tococpherols, tocotrienols and ubiquinone in indomethacin – induced experimental gastric lesion. *Int. J. Vitam. Nutr. Res.* 75 (2): 149–55.

Nesaretnam, K., Khor, H. T., Ganesan, J., Chong, Y. H., Sundram, K. and Gapor, A. (1992). The effect of vitamin E tocotrienols from palm oil on chemically-induced mammary carcinogenesis in female rats. *Nutr. Res.* 12: 63–75.

Nesaretnam, K., Guthrie, N., Chambers, A. F. and Caroll, K. K. (1995). Effect of Tocotrienols on the growth of human breast cancer cell line in culture. *Lipids* 30: 1139–43.

Nesaretnam, K., Stephen, R., Dils, R. and Darbre, P. (1998). Tocotrienols inhibit the growth of human breast cancer celss irrespective of estrogen receptor status. *Lipis* 33 (5): 461–9.

Nesaretnam, K. (2000). Antioxidant and anti cancer properties of Tocotrienols. *Palm Oil Dev.* 32: 10–17.

Nesaretnam, K., Radhakrishnan, A., Selvaduray, K. R., Reimann, K., Pailoor, J., Razak, G., Mahmood, M. M. and Dahliwal, J. S. (2002). Effect of palm oil carotene on breat cancer tumorigenicity in nude mice. *Lipids* 37 (6): 557–60.

Nesaretnam, K., Patricial, A. L., Kanga Rani, S. and Ghazali, A. R. (2007a). Tocotrienols and breast cancer – The outcome from a clinical trial. In Proceeding 2007 PIPOC International Palm Oil Congress: Empowering Change 26–30 August.

Nesaretnam, K. N., Wong, W. Y. and Mohd Basri, W. (2007b). Tocotrienols and cancer: Beyond antioxidant activity. *Eur. J. Lipid Sci. Technol.* 109 (4): 445–52.

Nesaretnam, K., Teoh, H. K., Sevaduray, K. R., Brono, R. S and Emily, H. (2008). Modulation of the cell growth and apoptosis response in human prostate cancer cells supplemented with tocotrienols. *Eur. J. Lipid Sci. Technol.* 111 (1): 23–31.

Ng, M. H., Choo, Y. M., Ma, A. N., Chuah, C. H. and Mohd Ali, H. (2004a). Isolation of palm Tocols using supercritical fluid chromatography. *J. Chrom. Sci.* 42 (10): 536–9 .

Ng, M. H., Choo, Y. M., Ma, A. N., Chuah, C. H. and Mohd Ali, H. (2004b). Separation of Vitamin E (Tocopherol, Tocomonoenol and Tocotrienol) from palm oil. *Lipids* 39: 1031–5.

Ng, M. H., Choo, Y. M., Ma, A. N., Chuah, C. H. and Mohd Ali. H. (2006). Separation of Coenzyme Q_{10} in palm oil by supercritical fluid chromatography. *Am. J. Appl. Sci.* 3 (7): 1929–32.

Ng, M. H., Choo, Y. M., Ma, A. N., Chuah, C. H. and Mohd Ali, H. (2009). Analyses of Coenzyme Q9 and Q10 in developing palm fruits. *J. Amer. Oil Chem. Soc.* 86: 201–5.

Nguyen T. L. et at. (2001). Effects of red palm oil supplementation on vitamin A and iron status of rural underfive chidren in Vietnam. Proceedings of Food Technology & Nutrition Conference, International Palm Oil Congress 2001, Kuala Lumpur, Malaysia.

Nicolson, G. (2014). Mitochondrial dysfunction and chronic disease: Treatment With natural supplements. *Integr. Med.* 13: 34–43.

Norazlina, M., Ng, F. W. and Ima-Nirwana, S. (2005). Gamma-tocotrienol is required for normal vitamin D metabolism in female rats. *Indian J. Pharmacol.* 37 (5): 309–14.

Ong, A. S. H., Choo, Y. M., Cheah, K. Y. and Bakar A. (1992). Production of Alkyl Esters from Oils and Fats. Australian Patent No. 43519/89.

Ong, A. S. H. and Goh, S. H. (2002). Palm oil: A healthful and cost effective dietary component. *Food Nutr. Bull.* 23 (1): 11–22.

Ooe, I. M. and Fujii, K. (2008). Mechanisms of inhibitory effects of CoQ10 on UVB-induced wrinkle formation *in vitro* and *in vivo*. *Biofactors* 32 (1–4): 237–43.

Osakabe, N., Yasuda, A., Natsume, M., et al. (2002). Catechins and their oligomers linked by C4 → C8 bonds are major cacao polyphenols and protect low-density lipoprotein from *oxidatioin vitro*. *Exp. Biol. Med.* (Maywood) 227: 51–6.

Osborne, A. E., Sanchez, J. A., Solomon, M., Stopa, A., Wangh, L. J., Sambanthamurthi, R. and Hayes, K. C. (2014a). Palm fruit juice mitigates AZT mitochondrial genotoxicity and dose- dependent cytotoxicity. *J. AIDS Clin. Res.* 5: 12.

Osborne, A. E., Wangh, L. J., Hayes, K. C. and Sambanthamurthi. R. (2014b). Treatment of DNA damage and mitochondrial dysfunction using palm fruit juice. Patent Application WO2014124140 A1.

Osborne, A. E., Sanchez, J. A., Wangh, L. J., Sambanthamurthi, R. and Hayes, K. C. (2015). Oxidative damage is not a major contributor to AZT-induced mitochondrial mutations. *J AIDS Clin. Res.* 6: 4.

Qureshi, A. A., Mo, H. B., Packer, L. and Peterson, D. M. (2000). Isolation and identification of novel tocotrienols from rice bran with hypochloesterolemic, antioxidant and antitumor properties. *J. Agric. Food Chem.* 48: 3130–40.

Raffoul, J. J., Wang, Y., Kucuk, O., et al. (2006). Genistein inhibits radiation-induced activation of NF-kappaB in prostate cancer cells promoting apoptosis and G2/M cell cycle arrest. *BMC Cancer* 6: 107.

Ranalli, A. and Demattia, G. (1997). Characterisation of olive oil produced with a new enzyme processing aid. *J. Amer. Oil Chem. Soc.* 49: 6026–32.

Ranalli, A., Demattia, G. and Ferrante, M. L. (1997). Comparative evaluation of the olive oil given by a new processing system. *Int. J. Food Sci. Tech.* 32: 289–97.

Ranalli, A., Ferrante, M. L., Demattia, G. and Costentini, N. (1999). Analytical evaluation of virgin olive oil of first and second extraction. *J. Agric. Food Chem.* 47: 417–24.

Ranheim, T. and Halvorsen, B. (2005). Coffee consumption and human health: beneficial or detrimental? Mechanisms for effects of coffee consumption on different risk factors for cardiovascular disease and type 2 diabetes mellitus. *Mol. Nutr. Food Res.* 49: 274–84.

Rao, G. and Achaya, K. T. (1970). Monohydroxylation of unsaturated oils: II. Autoxidation – Reduction. *J. Amer. Oil. Chem. Soc.* 47 (8): 289–91.

Rao, C. V., Newmark, H. L. and Reddy, B. S. (1998). Chemopreventive effect of squalene on colon cancer. *Carcinogenesis* 19 (2): 287–90.

Reddy, P. H. (2008). Mitochondrial medicine for aging and neurodegenerative diseases. *Neuromolecular Med.* 10 (4): 291–315.

Regnstrom, J., Nilsson, J., Tornvall, P., et al. (1992). Susceptibility to low-density lipoprotein oxidation and coronary athero-sclerosis in man. *Lancet* 339: 1183–6.

Rice-Evans, C. A., Miller, N. J. and Paganga, G. (1996). Structure – anti- oxidant activity relationships of flavonoids and phenolic acids. *Free Radic. Biol. Med.* 20: 933–56.

Rodríguez-Vaquero, M. J., Vallejo, C. V. A., Sosa, O., Saguir F. M. and Aredes-Fernandez P. A. (2015). Antimicrobial properties and potential usefulness of Caffeic acid. In *Caffeic Acid: Biological Properties, Structure and Health Effects* (Leanna Vaughn ISBN: 978-1-63483-159-8), 63–76.

Runnie, I., Nordin, M. M., Radzali, M., Azizah, H., Hapizah. N., et al. (2003). Antioxidant and hypocholesterolemic effects of Elaeis Guineensis leaves extract on hypercholesterolemic rabbits. *ASEAN Food J.* 12: 137–47.

Salah, N., Miller, N. J., Paganga, G., et al. (1995). Polyphenolic flavanols as scavengers of aqueous phase radicals and as chain-breaking antioxidants. *Arch Biochem. Biophys.* 322: 339–46.

Salazar-Martinez, E., Willett, W. C., Ascherio, A., Manson, J. E., Leitzmann, M. F., Stampfer, M. J. and Hu, F. B. (2004). Coffee consumption and risk for type 2 diabetes mellitus. *Ann. Intern. Med.* 140: 1–8.

Salleh, M. N., Runnie, I., Roach, P. D., Mohamed, S. and Abeywardena, M. Y. (2002). Inhibition of low-density lipoprotein oxidation and up-regulation of low-density lipoprotein receptor in HepG2 cells by tropical plant extracts. *J. Agric. Food Chem.* 50: 3693–7.

Sambanthamurthi, R., Sundram, K. and Tan, Y. (2000). Chemistry and biochemistry of palm oil. *Prog. Lipid Res.* 39 (6): 507–58.

Sambanthamurthi, R., Tan, Y. A., Sundram, K. inventors (2008). Malaysian Palm Oil Board, assignee. Treatment of vegetation liquors derived from oil-bearing fruit. United States patent US 7,387,802 B2.

Sambanthamurthi, R., Rha, C., Sinskey, A. J., Tan, Y. A., Sambandan, T. G., Wahid, M. B. (2010). Oil Palm Phenolics as a source of shikimic acid. MPOB Information Series ISSN 1511-7871.

Sambanthamurthi, R., Tan, Y. A., Sundram, K., Abeywardena, M., Sambandan, T. G., Rha, C., Sinskey, A. J., Subramaniam, K., Leow, S. S., Hayes, K. C. and Wahid, M. B. (2011). Oil palm vegetation liquor: a new source of phenolic bioactives. *Br. J. Nutr.* 106: 1655–63.

Sambanthamurthi, R., Tan, Y. A., Sundram, K., Hayes, K. C., Abeywardena, M., Leow, S. S., Sekaran, S. D., Sambandan, T. G., Rha, C., Sinskey, A. J., Subramaniam, K., Fairus, S. and Wahid, M. B. (2011). Positive outcomes of oil palm phenolics on degenerative diseases in animal models. *Br. J. Nutr.* 106: 1664–75.

Sekaran, S. D., Leow, S. S., Abobaker, N., Tee, K. K., Sundram, K., Sambanthamurthi, R. and Wahid, M. B. (2010). Effects of oil palm phenolics on tumor cells *in vitro* and *in vivo*. *Afr. J. Food Sci.* 4: 495–502.

Shults, C. W., Oakes, D., Kieburtz, K., Beal, M. F., Haas, R., Plumb, S., Juncos, J. L., Nutt, J., Shoulson, I., Carter, J., Kompoliti, K., Perlmutter, J. S., Reich, S., Stern, M., Watts, R. L., Kurlan, R., Molho, E., Harrison, M. and Lew, M. (2002), Effects of coenzyme Q10 in early Parkinson disease: evidence of slowing of the functional decline. *Arch Neurol.* 59 (10): 1541–50.

Sinatra, S. T. (1998). *The Coenzyme Q10 Phenomenon*. Keats Publishing, New Canaan.

Sylvester, P. W. and Sumit, S. (2003). Intracelluar mechanisms mediating tocotrienol-induced apoptosis in neoplastic mammary epithelial cells. In *PORIM Int. Conf.*, Putrajaya, Malaysia.

Smith, T. (2000). Squalene: potential chemopreventive agent. *Informa* 9 (8): 1841–8 .

Spiller, G. A. (2004). *The Trans Fats Dilemma and Natural Palm Oil*. Pascoe Publishing Inc., California.

Steinberg, D., Parthasarathy, S., Carew, T. E., Khoo, J. C. and Witztum, J. L. (1989). Beyond cholesterol. Modifications of low-density lipoprotein that increase its atherogenicity. *N Engl. J. Med.* 320: 915–24.

Stuijvenberg, M. E. et al. (2000). Red palm oil as a source of beta-carotene in a school biscuits used to address vitamin A deficiency in primary school children. *Int. J. Food Sci. Nutr.* 51: S43–50.

Sundram, K., Khor, H. T., Ong, A. S. H. and Pathmanathan, R. (1989). Effect of dietary palm oil on mammary carcinogenesis in female rats induced by 7,12-dimethylbenz(a)anthracene. *Cancer Res.* 49: 1447–51.

Sundram, K., Sambanthamurthi, R. and Tan, Y. A. (2003). Palm fruit chemistry and nutrition. *Asia Pacific J. Clin. Nutr.* 3: 355–62.

Suzuki, A., Fujii, A., Yamamoto, N., et al. (2006). Improvement of hypertension and vascular dysfunction by hydroxyhydroquinone-free coffee in a genetic model of hypertension. *FEBS Lett.* 580: 2317–22.

Swerdlow, R. H. (2011). Brain aging, Alzheimer's disease, and mitochondria. *Biochim. Biophys. Acta.* 1812 (12): 1630–9.

Takada, K., Terada, T., Yamanishi, H., Ashida, Y., Inomata, S., Nishiyama, T. and Amano, S. (2009). Coenzyme Q10 protects against oxidative stress-induced cell death and enhances the synthesis of basement membrane com-ponents in dermal and epidermal cells. *Muta- Biofactors.* 35 (5): 435–41.

Tan, B., Grady, C. M. and Gawrenowski, A. M. (1986). Hydrocarbon carotenoid profiles of palm oil processed fractions. *J .Am. Oil Chem. Soc.* 63: 1175–9.

Tan, B. (1987). Novel aspects of palm oil carotenoid analytical biochemistry. *Proc. O.P/P. O. Conference-Technology*: 370–6.

Tan, R. R. T., Mohamed, S., Samaneh, G. F., Noordin, M. M., Goh, Y. M. and Manap, M. Y. A. (2011). Polyphenol rich oil palm leaves extract reduce hyperglycaemia and lipid oxidation in STZ-rats. *Int. Food Res. J.* 18: 179–88.

Tan, Y. A. and Kuntom, A. (1993). Gas chromatographic determination of hydrocarbons in crude palm kernel oil. *J. AOAC Int.* 76: 371–6.

Tan, Y. A. and Kuntom, A. (1994). Hydrocarbons in crude palm kernel oil. *J. AOAC Int.* 77: 67–73.

Tan, Y. A., Sambanthamurthi, R., Sundram, K. and Wahid, M. B. (2007). Valorisation of palm by-products as functional components. *Eur. J. Lipid Sci. Tech.* 109: 380–93.

Tanaka, T., Nishikawa, A., Shima, H., Sugie, S., Shinoda, T., Yoshimi, N., Iwata, H. and Mori, H. (1990). Inhibitory effects of chlorogenic acid, reserpine, polyprenoic acid (E-5166), or coffee on hepatocarcinogenesis in rats and hamsters. *Basic Life Sci.* 52: 429–40 .

Tay, B. Y. P. and Choo, Y. M. (1999). Valuable constituent of commercial red palm olein: Carotenoids, Vitamin E, Ubiquinones and sterols. *Proceedings of the 1999 PORIM International Palm Oil Congress (Chemistry and Technology)*: 97–108.

Tay, B. Y. P. and Choo, Y. M. (2000a). Valuable minor consitituents of commercial red palm olein: Carotenoids, Vitamin E, Ubiquinones and sterols. *J. Oil Palm Res.* 12 (1): 14–24.

Tay, B. Y. P. and Choo, Y. M. (2000b). Practical guide to establishing palm carotenoids profiles by HPLC with three dimensional diode array detector. *Palm Oil Dev.* 22: 13–17.

Tay, B. Y. P., Choo, Y. M., Gwenoline Ee, C. L. and Goh, S. H. (2001). Geometrical isomers of the major provitamin A palm carotenes, a- and b- carotenes in the mesocarp oil of fresh and sterilized palm fruits, crude palm oil and palm carotene based products: red palm olein and carotene concentrates. *J. Oil Palm Res.* 13 (2): 23–32.

Tomeo, A. C., Geller, M., Watkins, T. R., Gapor, A. and Bierenbaum, M. L. (1995). Antioxidant effects of tocotrienols in patients with hyperlipidemia and carotid stenosis. *Lipids*: 1179–83 .

Touaibia, M., Jean-Francois, J. and Doiron, J. (2011). Caffeic acid, a versatile pharmacophore: an overview. *Mini Rev. Med. Chem.* 11: 695–713.

Tran, M. T., Mitchell, T. M., Kennedy, D. T. and Giles, J. T. (2001). Role of Coenzyme Q_{10} in chronic heart failure, Angina and Hypertension. *Pharmacotherapy* 21 (7): 797–806.

Trémolières, A. and Lepage, M. (1971). Changes in lipid composition during greening of etiolated pea seedlings. *Plant Physiol.* 47: 329–34.

Umegaki, K., Uramoto, H., Suzuki, J. and Esashi, T. (1997). Feeding mice palm carotene prevents DNA damage in bone marrow and reduction of peripheral leukocyte counts, and enhances survival following X-ray irradiation. *Carcinogenesis* 18 (10): 1943–7.

Van Het Hof. K. H. et al. (1999). Comparison of the bioavailability of natural palm oil carotenoids and synthetic beta-carotene in humans. *J. Agric. Food Chem.* 47 (4): 1582–6.

van Praag, H., Lucero, M. J., Yeo, G. W., Stecker, K., Heivand, N., Zhao, C., et al. (2007). Plant-derived flavanol (–)epicatechin enhances angiogenesis and retention of spatial memory in mice. *J. Neurosci.* 27: 5869–78.

Victor, V. M., Apostolova, N., Herance, R., Hernandez-Mijares, A. and Rocha, M. (2009). Oxidative stress and mitochondrial dysfunction in atherosclerosis: mitochondria-targeted antioxidants as potential therapy. *Curr. Med. Chem.* 16 (35): 4654–67.

Vitaglione, P., Donnarumma, G., Napolitano, A., Galvano, F., Gallo, A., Scalfi, L., et al. (2007). Protocatechuic acid is the major human metabolite of cyanidin-glucosides. *J Nutr.* 137: 2043–8.

Wan Nazaimoon, W. N. and Khalid, B. A. K. (2002). Tocotrienols- rich diet decrease advanced glycosylation end – product in non –diabetic rats and improves glycemic control in streptozotocin-induced diabetib rats. *Malays. J. Pathol.* 34 (2): 77–82.

Wang, J., Ho, L., Zhao, Z., Seror, I., Humala, N., Dickstein, D. L., et al. (2006). Moderate consumption of Cabernet Sauvignon attenuates Abeta neuropathology in a mouse model of Alzheimer's disease. *FASEB J.* 20: 2313–20.

Wang, X. L., Rainwater, D. L., Mahanev, M. C. and Stocker, R. (2004). Cosupplementation with vitamin E and coenzyme Q_{10} reduces circulating markers of inflammation in baboons. *Am. J. Clin. Nutr.* 80 (3): 649–55.

Wong, W. Y., Kanga Rani, S., Cheng, H. M. and Kalanitri, N. (2009). Suppression of tumor growth by palm tocotrinols via the attenuation of angiogenesis. *Nutr. Cancer* 61 (3): 367–73.

Wyman, M., Leonard, M. and Morledge, T. (2010). Coenzyme Q10: A therapy for hypertension and statin-induced myalgia? *Clevel. Clinic J. Med.* 77 (7): 435–42.

Yamamura, Y. (1985). *A Survey of the Therapeutic Uses of Coenzyme Q* (Lenaz, G. ed.). John Wiley & Sons Ltd, New York, 479–95.

Yang, T. T. and Koo, M. W. (2000). Inhibitory effect of Chinese green tea on endothelial cell-induced LDL oxidation. *Atherosclerosis* 148: 67–73.

Yankah, V. V. and Jones, P. J. H. (2001). Phytosterols and health implications – commercial products and their regulation. *Inform* 12: 1011–14, 1016.

Yap, S. C., Choo, Y. M., Ooi, C. K., Ong, A. S. H. and Goh, S. H. (1991). Carotenes in the oil from different palm species. *Elaeis* 8 (2): 369–78.

Yu F. L., Gapor M. T., Wanda B. and Katarzyna B. (2008). Palm oil tocotrienols as antioidants and chemopreventive agents. *J. Oil palm Res.* October (Special issue).

Yurttas, H. C. and Addis, P. B. (2001). Antioxidant activity of palm tocotrienols in meat systems. In *2001 PORIM Int. Conf.*, Kuala Lumpur, Malaysia.

Zagre, N. M. et al. (2002). Changes in vitamin A intake following the social marketing of red palm oil among children and women in Burkina Faso. *Sante* 12 (1): 38–44.

Ziegler, R. G. (1989). A review of epidemiological evidence that carotenoids reduce the risk of cancer. *J. Nutr.* 119: 116–22.

Ziegler, R. G., Colavito, E. A., Hartge, P., Mcadams, M. J., Schoenberg, J. B., Mason, T. J. and Fraumen, J. F. Jr. (1996a). Importance of α-carotene, β-carotene and other phytochemicals in the etiology of lung cancer. *J. Nat. Cancer Inst.* 88: 612–15.

Ziegler, R. G., Mayne, S. T. and Swanson, C. A. (1996b). Nutrition and lung cancer. *Cancer Causes Control* 7: 157–77.

Chapter 8

Palm oil and health

Jean-Michel Lecerf, Institut Pasteur de Lille, France

1 Introduction

Nutrition is a major factor in maintaining good health. A well-balanced diet is a key factor, with the total fat intake and the distribution of fatty acids being particularly important. Saturated and unsaturated fatty acids must be present in correct proportions. Palm oil is rich in saturated fatty acids, accounting for half its total fatty acid content. Western diets are high in saturated fatty acids (SFAs), which come mainly from animal sources (dairy and meat). Saturated fatty acids can increase total and LDL cholesterol in the blood stream (atherogenic lipoprotein fraction (Mensink et al., 2003). Unsaturated fatty acids, particularly polyunsaturated fatty acids, may decrease both total and LDL cholesterol (Mensink et al., 2003). It has been believed for half a century that saturated fatty acids are responsible for atherosclerosis, because of their effect on blood cholesterol (Vartiainen et al., 2010). However, during the last decade, this has been the subject of considerable controversy (Astrup et al., 2011).

This chapter examines the composition of palm oil, its exact content of fatty acids and their position on the glycerol molecule which may modify the biological effect. Several published studies of clinical trials on the effect of palm oil on blood cholesterol and lipoprotein are reviewed. The role of diet and palm oil on other biological markers of cardiovascular risk are discussed. This chapter will focus on the role of fatty acid intake (particularly of saturated fatty acids) on cardiovascular and coronary heart disease through the more recent epidemiological studies. Various epidemiological studies concerning the role of palm oil

http://dx.doi.org/10.19103/AS.2017.0018.22

on coronary heart disease morbidity and mortality are explored. The role of palm oil in dietary fat intake will be presented. Finally, its place in a well-balanced diet is considered, as the physical properties of palm oil have enabled the limitation of processes such as partial hydrogenation and thus the reduction of trans-fatty acids in Western diets.

2 Palm oil composition and properties

2.1 Fatty acids (May & Nesaretnam, 2014)

Palm oil belongs to the family of solid fats, together with kernel oil, coprah oil, cocoa butter and karite oil. It is a vegetable oil obtained from the fruit of the palm tree (*Elaeis guineensis*), while kernel oil is extracted from the seed of the palm fruit.

Crude (red) palm oil is obtained from the mesocarp (outer pulp) before the refining process. It is produced in palm oil mills by pressure. Fractionation and refining of the crude palm oil then takes place to obtain palm olein and superolein, which are fractions less rich in palmitic acid, the main palm oil fatty acid.

Crude palm oil may be directly used for culinary purposes in the countries where it is produced (mainly Western Africa and Brazilian Nordeste), but its global use is based on the fractionated and refined oils.

Palm oil is composed of saturated and unsaturated fatty acids in almost equal proportions. The main fatty acid is palmitic acid (C16:0), which accounts for 44% of the total fat content. Lauric acid (C12:0) and myristic acid (C14:0) are lacking. The second main fatty acid is oleic acid (C18:1n-9), which accounts for 39% of the total fat content and is a monounsaturated fatty acid. Polyunsaturated fatty acids account for 10% in the form of linoleic acid (C18:2 n-6). The typical fatty acid composition of palm oil is presented in Table 1.

The fatty acids are attached to a glycerol molecule and their distribution is specific to palm oil. Sn-1 and Sn-3 are located on the extremities of the glycerol molecule and Sn-2 is on the middle of the glycerol molecule. In palm oil, more than 87% of the unsaturated

Table 1 Fatty acid composition of palm oil

	g/100g
• Saturated fatty acids	44–55
< 10:0	0
Lauric acid C12:0	0
Myristic acid C14:0	0.5–2
Palmitic acid C16:0	39.5–47.5
Stearic acid C18:0	3.5–40
• Monounsaturated fatty acids	38–45
Oleic acid	36–40
• Polyunsaturated fatty acids	9–12
Linoleic acid	9–12
Alpha linolenic acid	< 0.5

fatty acids are located in the Sn-2 position while only 7–11% of the palmitic acid is located in the Sn-2 position.

In the digestive process, pancreatic lipase and colipase hydrolyse the Sn-1 and Sn-3 fatty acids which become free, while the Sn-2 fatty acid will remain attached to the glycerol as a 2-monoacylglycerol (2-MAG). In the presence of calcium, a fraction of the free long-chain saturated fatty acid (palmitic acid) precipitates as calcium soaps, which are excreted and not absorbed. However, the 2-MAG, which is mainly unsaturated, is well absorbed for re-synthesis of a new triacylglycerol (triglycerides) and incorporation into chylomicrons.

2.2 Other components (Sundram et al., 2003)

The minor components of palm oil are divided into two groups. The first group comprises fatty acid derivatives such as partial glycerides which are saponifiable. These do not occur in significant amount in good-quality oils. The second group is found in the unsaponifiable fraction. These consist mainly of sterols (phytosterols), triterpene alcohols, tocopherols and tocotrienols, phospholipids, chlorophylls, carotenoids and volatile flavour components such as aldehydes and ketones.

The pigmentation of palm oil depends on its maturity. Two classes of natural pigments occurring in crude palm oil are carotenoids and chlorophylls. Crude palm oil is the richest natural source of carotenoids. The carotenoid content of crude palm oil, which is responsible for its red colour, ranges from 500 to 2000 mg/kg. However after refining and de-colouration, this content will be much lower (10 mg/kg).

As with all vegetable oils, palm oil contains tocopherols and tocotrienols, which are fat-soluble vitamin E isomers. Together with rice bran and corn oil, palm oil is rich in tocotrienols. Crude palm oil contains 20–30% of tocopherols and 70–80% of tocotrienols, up to 600–1000 mg/kg. Refining reduces the level to 350–630 mg/kg. Heating and repeated deep frying can reduce 90% of the tocotrienols content.

2.3 Physical properties (Ong and Goh, 2002)

Due to its high level (50%) of saturated fatty acids (mainly palmitic acid), palm oil has specific physical properties. As it is solid at room temperature (20°C) it has increased utility in the food-processing industry. Its solidity is mainly due to its high melting point (36–38°C). Palm stearin has a higher melting point (50–55°C) while that of olein palm is lower (19–21°C). The uses for these different fractions may therefore differ.

Because of its low level of poly- and monounsaturated fatty acids, palm oil is stable and is not easily oxidized. This stability is advantageous in cooking or frying.

There are three processes for improving the hardness and stability of vegetable oils:

i) dividing into a solid fraction (soft stearin or hard palm stearin) and a non-solid fraction (olein) – this is a natural physical process; ii) interesterification, which is a process modifying the distribution of fatty acids on the glycerol molecule (which has been the most widely used for many years); and iii) hydrogenation – total hydrogenation adds hydrogen to all the double bonds of the unsaturated fatty acids, which then become totally saturated; partial hydrogenation adds hydrogen to a part of the double bonds. However, this process induces the generation of trans-fatty acids on the remaining double bounds. Unfortunately, trans-fatty acids have adverse effects on lipoprotein metabolism and therefore lead to cardiovascular risk (Mozaffarian and Clarke, 2009).

3 Effects of palm oil on cardiovascular risk markers

3.1 On fasting lipids and lipoproteins (Sundram, 1997; Lecerf, 2013; Fattore et al., 2014)

Because palm oil contains 50% saturated fatty acids (44% from palmitic acid), it has been considered harmful for the metabolism of both lipids and lipoproteins. However, lauric and myristic acids, which are the main cholesterol-raising fatty acids, are present only in trace amounts in palm oil, while palmitic acid has a much weaker cholesterol-raising potential. Moreover, in the three principal triglyceride species of palm oil, palmitic acid is located at the alpha-position (Sn-1 and Sn-3) on the glycerol molecule. Palm oil also consists of half unsaturated fatty acids (40% monounsaturated and 10% polyunsaturated fatty acids, mainly oleic acid and linoleic acid, respectively). It is therefore necessary to conduct good clinical randomized studies comparing palm oil, or its fractions, with other vegetable oils. Several studies have been made to assess the effect of different oils on cholesterol and lipoproteins (Bautista et al., 2001; Choudhury et al., 1995; Jensen et al., 1999; Müller et al., 1998; Ng et al., 1991; Pedersen et al., 2005; Sundram et al., 1992; Tholstrup et al., 2011; Vega-López et al., 2006; Wood et al., 1993; Zhang et al., 1997).

Total cholesterol is not a good marker of cardiovascular risk, except for severe dyslipidaemia. LDL cholesterol is clearly associated with atherosclerosis in epidemiological studies, and this has been confirmed through interventional studies with cholesterol-lowering diet or drugs. The underlying mechanisms are now clearly known: when LDL lipoprotein goes through the endothelial, the unsaturated fatty acids of the esterified cholesterol may be oxidized. The oxidation of LDL depends also on the antioxidant content of LDL particles, the size of LDL (small dense LDL is more readily oxidized), and finally the level of HDL cholesterol. A low HDL concentration is associated with higher cardiovascular risk. One of the explanations is the anti-oxidative role of para-oxonase, an enzyme carried by HDL.

It has been established that saturated fatty acids increase LDL cholesterol blood concentration and also HDL cholesterol concentration. However, there is no evidence of the protective effect of high HDL cholesterol, or of the increase of low HDL cholesterol. Finally, it is also important to consider the role of blood triglycerides concentration because each increase in triglycerides is inversely associated to lower HDL cholesterol and higher small dense LDL.

Several studies have been performed over many years comparing oils and fats, both hydrogenated and non-hydrogenated. The majority of these are well-conducted (double-blind, controlled-randomized) studies. However, there is a heterogeneity due to the level of the total fat intake and/or to the dietary cholesterol intake and/or to the linoleic acid intake. For this reason, these results were analysed, taking into account the percentage of fat in the global diet (Lecerf, 2013). Of course, the observed effects will also depend on the oils or fats compared.

In the case of a low or moderate fat diet (≤27–30%), as in Malaysia, palm oil or its fraction induces a lower LDL cholesterol (or apolipoprotein B) than other solid oils or saturated fatty acid rich fats. There are no differences in total, LDL and HDL cholesterol concentrations between palm oil and olive oil. When compared to peanut oil, the LDL cholesterol is lower and the HDL cholesterol is slightly higher. Some studies have been conducted in order to compare partially hydrogenated oils (rich in trans-fatty acids) with palm oil: it was found that palm oil induces a higher HDL cholesterol concentration and lower LDL cholesterol concentration. An interesting study (French et al., 2002) has shown that palmitic acid does not cause an increase in blood cholesterol when the linoleic acid intake is higher than 4,5% of total energy intake.

In the case of a normal fat intake (35–40%, which is the French recommendation), palm oil has a similar or better effect on HDL cholesterol and a similar or less favourable effect on LDL cholesterol than olive oil, and a more favourable effect than butter or partially hydrogenated fat. In the case of a very high fat diet (>40%) palm oil causes a higher LDL and total cholesterol concentration than sunflower oil.

Another review was published (Fattore et al., 2014) with a meta-analysis of dietary intervention trials. Comparison of palm oil diets with diets rich in stearic acid (C18:0), monounsaturated and polyunsaturated fatty acids showed significantly higher total cholesterol, LDL cholesterol, apolipoprotein B, HDL cholesterol and apolipoprotein A-I. However, the unfavourable effects of palm oil on total and LDL cholesterol compared with monounsaturated and polyunsaturated fatty acids disappeared when young people (≤30 years of age) were considered. Moreover, these effects were attenuated in normocholesterolaemic subjects and were abolished in patients whose level of total energy derived from dietary fat was lower. This is probably why the impact of palm oil may differ between Asian and Western countries. Most of the lipid and lipoprotein markers were significantly lower when compared with diets rich in myristic/lauric acid (butter for instance). Finally, the comparison of palm oil rich diets with diets rich in trans-fatty acids showed significantly higher concentrations of HDL cholesterol and apolipoprotein A-I and significantly lower concentrations of apolipoprotein B, triacylglycerols and total cholesterol/HDL cholesterol.

3.2 On post-prandial plasma lipids and lipoproteins (Lecerf, 2013; Mukherjee and Mitra, 2009)

It has been established that the post-prandial hyperlipidaemia is more atherogenic than the fasting state. But the body is more often in the post-prandial state (except during the second part of the night). Lipid and lipoprotein post-prandial metabolism has been widely studied. In some of these studies, the response was the same with palm oil, trans-fatty acid fat or animal fat (lard), or with palm olein, olive oil or coconut oil. Another study showed a lower response with lard than with palm or olive oil, and with palm oil than with olive oil. It seems that long-chain saturated fatty acids are less quickly absorbed than long-chain unsaturated fatty acids, mainly when these are located in the Sn-1 and Sn-3 position on the glycerol molecule, but the global area under the curve of plasma triglycerides is the same.

3.3 Effects on other cardiovascular risk markers (Sundram, 1997; Lecerf, 2013; Mukherjee and Mitra, 2009)

Cardiovascular diseases may not be reduced merely to the atherosclerosis process, which partially, but not totally, implies cholesterol and lipoprotein metabolism. Severe cardiovascular events may also be due to thrombosis, which occurs when there is platelet aggregation as soon as the rupture of a vulnerable inflammatory atherosclerosis plaque occurs. The post-prandial triglycerides-rich lipoproteins increase stimulates the factor

VII implied in the coagulation steps and in the thrombosis state. Probably because the post-prandial lipid response between tested oils are similar, the thrombotic activity of palm oil showed no difference when compared to other oils. A study showed a more favourable effect of palm oil when compared to a trans-fat diet on fibrinolysis. Other studies have shown that palm oil has a favourable or no effect on platelet aggregation.

Some studies have been conducted on fasting or post-prandial inflammatory markers such as C-reactive protein or cytokines, and these showed no difference between the tested oils.

Although the beneficial role of carotenoids is well known, they are present only at a low level in refined palm oils. Their benefits are more evident in crude red palm oil in which the tocotrienols content remains high after refining, but decreases dramatically after intensive heating or repeated frying. Tocotrienols have beneficial properties, partly due to having a more powerful antioxidant effect than tocopherols, and must be considered in the overall assessment of a healthy diet.

4 The health impacts of saturated fatty acids

4.1 Saturated fatty acids and coronary heart disease

Since ecological studies were conducted in the 1960s by Ancel Keys (Keys et al., 1986), it has been thought that saturated fatty acids were deleterious to health, leading to cardiovascular risk and coronary heart disease morbidity. However, several misinterpretations have been pointed out, particularly the fact that fatty acids are also a marker of dietary patterns, but may not be implied in the pathogenesis of the disease (Astrup et al., 2011). Highly saturated fatty acids have an effect, along with dietary cholesterol, although not exclusively, on blood cholesterol, elevating LDL cholesterol and HDL cholesterol. They differ from trans-fatty acids which increase both, and polyunsaturated fatty acids which decrease LDL cholesterol. Moreover, not all saturated fatty acids have the same effect on blood lipids. Stearic acid (C18:0) does not increase LDL cholesterol, while palmitic acid, myristic acid and lauric acid do. The Nurses' Health Study (Hu et al., 1999) has specifically addressed the relationship between dietary saturated fatty acids with differing carbon chain lengths and ischaemic heart disease. The cohort showed a moderately increased risk of ischaemic heart disease for all longer-chain saturated fatty acids (lauric acid through stearic acid), whereas for short to medium-chain saturated fatty acids (butyric through capric acid) no associations with ischaemic heart disease were observed. However, the relationship between cholesterol and cardiovascular disease is complex and also depends on overall diet and the lifestyle. In the 25-year follow-up from the Seven Countries Study (Kromhout, 1999), the relationship between blood cholesterol and coronary heart disease mortality was mainly observed in Western countries and not in Mediterranean societies. However, dyslipidaemia, particularly familial dyslipidaemia, and metabolic syndrome must be corrected by diet and drug therapy.

In 2010, a first meta-analysis of 16 prospective (not ecological) studies was published (Siri-Tarino et al., 2010a), which did not show any association between saturated fatty acids and ischaemic heart disease.

4.2 Prevention of cardiovascular disease and cerebro-vascular disease

An update of the meta-analysis, including four additional prospective cohort studies as well as a meta-analysis of a selection of 12 cohort studies, did not observe any association with a non-significant relative risk (de Souza et al., 2015). However, the negative role of trans-fatty acids on cardiovascular mortality and type 2 diabetes was confirmed.

There is controversy about secondary prevention studies. Few of these are recent and many show bias in their methodology. In most cases, the tested diets consisted of a decreased intake of saturated fatty acids and an increase of polyunsaturated fatty acids. Some, such as the Oslo Diet Heart Study (Hjermann et al., 1981), showed a spectacular reduction of the occurrence of new coronary events. However, a considerable number of the subjects also stopped smoking; the mean level of blood cholesterol was initially very high; and soy bean oil and fish, rich in omega-3 fatty acids, which are polyunsaturated-rich foods, were added to the diets. Three other studies (Strandberg et al., 1991; Ramsden et al., 2013; Ramsden et al., 2016) did not find evidence of any deleterious effect or increase in cardiovascular mortality or morbidity long after the end of the studies. Although post hoc analyses are interesting, they are mostly questionable: harmful effects occur when there is too high an increase of omega-6 polyunsaturated fatty acid intake with very high P/S levels (Polyunsaturated/Saturated). Where there is a reduction in saturated fatty acids combined with a moderate increase of polyunsaturated fatty acids, including omega 3 and omega 6 unsaturated fatty acids, there is a beneficial effect in secondary prevention (Ramsden et al., 2013).

4.3 The sources of saturated fatty acids

A new approach has recently been developed to aid in understanding the discrepancy between dietary intakes and predicted impairment with regard to cardiovascular risk, and the non-deleterious effect of saturated fatty acids. The focus was on the food sources of saturated fatty acids: first in the MESA study (de Oliveira Otto et al., 2012) and then in a subgroup of EPIC study (Praagman et al., 2016). These studies have clearly shown that saturated fatty acids provided by meat products are associated with high cardiovascular risk, whereas saturated fatty acids provided by dairy products are associated with a reduction of this risk. This is probably due to other nutrients or components which are directly involved in a protective or harmful effect in these foods, through specific mechanisms. These include the atherogenic TMAO production from the metabolism of carnitine-containing meat by the microbiote and the beneficial effect of calcium, certain specific fatty acids (conjugated linoleic acid or rumenic acid for instance, or short chain fatty acids), probiotics or functional peptides in dairy products. Finally, it should be noted that fats are not consumed separately.

There is therefore some interaction between fat and carbohydrate intakes while protein intake is almost stable. When saturated fatty acids decrease, carbohydrates and/or sugar increase. But it has been well known for more than fifteen years that carbohydrates induce a higher blood concentration of triglycerides, a lesser HDL cholesterol concentration and mainly small dense LDL particles which are oxidized and atherogenic (Siri-Tarino et al., 2010b) (Siri-Tarino et al., 2015).

4.4 Epidemiological studies on palm oil and cardiovascular risk

Unfortunately there are very few available studies in this area and none are of adequate quality.

The first is an ecological study (Zhang and Kesteloot, 2001) about mortality data in Hong Kong and Singapore between 1963 and 65, compared with 1993–5, using Spain, Japan and the United States as reference countries. Mortality and food consumption data were obtained from the WHO and FAO respectively. The authors observed a higher increase of ischaemic heart mortality in Singapore when compared to Hong Kong, and differences in dietary habits, including a higher consumption of coconut and palm oil in Singapore. However, no relationship can be drawn from ecological studies: the same mistake was made during the 1960s with the Seven Countries Study. These are not individual data, and significant differences exist between the lifestyles of the studied populations.

The second study was conducted by Campos and Baylin (Kabagambe et al., 2005). This is a large population-based incident case-control study. The studied cases were survivors of a first acute myocardial infarction (2111 subjects) and were matched to an equal number of randomly selected population controls. In order to validate the dietary intake, adipose tissue profiles of essential fatty acids were assessed. It was found that palm oil users were more likely to have a myocardial infarction than users of soybean oil (OR 1,33) or other cooking oils (OR = 1,23 non-significant), but these patients did not differ from users of soybean oil with a high trans-fatty acid content (OR 1,14). Compared with palm oil users, linoleic acid and alpha linolenic acid were higher among users of soybean oil. Unfortunately, case-control studies have significant bias and cannot prove any relationship between the diet observed and the disease due to a lack of chronology. However, it has been established that in secondary prevention for coronary heart disease, higher alpha linolenic and linoleic intakes are beneficial (Virtanen et al., 2014; Wu et al., 2014). There is no information about the amount of oil consumed.

The third study (Chen et al., 2011) is also an ecological study which analysed the deaths, stroke mortality and ischaemic heart mortality of adults aged 50 and older in 23 countries. The increased rates of mortality between 1980 and 1997 were examined, as well as the per capita consumption of palm oil and cigarettes, and the per capita Gross Domestic Product. In developing countries, for every additional kilogram of palm oil consumed per capita annually, the ischaemic heart mortality rates increased by 68 deaths per 100 000, whereas stroke mortality rates increased by 19 deaths per 100 000, but this was non-significant. For historically high-income countries, changes in ischaemic heart disease and stroke mortality were found to be smaller (17 and 5 respectively). Unfortunately, as with the Zhang and Kesteloot study, no link between a single fact and health can be drawn, although global lifestyle and dietary pattern changes cannot be ignored.

It is useful to remember that in countries such as Malaysia and Indonesia, where palm oil is traditionally used for cooking, the coronary heart disease mortality rate is still low, probably because the global diet is poor and the incidence of obesity is low. Given the globalization of food habits and the rapid rate of urbanization in such countries, it is of paramount importance that serious cohort studies on the health effect of the consumption of palm oil as the almost unique source of vegetable fat are undertaken.

The fourth study (Basu et al., 2013) is a calculation of the potential effect of a palm oil tax on hyperlipidaemia and cardiovascular disease mortality in India. A 20% excise tax on palm oil purchases was studied over the period 2014–23 through a micro-simulation model of mortality. The model was used to project future mortality due to myocardial infarction and stroke as well as the potential effect of a tax on food insecurity accounting for the effect of increased food prices. A 20% tax on palm oil purchases was expected to avert approximately 363 000 deaths from myocardial infarction and stroke over the 2014–23 period out of a total of 36.6 million expected deaths (a 1.3% reduction in cardiovascular deaths) if consumers did not substitute other oils to compensate for their reduced palm oil consumption. A total of 421 000 deaths were expected to be averted if there was a substitution with increased polyunsaturated fats. The calculation of the estimated relation between changes in the consumption of various oils and changes in total cholesterol was drawn from the Keys' formula (Keys et al., 1965). Although the relationship between LDL cholesterol (not total cholesterol) and cardiovascular disease is well established, the link between saturated fatty acids and cardiovascular risk and mortality is not. So this extrapolation cannot be considered as being valid.

5 Palm oil consumption

5.1 Estimating the consumption of palm oil

For many years there was no accurate data on palm oil dietary intakes, only indirect data about apparent consumption obtained through economic data (importation-exportation). However, it may be estimated at the level of 2.7 g/d/pers in France in 2009. Recently, the annual CCAF study conducted in France by CREDOC was made available (CREDOC, 2013). This is a dietary survey conducted on a sample of 2500 representative adults and 900 young adults and children (3–18 years) with a three-day dietary record. All information on food composition was accurately obtained from industry. The daily palm oil consumption was found to be 2.8 g/d/pers and is slightly higher in younger people (3.4 g/d/pers). For the 90th percentile, the intake may reach more than 6.5 g/d/pers in adults and more than 7.5 g/d/pers in children and young adults. This represents an average of 1.4 g of saturated fatty acids. It is useful to keep in mind that the total fat daily intake in France is 75 g/d and that the French recommended dietary intakes for saturated fatty acids is 12% of the total fat intake (ANSES, 2011), which represents 27 g of saturated fatty acids.

5.2 A place for palm oil in a balanced diet

The importance of a global well-balanced diet with healthy foods must be kept in mind. In Western countries, palm oil fractions are included in processed foods such as biscuits, snacks, confectionery, cakes, pastry, chips, fried foods, crackers, desserts and ice cream– consumption of which should be limited. These foods are also often rich in sucrose, fats and salt and poor in fibre. They replace healthier foods such as fruit, yogurt, legumes, vegetables and whole grains in the diet. Palm oil is not a 'bad' oil as it contains half saturated and half unsaturated fatty acids, but it is relatively poor in linoleic acid, and does not provide alpha linolenic acid. Although it contains tocotrienols (refined oil no longer contains carotenoids), its polyphenol content is much lower than that of olive oil. It is therefore advisable to use a wide variety of oils for cooking, salad dressing or baking.

Since its introduction 20 years ago, particularly in Western countries such as France, palm oil has made possible a dramatic reduction in the use of partially hydrogenated vegetable oils in the agrofood industry (and thus the consumption of trans-fatty acids). The subsidiarity principle should be applied to its use: if other solutions are possible, they should be pursued; if not, palm oil may be used.

Finally, it remains important to limit those food products which contain fat, sugar and salt in excess and to promote a varied diet.

6 Further trends in research

The negative role of saturated fatty acids is no longer recognized in all studies. This does not mean, however, that their consumption can be increased without adverse effects, nor that a good balance between fatty acids is unnecessary, but that a sufficient polyunsaturated linoleic and alpha linolenic acids intake is important for cardiovascular health.

Furthermore, two strands of information are lacking. The first is the specific role of saturated fatty acids according to their sources. We know that saturated fatty acids from meat products and from dairy products have opposite effects, probably due to the presence of protective components in dairy products and to negative components such as carnitine in meat. So one explanation is that saturated fatty acids are not implied in cardiovascular risk although they increase LDL (and HDL) cholesterol and that they are only a marker. Another explanation is that the total profile of fatty acids originating from food products are different despite some similarities.

The second is the lack of available prospective cohort studies designed to measure the impact of palm oil consumption on cardiovascular risk which take into account the whole diet, risk factors and lifestyle patterns.

7 Conclusion

Palm oil is increasingly used throughout the world. It is a solid oil due to its high content of saturated fatty acids, mainly palmitic acid. Half its fatty acids are unsaturated. Saturated fatty acids are specifically located on Sn-1 and Sn-3 (on glycerol), which diminishes their intestinal absorption. Crude palm oil is also rich in carotenoids, and has a high tocotrienols content, but the refining processes greatly decreases the concentration of carotenoids, while heating and frying lowers the concentration of tocotrienols. Fractioning makes it possible to obtain more or less solid fractions dedicated to various industrial uses. This has enabled the removal of trans-fatty acids through the interruption of partial hydrogenation of vegetable oils. Due to its fatty acid composition, palm oil is not only a solid oil but also stable and heat resistant with low oxidization ability.

Palm oil may increase blood cholesterol concentration by increasing LDL cholesterol and HDL cholesterol concentration. But these effects are modest and smaller than those observed in vegetable oils rich in trans-fatty acid. They are also close to that observed in olive oil. Because polyunsaturated fatty acids decrease LDL cholesterol, comparison with oils such as soy, sunflower or corn oil is less favourable. Palm oil has no clearly deleterious effect on other cardiovascular parameters involved in thrombosis or risk factors. Although the oldest ecological studies, undertaken in the 1960s, suggested that saturated fatty acids were responsible for coronary heart disease, the meta-analysis of prospective studies does not confirm that view, and show no relationship between the two. However, secondary prevention for ischaemic heart disease requires an increase in omega 3 and omega 6 polyunsaturated fatty acids.

No consistent epidemiological study has been performed on the effect of palm oil on cardiovascular risk. Poor methodology in the available studies makes it impossible to confirm a deleterious effect, and further good epidemiological studies are necessary. Finally, the consumption of palm oil remains very low in Western countries such as France and represents a small part of fat and saturated fatty acid intakes. However most of the foods containing palm oil are not particularly healthy and need caution in order to limit their consumption. Moreover the promotion of a diversity of oils for dressing and cooking and the encouraging of a well-balanced diet are important. In the agrifood industry the subsidiary principle must be applied.

8 Where to look for further information

Some scientific references have been mentioned in this chapter, however there is still a need to conduct a good epidemiological study on the role of palm oil in health. Indeed, there is a lack of data on the effect of palm oil consumption on cardiovascular disease in both developing and developed countries. Any study must be prospective with a large cohort of subjects, initially without cardiovascular disease, and with a follow-up of more than 10 years. All confounding factors must be taken in account. Dietary intakes must be determined at the beginning of the study and during the follow-up. It would therefore be necessary to know the exact food consumption and to be able to distinguish exactly the sources of fatty acids in the diet, and not only in the final composition. It will require some effort to obtain that information.

9 References

ANSES, (2011), 'Actualisation des apports nutritionnels conseillés pour les acides gras'. *Rapport d'expertise collective*. Available at: http://www.civ-viande.org/2013/08/26/actualisation-des-apports-nutritionnels-conseilles-pour-les-acides-gras/.

Astrup, A., Dyerberg, J., Elwood, P., Hermansen, K., Hu, F. B., Jakobsen, M. U., et al., (2011), 'The role of reducing intakes of saturated fat in the prevention of cardiovascular disease: where does the evidence stand in 2010?'. *The American Journal of Clinical Nutrition*, 93(4), 684–8.

Basu, S., Babiarz, K. S., Ebrahim, S., Vellakkal, S., Stuckler, D. and Goldhaber-Fiebert, J. D., (2013), 'Palm oil taxes and cardiovascular disease mortality in India: economic-epidemiologic model.' *BMJ (Clinical research ed.)*, 347, f6048.

Bautista, L. E., Herrán, O. F. and Serrano, C. (2001), 'Effects of palm oil and dietary cholesterol on plasma lipoproteins: results from a dietary crossover trial in free-living subjects.' *European Journal of Clinical Nutrition*, 55(9), 748–54.

Chen, B. K., Seligman, B., Farquhar, J. W. and Goldhaber-Fiebert, J. D. (2011), 'Multi-Country analysis of palm oil consumption and cardiovascular disease mortality for countries at different stages of economic development: 1980–1997.' *Globalization and Health*, 7, 45.

Choudhury, N., Tan, L. L. and Truswell, A. S. (1995), 'Comparison of Palmolein and Olive Oil – effects on Plasma-lipids and Vitamin-E in young-adults'. *American Journal of Clinical Nutrition*, 61(5), 1043–51.

CREDOC, (2013), 'Comportements et consommations alimentaires en France'. *CCAF*, Paris.

de Oliveira Otto, M. C., Mozaffarian, D., Kromhout, D., Bertoni, A. G., Sibley, C. T., Jacobs, D. R. and Nettleton, J. A. (2012), 'Dietary intake of saturated fat by food source and incident cardiovascular disease: the Multi-Ethnic Study of Atherosclerosis.' *The American Journal of Clinical Nutrition*, 96(2), 397–404.

de Souza, R. J., Mente, A., Maroleanu, A., Cozma, A. I., Ha, V., Kishibe, T., Uleryk, E., Budylowski, P., Schünemann, H., Beyene, J. and Anand, S. S. (2015), 'Intake of saturated and trans unsaturated fatty acids and risk of all cause mortality, cardiovascular disease, and type 2 diabetes: systematic review and meta-analysis of observational studies.' *BMJ (Clinical research ed.)*, 351, h3978.

Fattore, E., Bosetti, C., Brighenti, F., Agostoni, C. and Fattore, G. (2014), 'Palm oil and blood lipid-related markers of cardiovascular disease: a systematic review and meta-analysis of dietary intervention trials.' *The American Journal of Clinical Nutrition*, 99(6), 1331–50.

French, M. A., Sundram, K. and Clandinin, M. T. (2002), 'Cholesterolaemic effect of palmitic acid in relation to other dietary fatty acids.' *Asia Pacific Journal of Clinical Nutrition*, 11 Suppl. 7, S401–7.

Hjermann, I., Velve Byre, K., Holme, I. and Leren, P. (1981), 'Effect of diet and smoking intervention on the incidence of coronary heart disease. Report from the Oslo Study Group of a randomised trial in healthy men.' *Lancet (London, England)*, 2(8259), 1303–10.

Hu, F. B., Stampfer, M. J., Manson, J. E., Ascherio, A., Colditz, G. A., Speizer, F. E., Hennekens, C. H. and Willett, W. C. (1999), 'Dietary saturated fats and their food sources in relation to the risk of coronary heart disease in women.' *The American Journal of Clinical Nutrition*, 70(6), 1001–8.

Jensen, J., Bysted, A., Dawids, S., Hermansen, K. and Hølmer, G. (1999), 'The effect of palm oil, lard, and puff-pastry margarine on postprandial lipid and hormone responses in normal-weight and obese young women.' *The British Journal of Nutrition*, 82(6), 469–79.

Kabagambe, E. K., Baylin, A., Ascherio, A. and Campos, H. (2005), 'The type of oil used for cooking is associated with the risk of nonfatal acute myocardial infarction in costa rica.' *The Journal of Nutrition*, 135(11), 2674–9.

Keys, A., Anderson, J. T. and Grande, F. (1965), 'Serum cholesterol response to changes in the diet: IV. Particular saturated fatty acids in the diet.' *Metabolism: Clinical and Experimental*, 14(7), 776–87.

Keys, A., Menotti, A., Karvonen, M. J., Aravanis, C., Blackburn, H., Buzina, R., Djordjevic, B. S., Dontas, A. S., Fidanza, F. and Keys, M. H. (1986), 'The diet and 15-year death rate in the seven countries study.' *American Journal of Epidemiology*, 124(6), 903–15.

Kromhout, D. (1999), 'On the waves of the seven countries study: a public health perspective on cholesterol.' *European Heart Journal*, 20(11), 796–802.

Lecerf, J.-M. (2013), 'L'huile de palme: aspects nutritionnels et métaboliques. Rôle sur le risque cardiovasculaire'. *Oléagineux, Corps gras, Lipides*, 20(3), 147–59.

May, C. Y. and Nesaretnam, K. (2014), 'Research advancements in palm oil nutrition.' *European Journal of Lipid Science and Technology : EJLST*, 116(10), 1301–15.

Mensink, R. P., Zock, P. L., Kester, A. D. M. and Katan, M. B. (2003), 'Effects of dietary fatty acids and carbohydrates on the ratio of serum total to HDL cholesterol and on serum lipids and apolipoproteins: a meta-analysis of 60 controlled trials.' *The American Journal of Clinical Nutrition*, 77(5), 1146–55.

Mozaffarian, D. and Clarke, R. (2009), 'Quantitative effects on cardiovascular risk factors and coronary heart disease risk of replacing partially hydrogenated vegetable oils with other fats and oils.' *European Journal of Clinical Nutrition*, 63 Suppl. 2, S22–33.

Mukherjee, S. and Mitra, A. (2009), 'Health effects of palm oil'. *Journal of Human Ecology*, 26(3), 197–203.

Müller, H., Jordal, O., Kierulf, P., Kirkhus, B. and Pedersen, J. I. (1998), 'Replacement of partially hydrogenated soybean oil by palm oil in margarine without unfavorable effects on serum lipoproteins.' *Lipids*, 33(9), 879–87.

Ng, T. K., Hassan, K., Lim, J. B., Lye, M. S. and Ishak, R. (1991), 'Nonhypercholesterolemic effects of a palm-oil diet in Malaysian volunteers.' *The American Journal of Clinical Nutrition*, 53(4 Suppl.), 1015S–20S.

Ong, A. S. H. and Goh, S. H. (2002), 'Palm oil: a healthful and cost-effective dietary component.' *Food and Nutrition Bulletin*, 23(1), 11–22.

Pedersen, J. I., Muller, H., Seljeflot, I. and Kirkhus, B. (2005), 'Palm oil versus hydrogenated soybean oil: effects on serum lipids and plasma haemostatic variables.' *Asia Pacific Journal of Clinical Nutrition*, 14(4), 348–57.

Praagman, J., Beulens, J. W., Alssema, M., Zock, P. L., Wanders, A. J., Sluijs, I. and van der Schouw, Y. T. (2016), 'The association between dietary saturated fatty acids and ischemic heart disease depends on the type and source of fatty acid in the European Prospective Investigation into Cancer and Nutrition-Netherlands cohort.' *The American Journal of Clinical Nutrition*, 103(2), 356–65.

Ramsden, C. E., Zamora, D., Leelarthaepin, B., Majchrzak-Hong, S. F., Faurot, K. R., Suchindran, C. M., Ringel, A., Davis, J. M. and Hibbeln, J. R. (2013), 'Use of dietary linoleic acid for secondary prevention of coronary heart disease and death: evaluation of recovered data from the Sydney Diet Heart Study and updated meta-analysis.' *BMJ (Clinical research ed.)*, 346, e8707.

Ramsden, C. E., Zamora, D., Majchrzak-Hong, S., Faurot, K. R., Broste, S. K., Frantz, R. P., Davis, J. M., Ringel, A., Suchindran, C. M. and Hibbeln, J. R. (2016), 'Re-evaluation of the traditional diet-heart hypothesis: analysis of recovered data from Minnesota Coronary Experiment (1968-73).' *BMJ (Clinical research ed.)*, 353, i1246.

Siri-Tarino, P. W., Chiu, S., Bergeron, N. and Krauss, R. M. (2015), 'Saturated fats versus polyunsaturated fats versus carbohydrates for cardiovascular disease prevention and treatment.' *Annual Review of Nutrition*, 35, 517–43.

Siri-Tarino, P. W., Sun, Q., Hu, F. B. and Krauss, R. M., (2010a), 'Meta-analysis of prospective cohort studies evaluating the association of saturated fat with cardiovascular disease.'. *The American journal of clinical nutrition*, 91(3), 535–46.

Siri-Tarino, P. W., Sun, Q., Hu, F. B. and Krauss, R. M. (2010b), 'Saturated fat, carbohydrate, and cardiovascular disease.' *The American Journal of Clinical Nutrition*, 91(3), 502–9.

Strandberg, T. E., Salomaa, V. V, Naukkarinen, V. A., Vanhanen, H. T., Sarna, S. J. and Miettinen, T. A. (1991), 'Long-term mortality after 5-year multifactorial primary prevention of cardiovascular diseases in middle-aged men.' *JAMA*, 266(9), 1225–9.

Sundram, K. (1997), 'Modulation of human lipids and lipoproteins by dietary palm oil and palm olein: a review.' *Asia Pacific Journal of Clinical Nutrition*, 6(1), 12–6.

Sundram, K., Hornstra, G., von Houwelingen, A. C. and Kester, A. D. (1992), 'Replacement of dietary fat with palm oil: effect on human serum lipids, lipoproteins and apolipoproteins.' *The British Journal of Nutrition*, 68(3), 677–92.

Sundram, K., Sambanthamurthi, R. and Tan, Y.-A. (2003), 'Palm fruit chemistry and nutrition.' *Asia Pacific Journal of Clinical Nutrition*, 12(3), 355–62.

Tholstrup, T., Hjerpsted, J. and Raff, M. (2011), 'Palm olein increases plasma cholesterol moderately compared with olive oil in healthy individuals.' *The American Journal of Clinical Nutrition*, 94(6), 1426–32.

Vartiainen, E., Laatikainen, T., Peltonen, M., Juolevi, A., Männistö, S., Sundvall, J., Jousilahti, P., Salomaa, V., Valsta, L. and Puska, P. (2010), 'Thirty-five-year trends in cardiovascular risk factors in Finland.' *International Journal of Epidemiology*, 39(2), 504–18.

Vega-López, S., Ausman, L. M., Jalbert, S. M., Erkkilä, A. T. and Lichtenstein, A. H. (2006), 'Palm and partially hydrogenated soybean oils adversely alter lipoprotein profiles compared with soybean and canola oils in moderately hyperlipidemic subjects.' *The American Journal of Clinical Nutrition*, 84(1), 54–62.

Virtanen, J. K., Mursu, J., Tuomainen, T.-P. and Voutilainen, S. (2014), 'Dietary fatty acids and risk of coronary heart disease in men: the Kuopio Ischemic Heart Disease Risk Factor Study.' *Arteriosclerosis, Thrombosis, and Vascular Biology*, 34(12), 2679–87.

Wood, R., Kubena, K., Tseng, S., Martin, G. and Crook, R. (1993), 'Effect of palm oil, margarine, butter, and sunflower oil on the serum lipids and lipoproteins of normocholesterolemic middle-aged men', Elsevier.

Wu, J. H. Y., Lemaitre, R. N., King, I. B., Song, X., Psaty, B. M., Siscovick, D. S. and Mozaffarian, D. (2014), 'Circulating omega-6 polyunsaturated fatty acids and total and cause-specific mortality: the Cardiovascular Health Study.' *Circulation*, 130(15), 1245–53.

Zhang, J. and Kesteloot, H. (2001), 'Differences in all-cause, cardiovascular and cancer mortality between Hong Kong and Singapore: role of nutrition.' *European Journal of Epidemiology*, 17(5), 469–77.

Zhang, J., Ping, W., Chunrong, W., Shou, C. X. and Keyou, G. (1997), 'Nonhypercholesterolemic effects of a palm oil diet in Chinese adults.' *The Journal of Nutrition*, 127(3), 509S–513S.

Chapter 9

The nutritional value of red palm oil

Hélène Delisle, University of Montreal, Canada

1 Introduction

Palm oil is now the most widely consumed vegetable oil worldwide (Mba et al., 2015). For one, its cost is low compared to other oils. The main consumers of palm oil are China, India, Indonesia and the European Union; their demand is entirely met by imports since they do not produce palm oil. Additionally, the nearly solid state of palm oil at room temperature makes it a good substitute for hydrogenated oils widely used until recently in the food industry and which contain undesirable *trans*-fatty acids. The ban on *trans* fat in several countries including Canada and the United States, because of adverse cardiovascular effects similar to saturated fat, drives the rapid global shift in consumption from soybean oil to palm oil (Global Industry Analysts, 2015).

There is a great deal of confusion regarding the nutritional value and health effects of palm oil. The controversy and conflicting views still continue on whether or not palm oil is atherogenic. Based on current evidence, it would appear that palm oil has both favourable and unfavourable effects. The nutritional and health properties of palm oil depend not only on the amounts consumed and the other components of the diet, but also on the extent of processing and on the fractions considered. The crude (red) palm oil (RPO) is very distinct from the refined product and its high content of antioxidants including vitamin E and provitamin A may be responsible for health benefits that are no longer present in the refined oil as more than half its antioxidants have been destroyed. The health aspects of palm oil were discussed in previous chapters. It is suspected that several publications may have tended to be positively biased due to the fact that the palm oil industry has been very active at sponsoring research, as reported in two large systematic reviews and

http://dx.doi.org/10.19103/AS.2017.0018.33

meta-analyses (Fattore et al., 2014; Sun et al., 2015). The primary focus of the present chapter is on the nutritional value of crude RPO, primarily as a source of vitamin A.

2 Nutritional composition of palm oil

The nutritional composition of palm oil and its fractions is summarized in Table 1. First, palm mesocarp oil has to be distinguished from palm kernel oil, the latter being much more saturated than the former, 80% versus 40%–50%, respectively. We will only refer to palm fruit oil in this chapter. Palm (fruit) oil consists of 94%–98% triglycerides. Myristic acid (1%), stearic acid (4–5%) and palmitic acid (42–47%) make up the saturated fatty acid component in addition to monounsaturated oleic acid (37–41%) and polyunsaturated linoleic acid (9–11%). Although it is not as saturated as coconut oil, palm oil is still at the top of the list when it comes to saturation.

Although palm oil and animal fat have similar saturated fat content, the positional distribution of fatty acids in triglycerides is different: 70% of the palmitic acid in palm oil is in the sn-1 and sn-3 positions, whereas the majority of palmitic acid in animal fat is in the sn-2 position (Zhao et al., 2005). Fatty acids in the sn-2 position might have an enhanced absorption (Hunter, 2001), and thus some researchers have suggested that the palmitic acid in palm oil may be less hypercholesterolemic and atherogenic than that in animal

Table 1 Nutritional composition of crude and refined palm oil and its fractions (per 100 g)

	Palm kernel oil	Crude (red) palm oil	Refined palm oil	Olein	Stearin
Lauric acid (12:0)	47.8	0.0	0.2	0.2	–
Myristic acid (14:0)	16.3	1.2	1.0–1.5	1.0–1.5	1.0–2.0
Palmitic acid (16:0)	8.5	39.5	42.0–47.0	38–42	47–74
Stearic acid (18:0)	2.4	4.0–5.0	4.0–5.0	4.0–5.0	4.0–6.0
Oleic acid (18:1)	15.4	43.4	37.0–41.0	40.0–44.0	16.0–37.0
Linoleic acid (18:2)	2.4	12.2	9.0–11.0	10.0–13.0	3.0–10.0
Linolenic acid (18:3)	–	0	0.4	0.4	
Total saturated fatty acids	82.1	44.4	49.9	45.9	
Total monounsaturated fatty acids	15.4	55.6	39.2	54.1	
Total polyunsaturated fatty acids	2.4		10.5		
Total carotenes (μg/g)	–	500–700 (90% α- and β-carotene)	*	12	
Tocopherols and tocotrienols (vitamin E) (μg/g)	–	600–1000 (73% tocotrienols, 27% tocopherols)	~50% loss	643	

Sources: Sambanthamurthi et al., 2000; Edem, 2002; O'Brien, 2004; Boon et al., 2013.

*Physical or chemical refining entails a nearly total loss of carotenoids, but a technology exists to produce a refined RPO retaining 80% of carotenoids and 85% of tocols (Mayamol et al 2007).

fat (Ebong et al., 1999). A more recent study found that palmitic acid in the sn-2 position could decrease postprandial lipaemia in humans (Sanders et al., 2011).

The resulting two components of the palm oil fractionation is palm olein (liquid) and palm stearin (solid). The fatty acid composition of palm olein is approximately 45% saturated fat and 54% unsaturated fat. The main saturated fatty acids are 40–44% palmitic acid and 4–5% stearic acid. The unsaturated fatty acids are 43% oleic acid and 12% linoleic acid. Technological advances in palm oil fractionation have allowed to further separate olein fractions (Boon et al., 2013).

Palm oil tends to be perceived negatively because it is highly saturated (and because of the environmental impact of the large plantations). However, this is somewhat balanced by its high content of provitamin A carotenoids, vitamin E and other antioxidants as well as phytosterols, at least in the crude RPO. The antioxidants contribute to the oil stability, as well as to its nutritional and health benefits, alleged or real (Oyewole and Amosu, 2010).

One characteristic of crude palm oil is its high content in vitamin E (tocopherols and tocotrienols), with a total content ranging from 600 to 1000 ppm. It is actually the highest food source of tocotrienols. The tocopherol/tocotrienol ratio is usually around 20%, whereas it is roughly reversed in palm oil. Compared to other oils, palm oil also has a high proportion of tocopherols and tocotrienols in relation to its unsaturation. The ratio of total vitamin E (tocopherols and tocotrienols in ppm) to polyunsaturated fatty acids (PUFA expressed in %) is about 50, while it is only 19 for soybean and 12 for sunflower oils. The combined effects of high tocopherols and tocotrienols, and low PUFA, could explain why palm oil would present a greater oxidative stability, for instance, in frying (Gibon et al., 2007).

Crude RPO also represents the highest natural source of carotenoids (500–2000 ppm). β-carotene predominates and represents with α-carotene about 90% of the total carotenoids. Additionally, these provitamin A carotenoids of RPO are highly bioavailable because of the absence of a vegetable matrix and the presence of fat (Cottrell, 1991). In terms of vitamin A activity (expressed in retinol activity equivalents [RAE]), RPO provides 15 800 μg and carrots 1000 μg per 100 g (Scrimshaw, 2000). Unfortunately most of the carotenoids in palm oil are destroyed during the refining process, giving rise to colourless products (Gibon et al., 2007). Crude oils are refined to remove all impurities and undesirable odour, flavour and colour, but at the same time the process destroys more than half its natural antioxidants. One of the modified refining processes developed by the Palm Oil Research Institute of Malaysia procures a refined RPO (Carotino®) with a light pink colour but which has retained 80% of the carotenoids, 85% of the tocols and 65% of the phytosterols (Mayamol et al., 2007). It is unfortunate that this technology is not more widely applied, particularly in areas where vitamin A deficiency is a problem while palm oil is produced, notably in West Africa and in India. The cost of the technology is possibly among main deterrents.

Elaeis guineensis is the principal variety of oil palm. It is originally from tropical Africa and it is the most widely cultivated, not only in Africa but also in Asia and Indonesia. *E. oleifera* is native of South America. Hybrids could have increased oil unsaturation, carotene, tocopherol and sterol content (O'Brien, 2004), but they are little produced.

3 RPO as a source of provitamin A carotenoids

Vitamin A deficiency is still widespread in several African and Asian regions, where it is responsible for excess morbidity and mortality particularly among young children, even in the absence of the ocular manifestations of the deficiency. Preformed vitamin A or retinol can be obtained from animal products (such as liver, eggs, dairy products) or as provitamin A carotenoids in some plant products, including orange flesh fruits and vegetables. RPO is the highest plant source of provitamin A carotenoids, and, as mentioned previously, its vitamin A bioactivity is very high. The provitamins are absorbed and partly converted into retinol. Various combinations of fortification, supplementation and dietary diversification strategies are used to control vitamin A deficiency, but periodic supplementation with high-dose vitamin A capsules is still a major component of child survival strategies in African and Asia countries where the deficiency is widespread. Promoting the production and consumption of RPO where feasible and preventing its overheating appear as a logical, although neglected strategy to prevent vitamin A deficiency. In sub-Saharan Africa, where oil palms originated, RPO is traditionally consumed in some but not all areas, while vitamin A malnutrition persists.

Table 2 lists the safe level of vitamin A intake in various age-sex groups, expressed in RAE, the corresponding β-carotene needed, and, in the last column, the amount of RPO required to meet the recommended daily intake. It is seen that, roughly equating 1 g RPO with 1 ml, from two to four teaspoons of RPO would meet the daily recommended intake of vitamin A. Even consuming small amounts of RPO can therefore go a long way towards meeting daily requirements for vitamin A.

Table 2 The amount of RPO needed to meet the FAO/WHO recommended intakes of vitamin A

Subpopulation	Recommended safe intake (RAE, μg)[1]	β-carotene (μg)[2]	RPO (g)[3]
Children			
0–6 months	375	4500	9.0
7–12	400	4800	9.6
1–3 years	400	4800	9.6
4–6	450	5400	10.8
7–9	500	6000	12.0
Adolescents			
10–18 years	600	7200	14.5
Adult women			
19–65 years	500	6000	12.0
65+	600	7200	16.4
Pregnant women	800	9600	19.2
Lactating women	850	10200	20.4
Adult men			
19–65 years	600	7200	14.5
65+	600	7200	14.5

[1] Source: FAO/WHO.
[2] Based on a conversion factor of 12 μg β-carotene for 1 μg of RAE.
[3] Average value of 500 μg of β-carotene for 1 g of RPO.

Several trials and quite a few programmes showed vitamin A efficacy and effectiveness of RPO in children and in women. In a systematic review (Rice and Burns, 2010), ten key intervention trials of RPO used as daily supplement, in-home fortificant or in fortified food programmes were analysed. These trials were conducted in Indonesia, Honduras, Papua New Guinea, Tanzania, India, South Africa and Burkina Faso. Except for the studies in Papua New Guinea, Tanzania and Burkina Faso, the RPO was from Malaysia or Indonesia, the largest producers of palm oil in the world. All these studies reported positive effects, whether on serum or on milk retinol levels. In an Indian village, for example, 8 g of RPO daily for 15 days proved as effective as a single high-dosage retinol supplement on vitamin A status of schoolchildren (Mahapatra and Manorama, 1997). In South African poor schoolchildren aged 5–11 years, RPO- and β-carotene-fortified biscuits resulted in similar serum retinol increases (Van Stuijvenberg et al., 2001). The RPO-fortified biscuit is now commercialized and available on the South African market. In Tanzania (Lietz et al., 2001) and in Honduras (Canfield et al., 2001), RPO supplementation trials during pregnancy and lactation showed beneficial effects on vitamin A status of mothers and infants. In the Honduras study, RPO was found to be as effective as β-carotene supplements in this regard. The RPO studies that we conducted in Burkina Faso over a 10-year period are detailed below as case study.

There are at least two published reports of successful RPO fortification of staples. In a district of Tanzania, RPO was incorporated in cassava flour on a pilot basis in villages representing three agro-ecological zones with high rates of vitamin A deficiency (Mosha et al., 1999). Out of five villages, two served as controls. At the end of the 20-month programme, most mothers had reportedly adopted the innovation and were preparing the mix at home. The prevalence of low plasma retinol decreased from 53% to 8.5% in pilot village children, while remaining above 50% in the control villages. Plasma retinol of pregnant and lactating women increased by 12.6% in the pilot villages and only by 3.5% in the control villages. However, in the absence of meta-analysis of the data for all these studies, the average resulting effect on serum (or milk) retinol cannot be ascertained.

Other applications of RPO that were not assessed for their impact on vitamin A status are mentioned in the review by Rice and Burns (2010), including its use as ingredient in lipid-based ready-to-use therapeutic food (RUTF) and as fortificant in commercial baked foods, as well as in condiments or salad dressings. It was also tested when added to other cooking oils, for instance, in Brazil. In India, a blend of oil including 6–12% RPO was found to be palatable, with no objectionable taste or smell of the food, although the oil was coloured in red (Rao, 2000). In Burkina Faso, we also tested blends of RPO with local vegetable fats as described below. More recent advances have been the development of high-carotene varieties of maize, sweet potatoes and cassava, as well as the genetically modified 'golden rice'.

It is now common practice to fortify refined palm oil (or other vegetable oils) with retinyl palmitate as a source of preformed vitamin A to compensate for the loss of provitamin A carotenes during processing. Such fortification programmes are considered highly cost-effective and they have been used successfully in various parts of the world with sugar, monosodium glutamate or wheat flour as vector foods. The feasibility of using as natural fortificant β-carotene from RPO as alternative strategy to address vitamin A deficiency was reviewed (Souganidis et al., 2013). While β-carotene may be as effective as retinyl palmitate to improve vitamin A status, the challenges of fortification with RPO-derived β-carotene include its unfavourable colour, taste and odour, as well as the lower stability to high temperatures of β-carotene compared to retinyl palmitate. Additionally, the RPO fortification cost would be roughly 40 times higher than that of retinyl palmitate (Souganidis et al., 2013). It is also reminded here that RPO is not only a high source of vitamin A (and fat), but also of several other healthful antioxidants including non-vitamin A carotenoids (lutein and lycopene) and vitamin E (tocopherols and tocotrienols), which are lost during refining. The health effects of these antioxidants were discussed in previous chapters.

4 Case study: RPO in Burkina Faso

This case study describes the rationale, the objectives and the activities of the three distinct phases of the RPO project. The results, lessons learned and challenges are then discussed.

4.1 Project background and history

The role of RPO for reducing vitamin A deficiency in Africa was considered underexploited (Solomons, 1998), and the possible nutritional and economic benefits for Sahelian countries were, and continue to be, little recognized. The idea for our project stemmed from the observation that refined palm oil could be found even in remote locations of the Sahel. Burkina Faso is probably the only country where the feasibility and impact of introducing RPO on a commercial basis in non-consuming population groups could be tested, because some RPO is produced in the western part of the country, although in still limited quantities. Besides, preliminary market studies had shown that the production of RPO could be considerably increased in Burkina Faso and that it could indeed generate profit.

The studies that our team conducted with institutional partners in Burkina Faso between 1998 and 2008 were intended to demonstrate the feasibility and nutritional impact of stimulating the production and consumption of RPO. Increasing RPO production, distribution and consumption indeed bears potential as part of overall strategies to increase vitamin A intake, in Burkina Faso and in surrounding Sahelian areas as well. The RPO project in Burkina Faso was developed on a small scale, although the initial intent was to scale it up if the initial phases proved successful. It is unique in that a new demand for RPO was created by social marketing and was used as driver for local production. It involved the development of oil palm orchards by women, and improved oil extraction methods and the expansion of local markets to generate nutritional, women's income and even environmental benefits as palm trees are known to consolidate the land of river banks. RPO was posited as a food 'fortificant' for the nutritional benefit of mothers and children primarily, as RPO is costly in the Sahel and as it is highly concentrated in vitamin A. Several papers on this project were published (Delisle et al., 2001, 2003, 2004; Zagré et al., 2002, 2003, 2004; Zeba et al., 2006).

The three phases of the project are described below, before commenting on the results and perspectives.

4.1.1 Phase I: 1999–2002

The purpose was to demonstrate that it was feasible to introduce locally produced RPO on a commercial basis, and that it was effective in improving the vitamin A status of women and young children. The Burkina Faso Association of Home Economists (ABESF) was the main field partner and the research was conducted in collaboration with the French Institute for

Research and Development (IRD). Field promotional activities were conducted by home economists after appropriate training in social marketing, in a pilot zone (north-central part of the country) where RPO is not part of food habits. RPO was purchased by the project from women in south-western Burkina Faso, and other women were retailing it in the intervention sites. Promotional activities included using the media, plays and testimonies at village level, demonstrations and tasting of RPO-fortified dishes, and contests among villages. RPO promotion was also coupled with vitamin A capsule distribution during National Immunization and Micronutrient Days in order to bring the population to make the connection between food (and not only capsules) and the prevention of vitamin A deficiency. The impact was assessed in a random sample of 210 mother-child pairs whose vitamin A status was assessed based on serum retinol levels before the beginning of the project and two years later.

4.1.2 Phase II: 2002–2004

The second phase of the RPO project pursued the same initial goal, that is, to improve vitamin A status of vulnerable groups via the purchase and consumption of RPO produced right in Burkina Faso, but it was carried out on a larger scale and involved a wider range of partners. Additionally, more emphasis was given to strengthening the RPO production and commercial distribution system. The intervention zone was extended to five provinces, whereas Phase I took place in only one province. A school component was introduced, and school meals were fortified two to three times a week with 15 ml of RPO per child in a total of 41 schools in three provinces. In one province, RPO-fortified snacks were first offered in schools before school canteens were operational. The formulations were developed by ABESF and were very well appreciated. In all cases, teachers, parents and communities were closely involved, and training for appropriate handling of RPO was provided to cooks and to school managers.

In one of the provinces participating in the RPO fortification of school lunch, a randomized controlled trial was carried out, with eight primary schools randomly assigned to RPO, eight to supplementation with vitamin A capsules and eight served as controls. Serum retinol was measured before the intervention in October and in the same children exactly 12 months later (to avoid a seasonality effect). The vitamin A capsule group received a 200 000 IU capsule before the school break, in July of the same school year. Serum retinol was again measured when school resumed in October.

During Phase II, RPO production groups benefited from training, in Bénin and in Burkina Faso, in order to improve the extraction techniques, to learn simple management tools and to contribute to setting up the RPO commercial distribution system through direct connections with wholesale outlets established in the main city of the five provinces. Women's groups extracting the oil went from four at the beginning of Phase II to eight by the end with an average of 25 women per group. The groups formed a union of RPO producers around the pioneer group in the main producing village. A local organization involved in supporting rural enterprises funded the purchase of oil extraction equipment and the construction of a shed for RPO storage; it also provided support on technical and managerial matters.

4.1.3 Phase III: 2006–2008

This phase was limited to the RPO production zone of south-western Burkina Faso owing to limited funding. The activities were carried out in close collaboration with Oxfam-Québec, the NGO in charge of the intervention proper for this phase of the project. The objectives were to improve vitamin A status of mothers and children by promoting the consumption of RPO and to further strengthen the RPO production and marketing efforts as a means of procuring additional income for women (Oxfam-Québec, 2008).

The research activities conducted by our university team were: (1) to assess vitamin A status and RPO consumption and production among women of reproductive age in the RPO-producing area of Burkina Faso; (2) to assess the quality of RPO samples from Burkina Faso and samples coming from neighbouring countries as available in Burkina Faso; and (3) to test the technical feasibility and consumer acceptance of blends of RPO with either peanut oil or shea butter at various concentrations. The underlying hypotheses were that women's vitamin A status would be better in the RPO production area than in non-producing areas, that the RPO produced in Burkina Faso with project support would be of high quality, and that mixtures of RPO with other traditional fats would be well accepted by women and their family. The research components included baseline data on RPO consumption and vitamin A status of women as well as on the quality of the RPO from local production. A total of 150 mothers randomly selected in 15 villages of the RPO-producing areas were interviewed twice over a ten-month period. Their vitamin A status based on serum retinol was assessed in the first survey round. An additional ten women from the main production village were interviewed to better assess their experience with RPO production.

4.1.4 Phase IV: 2008–2011

This phase was planned to consolidate the previous achievements, to increase RPO production and consumption and to undertake a comprehensive evaluation of the health, income and environmental effects of the RPO project. This was to be a critical phase to ensure the sustainability and scaling-up of the production and consumption of RPO beyond the production zone to contribute to vitamin A status improvement and to women's income. The project was to be scaled up further to set the stage for a national RPO programme as part of the vitamin A strategy. The health, agriculture and social sectors of the government of Burkina Faso were all mobilized and gave their support to the approach. The expected duration of Phase III was three years. However, in spite of the positive results of the previous phases and after several proposals to various potential funding agencies, the funding was not secured. RPO apparently continues to be produced locally and palm orchards are growing but no evidence of the beneficial effects and costs could be generated and our dream of positing RPO as a major component of the vitamin A strategies in the whole region of sub-Saharan Africa had to be dropped.

Before discussing the effects of the RPO project, we summarize the main outputs below:

- A total population of 1.3 million was reached through community activities, the school component, R&D activities and media-based promotion of RPO in five provinces of Burkina Faso;
- The more intensive community-based RPO strategy in one province reached 26 villages, for an estimated direct target of 20 000 women and under-five children; RPO retailing committees were trained (150 persons);
- 6400 pupils aged 6–10 years in 41 schools of three provinces, the teachers and the parents were reached by the school component involving RPO-fortified meals and community mobilization. Some 76 school canteen cooks and at least 76 teachers were trained in adequate handling, storage and portioning of RPO in schools;

- The number of RPO production women's groups went from 4 to 8 (~25 women per group), and the production was around 4200 liters for 2003 and 2004; 330 women producing or selling RPO received training and were given an opportunity to generate some additional income. Information on the number of palm tree planted and on their oil yield once they began to be productive is unfortunately not available.

4.2 Effects on vitamin A status of women and children under five

We could demonstrate the feasibility and effectiveness of increasing production and consumption of local RPO. In the north-east pilot area where RPO was not previously available or consumed, nearly half the women purchased and consumed RPO after two years of social marketing of the product. This was based on a random sample of 210 mother-child pairs who were surveyed at the beginning of the project and then 24 months later. Highly significant improvements of the vitamin A status of women and children were observed: the prevalence of low serum retinol (<0.7 μmol/L), while remaining high, went from 85% to 67% in children, and from 62% to 28% in mothers. The prevalence of low serum retinol remained high particularly among children in spite of the fact that they had received a vitamin A capsule six months before the beginning of the RPO project. This shows the importance of mixed strategies for the control of the deficiency which can include supplementation, food-based approaches and public health measures to reduce infection. The reduction of the prevalence of vitamin A deficiency was more marked among women, which strongly suggests that other factors of the deficiency are at work in children, particularly recurrent infections. It was noted that in mothers as well as in children, the most deficient were also the ones whose vitamin A status improved the most, which tends to confirm that provitamin A carotenoids are better utilized in deficient subjects.

Among schoolchildren included in the controlled trial, the prevalence of low serum retinol was significantly reduced with both the RPO and the vitamin A capsule, as shown in Fig. 1. Although the baseline prevalence of vitamin A deficiency was much lower in the control schools, which was unexpected, mean serum retinol varied little in the control group, whereas it increased sharply in the RPO and the vitamin A capsule groups. This trial in schools confirmed that vitamin A deficiency was also widespread at school age and that

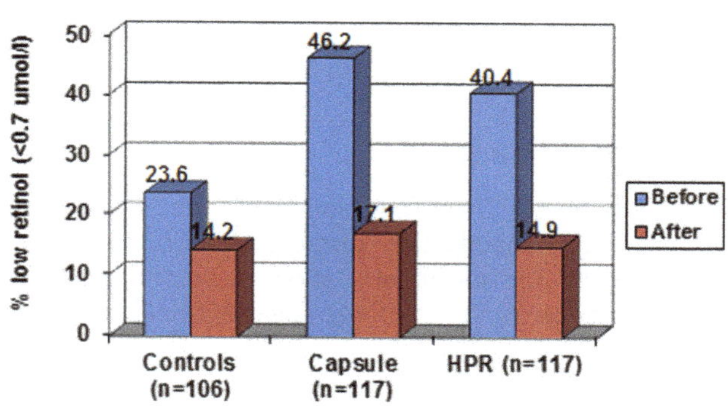

Figure 1 Serum retinol changes in primary schoolchildren exposed to HPR or high-dose vitamin A supplements compared with control subjects.

an average of 51 RPO-fortified meals over a school year was as effective as one vitamin A capsule to improve vitamin A status, besides RPO being well liked by children. Through RPO fortification of school meals two to three times/week, it can be assumed that 34% of the pupils were protected from vitamin A deficiency, based on the observed reduction of low serum retinol from 47% before the school intervention to 13% one year later.

RPO was also successfully integrated into a local infant cereal mix and in nutrition rehabilitation centres, although this was only done on a pilot basis and there was no assessment of impact.

At the beginning of Phase III of the project in the RPO production zones, it was found that the vitamin A status of RPO-producing women was significantly better, with only 5.9% showing low serum retinol (<0.7 µmol/L), compared with 20.8% in non-producing women (and 62% initially among women of the pilot intervention area of Phase I of the project). Although there was a tendency for RPO-producing women to report better socio-economic conditions, RPO production was the only significant variable in the logistic regression of vitamin A status. The odds ratio for low serum retinol was 0.28 (CI 0.09–0.79) in RPO-producing women compared to non-producers. Mean intakes of total vitamin A as well as preformed retinol were significantly higher among RPO-producing than non-producing women, which suggested that not only RPO but also other sources of vitamin A were consumed in higher amounts among the former compared with the latter.

4.3 Quality of the RPO

Throughout the project, producing high-quality RPO and using labour-saving equipment were sought. During Phase II, RPO samples from Burkina Faso (n = 4, including two from RPO project) and nine imported samples from Togo, Côte d'Ivoire and Ghana were analysed for chemical properties, microbiological quality and content of provitamin A. An expert tasting panel assessed the organoleptic quality of seven samples (four from Burkina and one randomly selected sample per import country). All 13 samples were again analysed 8 months later, and 5 were again tested 12 months after the initial analyses.

The results varied widely, whether in terms of chemical, microbiological or nutritional quality. There was no relationship between organoleptic quality rating by the taste panel and the other aspects of quality of the RPO samples, not even their acidity. All except one imported sample were adequate from a bacteriological standpoint, and the bacterial counts went down with time, likely owing to increasing acidity. The initial provitamin A carotenoid results confirmed that in spite of the high variance, RPO is a concentrated provitamin A source, providing more than 100 µg RAE per gram.

4.4 RPO as a food fortificant

Other approaches for promoting RPO consumption were tested and found very successful during Phase II. In a nutrition rehabilitation centre (CREN) close to the capital city of Ouagadougou, introducing RPO in the rehabilitation meals encouraged the mothers to purchase some RPO to continue using it at home (Delisle, unpublished results). RPO was also tested as a fortificant for Misola®, a locally prepared complementary food (weaning mix), with the local unit of fabrication in one province. It proved technically feasible to add 4–5% RPO without changing the process except for a slight reduction in groundnut content, and the mothers and children liked it very much. RPO fortification of Misola on a broader scale could therefore have been considered if the RPO project had been sustained.

Blends of RPO with either groundnut oil or shea butter were tested among women because mixing RPO with other local fats and oils may encourage its consumption without changing cooking practices. The response was positive for both mixes. The taste of RPO was found more acceptable when mixed with either fat and there was no technical problem with mixing RPO with these other fats and oils. The tests were first conducted in the RPO pilot project area and, for that reason, we suspected that exposure to the RPO project may have introduced some bias. The study was therefore repeated in a non-RPO area with 50 women (in charge of kitchen) randomly selected in the five neighbourhoods of a town located 20 km away from the capital city of Ouagadougou. Following the initial interview of all women on their perceptions and use of fats and oils in general and RPO in particular, 12 women were retained for two focus groups, one on RPO mixed with peanut oil and the other one on RPO mixed with shea butter. In each case, two different concentrations of RPO were presented for evaluation and tasting. The women were then given samples of the two mixtures (prepared with them) at the preferred RPO concentration for trial with their family. A final individual interview of these women was conducted 15 days later in their home. The results showed that the mixtures were feasible and acceptable in terms of colour, texture, taste and culinary use. Only one woman (out of 12) reported a negative response. The mixtures with a lower RPO concentration (10%) were preferred. The mixture of RPO with peanut oil was better rated than that with shea butter, and was considered preferable to using RPO alone in various preparations. The women appeared ready to prepare the mix again and they would even be willing to pay ~10% more for the mix already prepared. Blending RPO with peanut oil therefore appeared as an interesting strategy to promote RPO consumption.

4.5 Effects on women's income and empowerment

During Phase III, there was particular emphasis on capacity building of women as producers and consumers. RPO production increased, and women extracting the oil formed a producers' group (500 women in five subgroups); they had planted dwarf palm trees and were even given literacy and business training. Women producers were trained in palm orchard and RPO management as well as advocacy, along with literacy training. Several community workers were trained to carry social marketing campaigns for the promotion of RPO consumption.

According to the survey (Bougma, 2008), RPO-producing women perceived several advantages to the activity: RPO availability for consumption (88%), potential source of income (54%), and savings to attend to other needs (18%). The negative aspects were that it is physically demanding (49%) and time consuming (20%). In Tin, the village where the extraction unit is located, the 11 respondents (out of 20 in the unit) producing RPO were highly satisfied with the extraction material making the process easier; they wanted to remain members of the production group, and they reported that RPO production was a source of income, whether used for investment by the group or shared among members. Unfortunately, our plan to thoroughly assess the health and income effects of RPO production and distribution could not materialize since Phase IV was not funded.

4.6 Perspectives

The 'scaling-down' of the project rather than its scaling-up prevented an adequate assessment of the approach. It may even have had a negative effect in communities previously exposed to the pilot project. It would have been critical to thoroughly evaluate the process, outcomes and impact of the approach, as there are few projects based on local agriculture, focusing on women, and potentially benefiting micronutrient health and women's income. It is unfortunate and indeed difficult to understand that this natural oil, which is now even promoted as 'natural' or 'health food' in the West, has raised until now so little enthusiasm among international decision-makers and donors.

As underlined by Rice et al. (2010), there are several barriers to expanding and scaling up RPO-based interventions, including local food preferences, the availability and cost of RPO and policies, and programmes and resources to fund such vitamin A initiatives. Consumer acceptance may be an issue where populations are unfamiliar with RPO, but not in Africa. Even if RPO is not produced locally, most African populations know RPO and it is a prestige food in some cultures. The quality, shelf life and cost of RPO remain challenges, however, in non-producing areas. Sustained investments in local palm oil production in Africa have been advocated (Skurtis et al., 2010), and it should have been possible to encourage the use of the improved RPO-extracting technology in order for the oil to retain its vitamin A activity and other oxidants, thereby contributing to improved nutritional health and even contribution to income in Burkina Faso and neighbouring Sahelian countries. As aptly stated by an eminent nutritionist (Solomons, 1998), RPO should have become the logical substitute for vitamin A capsules in Africa.

5 Conclusion

Crude RPO is the highest and most bioavailable plant source of vitamin A in the form of carotenoids, primarily α- and β-carotenes. Palm oil is nearly 100% fat and it is 50% saturated, whether crude or refined; it may therefore contribute to overweight as well as nutrition-related chronic diseases such as cardiovascular disease or diabetes (Mensik, 2016). However, RPO may have important health benefits because of its high provitamin A, vitamin E and other antioxidant content. Furthermore, unlike Malaysian plantations, palm trees growing in the wild and small plantations in Africa are not detrimental to the environment. The project in Burkina Faso showed that RPO was effective in improving the vitamin A status of women, under-five children and school-age children, that it is well accepted by the population which is also willing to purchase the product, and that there is income generation for women producing RPO. The project also confirmed the high vitamin A content of RPO, its safety, the motivation of women to engage in RPO extraction as source of food and income and the prospective value of blending RPO with peanut oil as a means of increasing the use of RPO (and the intake of vitamin A). The project was deemed successful and the results impressive in spite of its small scale, but it proved impossible to get further funding to scale up the initiative and mount the distribution system.

6 Future trends

Implementation research would now be needed during a 'revitalization' phase of the RPO project in Burkina Faso. Additionally, various RPO commercial distribution strategies should be tested for their economic sustainability. Another area for promising research is on the adoption of the improved technology for processing palm oil without destroying its antioxidants – enabling factors and barriers.

Research on potential cost-effectiveness of RPO-based interventions has been advocated because such data are missing (Rice et al., 2010). However, cost-effectiveness comparisons between RPO (and other food diversification) interventions, vitamin A capsule supplementation and food fortification may not be appropriate and indeed work against RPO (and other food diversification schemes) as these are more expensive for less spectacular results on vitamin A status. RPO-based projects have benefits other than vitamin A status, and these benefits are not taken into account in cost-effectiveness studies, for instance, the income effects and the effects on local agriculture and the anti-erosion effects. For this reason, cost-utility evaluation or, even better, separated costs and consequences are to be preferred in order to also take into account the benefits other than nutritional. RPO production and distribution should be researched in inclusive evaluation schemes.

7 Where to look for further information

The Micronutrient Forum: www.micronutrientforum.org
Roundtable on Sustainable Palm Oil: http://www.rspo.org
Carotino® website: www.carotino.com/improving-vitamin-a-and-antioxidant-status-59.aspx

8 References

Boon, C. M., Ng, M. H., Choo, Y. M. and Mok, S. L. Super, red palm and palm oleins improve the blood pressure, heart size, aortic media thickness and lipid profile in spontaneously hypertensive rats. *Plos One* 2013; 8(2): e55908. doi:10.1371/journal.pone.0055908.

Bougma, K. [Red palm oil in Burkina Faso: Quality, consumption by women and impact on their vitamin A status]. Huile de palme rouge au Burkina Faso: Qualité et consummation par les femmes de la zone de production et impact sur leur statut en vitamin A. University of Montreal: M.Sc. essay, 2008.

Canfield, L. M., Kaminsky, R. G., Taren, D. L., Shaw, E. and Sander, J. K. Red palm oil in the maternal diet increases provitamin A carotenoids in breastmilk and serum of the mother-infant dyad. *Eur. J. Clin. Nutr.* 2001; 40: 30–8.

Cottrell, R. C. Introduction: nutritional aspects of palm oil. *Am. J. Clin. Nutr.* 1991; 53: 989S–1009S.

Delisle, H., Zagré, N., Ouedraogo, V. and Tarini, A. Marketing of red palm oil for vitamin A in Burkina Faso: A pilot project involving women's groups. *Food Nut. Bull.* 2001; 22: 388–94.

Delisle, H., Bakari, S. and Zagré, N. [Rolling back vitamin A deficiency in developing countries with food diversification strategies]. Faire reculer l'avitaminose A dans les pays en développement par des stratégies de diversification alimentaire]. *Rev Canadienne Économie Familiale* 2003; 52: 18–21.

Delisle, H. Food diversification strategies are neglected in spite of their potential effectiveness: Why is it so and what can be done? In Brouwer, I. D., Traoré, A. S. and Treche, S. (Eds), Proceedings of the 2nd International Workshop on food-based approaches for a healthy nutrition in West Africa, 23–28 November 2003. Ouagadougou, Presse Universitaire, 2004: 303–16.

Ebong, P. E., Owu, D. U., Isong, E. U. Influence of palm oil (*Elaesis guineensis*) on health. *Plant Foods Hum. Nutr.* 1999; 53: 209–22.

Edem, D. O. Palm oil: Biochemical, physiological, nutritional, hematological, and toxicological aspects: A review. *Plant Foods Hum. Nut.* 2002; 57: 319–41.

Fattore, E., Bosetti, C., Brighenti, F., Agostoni, C. and Fattore, G. Palm oil and blood lipid-related markers of cardiovascular disease: A systematic review and meta-analysis of dietary intervention trials. *Am. J. Clin. Nutr.* 2014; 99: 1331–50.

Gibon, V., deGreyt, W. and Kellen, M. Palm oil refining. *Eur. J. Lipid Sci. Technol.* 2007; 109: 315–35.

Global Inudstry Analysts Inc. The global palm oil market trends, drivers and projections, February 2015. http://www.strategyr.com/MarketResearch/Palm_Oil_Market_Trends.asp.

Hunter, J. E. Studies on effects of dietary fatty acids as related to their position on triglycerides. *Lipids* 2001; 36: 655–68.

Mahapatra, S. and Manorama, R. The protective effect of red palm oil in comparison with massive vitamin A dose in combating vitamin A deficiency in Orissa, India. *Asia Pacific J. Clin. Nutr.* 1997; 6: 246–50.

Mayamol, P. N., Balachandran, C., Samuel, T. and et al. Process technology for the production of micronutrient rich palm olein. *J. Amer. Oil Chemists. Soc.* 2007; 84: 587–96.

Mba, O. I., Dumont, M. J. and Ngadi, M. Palm oil: Processing, characterization and utilization in the food industry – A review. *Food Biosci.* 2015; 10: 26–41.

Mensik, R. P. Effects of saturated fatty acids on serum lipids and lipoproteins: a systematic review and regression analysis. Geneva: WHO, 2016.

Mosha, T., Laswai, H. and Mtebe, K. Control of vitamin A deficiency disorders through fortification of cassava flour with red palm oil: A case study of Kigoma district, Tanzania. *Ecol. Food Nutr.* 1999; 37: 569–93.

O'Brien, R. D. *Fats and Oils* (2nd edition). Boca Raton (Fla): CRC Press, 2004.

Osia, O. J., Ofori, S. and Maduka, O. Palm oil and the heart: A review. *World J. Cardiol.* 2015; 7(3): 144–9.

Oxfam-Québec. Red palm oil project in Burkina Faso Final technical report, 2008 (unpublished).

Oyewole, O. E. and Amosu, A. M. Public health nutrition concerns on consumption of red palm-oil (RPO): The scientific facts from literature. *Afr. J. Med. Med. Sci.* 2010; 39: 255–62; discussion 263–5.

Rao, B. S. N. Potential use of red palm oil in combating vitamin A deficiency in India. *Food Nut. Bull.* 2000; 21: 202–11.

Rice, A. L. and Burns, J. B. Moving from efficacy to effectiveness: Red palm oil's role in preventing vitamin A deficiency. *J. Amer. College Nut.* 2010; 29: 302S–13S.

Sambanthamurthi, R., Sundram, K. and Tan, Y. Chemistry and biochemistry of palm oil. *Prog. Lipid Res.* 2000; 39: 507–58.

Sanders, T. A., Filippou, A., Berry, S. E., Baumgartner, S. and Mensink, R. P. Palmitic acid in the sn-2 position of triacylglycerols acutely influences postprandial lipid metabolism. *Am. J. Clin. Nutr.* 2011; 94: 1433–41.

Scrimshaw, N. S. Nutritional potential of red palm oil for comabting vitamin A deficiency. *Food Nut. Bull.* 2000; 21: 195–201. https://doi.org/10.1177/156482650002100214

Skurtis, T., Aïnaché, G. and Simon, D. [Oil palm investiment: Why financial institutions for development must sustain their investments in Africa]. Le financement du secteur de l'huile de palme: Pourquoi les institutions financières de développement doivent continuer à investir en Afrique. *Oilseeds Fat Crops Lipids* 2010; 17: 400–3. https://doi.org/10.1051/ocl.2010.0346

Solomons, N. W. Plant sources of vitamin A and human nutrition: Red palm oil does the job. *Nut. Rev.* 1998; 56: 309–11.

Souganadis, E., Laillou, A., Leyvraz, M. and Moench-Pfanner, R. A comparison of retinyl palmitate and red palm oil β-carotene as strategies. *Nutrients* 2013; 5: 3257–71.

Sun Y. and Neelakantan, N. and Wu, Y. Palm oil consumption increases LDL cholesterol compared with vegetable oils low in saturated fat in a meta-analysis of clinical trials. *J. Nutr.* 2015; 145: 1549–58.

Van Stuijvenberg, M. E., Dhansay, M. A., Lombard, C. J., Faber, M. and Benade, A. J. The effect of a biscuit with red palm oil as a source of beta-carotene on the vitamin A status of primary school children: A comparison with beta-carotene from a synthetic source in a randomised controlled trial. *Eur. J. Clin. Nutr.* 2001; 55: 657–62.

Zagré, N., Delisle, H., Tarini, A. and Delpeuch, F. [Changes in vitamin A intake of children and women in Burkina Faso following the promotion of red palm oil consumption]. Évolution des apports en vitamine A à la suite de la promotion de l'huile de palme rouge chez les enfants et les femmes au Burkina Faso. *Cah Études Rech Francophones Santé* 2002; 12: 38–44.

Zagré, N. M., Delpeuch, F., Traissac, P. and Delisle, H. Red palm oil as a source of vitamin A for mothers and children: Impact of a pilot project in Burkina Faso. *Pub. Health Nutr.* 2003; 6: 733–42. https://doi.org/10.1079/phn2003502

Zagré, N. M., Delisle, H. and Delpeuch, F. [Red palm oil in Burkina Faso: One step toward integrated food diversification into the national vitamin A strategy]. L'huile de palme rouge au Burkina: Un pas pour la diversification alimentaire dans la stratégie nationale de lutte contre la carence en vitamine A. In Brouwer, I. D., Traoré, A. S., Trèche, S. (Eds), Proceedings of the 2nd International Workshop on food-based approaches for a healthy nutrition in West Africa, 23–28 November 2003. Ouagadougou, Presse Universitaire, 2004: 337–48.

Zeba, A. N., Martin-Prével, Y. and Delisle, H. F. The positive impact of red palm oil in school meals on vitamin A status: Study in Burkina Faso. *Nutrition J.* 2006; 5: 17. doi:10.1186/1475-2891-5-17.

Zhao, H., Lu, Z., Bie, X. and Lu, F. Analysis of positional distribution of fatty acids in triacylglycerols from lard by high performance liquid chromatography. *Se Pu* 2005; 23: 142–5.

Part 3

Sustainability and supporting smallholders

Chapter 10

Life cycle assessments of oil palm products

Cécile Bessou, CIRAD, France; Heinz Stichnothe, Thünen Institute of Agricultural Technology, Germany; Abdul-Manan, Saudi Aramco, Saudi Arabia; and Shabbir Gheewala, King Mongkut's University of Technology Thonburi, Thailand

1 Introduction

Quantifying environmental impact is becoming a requirement for agricultural commodity chains. Given the various pollution risks (e.g. eutrophication, global warming, ecotoxicity), and the opportunities to mitigate those risks (e.g. increasing nitrogen utilisation efficiency, nutrient recycling, carbon sequestration to reduce global warming), it is crucial to apply models and tools that allow for the identification of best practices in order to reduce the environmental impact of agriculture. It is particularly crucial for increasingly important crops such as oil palm that may impact the environment both during cultivation and due to land use change (LUC) for new plantations.

Over the last 20 years, the area of oil palm plantations has increased drastically. The total productive area reached 18.7 Mha in 2014 compared with 7.5 Mha in 1994, according to FAO[1]. This expansion was particularly remarkable in Indonesia and Malaysia, where the productive areas increased by a factor of two and seven, respectively, over the same time period[1]. Oil palms have the highest oil yield per hectare and palm oil can be used for various purposes. Given the growth of the world's population and the consequent growing demand for food and fuel, the increase in oil palm production is expected to continue, albeit at a slower pace than over the last decade (OECD and FAO, 2013). This increase is also expected to extend to other developing or emerging countries in Africa and Latin

1. http://www.fao.org/faostat consulted on 28 January 2017.

http://dx.doi.org/10.19103/AS.2017.0018.38

America, where governments are promoting palm oil development in order to alleviate poverty and increase energy security (Pirker et al., 2016).

Over the last decade, life cycle assessment (LCA) has become the worldwide standard for reporting environmental product declarations (ISO 14025 Type III Environmental Declarations) and the baseline model behind various greenhouse gas (GHG) calculators (BIOGRACE[2], GREET[3], CCaLC[4]) and GHG certification schemes (European Commission, 2009, BSI 2008, ISCC[5]). Initially developed in the 1980s to assess the environmental impact of industrial products and services, such as packaging, life cycle approaches were rapidly applied in increasingly diverse contexts urging for the development of harmonised guidelines. In the 2000s, the framework and methodological aspects of LCA were standardised through international norms (ISO 14040 series 2000–2006), particularly through the structuring and formalisation work led by SETAC[6]. LCA has been applied to agricultural commodities primarily for the purpose of assessing various environmental impacts and trade-offs, for example, bioenergy chains compared with fossil ones. Adaptation of the LCA framework to agricultural products requires scientific and methodological developments that are still ongoing and represent specific challenges for tropical perennial crops such as oil palm (Basset-Mens et al., 2010; Bessou et al., 2013; Bellon-Maurel et al., 2013).

In this chapter, first we briefly present LCA modelling principles and methodological steps, and then review the results from published LCA and GHG assessments of palm oil products. Finally, we discuss the available information on the environmental impact of palm oil and remaining challenges regarding LCA development and applications to palm oil products.

2 LCA principles and methodology

LCA is based on two fundamental principles. Firstly, environmental burdens are gathered throughout the commodity chain or 'life cycle', from raw material extraction ('cradle') to the end-of-life of products or services ('grave'). Secondly, environmental impacts are quantified with respect to a functional unit (FU), either a product quantity (one kilo, one car, etc.) or a usage or service [hours utilised, tonne-kilometre (tkm), etc.]. The entire life cycle of a product has to be taken into account so that local environmental improvements at one production stage or in one location do not result in a problem shifting to another stage or location (Jolliet et al., 2010). Similarly, the comparison of two or more products or services, based on the same FU, is paramount in order to identify all environmental impacts of every compared product, which enables decision-makers to avoid hidden problem shifting. Finally, LCA assesses environmental performance across numerous impact categories, such as climate change, acidification or ozone layer depletion. Such a multi-criteria approach does not focus on any one impact but rather pinpoints the relevant impacts and their origins at given production stages. This holistic approach enables identification of trade-offs and makes decision-making more transparent.

2. http://www.biograce.net/home consulted on 28 January 2017.
3. https://greet.es.anl.gov/consulted on 28 January 2017.
4. http://www.ccalc.org.uk/consulted on 28 January 2017.
5. http://www.iscc-system.org/consulted on 28 January 2017.
6. SETAC: Society of Environmental Toxicology and Chemistry, one of the most important international scientific organisations dealing with structural issues of life cycle assessment (Jolliet et al., 2010).

LCA employs a four-stage methodology (ISO 14040 series 2000–2006):

- definition of the objectives and boundaries of the system to be studied from the beginning to the end of the chain;
- inventory of mass and energy flows used within the system and those released into the environment;
- characterisation or modelling of impacts based on the inventory; and
- interpretation of the results.

Definition of the study objectives (stage 1) implies definition of the FU and the scope of the system processes to be assessed: for example, the LCA of FU = 1 t fresh fruit bunch (FFB) includes accounting for all burdens from all processes, from raw material extraction up to the harvest of FFB at the edge of the palm block, in the relative proportions needed to produce 1 t of FFB. The flows (resources used and substances emitted) are inventoried (stage 2) according to the technical specificities of the studied system. Effects of resource use and emissions generated are quantified and grouped into a limited number of impact categories (stage 3), which are expressed as problem-orientated indicators (global warming potential, eutrophication potential, etc.) or damage-orientated indicators (human health, biotic and abiotic resources, etc.). The respective indicators are calculated based on a linear model (Eq. 1):

$$I_P = \sum_{i}^{n} m_i . CF_{i,P} \tag{1}$$

where:
I_p is the indicator for the potential impact P,
m_i is the mass of the substance i contributing to the potential impact P,
$CF_{i,P}$ is the characterisation factor for the contribution of substance i to the potential impact P.

This linear model – a simplification of actual environmental impact mechanisms – does not usually account for local medium sensitivity or threshold effects; hence, LCA impacts are potential and not actual impacts. The interpretation of results (stage 4) is achieved considering uncertainties related to all the previous steps. LCA allows for the identification of environmental impact hot spots, process impact contributions and potential trade-offs between impact categories or process stages.

For example, the impact on climate change is calculated by taking into account an inventory of all GHG emissions per unit product. The emissions are then aggregated into a single impact indicator (global warming potential or climate change) using IPCC's model, which characterises what happens to GHGs in the atmosphere and their relative contributions to the global greenhouse effect. Characterisation factors in the case of climate change are expressed in CO_2 equivalent (CO_{2e}) based on mass.

Despite the intuitive methodological stages and well-documented guidelines, LCA implementation poses some problems because of insufficient data or scientific knowledge, which gives rise to a number of uncertainties, notably when inventorying field emissions and characterising final impacts. Several characterisation methods exist that provide varying environmental profiles, that is, a set of potential impact indicators. In the following section, we review palm oil LCA results, which are available in the literature, without further discussion regarding the underlying issues for LCA implementation. The challenges for LCA implementation to oil palm products are then discussed in detail, that is, stage by stage, in Section 4.

3 Results of LCA applied to oil palm products

3.1 Oil palm LCA studies

Several full or partial LCAs of oil palm products have been published over the last 20 years, with a drastic increase in publication rate over the last ten years (Fig. 1). A review of the Web of Science 1975–2017 database provided 106 publications related to palm LCA, with a large proportion of the published LCA studies focusing on palm oil-based bioenergy. Energy Fuels is the top research field covered, concerning almost 40% of the literature (Fig. 2), and *Biomass & Bioenergy* and *Applied Energy* are among the top five journals (Fig. 3). These publications were notably motivated by the debate on potential net advantages of biofuel compared with their fossil fuel equivalents and the subsequent release of the European Directive on Renewables (2009/28/EC), which details sustainability criteria including minimum GHG savings compared with the use of

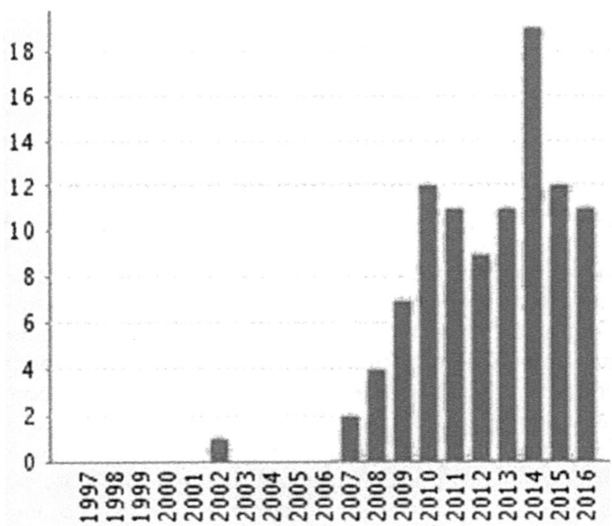

Figure 1 Published items during each year over the last 20 years from the Web of Science (February 2017). Searching terms were TOPIC: (palm NEAR/1 oil) *AND* TOPIC: (lca OR 'life cycle assessment' OR 'life cycle' OR 'lifecycle'); the total output included 248 items, some were then withdrawn due to a mistake in the Web of Science KeyWords Plus. The final item count was 106 publications.

fossil fuels. Hence, most of the published palm oil-based LCA studies focused on GHG (or climate change impact) and energy balance (or fossil resource depletion) (Manik and

Field: Web of Science Categories	Record Count	% of 106	Bar Chart
ENERGY FUELS	41	38.679 %	
ENVIRONMENTAL SCIENCES	38	35.849 %	
ENGINEERING ENVIRONMENTAL	29	27.358 %	
GREEN SUSTAINABLE SCIENCE TECHNOLOGY	21	19.811 %	
BIOTECHNOLOGY APPLIED MICROBIOLOGY	16	15.094 %	
ENGINEERING CHEMICAL	16	15.094 %	
FOOD SCIENCE TECHNOLOGY	12	11.321 %	
AGRICULTURAL ENGINEERING	11	10.377 %	
CHEMISTRY APPLIED	9	8.491 %	
CHEMISTRY PHYSICAL	6	5.660 %	

Figure 2 Published items classified according to their research topics, that is, Web of Science Categories.

Field: Source Titles	Record Count	% of 106	Bar Chart
INTERNATIONAL JOURNAL OF LIFE CYCLE ASSESSMENT	12	11.321 %	
JOURNAL OF CLEANER PRODUCTION	10	9.434 %	
JOURNAL OF OIL PALM RESEARCH	10	9.434 %	
BIOMASS BIOENERGY	8	7.547 %	
APPLIED ENERGY	6	5.660 %	
TENSIDE SURFACTANTS DETERGENTS	5	4.717 %	
RENEWABLE ENERGY	4	3.774 %	

Figure 3 Published items classified according to journal titles.

Halog, 2013; Bessou et al., 2013). A small number of published LCA have actually looked over the available panel of environmental impacts provided by LCA methodology. In the following sections, we first review environmental information on palm biofuel and then focus on palm oil LCA.

3.2 Environmental impact of palm oil-based bioenergy

Most LCA studies on palm oil-based bioenergy have been conducted in Malaysia and Thailand (with 29% and 12% of the total 106 recorded items, respectively); the few remaining predominantly cover Indonesia (more recent publications), Brazil, Colombia and Cameroon. The large majority of these studies assessed the cradle-to-grave (well-to-wheel) system boundary of palm methyl ester (PME), that is, including all processes from background input production (e.g. fertiliser manufacture) up to the vehicle tank, assuming total combustion or including engine efficiency to calculate final energy and GHG indicators.

The two main energy indicators commonly used are the Net Energy Ratio (NER = output/input) and the Net Energy Gain or Balance (output-input). Although the common LCA indicator for energy use is usually expressed in total used fossil resource equivalents, these indicators give an approximation of the environmental impact in terms of fossil resource depletion. Energy indicators may include or exclude co-products depending on the allocation ratios or whether system expansion was applied. Results vary greatly among studies (with a mean NER value of approximately 2.9) notably regarding yields, the handling of co-products, the inclusion or exclusion of capital goods (infrastructure) and discrepancies in terms of transport scenarios. Despite some differences, all studies highlight the great importance, in terms of energy costs, of both the agricultural production of palm oil feedstock and transesterification. The oil extraction stage at the mill shows low energy requirement in comparison due to the internal recycling of co-products for energy purposes. During the agricultural stage, the upstream production of fertilisers and fruit transport are the most energy-intensive steps. The upstream production of methanol is the main contributor to the energy costs of both industrial phases; however, if bioethanol replaced methanol, the NER could be improved up to ~3.6 (Papong et al., 2010).

GHG balances also vary greatly among studies and the main influencing factor is whether LUC is accounted for or not, as the type of previous land use determines the final GHG balance. The mean GHG balance (Fig. 4), accounting for various LUC scenarios, reaches 40 g CO_{2e}/MJ (9 g CO_{2e}/MJ without LUC), but is multiplied tenfold when peatland forest is converted to palm plantations (in the upper range of the min-max values). Net savings of GHG are possible when palms are planted on degraded lands or grasslands, and depend

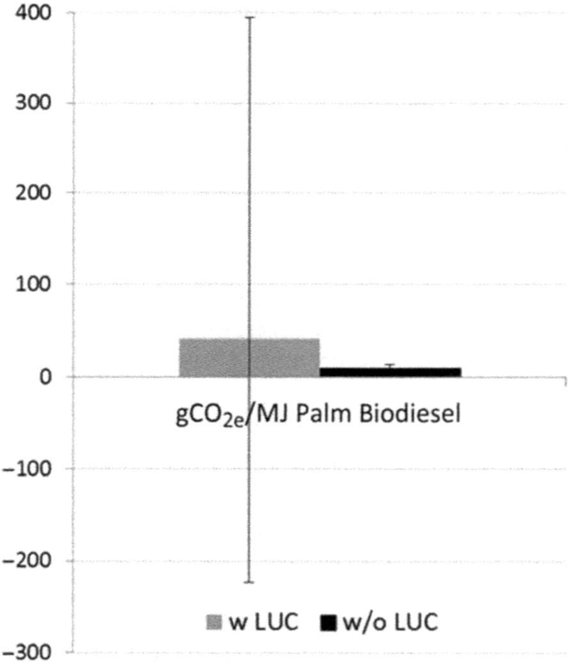

Figure 4 Comparison of LCA results on palm biodiesel (PME) based on data collected in Manik and Halog 2013: Mean GHG balance and minimum and maximum values with or without including land use change (LUC).

upon the existing carbon stock of previous land uses. Compared with fossil fuels, palm biodiesel is disadvantageous in terms of GHG if peatlands are converted or if tropical forests are cleared and the palm plantation lasts less than a century (Reinhardt et al., 2007). Otherwise, GHG savings ranging between 55% and 89%, compared with fossil diesel, can be achieved (Wicke et al., 2008; Pleanjai et al., 2009; Thamsiriroj and Murphy, 2009; Achten et al., 2010). Besides LUC, the main GHG sources are fertilisers (70–90% in field emissions, 10–30% emissions at the manufacturing site), methane emissions from the treatment of palm oil mill effluent (POME) when methane is not captured and the transesterification process (methanol and electricity) (Pleanjai et al., 2009; Thamsiriroj and Murphy, 2009; Achten et al., 2010; Choo et al., 2011).

Moreover, not all studies that include LUC use the same methodology to calculate GHG impact, which hinders any comparison. The major calculation parameters that vary are the carbon stocks accounted for (considered biomass compartments and amount of carbon released/stored) and the time frame for amortisation (Wicke et al., 2008; Hansen et al., 2014). Some of the studies that do not include LUC-related GHG emissions directly in the balance give information on the carbon debt or payback time[7] together with other results. This carbon debt varies between 8 and 169 years for palm biodiesel with mean and median values of 54 and 43 years, respectively (Fargione et al., 2008; Wicke et al., 2008; Pleanjai et al., 2009; Achten et al., 2010; de Souza et al., 2010; Harsono et al., 2012).

It is important to note that GHG accounting methodologies adopted by regulators within existing biofuel directives can also differ quite substantially. These regulations impose thresholds of minimum GHG emission reductions that biofuels must achieve, relative to fossil fuels, to show compliance. The adoption of a well-to-wheels LCA-based accounting perspective within regulations helps to ensure that such policies lead to actual reductions in global emissions as opposed to shifting the burden to a different economic sector or geographical region. Despite this, there are still many regulations today that continue to mandate the use of biofuels in transport without imposing a minimum GHG emissions reduction criterion, and therefore risk worsening global GHG emissions by forcing the substitution of fossil fuels with a biofuel that can potentially have higher emission intensity (Abdul-Manan et al., 2015).

Two of the most advanced biofuel regulations currently available are the EU's Renewable Energy Directive (RED) and the US Renewable Fuels Standard 2 (RFS2). The EU's RED requires biofuels to initially achieve a minimum reduction of 35% GHG emission, which is then increased to 50% for new plant installations operating from October 2015, and 60% for all biofuels effective from 2018 (European Commission, 2015). The US RFS2 stipulates that for a biofuel to be granted 'renewable fuel' status, it has to demonstrate a minimum GHG emission reduction of 20% (EPA, 2010). Although these regulations both adopt an LCA approach, the detailed GHG accounting methodologies they relied on are in reality very different, which prohibits any direct comparison.

An important distinction between the US RFS2 and the EU's RED is the way they take LUC into account. Presently, both regulations require the incorporation of direct LUC (dLUC) effects when accounting for biofuel GHG emissions. dLUC is the direct alteration of lands by the farmers themselves to produce biofuel crops. Indirect LUC (iLUC) is the unintended change of land use worldwide, typically from carbon-rich non-agricultural land to carbon-poor agricultural land, in response to economic pressures arising from the increasing

7. Years needed to recover the carbon loss due to LUC based on the annual GHG savings allowed by biofuel when displacing fossil fuel (Fargione et al., 2008; Gibbs et al., 2008).

demand for biofuels. iLUC requires a sophisticated economic modelling of global supply and demand of lands worldwide and how they respond to economic pressures. Today, both regulators acknowledge the importance of iLUC in terms of GHG emissions and its potential influence on reducing GHG from biofuels. The US RFS2 and the California Low Carbon Fuels Standard have included iLUC GHG penalties in their regulatory LCA, while policymakers in the EU opted for a virtual control of iLUC through limiting the maximum amount of conventional first-generation biofuels, like palm biodiesel, to be claimed under the EU's RED (Abdul-Manan, 2017), thus assuming that first-generation biofuel feedstocks are more likely to drive iLUC.

Using their respective methodologies, both the US RFS2 and the EU's RED have provided estimated reduction potentials of GHG emissions for palm biodiesel. Under the EU's RED framework, the GHG savings potential for typical palm biodiesel processes without and with methane capture are 36% (54 g CO_{2e}/MJ) and 62% (32 g CO_{2e}/MJ), respectively. These values exclude dLUC, which operators need to estimate to show compliance. However, according to the EU's RED sustainability criteria, no dLUC should occur after 1 January 2018 at the expense of lands with high carbon stock or high biodiversity value. In comparison, the GHG emissions reduction potential for palm biodiesel under RFS2 has been estimated to average approximately 17% (76 g CO_{2e}/MJ); this value includes both dLUC and iLUC. The large difference between the regulatory values in the EU and the United States are attributable to the methodological distinctions, including the treatment of LUC but also allocation, fossil references and so on.

Although the control of GHG emissions is a major issue in biofuel regulations, they also include further pass-or-fail sustainability criteria such as elements relating to the protection of land with high conservation value, prevention of habitat loss, fair and equitable treatment of workers and communities and so on. Only GHGs are quantified using an environmental LCA approach in spite of the much wider potential for the use of LCA techniques in biofuel sustainability impact assessments. The scientific literature details many studies which evaluate other environmental impacts of biofuel production (Achten et al., 2010; Puah et al., 2010; Arvidsson et al., 2011; Silalertruska and Gheewala, 2012). These studies concomitantly highlight the important contribution of the agricultural phase to other impact categories, for example, eutrophication and acidification potentials, carcinogens and respiratory inorganics. Fertilisers which leak into the environment contribute significantly to eutrophication and acidification. The use of biodiesel in engines also adds to the potential impact of eutrophication and acidification (Arvidsson et al., 2011), and contributes significantly to the impact category of respiratory inorganics (Puah et al., 2010).

3.3 Environmental impact of oil palm fruits and palm oil

LCA studies on palm fruits and oil are less numerous than those focusing on palm biodiesel, but they globally cover more impact categories and provide more details on the agricultural phase (Yusoff and Hansen, 2005; Reijnders and Huijbregts, 2008; Zulkifli et al., 2009; Vijaya et al., 2010; Schmidt, 2010; Stichnothe and Schuchardt, 2011). A few studies also focus on GHG assessment (Chuchuoy et al., 2009; Choo et al., 2011; Kaewmai et al., 2012; Bessou et al., 2014).

As expected, the main contributors to the GHG balance of crude palm oil (CPO) are the same as for palm biodiesel, except transesterification, with LUC and peat oxidation being critical and potentially overwhelming drivers (Schmidt, 2007; Reijnders and Huijbregts, 2008; Zulkifli et al., 2009), followed by methane emissions from the treatment of POME and fertiliser-related emissions, notably N_2O field emissions (Schmidt, 2007; Choo et al., 2011; Chase et al., 2012; Bessou et al., 2014). Nevertheless, the impact of POME can be significantly reduced if biogas is captured at the mill level (Chavalparit et al., 2006; Choo et al. 2011; Bessou et al., 2014; Harsono et al., 2014) or, to a lesser extent, if raw or partially treated POME are injected into a composting process for organic residues (Singh et al., 2010; Stichnothe and Schuchardt, 2010).

In a pilot application of palm GHG (RSPO GHG calculator, Chase et al., 2012) on mills in Southeast Asia and Latin America, the average GHG balance was 1.67 t CO_{2e}/t CPO and ranged from −0.02 to +8.32 t CO_{2e}/t CPO (Bessou et al., 2014). Of the mills not supplied by a peat area, land clearing, POME methane emissions and fertiliser-related emissions accounted for 41–80%, 15–35% and 3–19% of total GHG emissions, respectively. The impact of fossil fuel use was not significant (0–5% and 0–2% of total emissions at the field and mill levels, respectively). Such a low impact was due to the low mechanisation level in the plantations and the recycling of numerous residues providing heat and power to operate the mill (with the potential production of excess electricity). Most field fuel use is dedicated to FFB transport; hence, the impact of fuel use may vary greatly according to FFB harvesting logistics.

Published GHG balances (or the climate change impact indicator) range between −0.55 and 24 t CO_{2e}/t CPO with median values around 1–2 t CO_{2e}/t CPO when LUC is applied to mixed previous land uses and less than 10% peatland, and methane is not captured (Reijnders and Huijbregts, 2008; Schmidt, 2010; Choo et al., 2011; Bessou et al., 2012).

Looking at the other impact categories, the agricultural phase remains the main contributor, except for human toxicity or respiratory inorganics impact categories, which are mainly caused by boiler emissions (Stichnothe and Schuchardt, 2011; Bessou et al., 2012). Mill emissions can also contribute to eutrophication which is driven by the emission of nitrogen and phosphorus compounds. The main eutrophication factors at the agricultural stage are nitrate leaching, and phosphorus and nitrate run-off. Other N-compound emissions also contribute to acidification and photochemical ozone impact categories. While palm oil generally performs worse than other oil crops on climate change impact, when LUC occurs and leads to carbon loss from previous land use (e.g. in the case of deforestation or peat oxidation), palm oil can perform better than rapeseed oil with regard to eutrophication, acidification, ozone depletion and photochemical ozone impacts when effective management is in place (Schmidt, 2010).

4 Challenges in building LCA of oil palm products

4.1 Issues at the Goal and Scope level

The Goal and Scope steps of the first stage of LCA are critical as they define the validity domain of the final outputs. The boundary of the studied system must be delineated in order to ensure that all potential environmental impacts linked to the investigated product or service are taken into account. At the same time, there might be trade-offs needed between an exhaustive system assessment and gathering representative and consistent data. Iterative adjustments from stage-to-stage are often needed to carry out a robust LCA.

Being a perennial crop, oil palms last for at least 25 years in the field, during which time the crop stand goes through different development phases. The whole life cycle of oil palms includes the nursery stage (three months in pre-nursery and nine months in the main nursery), the early growing stage of immature non-productive palms (2–3 years) in addition to the productive harvest period (Stichnothe et al., 2014). Palm trees older than 25–28 years old (depending on planting material and site conditions) are often too high for harvesting to be kept longer in the field. The early growing stages account for 10–15% of the entire plantation cycle. These long and partitioned cycles require specific management, which usually combines long-term management strategies and short- or medium-term adjustments. Moreover, it also implies complex and evolving interactions with the ecosystem, which can affect the potential performance of the crop and management efficacy. Nevertheless, in most published studies, only the productive area of the plantation and the associated FFB yield are considered. Given the potential significant contribution of the early stages and the variability in practices and performances throughout the long productive period, the modelling choices to account or not for the whole perennial cycle can influence LCA results (Bessou et al., 2016). Hence, when defining the goal and scope of an oil palm product LCA, attention should be paid to the whole perennial cycle in order to produce representative results. Considering the whole growing cycle is particularly relevant for nitrogen losses (Pardon et al., 2016a) and hence for the life cycle inventory (LCI).

Another peculiar aspect of perennial compared with annual crops is the potential importance of changes in carbon stocks (Mithraratne et al., 2008). Henson showed that mature oil palms on coastal soil in Malaysia generated a net carbon fixation of 11 t ha^{-1} y^{-1} based on the eddy covariance technique (Henson, 1999). This fixation rate varies depending on the plantation age and management, and it does not represent an actual net carbon fixation in the biosphere. Indeed, a large proportion of the assimilated carbon is exported to the oil mill (Melling et al., 2010).The temporary storage of carbon in oil palm stipes might improve the GHG balance of palm plantations (Lam et al., 2009), but there is no generally accepted method for quantifying temporary carbon storage (Levasseur et al., 2012). The most generally used and reproduced guidelines are those from IPCC (IPCC, 2006). Further guidelines developed on the same basis, such as PAS2050 (BSI, 2011) or the European Renewable Directive (European Commission, 2009), all consider potential carbon storage in biomass as long as it represents a stable stock at equilibrium for at least 20 or 25 years. The way stocks are calculated and changes are modelled varies considerably across methods and published studies. Whether or not oil palm plantations are a net sink or source of carbon depends on the soils, climate, cultivation and residue management practices; however, the history of the site, especially LUCs (Melling et al., 2005, Melling et al., 2010), may significantly affect the GHG balance of end products such as palm biodiesel (see Section 3.1). Defining if and how LUC should be included in the LCA is a crucial parameter in the goal and scope definition of the LCA of palm products.

4.2 Issues related to LCI data collection

Specific quantified LCI data, for example, history of LUCs, influence of plantation management practices, nitrogen budget of oil palms, residue treatment, etc., are frequently missing, which is a current issue in tropical crop LCA (Basset-Mens et al., 2010). In the oil palm sector, the lack of representative data is accentuated by the concomitant lack of detailed institutionalised agricultural census for certain key producing countries and the great diversity in oil palm practices observed in the field (Lee et al., 2014; Moulin et al., 2016). Current knowledge regarding the influence of different management practices on the plantation and/or the palm oil mills varies from fragmented to non-existent. Examples include nutrient management, water level management on peat soils, pest control, residue treatment (empty fruit bunches (EFB), POME and nutshells), energy efficiency in oil mills, to name just a few. This critical lack of data persists despite the recent growing number of LCA studies driven by environmental concerns notably due to the expansion of oil palm areas.

LUC and peat oxidation lead to severe damage to the environment in terms of both biodiversity loss and GHG emissions. The proper identification of LUCs, from the type and extent of land cover, and subsequent land use fluxes and related emissions is therefore critical. Assessing the impact of oil palm area expansion requires the identification of LUCs and LUC impacts, as well as the impact of oil palm land use, for example, the impact on soil or carbon sequestration. Impacts of land use and LUCs are highly sensitive to soil type and climate conditions so that site or region-specific assessment is required to adequately cover this aspect. The development of region-specific LCI methods is hampered by the lack of regional and site-specific data. Moreover, there is still a lack of consensus on the methodology to address LUC history, carbon stock accounting, fluxes and therefore a lack of adequate and representative site-specific data sets.

Over the past 20 years, 95% of the Indonesian palm oil production area has been located in Sumatra and Kalimantan, and palms have been increasingly cultivated on peatlands (Afriyanti et al., 2016). Tropical peatlands store a huge amount of carbon, roughly 7000 t C ha^{-1} in below-ground biomass (Moore et al., 2013) and are highly vulnerable to natural and human disturbance. Under normal weather conditions, peatland in Indonesia is almost entirely waterlogged, which must be drained via hydrological engineering prior to oil palm planting. The water level is the main control for GHG fluxes from tropical peat soils. Crouwenberg et al. (2010) calculated emissions of at least 9 t CO_2 ha^{-1} y^{-1} and considered that to be a conservative estimate, because the role of oxidation in subsidence and the increased bulk density of the uppermost drained peat layers are insufficiently quantified (Couwenberg et al., 2010). The decomposition of biomass due to the lowering of the water table levels also goes along with nitrous oxide emissions. Despite dedicated research (Melling et al., 2007; Jauhiainen et al., 2012a,b) and recent guidelines (IPCC, 2013), there is still considerable uncertainty on the impact of various water level management practices on peat emissions and on the various direct and indirect fluxes and impacts of peat cultivation; hence, LCI for oil palm plantations on peat soil are not comprehensive.

Nitrogen losses in agroecosystems are a major environmental and economic issue. Indeed, agroecosystems receive approximately 75% of the reactive nitrogen created by human activity (Galloway et al., 2008; Galloway et al., 2013). In oil palm plantations, nitrogen fertilisation is a common practice that is associated with water pollution risks and GHG emissions (Corley and Tinker, 2008; Choo et al., 2011; Comte et al., 2012), notably nitrous oxide, a very potent GHG[8]. Furthermore, fertilisers constitute 46–85% of plantation field costs (Caliman et al., 2001; Goh and Härdter, 2003; Silalertruksa et al., 2012). Oil palm plantations have three main peculiarities affecting nitrogen dynamics in a way that differs from other cropping systems: the long duration of the growing cycle, the marked spatial heterogeneity and the large internal fluxes and pools of nitrogen. Substantial losses of reactive nitrogen can occur during the immature phase, when palms are still young and legume cover is vigorous; as well as during the mature phase in areas with sparse or no soil cover; or where high amounts of organic and mineral fertilisers are applied (Pardon et al.,

8. Nitrous oxide has a global warming potential 298 times greater than carbon dioxide on the same mass basis (IPCC, 2007).

2016a). Pardon et al. investigated several models to estimate nitrogen losses of oil palm plantations; most of the models indicated substantial losses at the early growing stage of oil palms. On average, 31% of nitrogen losses occur during the immature growing phase (Pardon et al., 2016b), which is frequently not taken into account in an LCA of oil palm plantations (see Section 4.1). The greatest uncertainty involves the loss of nitrogen via the emission of gaseous nitrogen compounds (N_2O, NO_x, N_2, NH_3) (Pardon et al., 2016a). Reactive nitrogen emissions contribute to several environmental problems, such as climate change, eutrophication or acidification. The lack of precise estimation of the nitrogen compounds released into the environment thus causes a great deal of uncertainty in the associated impact categories, emphasising the need for representative and robust LCI data on nitrogen fluxes as far as possible.

The management of organic residues from palm oil mills is paramount to emission reduction and nutrient recycling (Stichnothe and Schuchardt, 2010, Kaewmai et al., 2013). Given the diversity of residues generated by the production of palm oil (EFB, fibres, shells, etc.) and their respective large amounts (e.g. POME), there are very diverse ways to reuse these products via various processes and potential impacts. One cubic metre of POME treated in conventional open ponds can generate up to 12 m^3 of methane emission, equivalent to approximately 200 kg CO_{2e}. Biogas production from improved POME treatment is associated with a highly favourable GHG budget (Bessou et al., 2014). A worst-case scenario is dumping EFB, causing GHG emissions equivalent to 1000 kg CO_{2e} t^{-1} (Stichnothe and Schuchardt, 2011; Langeveld et al., 2016). POME and EFB can also be co-composted, which can lead to emission reductions as well as benefits to soil quality (Stichnothe and Schuchardt, 2010). Indeed, EFB are generally applied back to the plantation to maintain soil fertility through increasing the organic matter content (Saletes, 2004, Carron et al., 2015). The application of palm oil mill residues back to the field may not only reduce GHG emissions but also preserve resources as it reduces the demand for mineral fertilisers. The impact of compost or EFB on soil quality, as well as upstream emissions during the various composting processes, are still poorly quantified. Further data collection is needed to better account for these practices within both LCI and life cycle impact assessment (LCIA).

4.3 Challenges in impact pathway characterisation

The development of several LCIA methodologies has created confusion partly due to differing results even for some midpoint or endpoint indicators. Several areas/indicators (soil property change, ecotoxicity, biodiversity, etc.) are still under development and consequently not fully ready for general use.

Land use causes various chemical, physical and biological changes to soil properties and functions such as life support or nutrient cycling. Despite recent developments by the LCA community (Milà i Canals et al., 2007; Oberholzer et al., 2012; Garrigues et al., 2013; Saad et al., 2013; Bos et al., 2016), there is currently no comprehensive impact assessment of the various branches of the cause–effect chains implemented in LCIA. In particular, impacts related to co-variations in the associated physico-chemical and biological soil properties and soil functions are hardly addressed in LCIA. Moreover, physical and chemical changes of surface and soil have further effects on flora and fauna and hence affect biodiversity within and above the soil. The accounting of land use and LUC impacts is critical for oil palms given the issue of area expansion and the peculiarities of oil palm as a perennial crop, that is, the long-term cultivation cycle with constant land cover and biomass accumulation, and the deep rooting system of the plant (see Section 4.1). In addition, practices related to the recycling of residues back to the field may also influence soil quality (see Section 4.2). Comprehensive impact pathways to relate the long-term trends and the influence of practices on the temporary storage of soil carbon, improvement of soil quality and protection from soil erosion are not currently part of the LCIA (Stichnothe and Schuchardt, 2011) as the existing level of knowledge impedes the modelling of all potential correlated processes and impacts. To design the best environmentally friendly scenarios of residues and global plantation management, the proper modelling of impact on soil is crucial.

The modelling of land use impacts on biodiversity is also considered a priority in LCA. Biodiversity can be considered at different levels, namely ecological diversity (ecosystems), population diversity (species) and genetic diversity (genes). The quantification is complex and many diverging approaches have been proposed in an expanding literature on the topic (Curran et al., 2016). Biodiversity loss can be linked to four midpoint indicators (land use, ecotoxicity, acidification and eutrophication) but also to the endpoint indicator 'Natural Environment'. Curran et al. (2016) evaluated the performance of 31 models to assess the biodiversity loss from both the LCA and the ecology/conservation literature. The authors concluded that there is room for improvement and suggested working on a 'consensus model' by the weighted averaging of existing information to complement future development (Curran et al., 2016). Currently, there is no agreed and harmonised approach which addresses how to quantify the spatially distinct environmental impacts of LUC in palm oil-producing countries.

Spatially explicit methods are needed in LCA in order to accurately quantify impacts of products and processes. Chaudhary et al. (2015) used the countryside species–area relationship to quantify regional species loss due to land occupation and transformation (Chaudhary et al., 2015; de Baan et al., 2015). These authors combined regional characterisation factors with vulnerable scores to calculate global characterisation factors. Oil palms grow in tropical areas and tropical biomes have higher characterisation factors than those of boreal biomes mainly because of their higher species richness per area.

Finally, dry peat soils are prone to subterranean fires, which smoulder and emit thick white smoke laden with hazardous particles (Goldstein, 2016). Such fires in Indonesia became an international health concern in 2015, enhanced by long and intense drought periods related to a severe El Niño episode occurring in the region; a similar catastrophe occurred in 1997. Such fires cause smog, haze and respiratory problems as far away as Malaysia, Singapore and the Philippines. Those were obviously extreme events that, by definition, have the potential to cause considerable health and other environmental impacts but whose occurrence is rare. The frequency, intensity and persistence of such extreme events are still important characteristics for deriving characterisation factors, for example, for human toxicity. Such information requires dedicated modelling work in combination with LUC and climate models.

4.4 The challenge of interpreting results

Results have to be discussed with respect to the particular goal and scope of the study, which in turn also define data requirements but also the limitation of the analysis. Describing the consequences of modelling choices, such as total or productive plantation area, LCI models (IPCC, crop model, etc.), time period (year, plantation cycle or several plantation cycles) considered and so on, is crucial, as all such factors can influence the results. The spatial dimension is given by the scope of the study, for example, a specific plantation, a particular region or national production. The obtained results are only valid for the system under investigation. Although this may seem obvious, results are frequently generalised without proper evidence.

Palm oil mills are multi-output systems and the difficulty is quantifying and identifying which product contributes to the emissions. System subdivision is not feasible as palm kernels cannot be obtained separately. System expansion is possible but difficult to interpret and the substitution method is prone to arbitrary choices for co-product substitutes, for example, can kernel meal be a substitute for soya meal or wheat? In attributional LCA, emissions can be allocated among the various products, for example, CPO, nuts or other downstream products such as palm kernel oil and palm kernel meal, while using physical (mass, energy content, nutrient content, etc.) or economic relationships. Obviously, all these choices will alter the results for a particular product (Wiloso et al., 2015). It is highly recommended to conduct a sensitivity analysis for the different options as well as an uncertainty analysis before discussing results. The epistemic uncertainty analysis is particularly crucial for LCI field emission models that are not well parameterised for tropical perennial crops such as oil palm, and for cause–effect processes, notably those related to soil functions, which are still not fully understood and modelled.

5 Oil palm LCA improvement tracks

5.1 The search for representative data sets

LCA studies of oil palm systems and their derived products are frequently restricted by data gaps. Consequently, the principal challenge is to build a consensus-based modelling framework, to gather regional- and management-specific inventory data and to define inventory models in order to estimate emissions and temporary carbon storage effects. Building a national LCI database for oil palm plantations and subsequent conversion processes would be a valuable asset.

Independent to the system boundaries studied, the agricultural phase, in particular fertiliser input, plays a key role in determining the final environmental profile. It is hence paramount to adjust fertiliser input to enhance productivity while limiting loss to the environment. To do so, there is an urgent requirement for adapted models (mechanistic or operational models) that allow for more precise estimation of field emissions linked to fertilisers. Indeed, the great majority of LCAs use IPCC emission factors to estimate nitrate leaching and run-off as well as ammoniac or nitrous oxide emissions. These emission factors are poorly calibrated for tropical regions (Bouwman et al. 2002a,b; Stehfest and Bouwman, 2006) and they do not take into account the specificities of perennial cropping cycles such as palm plantations. A recent review emphasised that the combined initial structure and long-term evolution of oil palm plantations induce specific spatio-temporal patterns in nitrogen fluxes that are poorly quantified and thus need further research. This review also highlighted that nitrogen losses through leaching and volatilisation may be important and all nitrogen gaseous losses remain unknown (Pardon et al., 2016). More field measurements are needed to establish more relevant emission factors.

Research projects are ongoing that will shed some light on ways to reduce uncertainty in the LCA results. Development work on other approaches, such as agro-ecological indicators, are complementary as they enable a better account of local conditions and practices to build up the LCA inventories. New knowledge and model developments are also expected to accurately account for the comprehensive role of organic fertilisers in soil quality and potential field emissions.

5.2 The need for comprehensive impact assessments

There are 13–18 impact category indicators in the current standard LCA methods (ILCD, ReCiPe; respectively). Nevertheless, a great proportion of published LCA studies on oil palm products solely focus on GHG and energy balances. Many LCA impact indicators need to be more widely explored across palm oil production systems such as the impacts of pesticides (e.g. paraquat or glyphosate) on terrestrial or freshwater ecotoxicity, or the impact of irrigation systems on water depletion (Nilsalab et al., 2016; Silalertruska et al., 2016). Given the important contribution of fertilisers to environmental impact during the agricultural phase, the eutrophication and acidification impacts related to nitrogen and phosphate inputs would also need to be further investigated. Several other environmental impact indicators (ecotoxicity, biodiversity, etc.) are still under development and consequently not ready for use.

The accounting of land use impacts on soil is critical for oil palm given i) the important challenge related to oil palm expansion and related LUCs and ii) the peculiarities of oil palm as a perennial compared with annual crops and due to the various and abundant recycled residues. Particular impacts related to co-variations in the associated physico-chemical and biological soil properties and soil functions are hardly addressed in LCIA due to limited knowledge. The impact of peat drainage on soil quality and causal relationship with increased risk of peat fires also need to be further investigated, and modelled within the LCA framework.

Finally, the limits of the linear globalised model may be overcome by developing regional characterisation factors that can be used to adapt the linear model to the sensitivity of the local environment. Such factors are particularly critical in the case of very local impacts that are more sensitive to changes in the immediate environment – such as eutrophication – or resources unequally distributed on the global scale, such as water in dryland areas. Such regional factors have not yet been highly developed in regions where palm plantations are established and in the context of LUC may affect the medium sensitivity during the transition phase in particular.

6 Conclusion

LCA is a very useful tool for assessing the environmental impacts of oil palm products as it helps to identify the hot spots across the whole commodity chain while avoiding hidden trade-offs between different environmental impacts. Studies show that the oil palm cultivation stage contributes to a major share of several impacts including climate change, acidification and eutrophication, though the palm milling stage can also release significant GHGs if wastewater is not properly managed. Palm mills also contribute significantly to toxicity effects due to the particulate emissions from boilers.

The assessment of GHGs that contribute to climate change has been widely carried out and a large contribution is from LUC and nitrogen fertiliser production and application. For palm LCAs, delineation of the system boundaries is critical to provide consistency to studies, which can be particularly confounded by the challenge of accurately defining and calculating the impacts from LUC and carbon stocks. This is in part linked to a lack of consensus on the methodologies for the calculation and also to the limited availability of robust and reliable inventory data on the history of LUCs, influence of plantation management practices, nitrogen budget of oil palms and residue treatment.

In the oil palm sector, the lack of representative data is accentuated by the great diversity in cultivation practices observed in the field. Currently, specific data and consequently knowledge on the influence of different management practices on the plantation and the palm oil mills are limited. The nitrogen dynamics of oil palm are complicated due to the long growing cycle, spatial heterogeneity and large internal fluxes and pools of nitrogen, which lead to great uncertainty in the assessment of climate change, acidification and eutrophication impacts. There is also a considerable lack of data on the various options of residue management which can affect the assessment results.

Significant challenges also remain when selecting impact assessment methods for characterising land use, biodiversity and including the effects of temporary carbon storage. Land use causes various chemical, physical and biological changes to soil properties and functions such as life support or nutrient cycling. There is currently no comprehensive impact assessment of the various branches of the cause–effect chains implemented in LCIA. In particular, impacts related to co-variations in the associated physico-chemical and biological soil properties and functions are insufficiently addressed. Moreover, physical and chemical changes of surface and soil have further effects on flora and fauna and hence affect biodiversity. Comprehensive impact pathways to relate long-term trends and the influence of practices on the temporary storage of soil carbon, improvement of soil quality and protection from soil erosion are not currently part of the LCIA. To design the best environmentally friendly scenarios of residues and global plantation management, correct modelling of impact on the soil is crucial.

To address the challenges of conducting palm LCAs, a consensus-based modelling framework is needed which can consistently define the inventory data needs for estimating emissions from fertiliser application and temporary carbon storage. Regionalised land use models need to be developed along with complementary agro-ecological indicators for better characterisation of the effects of oil palm cultivation. Finally, the obtained results have to be discussed with respect to the particular goal and scope of the study, including model, allocation and other methodological choices as well as data quality assessment, which when combined, define the validity domain of the results and hence the application limitation of the analysis. Such limitations should be estimated by scenario analysis. It is highly recommended to support the final interpretation of results by sensitivity and uncertainty analyses.

7 Where to look for further information

Key organisations for LCA development:

SETAC: http://www.setac.org/
UNEP SETAC Life Cycle Initiative: http://www.lifecycleinitiative.org/

LCA networks in Southeast Asia:

Indonesian LCA Network (ILCAN): http://www.ilcan.or.id/
Thai LCA Agri Food Asia Network: http://www.lcaagrifoodasia.org

LCA networks in Europe:

European Platform: http://eplca.jrc.ec.europa.eu/
French Environmental Lifecycle and Sustainable Assessment: http://www.elsa-lca.org/?lang=en
German LCA network: http://www.lcanet.de/en/

Dedicated journals and conferences:

International Journal of LCA: http://link.springer.com/journal/11367
Journal of Cleaner Production: https://www.journals.elsevier.com/journal-of-cleaner-production
Indonesian Journal of LCA and Sustainability: http://ijolcas.ilcan.or.id/index.php/IJoLCAS

Environmental Toxicology and Chemistry:

http://onlinelibrary.wiley.com/journal/10.1002/(ISSN)1552-8618

Integrated Environmental Assessment and Management:

http://onlinelibrary.wiley.com/journal/10.1002/(ISSN)1551-3793
LCA-Food conferences: http://lcafood2016.org/
SETAC Europe LCA Case Study Symposium: http://events.setac.eu/?contentid=179

8 References

Abdul-Manan, A. F. N., Baharuddin, A. and Chang, L. (2015). Application of theory-based evaluation for the critical analysis of national biofuel policy: A case study in Malaysia. *Evaluation and Program Planning*, 52: 39–49.

Abdul-Manan, A. F. N. (2017). Lifecycle GHG emissions of palm biodiesel: Unintended market effects negate direct benefits of the Malaysian Economic Transformation Plan (ETP). *Energy Policy*, 104: 56–65.

Achten, W. M. J., Vandenbempt, P., Almeida, J., et al. (2010) Life cycle assessment of a palm oil system with simultaneous production of biodiesel and cooking oil in Cameroon. *Environmental Science and Technology* 44: 4809–15. doi:10.1021/es100067p.

Afriyanti, D., Kroeze, C. and Saad, A. (2016). 'Indonesia palm oil production without deforestation and peat conversion by 2050'. *Science of the Total Environment* 557–558: 562–70.

Arvidsson, R., Persson, S., Fröling, M. and Svanström, M. (2011). Life cycle assessment of hydrotreated vegetable oil from rape, oil palm and Jatropha. *Journal of Cleaner Production* 19: 129–37. doi:10.1016/j.jclepro.2010.02.008.

Basset-Mens, C., Benoist, A., Bessou, C., Tran, T., Perret, S., Vayssières, J. and Wassenaar, T. (2010). Is LCA-based eco-labelling reasonable? The issue of tropical food products. In: *7th International Conference on Life Cycle Assessment in the Agri-Food Sector.*

Bellon-Maurel, V., Aissani, L., Bessou, C., Lardon, L., Loiseau, E., Risch, E., Roux, P. and Junqua, G. (2013). What scientific issues in Life Cycle Assessment applied to waste and biomass valorization? Editorial. *Waste Biomass Valorization* 4(2): 377–83.

Bessou, C., Vélu, A. and Caliman, J. P. (2012). LCA of palm oil in sumatra, comparison of cropping systems. In: *International Conference on Oil Palm and Environment (ICOPE 2012)*. Bali, Indonésie, 22–4 février 2012.

Bessou, C., Basset-Mens, C., Tran, T. and Benoist, A. (2013). LCA applied to perennial cropping systems: A review focused on the farm stage. *The International Journal of Life Cycle Assessment* 18(2): 340–61.

Bessou, C., Chase, L. D. C., Henson, I. E., et al. (2014). Pilot application of PalmGHG, the Roundtable on Sustainable Palm Oil greenhouse gas calculator for oil palm products. *Journal of Cleaner Production* 73: 136–45. doi:10.1016/j.jclepro.2013.12.008.

Bessou, C., Basset-Mens, C., Latunussa, C., Vélu, A., Heitz, H., Vannière, H. and Caliman, J.-P. (2016) Partial modeling of the perennial crop cycle misleads LCA results in two contrasted case studies. *The International Journal of Life Cycle Assessment* 21(3): 297–310.

Bos, U., Horn, R., Beck, T., Lindner, J.P. and Fischer, M. (2016). *LANCA® - Characterisation Factors for Life Cycle Impact Assessment, Version 2.0.* Fraunhofer-Verlag, Stuttgart, Fraunhofer.

Bouwman, A. F., Boumans, L. J. M. and Batjes, N. H. (2002a). Emissions of N_2O and NO from fertilized fields: Summary of available measurement data. *Global Biogeochemical Cycles* 16: 1058.

Bouwman, A. F., Boumans, L. J. M. and Batjes, N. H. (2002b). Modeling global annual N_2O and NO emissions from fertilized fields. *Global Biogeochemical Cycles* 16: 11. doi:10.1029/2001GB001812.

BSI (2011) *PAS2050: Specification for the Assessment of the Life Cycle Greenhouse Gas Emissions of Goods and Services*, British Standards Institution, London.

Caliman, J. P., Hardianto, J., Ng, M., et al. (2001). Strategy for fertilizer management during low commodity price. In: *Cutting-edge Technologies for Sustained Competitiveness: Proceedings of the 2001 PIPOC International Palm Oil Congress, Agriculture Conference, Kuala Lumpur, Malaysia, 20–22 August 2001*. Malaysian Palm Oil Board (MPOB), Kuala Lumpur, Malaysia, pp. 295–312.

Carron, M. P., Auriac, Q., Snoeck, D., Villenave, C., Blanchart, E., Ribeyre, F., Marichal, R., Darminto, M. and Caliman, J. P. (2015). Spatial heterogeneity of soil quality around mature oil palms receiving mineral fertilization. *European Journal of Soil Biology* 66: 24–31.

Chase, L. D. C., Henson, I. E., Abdul-Manan, A. F. N., Agus, F., Bessou, C., Mila i Canals, L. and Sharma, M. (2012). *The PalmGHG Calculator: The RSPO Greenhouse Gas Calculator for Oil Palm Products, Betaversion*. RSPO, Kuala Lumpur.

Chaudhary, A., Verones, F., de Baan, L. and Hellweg, S. (2015). Quantifying land use impacts on biodiversity: Combining species–area models and vulnerability indicators. *Environmental Science & Technology* 49(16): 9987–95.

Chavalparitm, O., Rulkensm, W. H., Molm, A. P. J. and Khaodhairm, S. (2006). Options for environmental sustainability of the crude palm oil industry in Thailand through enhancement of industrial ecosystems. *Environment, Development and Sustainability* 8: 271–87. doi:10.1007/s10668-005-9018-z.

Choo, Y. M., Muhamad, H., Hashim, Z., et al. (2011). Determination of GHG contributions by subsystems in the oil palm supply chain using the LCA approach. *The International Journal of Life Cycle Assessessment* 16:669–81. doi:10.1007/s11367-011-0303-9.

Chuchuoy, K., Paengjuntuek, W., Usubharatana, P. and Phungrassami, H. (2009). Preliminary study of Thailand carbon reduction label: A case study of crude palm oil production. *Europial Journal of Science Research* 34: 252–9.

Comte, I., Colin, F., Whalen, J. K., et al. (2012). Agricultural practices in oil palm plantations and their impact on hydrological changes, nutrient fluxes and water quality in Indonesia. In: D. L. Sparks, *Advances in Agronomy*. Elsevier, Newark, USA, pp. 71–124.

Corley, R. H. V. and Tinker, P. B. H. (2008). *The Oil Palm*. Fifth Edition. Blackwell Science Ltd, UK.

Couwenberg, J., Dommain, R. and Joosten, H. (2010) Greenhouse gas fluxes from tropical peatlands in south-east Asia. *Global Change Biology* 16(6): 1715–32.

Curran, M., Maia de Souza, D., Antón, A., Teixeira, R. F. M., Michelsen, O., Vidal-Legaz, B., Sala, S. and Milà i Canals, L. (2016) How well does LCA Model land use impacts on biodiversity?—a comparison with approaches from ecology and conservation. *Environmental Science & Technology* 50(6): 2782–95.

de Baan, L., Curran, M., Rondinini, C., Visconti, P., Hellweg, S. and Koellner, T. (2015). High resolution assessment of land use impacts on biodiversity in Life Cycle Assessment using species habitat suitability models. *Environmental Science & Technology.*, 49(4) : 2237–44.

de Souza, S. P., Pacca, S., de Ávila, M. T. and Borges, J. L. B. (2010). Greenhouse gas emissions and energy balance of palm oil biofuel. *Renew Energy* 35:2552–61. doi:10.1016/j.renene.2010.03.028.

European Commission (2015). Directive (EU) 2015/1513 of the European Parliament and of the Council of 9 September 2015 amending Directive 98/70/EC relating to the quality of petrol and diesel fuels and amending Directive 2009/28/EC on the promotion of the use of energy from renewable. *Official Journal of the European Union.*

EPA (2010). Part II Environmental Protection Agency. 40 CFR Part 80. Regulation of Fuels and Fuel Additives: Changes to Renewable Fuels Standard Program; Final Rule. National Archives and Records Administration, Washington.

European Commission (2009). Directive 2009/28/EC of the European parliament and of the council of 23 April 2009 on the promotion of the use of energy from renewable sources. *Official Journal of the European Union*, June 5th.

Fargione, J., Hill, J., Tilman, D., et al. (2008). Land clearing and the biofuel carbon debt. *Science* 319: 1235–8. doi:10.1126/science.1152747.

Galloway, J. N., Townsend, A. R., Erisman, J. W., et al. (2008). Transformation of the nitrogen cycle: Recent trends, questions, and potential solutions. *Science* 320: 889–92. doi:10.1126/science.1136674.

Galloway, J. N., Leach, A. M., Bleeker, A. and Erisman, J. W. (2013). A chronology of human understanding of the nitrogen cycle. *Philosophical Transactions of the Royal Society B: Biological Science* 368:20130120. doi:10.1098/rstb.2013.0120.

Garrigues, E., Corson, M., Angers, D., Werf, H. G. and Walter, C. (2013). Development of a soil compaction indicator in life cycle assessment. *The International Journal of Life Cycle Assessment* 18(7): 1316–24.

Gibbs, H. K., Johnston, M., Foley, J. A., Holloway, T., Monfreda, C., Ramankutty, N. and Zaks, D. (2008) Carbon payback times for crop-based biofuel expansion in the tropics: The effects of changing yield and technology. *Environemental Research Letters* 3(3): 034001. doi:10.1088/1748-9326/3/3/034001.

Goh K.-J. and Härdter R. (2003). General oil palm nutrition. In: F. Härdter (Ed.) *Oil Palm Manag Large Sustain Yields*. PPIPPIC-IPI, Singapore, pp. 191–230.

Goldstein, J. E. (2016). Knowing the subterranean: Land grabbing, oil palm, and divergent expertise in Indonesia's peat soil. *Environment and Planning A* 48(4): 754–70.

Hansen, S. B., Olsen, S. I. and Ujang, Z. (2014). Carbon balance impacts of land use changes related to the life cycle of Malaysian palm oil-derived biodiesel. *International Journal of Life Cycle Assessment* 19: 558–66. doi:10.1007/s11367-013-0672-3.

Harsono, S. S., Prochnow, A., Grundmann, P., et al. (2012). Energy balances and greenhouse gas emissions of palm oil biodiesel in Indonesia. *GCB Bioenergy* 4:213–28. doi:10.1111/j.1757-1707.2011.01118.x.

Harsono, S. S., Grundmann, P. and Soebronto, S. (2014). Anaerobic treatment of palm oil mill effluents: Potential contribution to net energy yield and reduction of greenhouse gas emissions from biodiesel production. *Journal of Cleaner Production* 64: 619–27. doi:10.1016/j.jclepro.2013.07.056.

Henson, I. E. (1999). Comparative ecophysiology of oil palm and tropical rain forest. In: *Oil Palm and the Environment*. Malaysian Oil Palm Growers' Council, Kuala Lumpur, Malaysia.

IPCC (2006). *Guidelines for National Greenhouse Gas Inventories*. Vol 4. Agriculture, Forestry and Other Land Use, WMO/UNEP. http://www.ipcc-nggip.iges.or.jp/public/2006gl/index.html.

IPCC (2007). Summary for Policymakers. In: S. Solomon, D. Qin, M. Manning, Z. Chen, M. Marquis, K. B. Averyt, M. Tignor and H. L. Miller (Eds) *Climate Change 2007: The Physical Science Basis. Contribution of Working Group I to the Fourth Assessment Report of the Intergovernmental Panel on Climate Change*. Cambridge University Press, Cambridge, United Kingdom and New York, NY, USA. p. 18.

IPCC (2013). In: T. Hiraishi, T. Krug, K. Tanabe, N. Srivastava, J. Baasansuren, M. Fukuda and T. G. Troxler (Eds), *Supplement to the 2006 IPCC Guidelines for National Greenhouse Gas Inventories: Wetlands*. Intergovernmental Panel on Climate Change, Switzerland.

Jauhiainen, J., Hooijer, A. and Page, S. E. (2012a). Carbon dioxide emissions from an Acacia plantation on peatland in Sumatra, Indonesia. *Biogeosciences* 9(2): 617–30.

Jauhiainen, J., Silvennoinen, H., Hämäläinen, R., Kusin, K., Limin, S., Raison, R. J. and Vasander, H. (2012b). Nitrous oxide fluxes from tropical peat with different disturbance history and management. *Biogeosciences* 9(4): 1337–50.

Jolliet, O., Saadé, M., Crettaz, P. and Shaked, S. (2010). *Analyse du cycle de vie. Comprendre et réaliser un écobilan*. 2e édition. Presses Polytechniques et Universitaires Romandes, Lausanne, Suisse. ISBN 978-2-88074-886-9. 302p.

Kaewmai, R., H-Kittikun, A. and Musikavong, C. (2012). Greenhouse gas emissions of palm oil mills in Thailand. *International Journal of Greenhouse Gas Control* 11:141–51. doi:10.1016/j.ijggc.2012.08.006.

Kaewmai, R., H-Kittikun, A., Suksaroj, C. and Musikavong, C. (2013). Alternative technologies for the reduction of greenhouse gas emissions from palm oil mills in Thailand. *Environmental Science & Technology* 47 (21) : 12417–25.

Lam, M. K., Lee, K. T. and Mohamed, A. R. (2009). Life cycle assessment for the production of biodiesel: A case study in Malaysia for palm oil versus jatropha oil. Biofuels, *Bioproducts and Biorefining* 3(6): 601–12.

Langeveld, J. W. A., Guisson, R. and Stichnothe, H. (2016). Mobilising Sustainable Supply Chains – Biogas Cases, IEA Bioenergy Task 43.

Lee, J. S. H., Abood, S., Ghazoul, J., Barus, B., Obidzinski, K. and Koh, L. P. (2014). Environmental impacts of large-scale oil palm enterprises exceed that of smallholdings in Indonesia. *Conservation Letters* 7(1): 25–33.

Levasseur, A., Lesage, P., Margni, M. and Samson, R. (2012). Biogenic carbon and temporary storage addressed with dynamic Life Cycle Assessment. *Journal of Industrial Ecology* 17(1): 117–28.

Manik, Y. and Halog, A. (2013). A meta-analytic review of life cycle assessment and flow analyses studies of palm oil biodiesel. *Integrated Environmental Assessessment Management* 9(1): 134–41.

Melling, L., Hatano, R. and Goh, K. J. (2005). Soil CO_2 flux from three ecosystems in tropical peatland of Sarawak, Malaysia. *Tellus Series B-Chemical and Physical Meteorology* 57(1): 1–11.

Melling, L., Hatano, R. and Goh, K. J. (2007). Nitrous oxide emissions from three ecosystems in tropical peatland of Sarawak, Malaysia. *Soil Science and Plant Nutrition* 53(6): 792–805.

Melling, L., Goh, K. J., Bouvais, C. and Hatano, R. (2010). Carbon flow and budget in a young mature oil palm agroecosystem on deep tropical peat. From: http://www.geog.le.ac.uk/carbopeat/media/pdf/yogyapapers/p43.pdf.

Milà i Canals, L., Romanyà, J. and Cowell, S. J. (2007). Method for assessing impacts on life support functions (LSF) related to the use of 'fertile land' in Life Cycle Assessment (LCA). *Journal of Cleaner Production* 15(15): 1426–40.

Mithraratne, N., McLaren, S. and Barber, A. (2008). Carbon footprinting for the Kiwifruit supply chain—report on methodology and scoping study. Landcare research Contract Report LC0708/156, prepared for New Zealand Ministry of Agriculture and Forestry, p 61.

Moore, S., Evans, C. D., Page, S. E., Garnett, M. H., Jones, T. G., Freeman, C., Hooijer, A., Wiltshire, A. J., Limin, S. H. and Gauci, V. (2013). Deep instability of deforested tropical peatlands revealed by fluvial organic carbon fluxes. *Nature* 493(7434): 660–3.

Moulin, M., Wohlfahrt, J., Caliman, J.-P. and Bessou, C. (2016). Deciphering agricultural practices and environmental impacts in palm oil plantations in Riau and Jambi provinces, Indonesia. *International Journal of Sustainable Development & World Ecology*: 1–12.

Nilsalab, P., Gheewala, S. H., Mungkung, R., Perret, S. R., Silalertruksa, T. and Bonnet, S. (2016). Water demand and stress from oil palm-based biodiesel production in Thailand. *The International Journal of Life Cycle Assessment* 1–12. Available at http://link.springer.com/10.1007/s11367-016-1213-7 (verified 6 March 2017).

Oberholzer, H.-R., Freiermuth Knuchel, R., Weisskopf, P. and Gaillard, G. (2012). A novel method for soil quality in life cycle assessment using several soil indicators. *Agronomy for Sustainable Development* 32(3): 639–49.

OECD, FAO (2013). *Oilseeds and Oilseed Products*. OECD-FAO Agricultural Outlook 2013. OECD Publishing, Paris.

Papong, S., Chom-In, T., Noksa-nga, S. and Malakul, P. (2010). Life cycle energy efficiency and potentials of biodiesel production from palm oil in Thailand. *Energy Policy* 38: 226–33. doi:10.1016/j.enpol.2009.09.009.

Pardon, L., Bessou, C., Nelson, P. N., Dubos, B., Ollivier, J., Marichal, R., Caliman, J.-P. and Gabrielle, B. (2016a). Key unknowns in nitrogen budget for oil palm plantations: A review. *Agronomy for Sustainable Development* 36(1): 1–21.

Pardon, L., Bessou, C., Saint-Geours, N., Gabrielle, B., Khasanah, N., Caliman, J.-P. and Nelson, P. N. (2016b) Quantifying nitrogen losses in oil palm plantations: Models and challenges. *Biogeosciences* 13(19): 5433–52.

Pirker, J., Mosnier, A., Kraxner, F., Havlík, P. and Obersteiner, M.. 2016. What are the limits to oil palm expansion? *Global Environment Change* 40: 73–81.

Pleanjai, S., Gheewala, S. H. and Garivait, S. (2009). Greenhouse gas emissions from the production and use of palm methyl ester in Thailand. *International Journal of Global Warming* 1:418–31.

Puah, C. W., Choo, Y. M. and Ma, A. N. (2010). Life cycle assessment for the production and use of palm biodiesel (part 5). *Journal of Oil Palm Research* 22.

Reijnders, L. and Huijbregts, M. A. J. (2008). Palm oil and the emission of carbon-based greenhouse gases. *Journal of Cleaner Production* 16: 477–82. doi:10.1016/j.jclepro.2006.07.054.

Reinhardt, G., et al. (2007). *Conclusive Evaluation of Studies Assessing the Environmental Impact of the Use of Palm Oil as a Bioenergy Carrier*. IFEU, Germany.

Saad, R., Koellner, T. and Margni, M. (2013). Land use impacts on freshwater regulation, erosion regulation, and water purification: A spatial approach for a global scale level'. *The International Journal of Life Cycle Assessment* 18(6): 1253–64.

Saletes, S., Caliman, J. P. and Raham, D. (2004). Study of mineral nutrient losses from oil palm empty fruit bunches during temporary storage. *Journal of Palm Oil Research* 16(1): 11–21.

Schmidt, J. H. (2007). Life assessment of rapeseed oil and palm oil. Ph.D. thesis, Part 3: Life cycle inventory of rapeseed oil. Aalborg University.

Schmidt, J. H. (2010). Comparative life cycle assessment of rapeseed oil and palm oil. *The International Journal of Life Cycle Assessessment* 15: 183–97. doi:10.1007/s11367-009-0142-0.

Silalertruksa, T. and Gheewala, S. H. (2012). Environmental sustainability assessment of palm biodiesel production in Thailand. *Energy* 43(1): 306–14.

Silalertruksa, T., Gheewala, S. H., Pongpat, P., Kaenchan, P., Permpool, N., Lecksiwilai, N. and Mungkung, R. (2016). Environmental sustainability of oil palm cultivation in different regions of Thailand: Greenhouse gases and water use impact. *Journal of Cleaner Production*. Available at http://linkinghub.elsevier.com/retrieve/pii/S0959652616319084 (verified 6 March 2017).

Silalertruksa, T., Bonnet, S. and Gheewala, S. H. (2012). Life cycle costing and externalities of palm oil biodiesel in Thailand. *Journal of Cleaner Production* 28: 225–32. doi:10.1016/j.jclepro.2011.07.022.

Singh, R., Ibrahim, M., Esa, N. and Iliyana, M. (2010). Composting of waste from palm oil mill: A sustainable waste management practice. *Reviews in Environmental Science and Biotechnology* 9(4): 331–44. doi:10.1007/s11157-010-9199-2.

Stehfest, E. and Bouwman, L. (2006). N_2O and NO emission from agricultural fields and soils under natural vegetation: Summarizing available measurement data and modeling of global annual emissions. *Nutrient Cycling Agroecosystems* 74: 207–28. doi: 10.1007/s10705-006-9000-7.

Stichnothe, H. and Schuchardt, F. (2010). Comparison of different treatment options for palm oil production waste on a life cycle basis. *The International Journal of Life Cycle Assessment* 15(9): 907–15.

Stichnothe, H. and Schuchardt, F. (2011). Life cycle assessment of two palm oil production systems. *Biomass Bioenergy* 35: 3976–84. doi:10.1016/j.biombioe.2011.06.001.

Stichnothe, H., Schuchardt, F. and Rahutomo, S. (2014). European renewable energy directive: Critical analysis of important default values and methods for calculating greenhouse gas (GHG) emissions of palm oil biodiesel. *The International Journal of Life Cycle Assessment* 19: 1294–304.

Thamsiriroj, T. and Murphy, J. D. (2009). Is it better to import palm oil from Thailand to produce biodiesel in Ireland than to produce biodiesel from indigenous Irish rape seed? *Applied Energy* 86: 595–604. doi:10.1016/j.apenergy.2008.07.010.

Vijaya, S., Choo, Y. M., Halimah, M., Zulkifli, H., Yew, A. T. and Puah, C. W. (2010). Life cycle assessment of the production of crude palm oil (part 3). *Journal of Oil Palm Research* 22, 895–903.

Wicke, B., Dornburg, V., Junginger, M. and Faaij, A. (2008). Different palm oil production systems for energy purposes and their greenhouse gas implications. *Biomass Bioenergy* 32: 1322–37. doi:10.1016/j.biombioe.2008.04.001.

Wiloso, E. I., Bessou, C. and Heijungs, R. (2015). Methodological issues in comparative life cycle assessment: Treatment options for empty fruit bunches in a palm oil system. *The International Journal of Life Cycle Assessment* 20(2): 204–16.

Yusoff, S. and Hansen, S. (2005). Feasibility study of performing an life cycle assessment on crude palm oil production in Malaysia (9 pp). *International Journal of Life Cycle Assessessment* 12: 5058. doi:10.1065/lca2005.08.226.

Zulkifli, H., Halimah, M., Mohd Basri, W. and Choo, Y. M. (2009). Life cycle assessment for FFB production. In: *PIPOC 2009 Palm oil - Balancing Ecologies with Economics*. Kuala Lumpur, Convention Center, Malaysia.

Chapter 11

Life cycle assessment (LCA) of palm oil in practice: the example of Malaysia

Vijaya Subramaniam, Zulkifli Hashim and Halimah Muhamad, Malaysian Palm Oil Board, Malaysia

1 Introduction

Since the late 1990s there has been a widespread emergence of eco-labelling criteria and the Environmental Management System, which have extended to the agriculture products and processing sectors. Eco-labelling is slowly evolving to become a market-based voluntary mechanism in the greening of the agriculture products supply chain. Eco-label Type I is a voluntary third-party programme that awards a licence that authorizes the use of environmental labels on a product indicating its better environmental performance. Eco-label Type II is a self-declaration label on environmental performance, while Eco-label Type III specifically requires a life cycle assessment (LCA) study to be conducted on the product before certification. LCA is a tool for evaluating the environmental impact of a product or process throughout its entire life cycle (SETAC, 1993). It was first developed as a method for evaluating the impact of a product over its entire life cycle, the so-called 'cradle-to-grave approach', but it has since grown for use in evaluating the environmental impact of business activities, and finally of all the economic activities of a nation as well. Over the last decade there has been a rapid expansion in the demand for LCA studies to chart the environmental performance of products. LCAs have thus become a common environmental management tool and an analytical method for assessing and optimizing the environmental quality of a system over its whole life cycle (Stalmans et al., 1995).

The oil palm industry is continually expanding to meet the demand for its products. However, as producing countries run out of land, the key to any future increase in production must come from higher yields, whether by better planting materials or by better agricultural practices or by both. This would avoid further clearing of forest

http://dx.doi.org/10.19103/AS.2017.0018.23

and peat swamps resulting in loss of biodiversity, erosion and greenhouse gas (GHG) emissions. Sustainable palm oil production must maintain a long-term balance between inputs and outputs across the system boundary during its life cycle. There is a need to quantify the environmental impacts associated with the processes, products or activities carried out along the oil palm supply chain. LCA is a decision support tool for mitigation of ecological concerns in which the environmental impacts are quantified by auditing the inputs and outputs to assess the environmental cost of producing palm oil. It is the most holistic and comprehensive method known to assess the environmental burden of any production.

The objective of this study was to chart the environmental performance of the production of crude palm oil (CPO) and to suggest mitigation measures to overcome hot spots if any.

2 Life cycle assessment (LCA) methodology

The system boundary for the LCA of palm oil production was divided into three subsystems (Fig. 1), namely the agricultural phase during which oil palm seedlings are grown in the nursery and then transplanted to the plantation for the cultivation of oil palm fresh fruit bunch (FFB) and extraction of CPO from FFB at the palm oil mill.

2.1 Nursery subsystem (production of palm seedlings)

The productivity of an oil palm plantation depends on many factors, the most important of which is the quality of seedlings derived from cross-pollination of selected parent palms. Besides the genetic characteristics of selected seeds, the production of high-quality oil palm seedlings is also dependent on good nursery management and practices (Halimah et al., 2010).

At nursery stage, the seedlings are grown in polybags and are given accurate treatment during the first 10 to 12 months of their development. This ensures that well-developed

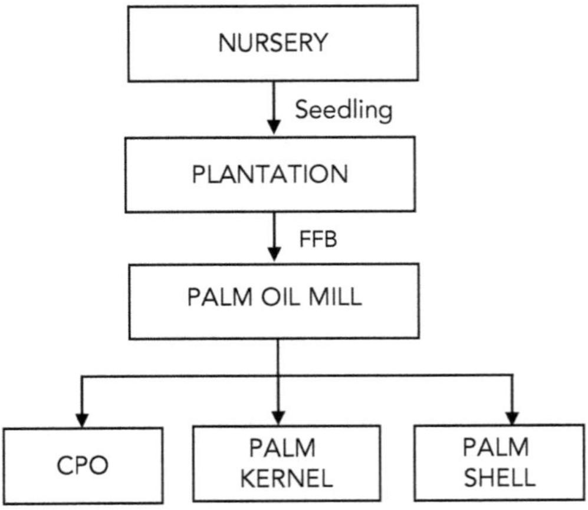

Figure 1 System boundary for LCA of palm oil production.

seedlings with optimum vigour at the time of field planting can be quickly established in the plantation (Corley and Tinker, 2003). Inventory data were collected from 21 oil palm nurseries in Malaysia.

2.2 Plantation subsystem (production of FFB)

Oil palms from the nursery are transferred and planted in the oil palm plantation when they are approximately 12–13 months old. The palms are planted at a density of 136–48 palms per hectare on mineral soils. Before the palms are planted, the soil is ploughed, compacted and a leguminous cover crop is sown. Oil palm trees bear their FFB within two to three years and continue to do so for the next 20 to 25 years. Detached FFB are placed by the roadside and collected later by 5- to 10-tonne lorries before being transported to a nearby mill within 24 hours (Corley and Tinker, 2003).

Replanting of oil palms is carried out when palms are 25–30 years old because of the difficulty in harvesting tall palms. Old palms also give low FFB yield. The palms are felled, chipped and left in the plantation as a nutrient source for replants. The felled palms contain about 95 t of dry weight per hectare (Khalid et al., 2000, 2009) which would decompose within two years. Our inventory data were collected from over 281 plantations in Malaysia. Two land-use change scenarios were chosen for these studies which were land-use change from logged-over forests and the continued land-use change from oil palm to oil palm. The reason for this was that most oil palm plantations that have been developed in peninsular Malaysia have continued land use while the more recently developed in Sabah and Sarawak are from logged-over forest.

2.3 Palm oil mill subsystem (production of CPO)

The FFB, delivered to the palm oil mill, are received at the FFB hoppers to be processed into CPO which is pressed out of the mesocarp fibre. The nuts with the pressed mesocarp fibres are separated at the fibre cyclone and then cracked to produce kernels and shells. The kernels are shipped to kernel-crushing plants to be processed into palm kernel oil while the shell and pressed mesocarp fibre are used as boiler fuel (Vijaya et al., 2010).

The main solid waste from the milling process is empty fruit bunch (EFB), pressed mesocarp fibre, shell and boiler ash while the liquid waste is palm oil mill effluent (POME). The gaseous emissions are from the boiler stack and biogas from the effluent treatment ponds (Vijaya et al., 2010). A contribution to soil fertility is carried out by composting EFB on oil palm plantations while the boiler ash is used for land application. Biogas from the anaerobic digestion from the POME is captured for use as an energy source. It can be either burnt directly in the boiler or purified and run through a gas engine to produce electricity for the grid (Vijaya et al., 2010).

The pressed mesocarp fibre and shells are burnt as fuel in the palm oil mill boiler while the excess shells are sold to other biomass boilers. However, the credits from the use of shells elsewhere are not included in this study as it is out of the system boundary. Palm kernels and shells are the co-products of the milling process and are accounted for by using a weight allocation of 61% CPO, 25% palm kernel and 14% palm shell (Vijaya et al., 2010). The inventory data for this study was collected from 45 palm oil mills located all over Malaysia. At the palm oil mills two scenarios were examined. One with biogas emissions and the other with biogas capture.

3 Life cycle impact assessment (LCIA) case studies

The life cycle impact assessment (LCIA) was carried out using the SimaPro Software version 7.1 with the Eco-indicator 99 methodology.

LCIA was conducted for the system boundary of nursery plantation – palm oil mill with allocation for three scenarios.

3.1 Scenario A

LCIA was conducted for one tonne of CPO produced at the palm oil mill for the system boundary with land-use change from logged-over forest at the plantation and with biogas emissions at the palm oil mill with allocation to co-products. The weighted results are shown in Fig. 2.

The weighted results show that the impact categories with significant impacts are from land use which shows the highest impact followed by fossil fuels respiratory inorganics and climate change. The impact under the land-use impact category is due to the conversion and occupation of the land by the oil palm plantation. The impact from the fossil fuels category came from the production of the various fertilizers as well as diesel usage for transportation and harvesting which were used in the nursery and plantation phases. The parameters that contributed towards the impact categories of climate change and respiratory inorganics were from the application of N fertilizer and energy use through transportation, traction and production of raw materials while the big effect on climate change impact category comes from land-use change upstream and POME at the palm oil mill. Both these impact categories are related to air emissions. The main air emission

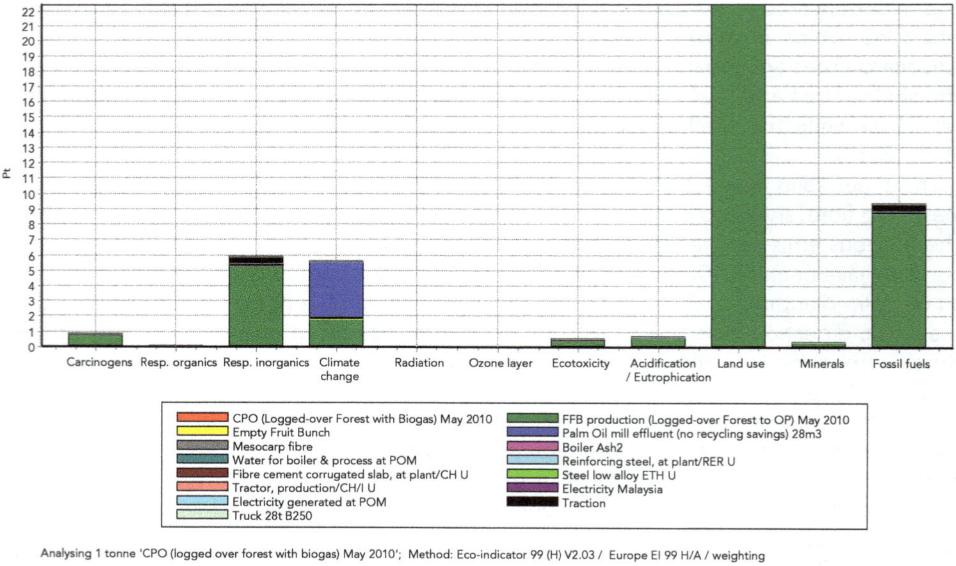

Figure 2 Weighting for LCIA of 1 tonne of CPO – land-use change 25 years logged-over forest, with biogas emissions.

from the POME ponds during anaerobic digestion was the biogas which consisted of methane, carbon dioxide and traces of hydrogen sulphide. The un-harvested biogas is a greenhouse gas which harms the ozone layer and contributes to global warming. The impact from biogas falls under the climate change impact category. The impact under respiratory inorganics and climate change from upstream activities was caused by the application of nitrogen fertilizers in the plantation as well as in the nursery.

Under the Economic Transformation Programme (ETP) which was launched in 2011 by the Malaysian government, palm oil was identified as one of the National Key Economic Areas (NKEAs). Under this NKEA entry point project 5 (EPP5) envisages that all palm oil mills in Malaysia will have a biogas capture facility in place by the year 2020 (Pemandu, 2011). In view of this the second scenario was to access the LCIA when biogas was captured at an 85% capture efficiency.

3.2 Scenario B

LCIA was conducted for one tonne of CPO for the system boundary with land-use change from logged-over forest at the plantation and with biogas capture at the palm oil mills with allocation to co-products. The weighted results are shown in Fig. 3.

The weighted results show that when biogas is captured, the impact from climate change impact category under POME is reduced significantly. The only remaining impacts are again from the upstream in green under land-use impact category due to the occupation of land; fossil fuel impact category due to the production of fertilizers and for climate change; and respiratory inorganics impact category from the application of fertilizers and land-use change. To date, about 20% of the palm oil mills in Malaysia have biogas capture facilities. Another 35 % of the palm oil mills are in the various stages from planning to construction to commissioning of their biogas plants. This shows a huge change in the practices of the Malaysian oil palm industry making the industry more sustainable. A third scenario was

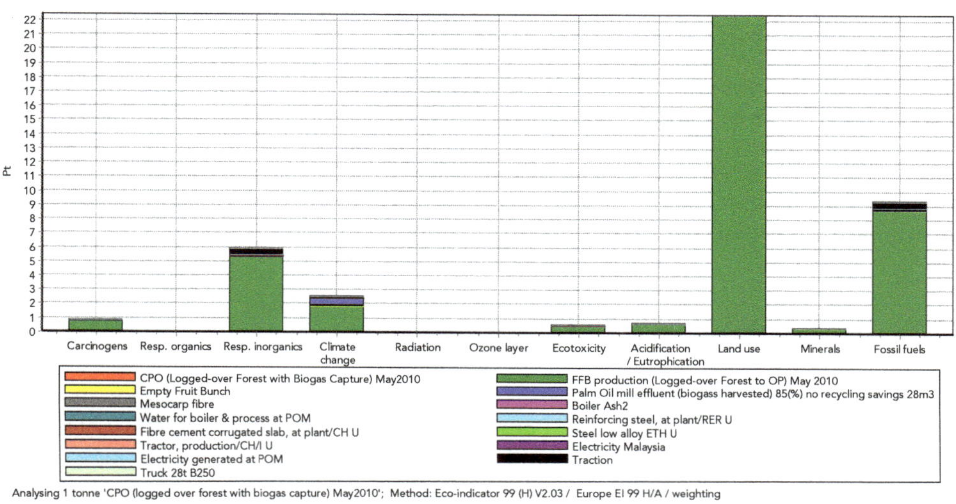

Figure 3 Weighting for LCIA of 1 tonne of CPO – land-use change 25 years, logged over forest with biogas capture.

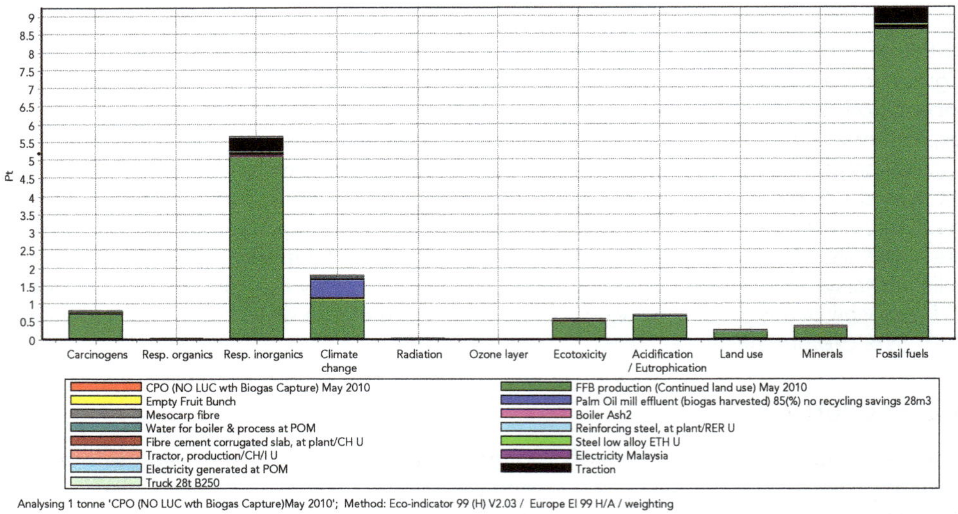

Figure 4 Weighting for LCIA of 1 tonne of CPO – continued land use, biogas capture.

conducted in order to examine the situation when the land transformation was from oil palm to oil palm which is the case of most plantations in peninsular Malaysia which are going into the second and third cycle of replanting.

3.3 Scenario C

LCIA was conducted for one tonne of CPO for the system boundary with continued land use from oil palm to oil palm with biogas capture at the palm oil mills with allocation to co-products. The weighted results are shown in Fig. 4.

When the land-use change scenario is from oil palm to oil palm which has a continued land use, the impact under the land-use impact category is reduced significantly. This is because with the continued land use there is no change in the land occupation. The climate change impact category from FFB production in green is also reduced as there is no land-use change. When biogas is captured, the impact from POME in blue under the climate change impact category has reduced and only the impact from the upstream for fertilizer application is still remaining under this impact category.

4 Greenhouse gas (GHG) emissions

The GHG emissions for the production of one tonne of CPO were calculated, and are shown in Fig. 5. This is the GHG right from the nursery to the plantation with continued land-use change up to the palm oil mill with allocation to co-products.

The main parameter causing the contribution to the GHG emissions is the biogas from the anaerobic treatment of the POME which totals to 546.85 kg CO_2 eq/tonne CPO in the continued land-use change scenario. The biogas emission alone accounts for about 56% of the total CO_2 emissions of 970.58 kg CO_2 eq/tonne CPO. When biogas is captured, the

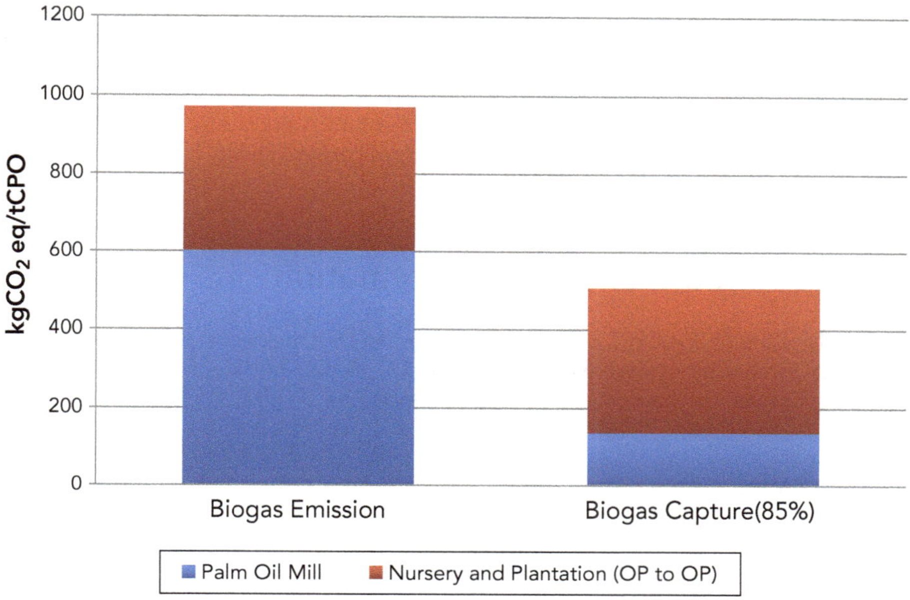

Figure 5 The GHG emissions for 1 tonne of CPO (nursery-plantation continued land use from oil palm to oil palm and palm oil mill with allocation).

total drops to 505.76 kg CO_2 eq/tonne CPO. This shows the urgency and need for the palm oil mills to capture their biogas and use it as renewable energy.

5 Future trends and conclusion

To date, environmental management has been more for image purposes. However, in recent developments, with a shift towards wanting a 'greener' earth, environmental demands are becoming marketing tools. Consideration for the environment is becoming a determining factor for the use of products. In view of the current shift towards higher environmental demands from customers as well as the emergence of eco labels, the need for the oil palm industry to also shift in parallel with the current trend is unavoidable. The main hotspots along the system boundary are land use at the plantation and biogas emissions at the palm oil mill. Land use in Malaysia is rather complicated as land issues are governed by the state government of the respective states. Even though most of the oil palm plantations in peninsular Malaysia have continued land-use change, new developments in Sabah and Sarawak are cleared from logged-over forest. The oil palm industry must somehow strive to strike a balance by increasing production and productivity rather than land expansion. This is easier said than done. The oil palm industry is also moving towards biogas capture or methane avoidance by the year 2020 under the efforts by the Malaysian government which shows a very positive step towards overcoming an important environmental burden. LCA is a tool that helps not only to identify the environmental burdens along the supply chain but also to improve the environmental performance of the product. This case can be seen in the Malaysian oil palm industry moving towards methane capture or avoidance based on the findings of the national LCA study conducted by MPOB. Under the Economic Transformation Programme (ETP) which was launched by the Government of Malaysia in 2011, palm oil was identified as one of the National Key Economic Area (NKEA). Under this NKEA entry point project 5 (EPP5) envisages that all POMs in Malaysia will have a biogas capture facility in place by the year 2020 (Pemandu 2011).

6 Where to look for further information

More details of the NKEA EPP5 on biogas capture can be obtained in the ETP website http://etp.pemandu.gov.my/Palm_Oil-@-Palm_Oil_-%E2%97%98-_Rubber_-_EPP_5-;_Developing_Biogas_Facilities_at_Palm_Oil_Mills.aspx

7 References

Corley, R. H. V. and Tinker, B. (2003). *The Oil Palm*. Blackwell Science Ltd, Oxford, 562p.

Halimah, M., Zulkifli, H., Vijaya, S., Tan, Y. A., Puah, C. W. and Choo, Y. M. (2010). Life cycle assessment of oil palm seedling production (Part 1). *J. Oil Palm Res.* 22: 878–86.

Khalid, H., Zin, Z. and Anderson, J. M. (2000). Decomposition processes and nutrients release patterns of oil palm residues. *J. Oil Palm Res.* 12 (1): 46–63.

Khalid, H., Chan, K. W. and Ahmad, T. (2009). Nutrient cycling and residue management during oil palm replanting in Malaysia. Paper presented at PIPOC 2009, 9–12 November 2009, 27p.

Pemandu (2011). Economic Transfoirmation Programme. http://etp.pemandu.gov.my/annualreport2011/12_National_Key_Economic_Areas-@-12_National_Key_Economic_Areas.aspx

SETAC. (1993). Guidelines for Life-Cycle Assessment: A Code of Practice. Pensacola: SETAC Foundation for Environmental Education.

Stalmans, M., Berenbold, H., Berna, J., Cavali, L., Dillarstone, A., Franke, M., Hirsinger, F., Janzen, D., Kosswig, K., Postlethwaite, D., Rappert, T., Renta, C., Scharer, D., Schick, K. P., Schul, W., Thomas, H. and Van Sloten, R. (1995). European Life- Cycle inventory for Detergent Surfactants Production. Carl Hanser Verlag, Munchen. *Tenside Surf. Det.* 32: 84–5.

Vijaya, S., Choo, Y. M., Halimah, M., Zulkifli, H., Tan, Y. A. and Puah, C. W. (2010). Life cycle assessment of the production of crude palm oil (Part 3). *J. Oil Palm Res.* 22: 895–903.

Zulkifli, H., Halimah, M., Chan, K. W., Choo, Y. M. and Mohd Basri, W. (2010). Life cycle assessment for oil palm fresh fruit bunch production from continued land use for oil palm planted on mineral soil (Part 2). *J. Oil Palm Res.* 22: 887–94.

Chapter 12

Modelling environmental impacts of agriculture, focusing on oil palm

Paul N. Nelson, James Cook University, Australia; Neil Huth, CSIRO, Australia; Marcus Sheaves, James Cook University, Australia; Cécile Bessou, CIRAD, France; Lénaïc Pardon, CIRAD, France; Han She Lim, James Cook University, Australia; and Rai S. Kookana, CSIRO, Australia

1 Introduction

Most people involved in the production, use and consumption of agricultural products, including palm oil, are interested in reducing or eliminating adverse environmental impacts of cultivation while maintaining or improving productivity. To do that we need to know what the impacts are, and predict how they will respond to changes in land use and management. Environmental impacts occur through movement and transformations of energy and materials. In an ideal world, we might monitor all these processes, but that is simply not feasible. We therefore need to estimate them, and this involves models.

Models are simple abstractions of complex systems and always involve a trade-off between complexity and comprehensibility. All of our understanding and knowledge of the environment is in fact in the form of models, because we must simplify the system to make sense of it. The better we understand systems, that is, the better our conceptual models, the more likely we can quantify them and describe them with mathematical relationships. A theoretically ideal model for a question might describe all the important processes mechanistically across space and time, be initialised with easily measured parameters and

http://dx.doi.org/10.19103/AS.2017.0018.24

be simple to process. However, that ideal is not realistic because of the complexity of the environment.

What we have is a large and growing collection of models that are useful for many purposes. Available models simulate some but not all processes and ignore, simplify or assume others. They also overlap in diverse ways. Many can serve as useful tools for tackling scientific and practical questions, if their scope and limitations are understood. The optimal model for any particular purpose is one that includes sufficient complexity to explain the processes of interest, but no more. The nature of models reflects the purpose for which they were designed. There are various typical but overlapping purposes related to environmental impact: developing a fuller understanding of environmental processes, helping managers make decisions, producing indicators for reporting against environmental certification criteria, informing policy to help officials plan and regulate industry, and communication.

Here, we look at modelling of the ways in which cultivation of oil palm influences the biophysical state of the environment within the field itself and in the wider environment, in particular via exchanges with the atmosphere and hydrosphere (Fig. 1). Understanding the environmental impacts of oil palm is essential because it is a globally important crop, especially in the humid tropics, and is rapidly expanding (Sayer et al., 2012). The area of oil palm plantations grew by 680 000 ha/year worldwide over the 2005–2013 period (FAOSTAT, Accessed 2016) and further expansion is expected (Corley, 2009). To put that growth in context, it is slower than soybean and rice (2 311 000 and 1 306 000 ha/year over the same period), but faster than rubber and cocoa (224 000 and 220 000 ha/year), which are crops grown in similar areas.

The key issues for modelling environmental impacts of agriculture, including oil palm cultivation, are choice of modelling approach, definition of the system and key parameters (Section 2), providing useful outputs through integration of our understanding (Sections 3 and 4) and accurate modelling of the causal processes (Section 5). Finally, we summarise the information and make some suggestions about future trends in research and where to look for further information (Sections 6 and 7). We focus on process-based models that have been used in oil palm, but also touch on conceptual and empirical models and relevant models that have not yet been applied to oil palm but could be.

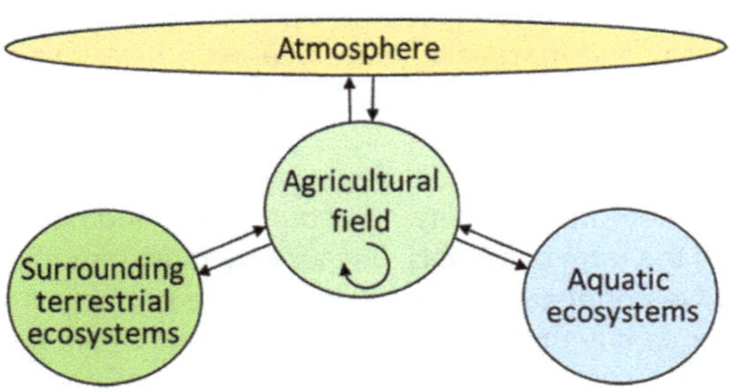

Figure 1 Cultivation of crops, including oil palm, impacts upon the environment through movement of energy, materials and organisms (arrows) between the field and the atmosphere, aquatic ecosystems and surrounding terrestrial ecosystems, and within the field itself.

2 Characteristics of models and the system

Environmental impacts of agriculture, and oil palm cultivation in particular, are diverse. Therefore, when modelling, it is important to be clear about the question being asked. Different questions must be tackled in different ways. Modellers have considered different dimensions and scales of time and space, and used different conceptual and numerical approaches, depending on their emphasis, the availability of data and the understanding of the processes. Several authors have provided useful procedures for selecting environmental models and evaluating their performance (Jakeman et al., 2006; Makowski et al., 2009; Bennett et al., 2013; Kelly et al., 2013; Harmel et al., 2014).

There are two main types of numerical model. On the one hand, there are statistical or empirical models, which describe relationships between data sets, and range from simple regressions to neural networks. They are useful for making predictions but because they are a 'black box', not relying on mechanisms or causality, they are limited in their predictive capacity outside of the conditions under which they were developed. On the other hand, there are 'physically based', 'deterministic', 'mechanistic' or 'process-based' models, which are usually also 'dynamic', having a temporal dimension. They are based on understood causal relationships. The predictive capacity of models generally increases as they become more complex and incorporate more parameters (Fig. 2). The two must go together. There is no point in measuring many parameters if they cannot be incorporated into a model, nor is there any point having a complex model if there are insufficient data to run it.

For problems involving a high degree of uncertainty, models that incorporate probability functions can be useful (Aguilera et al., 2011). For example, Bayesian networks, which are based on probability of one event leading to another, have been used to model ecosystem services (Landuyt et al., 2013), nutrient exports from agricultural areas (Nash et al., 2013;

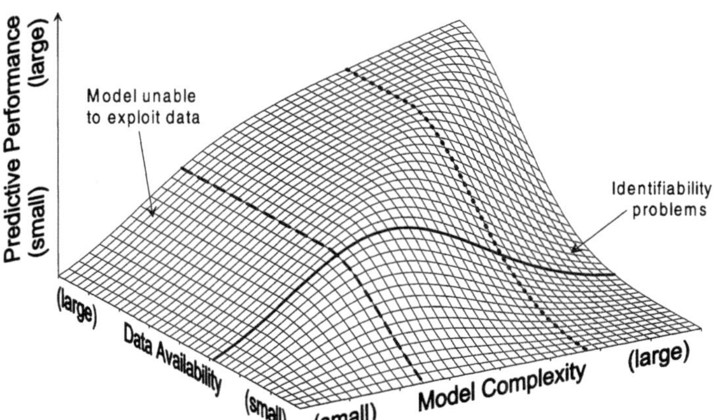

Figure 2 Schematic diagram of the relationship between model complexity, data availability and predictive performance (reproduced from R. Grayson and G. Blöschl, *Spatial Patterns in Catchment Hydrology*, Cambridge University Press, 2001, with permission from Cambridge University Press).

Lucey et al., 2014) and habitat suitability (Hamilton et al., 2015). However, in this chapter we focus on numerical rather than probability-based models.

No matter how sophisticated numerical process-based models are, or how well they model a particular set of circumstances, they are never a perfect representation of reality. Process-based models commonly incorporate relationships that were determined empirically. There is also error involved in determining correct values for parameters. Therefore, for any particular problem and situation, we might expect an optimal model complexity (Fig. 3). Complex models are constructed by combining simpler models, and it follows that model accuracy will be limited by the weakest component.

Models can operate in one, two or three dimensions and at different temporal resolutions. Those operating in two or three dimensions are often called 'distributed' or 'lumped'. Distributed models explicitly model lateral or vertical processes, whereas lumped models assume processes are uniform within given areas. For both, the scale should match the problem being tackled. As for time, some problems might be tackled adequately at coarse timescales such as a year, but others need daily or finer resolution to provide a useful representation of reality.

Impacts on the atmosphere are felt at a global scale because of the relatively well-mixed nature of the atmosphere. Consequently, global climate models and earth system models operate in two or three dimensions across the earth's surface. However, they derive inputs from, or link to, agricultural systems via one-dimensional models of the soil–plant–atmosphere continuum operating with lateral resolution of fields to regions.

Impacts on aquatic ecosystems occur in the field, downstream of it and perhaps also upstream, due to effects on the movement of organisms. Thus, aquatic ecosystem impact models may operate at various scales and dimensions, from one-dimensional to three-dimensional. In general, the effects of agricultural land use tend to accumulate downstream due to an increase in proportion of water that has passed through agricultural fields. However, the impact of disturbance at points (e.g. road crossings and mill outfalls) may diminish downstream and, once the estuary is reached, the effect of the sea becomes important. The direct effect of shading is essentially one-dimensional,

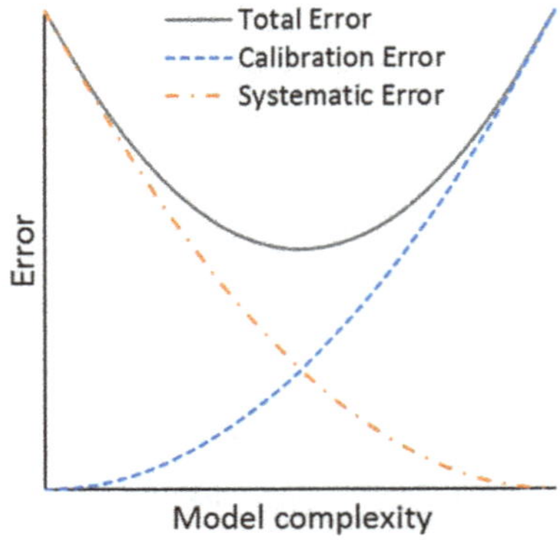

Figure 3 As models become more complex, systematic error (i.e. error from the assumptions made) tends to decrease and calibration error (resulting from limited knowledge about the necessary parameters) tends to increase.

although distributed along stream reaches, whereas nutrient movements and effects must be modelled in two or three dimensions due to downward leaching through the soil, lateral movement through groundwater and discharge into surface water bodies. Quality and ecology of the groundwater may also be important, especially when it is used for drinking water.

Terrestrial impacts are felt in surrounding ecosystems and in the field itself (Fig. 1). Impacts on the surrounding ecosystems tend to be proportional to size and distribution of agricultural fields, reaching negligible levels at distances from fields that are beyond the movement range of organisms and groundwater. Therefore, such models tend to be two- or three-dimensional. Models of in-field soil processes and quality tend to operate in the one-dimensional soil–plant–atmosphere continuum, but lateral variability and movement of materials and organisms may also be taken into account.

Models need parameters that adequately describe the characteristics of the environment, crop and management practices. The greater the availability of data and the more complex the model, the greater the predictive power (Fig. 2). To satisfactorily define parameters, modellers need to know and understand the key factors controlling the processes of interest. Or, on the other hand, models may help researchers find out which are the most important parameters. Models differ in the types of parameters they require, the ways in which the parameters are measured or estimated, and the ways in which they have been calibrated and evaluated (Mulligan and Wainwright, 2013a).

To define characteristics of the oil palm system, it is helpful to recognise that it has some characteristics in common with other cropping systems and others that are unique. Characteristics of oil palm systems are described in detail by Corley and Tinker (2016), but we briefly outline some key features here.

Key climatic characteristics of the system are related to the high temperature and water requirements of oil palm. Radiation tends to be high due to low latitude, although cloud cover can reduce it significantly. Day length varies little, but may affect flowering. Seasonal and daily temperature ranges are also limited, but are important for regulating the rate of processes (e.g. Wang et al. 2014), especially at higher altitudes or latitudes. Rainfall is usually greater than evapotranspiration for much of the year, leaving considerable surplus water available for surface run-off or deep drainage. High-intensity rainfall events mean timescales for modelling generally need to be short to adequately model water-related processes. Irrigation is practised in some places where long periods of water deficit impact on production. While appropriate historical climate data is an essential requirement for modelling, the availability and quality of such data is often limiting.

Key physical characteristics of the system are mostly determined by the climate and topography of tropical lowlands chosen for plantations (Paramananthan, 2011). Topography tends to be flat, often with shallow water table. However, steeper areas are also planted and topography is important for processes such as erosion. The maximum recommended slope is around 20–25°, although steeper slopes have been planted. Terracing is generally practised on slopes steeper than 6–10°.

Soils are typically deep, highly weathered and acidic, with low nutrient holding capacity and high permeability. However, there is also considerable variation, with excessively or poorly drained, shallow, stony, saline, sodic or acid sulphate conditions all existing in plantations. These soil characteristics all affect environmental processes. Most soils are 'mineral' soils with topsoil organic matter contents typically <10%. However, organic soils or peats, which occur in waterlogged areas and have organic matter contents >65%, are also important because of the extensive areas planted with oil palm in Malaysia and Indonesia. Koh et al. (2011) demonstrated that 6% (or ≈880 000 ha) of tropical peatlands in the region had been converted to oil palm plantations by the early 2000s. Appropriate geomorphological and soil data are an essential requirement for modelling many processes.

Infrastructure, especially drainage and road systems, is also an important feature of the physical environment. Networks of earth and gravel roads are relatively dense in oil palm systems, and have a large influence on the movement of water and wildlife, as do drains. Road density and related erosion tends to increase with slope.

The oil palm crop itself has particular key features. It comprises palms at a density of 120–150 ha^{-1}, groundcover vegetation and areas kept bare for harvesting and access. The palms are planted as seedlings, grown to maturity (full canopy cover) by about five years of age and then continued to grow until they are cut down at about 25 years of age. Most oil palm is the 'tenera' type of *Elaeis guineensis* Jacq., a hybrid between 'dura' and 'pisifera' types. However, hybrids between *E. guineensis* and *E. oleifera* are increasingly planted in America (Corley and Tinker, 2016). Corporate plantations are generally managed in field blocks of 25–30 ha, surrounded by roads, whereas smallholder fields may be as small as 2 ha. Intercropping is common in smallholder fields during the immature phase but is not common in corporate plantations. The land use preceding oil palm is also important for many environmental processes. The previous crop may be oil palm, but due to the rapid expansion of the industry, most current oil palm was preceded by a different land use, typically forest, grassland, food crops or other commodity crops such as cocoa, rubber or coconuts.

Finally, models must consider key management practices. The most intensive management is undertaken during the establishment phase, which involves clearing and windrowing (sometimes burning) the existing vegetation, and forming of roads, drains and terraces. Considerable amounts of soil and surface organic matter can be eroded during this phase. The fields are then planted with palms and legume cover crop. The main materials applied are fertiliser (mostly urea or ammonium-based fertilisers and potassium chloride), the palm oil mill by-products–empty fruit bunches (especially around young palms and in fields close to the mill), liquid effluent and increasingly compost, and herbicides (to the harvest paths and weeded circles). Fruit bunches are removed in regular harvests, and old fronds are pruned regularly and placed in heaps between planting rows.

3 Integrated environmental impact modelling approaches

The main approach to quantifying all environmental impacts of the production of a particular agricultural product such as palm oil is life cycle assessment (LCA) (Fig. 4). LCA is, in a sense, an integrated model of environmental impacts, so we discuss it here. As for dynamic process-based models, 'earth system models' are in principle the most integrated. However, earth system models tend to focus on the atmosphere and do not yet incorporate many of the processes that are important in agricultural systems. Models that do so can be classified as 'agricultural (eco)system' models, 'catchment process' models or 'ecological' models. All these integrated approaches incorporate or couple

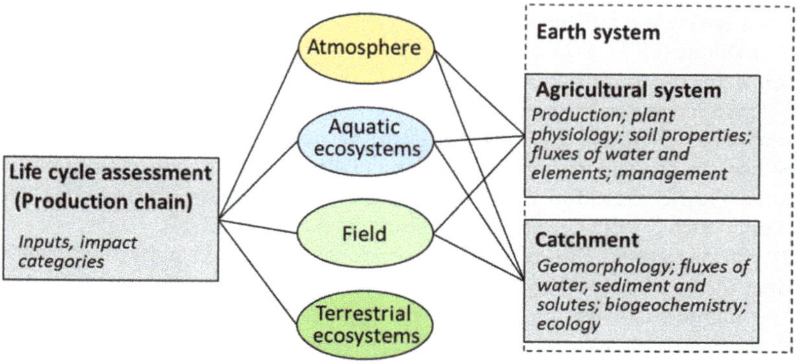

Figure 4 Integrated biophysical modelling of environmental impacts of agriculture can be categorised into different approaches: life cycle assessment on the one hand, or dynamic process-based modelling of the earth system, agricultural systems or catchment processes on the other. Boxes represent typical system boundaries and modelling focus, and lines represent the principal environmental impacts modelled.

simpler models, which we discuss further in subsequent sections. Ecological models tend to focus on the surrounding terrestrial ecosystems or the field itself so we also discuss them later.

The challenge of integrating diverse data types, approaches and models increases even more if social dimensions are included (Granell et al., 2013). Examples include the SEAMLESS and Forest Land Oriented Resource Envisioning System (FLORES) projects. The SEAMLESS modelling project, developed in Europe using the MODCOM framework, is designed to facilitate development of policy (Brouwer and Ittersum, 2010; Ewert et al., 2011). Agricultural systems are simulated using its Agricultural Production and Externalities Simulator (APES) approach (Donatelli et al., 2010). The FLORES model also analyses complex spatial, environmental and social processes (Vanclay et al., 2003). FLORES was developed within the SIMILE graphical modelling environment (Muetzelfeldt and Massheder, 2003) to assist in model development and transparency. Within FLORES, land use decisions are made by 'actors' who can be individuals or groups of individuals who collaborate as families, clans, associations and corporations. These actors can make decisions to maximise benefits or minimise risks. Outputs are calculated from models within a spatial context, allowing users to explore impacts at a landscape scale. Socio-ecological systems have a high degree of uncertainty, so they lend themselves to modelling approaches based on probability and expert judgement, such as Bayesian networks (Ropero et al. 2016).

3.1 Life cycle assessment

LCA involves assessment of a suite of environmental impacts throughout the commodity chain, from raw material production to end-of-life of the product or service, with respect to a functional unit (e.g. 1 kg of palm oil). As LCA assesses environmental performance across multiple impacts, such as climate change, acidification and ozone layer destruction, the weightings can be documented and trade-offs assessed. LCA has become a worldwide standard method (ISO 14040 series 2000–2006) for reporting on environmental impacts of production, and is the baseline modelling approach of various greenhouse gas (GHG) inventory guidelines (IPCC, 2006; European Commission, 2009).

LCA of oil palm products enables assessment of palm cultivation together with other parts of the production chain. It also facilitates comparisons between production of palm oil and alternatives (Schmidt, 2015). While palm oil generally performs worse than other oil crops on climate change impact, due in particular to land use change from high carbon stock systems, it performs better than rapeseed oil regarding eutrophication, acidification, ozone depletion and photochemical ozone impacts (Schmidt, 2010).

About 70 full or partial LCAs of palm oil products have been published over the last decade. Most of the published studies were on palm oil as biofuel, and focused on GHG emissions (or climate change impact) and energy balance (or fossil resource depletion) (Bessou et al., 2013; Manik and Halog, 2013). They mostly assessed the life cycle for palm methyl ester or biodiesel, that is, including all processes from background input production (e.g. fertiliser manufacture) up to the vehicle tank, assuming total combustion or including engine efficiency, to calculate final energy and GHG indicators.

Palm oil LCAs have indicated that the agricultural phase is the main contributor to most of the impact, except for human toxicity or respiratory impacts, to which boiler emissions are the main contributor (Stichnothe and Schuchardt, 2011). Climate change impact is strongly influenced by fertilisers, especially nitrogen fertilisers, and also by carbon stock

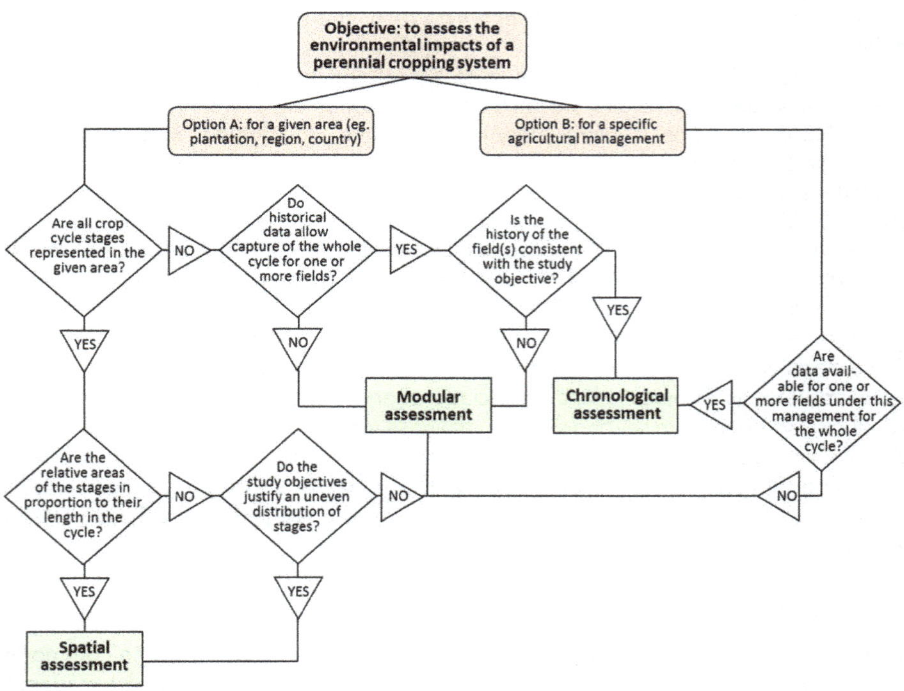

Figure 5 Decision tree for life cycle assessment of a perennial crop system. 'Field' means a unit of plantation with a unique and homogenous planting date, genotype, climate and soil type. 'Crop cycle stages' generally encompass immature phase, most productive phase and declining production phase. From Bessou et al. (2013).

changes in case of land use change or peat drainage (Bessou et al., 2014). Eutrophication impact is driven by nitrogen- and phosphorus-compound emissions, although mill emissions can also contribute. The main eutrophication factors at the agricultural stage are nitrate leaching, and phosphorus and nitrate run-off. The acidification and photochemical ozone impact categories are also influenced by fertilisers, especially due to their influence on nitrogen compound emissions. The largest uncertainties relate to cultivation of peat and emissions of N_2O (Schmidt, 2015).

Impacts originating outside the field are also influenced by in-field management practices, in particular the use of fertiliser. Significant GHG emissions are generated in the production of nitrogen fertiliser (10–30% of total emissions from fertilisers, compared to 70–90% generated in the field), in palm oil mill effluent treatment ponds, especially CH_4 (Wicke et al., 2008; Bessou et al., 2014). Emission factors for the manufacture and transport of inputs are usually taken from databases such as ecoinvent (Nemecek and Kägi, 2007, http://esu-services.ch/data/ecoinvent/).

LCA uses various process models to evaluate impacts, and these are discussed below. Nitrogen management is critical for several impact categories. Perennial cropping systems can be evaluated in different ways, depending on the objectives and the data available (Fig. 5).

Several LCA impact indicators remain to be more widely explored in palm oil production systems. Given the large contribution of fertilisers to environmental impact of the agricultural phase, the eutrophication and acidification impacts related to nitrogen and phosphate inputs need to be further investigated. Other indicators that are little studied include the impacts of herbicides on terrestrial or freshwater ecotoxicity, and the impact of irrigation systems on water depletion.

3.2 Earth system modelling

Earth system models generally focus on the atmosphere and climate, but are increasingly including soil–plant–atmosphere and catchment processes (e.g. Krinner et al., 2005; Fatichi et al., 2012a,b; Oleson et al., 2013; Shen et al., 2013). For example, the Community Earth System Model (CESM) models climate together with land use, vegetation and catchment processes in its Community Land Model (CLM) component (Hurrell et al., 2013; Oleson et al., 2013). A palm growth module incorporating carbon and water cycling was recently developed for the CESM and tested on oil palm (Fan et al., 2015).

Integrating process-based models to larger scales involves a range of challenges. Such integration is aided by software platforms that facilitate data exchange between models while maintaining model transparency for developers and users. For example, The Invisible Modelling Environment (TIME) (Rahman et al., 2003) uses the features of modern software techniques (Rahman et al., 2004) to simplify the communications interface between model components within its spatial modelling framework for issues such as catchment hydrology.

3.3 Agricultural system modelling

Dynamic agricultural system models provide a means for exploring environmental impacts related to water, carbon and nutrient cycles and agrochemicals, especially where several drivers of the system may interact in complex ways. A series of dynamic simulation models are used extensively around the globe for various cropping systems. Examples include Decision Support System for Agrotechnology Transfer (DSSAT-CSM) (Jones et al., 2003),

CropSyst (Stockle et al., 2003), Simulateur mulTIdisciplinaire pour les Cultures Standard (STICS) (Brisson et al., 2003), APES (Donatelli et al., 2010) and Agricultural Production Systems Simulator (APSIM) (Holzworth et al., 2014).

These agricultural system models have been developed over several decades to capture an increasing number of environmental interactions. This development involved a shift of focus from the crop to the soil resource base and the cropping system. The shift allowed individual crops to come and go, finding the soil in one state and leaving it in another (McCown et al., 1995). This move towards a stronger focus on soils was a major step from plant models to environmental impact models and application to more complex systems, including perennial crops.

New applications required the modelling of crop management decisions (Moore et al., 2014). These capabilities also allowed the inclusion of livestock models, and simulations of farms with multiple fields (Holzworth et al., 2014). This journey has seen a change in focus such that model use was no longer aimed just at research, but also for management of production systems and exploration of options for farming systems at various scales (van Ittersum and Donatelli, 2003).

The APSIM model (Holzworth et al., 2014) is an example of a modern integrated environmental model that has been applied to oil palm systems. APSIM is developed and maintained within the APSIM community source framework, which provides a freely available open source platform for collaborative development. The software design allows models from different modelling teams and problem domains to be integrated within a single modelling environment. Furthermore, the framework includes software and processes, such as the APSIM Plant Modelling Framework (Brown et al., 2014), that assist in the development of new models.

The APSIM-Oil Palm model (Huth et al., 2014) was developed to communicate with previously developed and tested models for water, nitrogen and carbon cycling (Probert et al., 1998), GHG emissions (Thorburn et al., 2010), surface organic matter dynamics (Probert et al., 1998; Probert et al., 2005) and agricultural management (Moore et al., 2014), in one dimension. The addition of an oil palm model within existing model capabilities improved model reliability and facilitated the capturing of complex environmental issues. For example, soil carbon stocks and nitrogen losses (leaching of nitrate and gaseous emission of N_2O), along with fruit production, can be estimated in a dynamic manner, taking into account soil type, climate and management (Fig. 6). This capacity allows trade-offs between production and environmental impacts to be assessed (Pardon et al., 2017). APSIM-Oil Palm has been validated for growth and yield at several sites (Huth et al., 2014).

Another agricultural system model that has been applied to oil palm is WaNuLCAS (Noordwijk et al., 2011), which used a different approach to the APSIM-Oil Palm model. WaNuLCAS was formulated within the STELLA graphical modelling environment. Development of the oil palm module benefitted from the group's experience in modelling agroforestry systems. WaNuLCAS focuses on interactions between crop species and can incorporate lateral variability in the form of up to four zones.

Adding complexity to dynamic simulation models comes with costs associated with increased data requirements and calibration error, and decreased ease of use and model transparency (Soltani and Sinclair, 2015). Many of these concerns are now being addressed using modern modelling and software engineering processes that ensure model integrity during ongoing development (Holzworth et al., 2011), encourage collaboration by simplifying model reuse (Holzworth et al., 2010) and assist model accessibility and

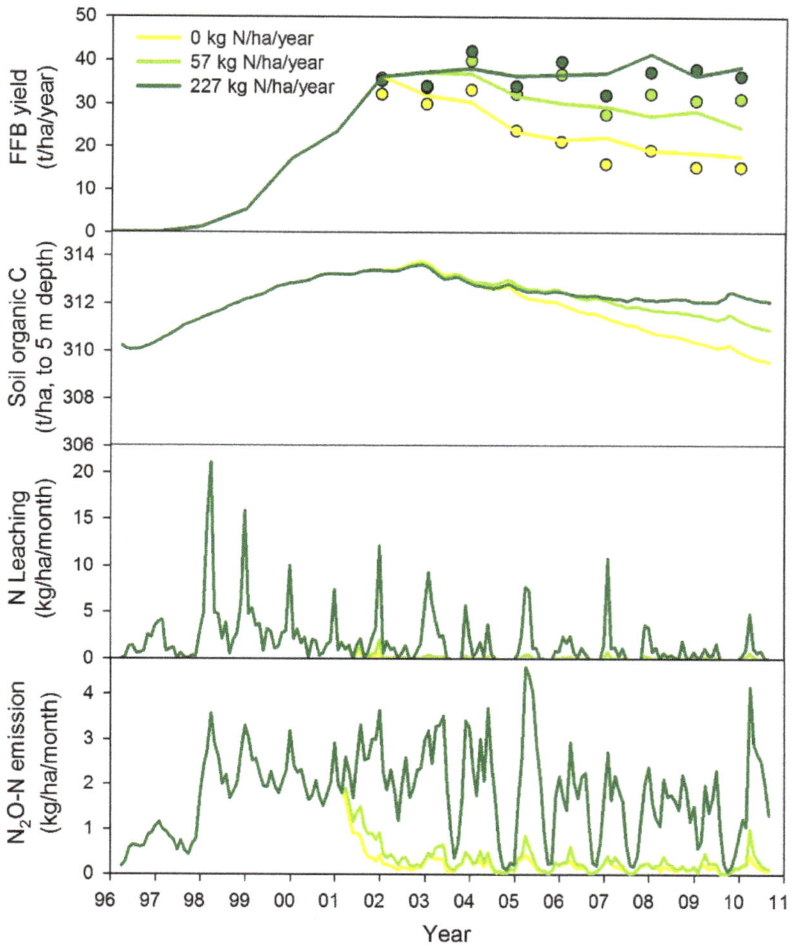

Figure 6 Measured fresh fruit bunch (FFB) yield (points), modelled yield, soil C stock, N leaching and N_2O emission (lines) of oil palm at three nitrogen fertiliser rates, over the course of a crop cycle in Sangara, Papua New Guinea. Modelling was carried out using the agricultural system model APSIM-Oil Palm.

transparency through well-designed user interfaces and automated documentation (Dietze et al., 2011; Brown et al., 2014).

3.4 Catchment process modelling

Catchment (or 'drainage basin' or 'watershed') process models are important tools for examining environmental impacts related to movement water and solutes (e.g. nutrients and pesticides) and erosion of soil. These processes directly impact on conditions of the field, groundwater and downstream aquatic ecosystems. These models couple hydrological models with sediment and solute/contaminant models, and in some cases biogeochemical and ecological models.

The hydrology of oil palm plantations differs to that of other vegetation types, and changes through the crop cycle. Hydrological changes are particularly large during establishment, due to clearing of vegetation, use of heavy machinery, compaction, loosening and creation of bare areas during formation of terraces and roads. Changes are also large, although less dramatic, during the immature phase, when palm canopies and root systems are growing. The presence of roads and oil palm waste from factories and human settlements in plantations also contribute to water quality degradation in these areas (Comte et al., 2012).

Physically based catchment-scale models are 'distributed', dividing the catchment into elements. This distribution is discussed below, in the discussion of the underlying hydrological models. A variety of integrated catchment models have been developed and used (Table 1), and the field is rapidly developing (Robson, 2014a; Paniconi and Putti, 2015). Most of the models do not consider redistribution of soil within the landscape, although some do like LandSoil (Ciampalini et al., 2012).

An integrated surface water and groundwater model such as MIKE SHE/MIKE 11 simulates the major hydrological processes at different levels of spatial distribution and complexity within the catchment. Groundwater, surface water, recharge and evapotranspiration are represented by MIKE SHE, whereas channel flow is represented by the MIKE 11 component of the model. The model also predicts solute transport in the unsaturated and saturated zones. This involves modelling advection and dispersion of solutes in the unsaturated and saturated zone, sorption and desorption processes, erosion and sediment transport (coupled with a sediment transport model), geochemistry reactions for groundwater transport (Geochemistry module) and agricultural applications such as crop yield and nitrogen consumption (MIKE SHE DAISY module).

Complex process-based catchment models such as MIKE SHE require large amounts of data, which is often not available, so simpler catchment models are often used. For example, the SedNet/ANNEX model, developed for catchments of the Great Barrier Reef in Australia, models flux of sediment, N, P and herbicides using a modular empirically based approach (McKergow et al., 2005; Waters et al., 2014; Wilkinson et al., 2014). Sediment sources are apportioned to hill slope erosion (using the Revised Universal Soil Loss Equation, RUSLE), gully erosion and stream bank erosion, and solutes are apportioned to point and diffuse dissolved sources. Budgets are calculated for subcatchments and stream links. SedNet/ANNEX is being used to model the effects of changed agricultural management practices on pollutant loads (Hateley et al., 2014).

So far, published literature on hydrological and solute/sediment modelling of oil palm plantations are few and limited to hydrological impacts. Modelled oil palm run-off yields have been compared with tropical rainforest run-off yields using the HEC-HMS model (Yusop et al., 2007). The Soil and Water Assessment Tool model (SWAT) was used to model run-off yield changes in oil palm at different stages of growth (Majid and Rusli, 2014) and to compare run-off with other biofuel crops (Babel et al., 2011). Integrated hydrological and sediment/solute models such as SWAT and MIKE/SHE provide opportunities to better understand the hydrological and water quality impacts of oil palm plantations, and there is potential to couple them with agricultural system models such as APSIM to model impacts of different management practices.

Table 1 Commonly used field-to-catchment-scale erosion and coupled hydrological sediment/solute models

Model	Spatial scale	Process representation	Spatial representation	Temporal scale	Water quality[1]	GIS[2]	Comments
RUSLE2	Plot	Empirical	Lumped	Annual	S		Improved version of the USLE
WEPP	Catchment	Physical	Distributed	Continuous	S	✓	GIS version (GEOWEPP)
EUROSEM	Field + small catchment	Conceptual	Distributed	Event	S		Linked to hydrological models; KINEROS and MIKE SHE
SHESED	Catchment	Physical	Distributed	Continuous	S		Linked to SHE model (MIKE SHE)
EPI C/APEX	Field	Empirical	Distributed	Continuous	SNP		APEX is the updated version of EPIC
AGNPS/ AnnAGNPS	Catchment	Physical	Lumped/Distributed	Event-based/ Continuous	SNP	✓	Includes stream routing, stream temperature and so on
ANSWERS	Small catchment (<100 km^2)	Physical	Distributed	Continuous	SN	✓	
CREAMS	Field	Physical	Lumped	Continuous	SNP		And GLEAMS (groundwater version)
SWAT	Small to large catchments	Physical	Distributed	Continuous	SNP	✓	Also includes capability for water quality parameters such as DO, BOD, pathogens. Can be linked to MODFLOW for surface groundwater modelling
MIKE SHE/ MIKE 11	Small to medium-sized (up to 5000 km^2)	Physical	Distributed	Continuous	SNP	✓	See text

S= sediment, N=nutrients, P=pesticides.
GIS = geographic information system-based.

4 Modelling impacts of cultivation on components of the environment

Integrated environmental impact models usually couple models focused on particular components of the environment (Fig. 7). Concerns about climate change have driven development of increasingly comprehensive and sophisticated models of atmospheric impacts, highly integrated in terms of the processes modelled. Modelling of impacts on aquatic ecosystems, surrounding terrestrial ecosystems and the field itself is not so developed, but our understanding of the processes is rapidly improving.

4.1 Modelling impacts on the atmosphere

Atmospheric impacts are felt via the net emission of the greenhouse gases CO_2, N_2O and CH_4; ozone; and haze-causing volatile organic compounds, smoke, dust and other particulates. Here, we do not consider ozone or haze, which originates mostly from burning during forest clearing, but focus on GHGs. Atmospheric impacts are modelled so as to better understand effects on climate. Conversion of land to agriculture may directly influence climate through decreased evapotranspiration and hence reduced rainfall. However, this effect is unlikely to be significant with oil palm because its transpiration rate is similar to intact forest (Henson, 1994), so the main concerns are related to GHG emissions.

To meet the requirements of the United Nations Framework Convention on Climate Change (UNFCCC) treaty, countries must report their annual GHG emissions. That accounting is based on the Intergovernmental Panel on Climate Change (IPCC) methods and thus involves various models. Most of the countries with an oil palm industry are classed

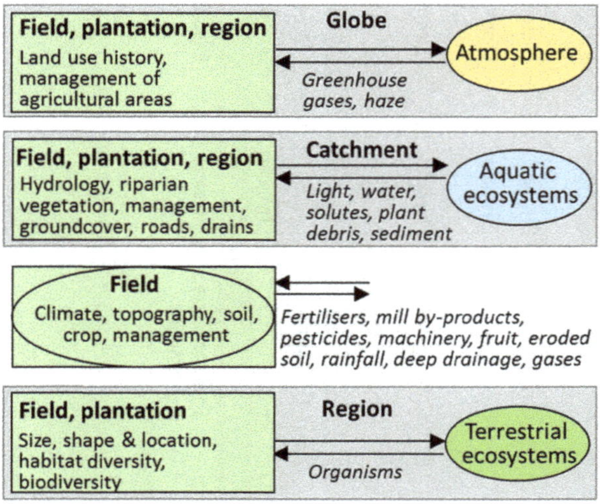

Figure 7 Modelling may simulate processes impacting upon a particular component of the environment (ellipses). That involves different system boundaries and characteristics (rectangles) and different types of exchanges (arrows and italic labels).

as developing economies and are given special status under the treaty as 'non-annex' parties. Nevertheless, palm oil producing countries such as Indonesia and Malaysia are developing carbon accounting procedures (INCAS and MYCarbon, respectively).

Parallel to, and sometimes contributing to, the development of global accounting systems, numerous process-based models have also been developed to simulate effects of land use on the atmosphere. The global earth system models have been compared in the Climate Modelling Intercomparison Project Phase 5 (CMIP5) and reviewed in the fifth Assessment Report of the IPCC (Flato et al., 2013). In some cases they assess management practices as well as land use, by incorporating various crop system, carbon and nitrogen models, discussed in subsequent sections.

GHG emissions are usually modelled using the guidelines of the IPCC (IPCC, 2003, 2006, 2014). The IPCC approach offers several methodological approaches, which differ in complexity. The 'Tier 1' methodology, the one most commonly employed in oil palm LCAs, uses a coarse spatial scale for land use, specified emission factors and a specified timeframe for amortisation if land use changes, that is, an empirical linear model. 'Tier 2' involves locally relevant emission factors, and 'Tier 3' methods involve the use of more sophisticated models (discussed in Section 5, under carbon and nitrogen models), if data are available.

The two main considerations in GHG emission models are whether or not land use change is incorporated and whether cultivation occurs on drained peat soils or mineral soils. Net CO_2 and N_2O emission take place when there is a net decrease in organic matter in the soil–plant system, which occurs most dramatically where oil palm replaces vegetation with higher biomass and net primary production, such as forest, or when it is established on peat soils. The amount of carbon in above-ground biomass is about 3–10 t ha^{-1} in annual crops and grasslands, 37–42 t ha^{-1} in oil palm (average over cycle) and 50–300 t ha^{-1} in tropical forests (Lewis et al., 2013; Lucey et al., 2014; Khasanah et al., 2015a). Establishment on peat entails drainage, which results in net mineralisation of the peat and release of huge quantities of CO_2. On the other hand, there is net sequestration of CO_2 when oil palm replaces vegetation with lower net primary production, such as grassland (Goodrick et al., 2015).

The effects of land use change on GHG emissions have been modelled in different ways. Some studies that do not model it directly nevertheless provide information on the 'carbon debt' or 'payback time'. This carbon debt, initially developed in the context of bioenergy, considers the time needed for a bioenergy value chain to compensate for the initial land use change-related GHG emissions, given its GHG savings compared to the fossil equivalent (Fargione et al., 2008; Gibbs et al., 2008). It ranges from 8 to 169 years for palm biodiesel, with mean and median values of 54 and 43 years, respectively (Fargione et al., 2008; Wicke et al., 2008; Pleanjai et al., 2009; Achten et al., 2010; de Souza et al., 2010; Harsono et al., 2012). The type of previous land use can determine the final GHG balance.

The IPCC approach to modelling effects of land use change assumes that the net flux of CO_2 to the atmosphere equals the difference between the original stock of carbon in biomass and soil and the stock in the new land use. The first-order approximation (Tiers 1 and 2) is to multiply the original stock, which depends on soil and climate type, by 'emission factors' specific to the type of land use and land management. The method operates on the annual time scale required for national reporting of GHG emissions, but measurements or estimates might be done less frequently. There are additional guidelines for calculating N_2O and CH_4 emissions.

GHG emissions from conversion of peatlands to plantations consist of both CO_2 and N_2O emissions, due to the large amounts of organic carbon and nitrogen that are microbially mineralised when oxygen enters the drained peat. Changes in aeration may also affect methanogenic processes. The IPCC provided guidelines for modelling emissions of GHGs from converted peats and other wetlands in its 2013 supplement (IPCC, 2014). There is a linear relationship between water table depth and carbon loss from drained peats, although other factors are also important (Carlson et al., 2015).

After land use change, the second largest contributor to GHG emissions are N_2O emissions. They are largely related to nitrogen inputs to the soil–plant system, especially fertiliser. For emission of N_2O unrelated to land use change, the IPCC (2006) provides Tier 1 factors for 'direct emissions' from the field and 'indirect emissions', which occur elsewhere, after the nitrogen has left the field. For direct emissions the factor of 1% is multiplied by the amount of nitrogen applied in fertiliser or crop residues. The same emission factor is applied irrespective of soil type or other factors. This emission factor was derived from a statistical analysis based on several hundreds of field measurements across the world and it hides large variability in practices and emissions.

However, tropical and perennial crops were poorly represented within the field measurements used for the statistical analysis (Bouwman et al., 2002a,b; Stehfest and Bouwman, 2006), so the factor does not take into account important characteristics of perennial cropping systems such as oil palm (Bessou et al., 2013; Pardon et al., 2016a).

For indirect N_2O emissions the factors are modified to account for nitrogen lost from the field via volatilisation and leaching. The proportion of nitrogen volatilised is set at 10% for synthetic fertiliser and 20% for organic fertiliser. The proportion of nitrogen leached is set at 30% unless specific conditions of water stress or precise irrigation can be demonstrated. The indirect N_2O emissions from these sources are then calculated as 0.75% of the leached nitrogen and 1% of the volatilised nitrogen.

In PalmGHG, a GHG calculator designed especially for the oil palm system, GHG emissions are calculated along the value chain according to the LCA approach, that is, emissions from fertiliser production and transport, land use change and peat drainage, field and mill fuel and electricity use, fertiliser application and palm oil mill effluent (Chase et al., 2012; Bessou et al., 2014). Land use change and fertiliser application-related emissions, as well as N_2O emissions from peat drainage, are calculated based on the IPCC Tier 1 method. In contrast, CO_2 emission from peat drainage is based on an emission factor related to the water table level (Hooijer et al., 2010), which allows more sensitivity to drainage practices than the IPCC Tier 1 method.

Although we focus here on the effects of oil palm cultivation, it is worth briefly mentioning other atmospheric impacts in the palm oil production chain. The impact of palm oil mill effluent treatment on the overall GHG emissions has been investigated by several authors (Yacob et al., 2005; Vijaya et al., 2009; Basri et al., 2010). Methane emissions vary greatly depending on seasonal variations in fruit quality and technical features of mill operations and ponds. GHG assessments mostly rely on derived emission factors based on a few published studies (Yacob et al., 2005; Schmidt, 2007). Based on UNFCCC equations, methane emissions may also be derived from on-site measurements of the chemical oxygen demand reduction during treatment. Considerable emphasis has been placed on reducing these emissions by capturing the biogas at the mill (Choo et al., 2011; Harsono et al., 2014) or by using raw or partially treated effluent in the composting process (Singh et al., 2010; Stichnothe and Schuchardt, 2010).

The impact of fossil fuel use is generally not significant, being 0–5% and 0–2% of total emissions at the field and mill levels, respectively (Bessou et al., 2014). This low impact is due to a low level of mechanisation in the plantations and the use of mill by-products for generating heat and electricity. Most field fuel use is dedicated to fruit transport from plantations to the mill, and its relative contribution varies according to logistics and geography. The high ratio of energy output:input in oil palm systems is matched by very few other agricultural systems (Henson, 1994).

4.2 Modelling impacts on aquatic ecosystems

Impacts on aquatic ecosystems are challenging to model because they involve a complex of processes and events that occur in both nearby and remote ecosystems (Robson, 2014a; Davis et al., 2017). They are influenced by conditions and activities upstream, which are moderated by normal stream discharge and flood events. They are also influenced by connectivity to downstream ecosystems and the marine environment via the migrations of fish and invertebrates up- and downstream. These biological and physical connections between aquatic systems need to be considered before impacts from surrounding terrestrial environments can be modelled.

The riparian zone along the banks of watercourses directly influences the physical characteristics of streams. Natural riparian zones are generally heavily vegetated, due to reliable water supply, and they provide many key services to aquatic ecosystems (Fig. 8), so maintaining their biodiversity and biomass is critical to maintaining near-natural functioning of streams in oil palm landscapes (Singh et al., 2015). Riparian vegetation modifies the flow of water from the land to produce a complex of abiotic and biotic outcomes (Gurnell et al., 2012; Zhang et al., 2012). Sediment and nutrient transport and their delivery to streams are modified (Cooper et al., 1987; Zhang et al., 2003), and plant root systems help bind stream banks and reduce erosion (Kui et al., 2014). The shade provided by overhanging vegetation acts as a major moderator of stream temperature (Moore et al., 2005; Kristensen et al., 2013) and influences the pattern of light and temperature in the water column, which in turn increases in-stream habitat diversity (Naiman et al., 1993) by producing a mosaic of water conditions.

Riparian vegetation also provides vital sources of food and habitat for in-stream fauna via detritus, fruit and insects that fall into streams (Gregory et al., 1991) and through

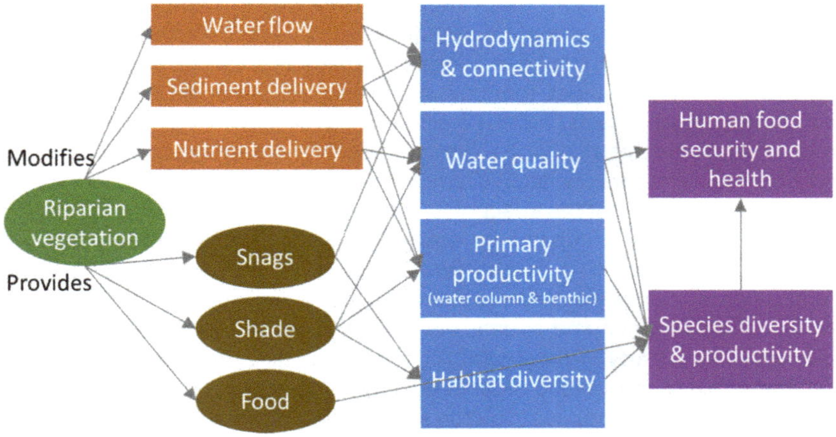

Figure 8 Summary of the services provided by riparian zones to aquatic ecosystems.

the feeding migrations of aquatic fauna into riparian areas during overbank flooding (Welcomme, 1988; De Graaf, 2003). This provision of food helps support high faunal diversity (Gregory et al., 1991). Fallen trees from the riparian zone are vital contributors to in-stream structure (Acuña et al., 2013), substantially enhancing the health and diversity of stream ecosystems (Boyer, 2003). Fallen timber provides structural complexity in streams (Caddy, 2008) and furnishes vital habitat for structure-dependent species (Sheaves, 1996; Valente-Neto et al., 2015), and traps and retains detritus in high-flow streams (Wantzen and Junk, 2000).

While intact and sufficient riparian buffer zones are critical to isolating aquatic systems from many of the impacts of oil palm, their value is circumvented where they are too damaged or narrow to be effective (Singh et al., 2015), or where drains flow directly to streams from plantation mill effluent outfall systems. Similarly, their value is reduced where drainage works lead to the loss of swamps, peatlands and ephemeral wetlands (Anderson, 2008; Comte et al., 2012).

While there have been relatively few investigations of physical processes specific to the oil palm–aquatic ecosystem link, individual physical processes have been quantified in particular locations. For instance, Carlson et al. (2014, 2015) showed that some streams draining oil palm plantations have elevated temperatures and increased total suspended solids concentrations. Such quantitative data are amenable to modelling, at least on a parameter-by-parameter basis. However, such results are likely to be very site- and situation-specific (Chew and Goh, 2015), limiting the value of any models constructed.

Most quantitative models of physical processes that are relevant to the oil palm–aquatic ecosystem situation have been developed in other systems or other parts of the world. Where oil palm-specific studies have been conducted, the majority have been at the local scale (Comte et al., 2012), with a few studies conducted at a watershed scale (but see Yusop et al., 2008 for an example of a small-scale watershed study). Relevant studies generally focus on the volume and quality of water moving into surface water bodies and the physico-chemical nature of those water bodies (e.g. Comte et al., 2012).

Modelling becomes more complex with movement away from simple parameter-by-parameter models to more realistic modes that include the diverse interactions among physical processes, and between physical and biological factors (Robson, 2014a,b; Janssen et al., 2015). Examples of these interactions include the simple effect of riparian vegetation shading on water temperature (Moore et al., 2005; Kristensen et al., 2013), and the impact of converting riparian zones from natural forest to oil palm on surface hydrology, which in turn changes the type and amount of sediment, woody debris and terrestrial-derived food (insects, fruit, leaves) and nutrients entering aquatic food webs (Bruijnzeel, 2004; Comte et al., 2012). This complexity is exacerbated by substantial and unquantified spatio-temporal variation (Comte et al., 2012; Chew and Goh, 2015).

Biological understanding of the effect of oil palm on aquatic ecosystems is in its infancy. There have been studies of impacts on aquatic insects in Malaysian rainforest streams (Mercer et al., 2014) and fish in Papua New Guinea (Nelson et al., 2010b), but little or no generalisation across oil palm areas. Therefore, ecological models of the relationship between oil palm cultivation and aquatic environments are qualitative and mostly based on studies from other parts of the world. Although tropical systems are unique in many respects (Gupta, 2011), most of the broader ecological understanding of streams and their riparian zones probably still applies (Boulton et al., 2008). Holistic models are needed that take into account characteristics of the environment, characteristics of the oil palm, the human dimension and management practices.

There are aquatic ecosystem impacts of palm oil production other than those resulting from cultivation practices. They include direct impacts, such as degraded water quality at mill outfalls and eutrophication in streams resulting from excess nutrient inputs, and less direct effects such as road crossings and gravel extraction for road building, which can greatly increase turbidity in streams.

These impacts reduce the habitat quality of streams for key parameters such as turbidity (Henley et al., 2000) and dissolved oxygen (Wannamaker and Rice, 2000). Other indirect impacts stem from the food and space requirements of local human populations, which are increased by oil palm developments. These impacts are manifested in harvesting of edible aquatic organisms and clearing of riparian zones for cropping.

To effectively model the impacts of oil palm cultivation on aquatic ecosystems it is critical that more extensive, more spatially representative and more holistic data are amassed. Developing empirical models without substantive underpinning data is dangerous because there is no way to understand either sources of error or spatio-temporal variability (Harris and Heathwaite, 2012) or identify tractable components of the causal thicket underpinning outcomes (Levin and Stunz, 2005). In the data-poor meantime, probability-based models show promise for modelling some impacts on aquatic ecosystems (Death et al., 2015).

4.3 Modelling impacts on surrounding terrestrial ecosystems

Oil palm plantations influence the ecology of nearby terrestrial ecosystems, mainly through their effects on populations and movement of organisms (Fig. 1). The degree of effect is related to the relative size and positions of plantations, natural areas and human activity (Foster et al., 2011), which determine size, quality and connectivity of habitat. There is much debate about how conservation and agricultural productivity can be optimised within landscapes (Sayer et al., 2012). Modelling of the interactions suggests that fewer more contiguous areas of oil palm and natural vegetation have a less detrimental effect on biodiversity than finer-scale mosaics of agricultural land and natural habitat (Phalan et al., 2011; Lee et al., 2014).

At a global level, several conceptual models have been proposed to understand the nature of these types of interactions (Tscharntke et al., 2012). It is clear that effects differ between types of organisms (Phalan et al., 2011; Newbold et al., 2014). Such modelling should also include the effects of increased human populations associated (directly or not) with oil palm developments, especially their hunting, fishing and clearing of surrounding forest for food crop cultivation.

Oil palm plantations can also affect nearby terrestrial ecosystems via their effect on movement of groundwater and gases. Lowering of the water table with drains in low-lying plantations can lower the water table in neighbouring ecosystems, but mitigation of the effects using dams is possible. The effects can be modelled using standard hydrology models (Jaenicke et al., 2010), but the high permeability of peat must be taken into account (Baird et al., 2017). Concern has also been raised that terrestrial ecosystems downwind of oil palm plantations might be adversely affected by ozone produced by the reaction of N_2O and volatile organic compounds produced in the plantations (Hewitt et al., 2009).

4.4 Modelling impacts on in-field environmental quality

In-field environmental quality can be described as the abundance, diversity and nature of organisms (above- and below-ground) and the physico-chemical condition of the soil. High quality can be defined as soil conditions and organisms that are conducive to abundant and diverse plant growth and have no negative impact on other components of the environment (Fig. 9). There are no models that integrate all the effects of agricultural management on in-field environmental quality, but there are many modelling approaches for various processes.

As for all agricultural systems, oil palm plantations have less biodiversity than the natural systems they have replaced. Species richness and biomass at all trophic levels is lower than primary forest due to lower net primary productivity and plant diversity (Henson, 1994; Sayer et al., 2012; Barnes et al., 2014). Empirical modelling of bird, beetle and ant populations showed that the larger-bodied species of higher trophic levels are most reduced in oil palm compared to forest, whereas the relatively few species that are more abundant in plantations tend to be smaller and from lower trophic levels (Senior et al., 2013). However, ant populations and species numbers are also much reduced (Fayle et al., 2010).

The abundance and diversity of organisms are determined by energy inputs and habitat. For above-ground organisms, such as insects and arthropods, habitat diversity within the plantation, especially epiphytes and groundcover, appears to be a significant driver of diversity (Koh, 2008; Fayle et al., 2013). The amount and proximity of natural forest in the surrounding landscape is important for more mobile organisms like butterflies, birds and mammals (Koh, 2008; Jennings et al., 2015; Prescott et al., 2016). The ecology and management of weeds, pests and diseases, which is much better studied, is discussed further in this book.

Abundance and diversity of soil organisms are determined by inputs of plant residues and physico-chemical properties of the soil (Carron et al., 2015a,b, 2016; Wakelin et al.,

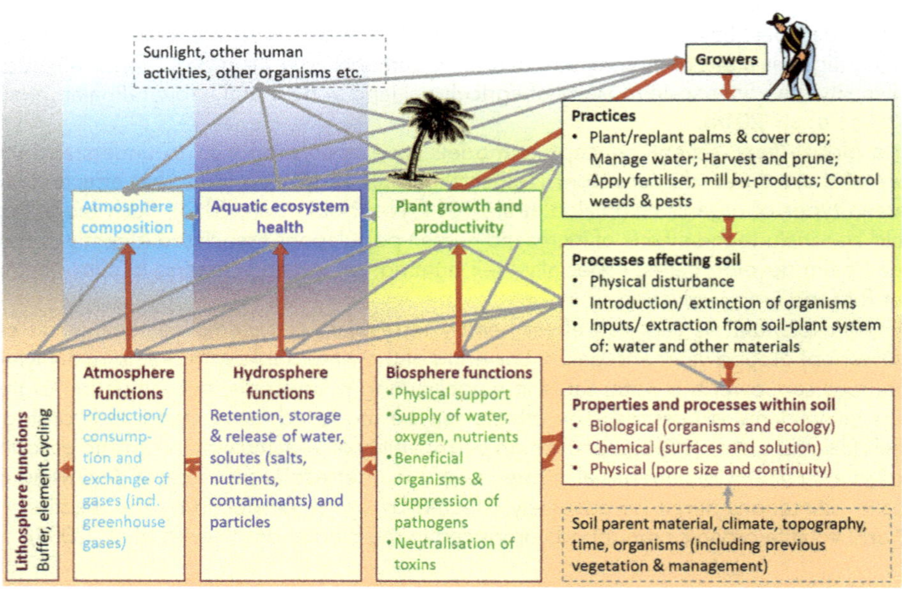

Figure 9 Key interactions (dark red arrows) between oil palm growers, cultivation practices, soil properties and soil functions. Soil is the place where there is a maximum degree of interaction between the lithosphere, hydrosphere, atmosphere and biosphere.

2016). Soil physico-chemical condition can be quantified by fairly stable properties such as texture and mineralogy, and by more dynamic characteristics such as the organic matter content, pH, structure and content of toxic contaminants. Soil organic matter (or carbon) content, accumulation of pesticides and other toxic materials and erosion are discussed in other sections, so here we focus on soil structure and pH. Measurements of soil condition in oil palm plantations must take into account the large spatio-temporal variability inherent in them, and methods are now available to do that efficiently (Nelson et al., 2015a).

Soil structure is enhanced by the faunal activity associated with high inputs of organic matter and is degraded by erosion and compaction. Compaction processes have been modelled but the models have limited applicability in complex heterogeneous systems (Hamza and Anderson, 2005), and simple, well-known empirical relationships can be applied to management. Compaction and sealing can be minimised by minimising the number of passes by machinery; reducing pressure on soil either by decreasing axle load and/or increasing the contact area of wheels with the soil; confining traffic to fixed tracks (using uniform axle widths); and maximising plant growth, including groundcover, and inputs of residues.

Soil pH tends to decline naturally in wet climates but the decline can be accelerated by agricultural management practices. Oil palm is tolerant to low pH, and the associated high availability of aluminium and low availability of calcium, magnesium and potassium. Therefore, accelerated acidification is not necessarily noticed or of concern in oil palm plantations. Nevertheless, it can be considered a detrimental environmental impact. The main drivers of soil acidification in oil palm are nitrate leaching and the off-take of non-acidic cations in harvested fruit (Kee et al., 1995; Nelson et al., 2010a; Dubos et al., 2016). Crop system models simulate these processes, so they are potentially useful for modelling soil acidification (Nelson et al., 2015b).

5 Modelling causal processes

Environmental impact models are based on models of causal processes. Energy enters the field as light, which is used to convert CO_2 and water to organic compounds, and eventually dissipates as heat. Energy transfer processes are discussed in the sections on crop growth models and carbon models. Energy also enters the field in the form of potential energy in water, which is mostly converted to kinetic energy (movement) of water. The water cycle and the associated movement of solutes and soil particles (erosion) are discussed in the section on hydrology models. Apart from carbon and water, a material of particular environmental interest is nitrogen, which is discussed in the section on nitrogen models. Finally, the input of pesticides has particular environmental interest.

Models of elements with less environmental impact than carbon or nitrogen are not considered here, although a range of other elements may be environmentally significant. For example, phosphorus is important in terms of resource depletion and eutrophication (Robson, 2014b), but it tends not to be a major environmental problem in oil palm growing areas, partly because tropical soils tend to have high phosphate retention capacity and also because the amounts of phosphorus fertiliser applied are relatively small. Potassium and chloride are important for oil palm because the quantities taken up are similar to nitrogen. Potassium and chloride have little direct environmental impact, but modelling their cycling in oil palm systems may become important in the future because of their large effect on palm growth.

5.1 Crop growth models

Crop models are not directly focused on environmental impacts, but can be used as a component of integrated models. Early development of OPSIM, an oil palm growth model (van Kraalingen et al., 1989), provided a model of photosynthesis, growth of organs and canopy development, and how these responded to climate and basic management actions such as planting population. After a conspicuously long delay compared to other agricultural domains, other models were developed to capture a wider range of impacts.

Models such as OPRODSIM (Henson et al., 2007), ECOPALM (Combres et al., 2013) and PALMSIM (Hoffmann et al., 2015) added the capacity to model the effect of water supply on crop growth through the incorporation of relatively simple soil water balance models (Table 2). Perhaps more importantly, these models brought a stronger physiological basis to the modelling of oil palm development. The more recent models provided a more detailed description of frond and inflorescence production, which, in the case of ECOPALM, has allowed exploration of the impact of water supply and photoperiod on fruit dynamics. PALMSIM (Hoffmann et al., 2014) is somewhat simpler in some aspects, but this has allowed it to be used in spatial analyses of potential production across climatic regions. CLM-Palm (Fan et al., 2015) is also quite simple but is designed to integrate, via the CLM (Oleson et al., 2013), with the global climate-focused CESM (Hurrell et al., 2013).

A wider range of crop and soil processes can be simulated by the models APSIM-Oil Palm (Huth et al., 2014) and WaNuLCAS (Van Noordwijk et al., 2011). These more detailed models provide opportunities for studying environmental impacts such as GHG emissions, hydrology, erosion and nutrient leaching and understorey management, as discussed in this chapter.

However, crop growth models do not yet represent many characteristics of the oil palm system, such as the presence of shallow groundwater or other soil constraints, or pests and diseases.

5.2 Hydrology, erosion and solute/sediment transport models

Water is the driver of sediment and solute/contaminant transport. Modelling the hydrological cycle and processes (Fig. 10) is important for applications such as water use and management, and water pollution. Hydrological models range in their complexity

Table 2 Oil palm crop models and the processes captured within them

	Model components					
Model	Growth and yield	Water	Soil C&N	Understorey	Management	Reference
OPSIM	•				•	van Kraalingen et al. (1989)
OPRODSIM	•	•			•	Henson et al. (2007)
WaNuLCAS	•	•	•	•	•	van Noordwijk et al. (2011)
ECOPALM	•	•				Combres et al. (2013)
PalmSIM	•	•			•	Hoffman et al. (2014)
APSIM	•	•	•	•	•	Huth et al. (2014)
CLM-Palm	•	•				Fan et al. (2015)

from simple 'black box' models developed from empirical data to powerful process-based computer models that are based on laws of mass, momentum and energy conservation (e.g. Richards equation and Darcy's equation). These models require huge data inputs but can provide distributed predictions of hydrological processes on a continuous timescale.

The development of hydrological models incorporating the interaction between surface water and groundwater provides opportunities to study water flows and contaminant transport between various parts of the hydrological system. Surface and subsurface flows can be modelled in one to three dimensions depending on the type of model. The MIKE SHE model simulates unsaturated flow in one dimension (Richards equation), surface flows in two dimensions (St. Vernant's equation) and groundwater flows in three dimensions (Boussinesq's equation). In addition, many hydrological models now include components to model erosion and solute transport at various levels of complexity (Table 1). Good examples of these coupled hydrological models include the SWAT and the MIKE models.

Erosion of soil is one of the most destructive environmental impacts of poor cultivation practices (Corley and Tinker 2016). Soil erosion is generally low in oil palm plantations due to low slope gradients, permeable soils and permanent ground cover (Lal, 1990; Labrière et al., 2015; Corley and Tinker, 2016). However, erosion can be large in times and places when the soil is exposed, especially during establishment and on steep slopes.

Erosion models simulate the detachment, transport and deposition of sediment in sheetflow, rills, gullies and within stream channels. The simplest empirical sediment transport model is the Universal Soil Loss Equation and its more recent modification, the Revised Universal Soil Loss Equation (Morgan and Nearing, 2011). Other commonly used soil erosion and sediment transport models are process-based, mostly modelling the erosion process in one or two dimensions where the erosion process and resultant

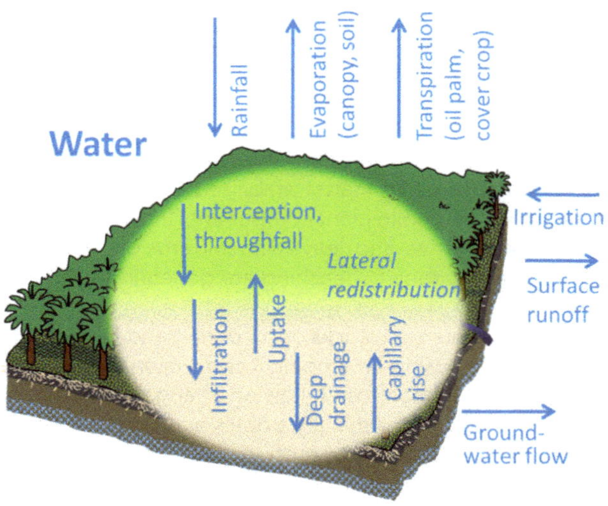

Figure 10 Fluxes of water into, out of and within the field. Water movement has direct impacts on aquatic ecosystems, as well as carrying other materials with their own impacts. Rainfall embodies a significant input of kinetic and potential energy to the system, driving processes such as erosion. Some water molecules are split during photosynthesis and created during mineralisation of organic matter but the amounts are insignificant compared to the fluxes shown.

sediment loads can be predicted either on an event or on a continuous basis. De Vente et al. (2013, 2014) recently made a useful evaluation of regional scale erosion models.

Solute transport models predict the transport of solutes through advection and/or dispersion, sorption/desorption and other geochemical reactions such as aqueous complexation, ion exchange, precipitation/dissolution and oxidation/reduction reactions in subsurface transport. Most of these models simulate solute transport processes in one or two dimensions, although three-dimensional modelling for the saturated zone is increasingly possible with improved computational power. The HYDRUS model simulates water, heat and solute movement in one to three dimensions in variably saturated media and can be linked to the groundwater model MODFLOW. Other examples of solute transport models include MT3DMS (3D multi-species transport model), SUTRA and HST3D (Maliva and Missimer, 2012).

5.3 Carbon models

The carbon cycle is fundamentally important for productivity, sustainability and environmental impacts of agricultural systems. Those modelling the cycle are concerned mostly with simulating the amounts of carbon in the atmosphere (as CO_2 and CH_4), plant biomass and soil (as organic matter), and the fluxes between those three pools. We considered modelling of atmosphere–plant exchanges above ('Atmospheric impact models' and 'Crop growth models'). Therefore, in this section we focus on the cycling of carbon through soil, the largest pool of the three.

Carbon cycling through soil depends on dynamic and incompletely understood interactions between organisms (all Kingdoms), substrates, availability and movement of water and oxygen, temperature, supply of other elements, especially nitrogen, and surface and solution chemistry (mineralogy and pH in particular). Carbon enters the soil mostly as plant-derived organic matter, which then decomposes (Fig. 11). Decomposition involves eventual loss of most or all of the carbon as CO_2, but along the way biomass and organic compounds are formed, especially by microorganisms.

Some organic materials decompose slower than others due to chemical recalcitrance, interactions with inorganic materials or physical inaccessibility to decomposing organisms and enzymes. Overall, if there are no fresh inputs, soil carbon becomes less biologically available over time, although organic matter stabilised by interactions with the mineral phase may originate from initially labile plant components (Haddix et al., 2016). Significant amounts of carbon may also arrive and leave via soil erosion and deposition at some locations and times. Losses of carbon by leaching are generally insignificant compared to the other fluxes.

Large changes in soil organic carbon stocks occur during and following land use change, and these have been modelled using an empirical conversion factor approach. The IPCC default values for soil organic carbon loss following change from forest to cropland are 0.69 (31% soil organic carbon loss) for the dry tropics and 0.58 (42% soil organic carbon loss) for the wet tropics (IPCC, 2003). The meta-analysis of Don et al. (2011) indicated about 30% loss following conversion of primary forest to perennial cropland, but more recent studies have suggested up to half may be lost after conversion (Bruun et al., 2013; Straaten et al., 2015). Much of the loss happens very quickly. However, within the oil palm crop, soil organic carbon stocks have shown no consistent decline, or even an increase over time (Henson, 1994; Haron et al., 1998; Law et al., 2009; Smith et al., 2012; Frazão et al., 2013; Goodrick et al., 2015; Khasanah et al., 2015a). Bahr et al. (2014) showed a

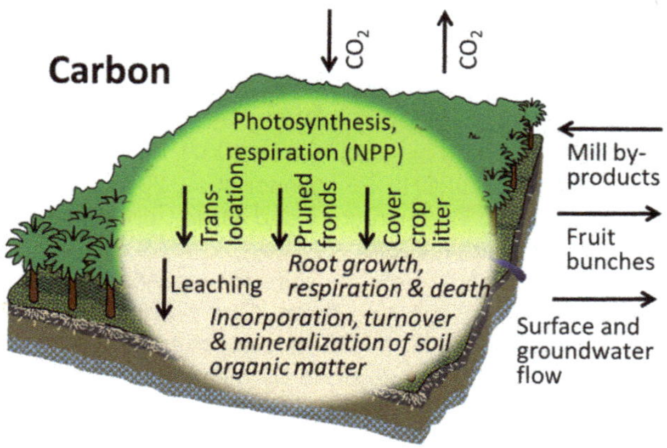

Figure 11 Fluxes of carbon into and out of the field, and transformations of carbon compounds. Carbon represents the main flow of energy through the system. Energy from radiation is converted to chemical bonds during photosynthesis, and used to drive all the biological processes in the system. Net primary production (NPP) is the net amount of organic matter produced in the system, or photosynthesis minus plant respiration.

pattern of decrease immediately after conversion, increase during the perennial cropping phase and then a decrease towards the end of the phase. Modelling with APSIM-Oil Palm shows a similar pattern (Fig. 6) and suggests it is due to inputs to the soil being high initially (from the preceding oil palm crop in the case of the modelling) and then reaching a steady state or declining depending on nitrogen supply.

Process-based models of change in soil carbon during land use change should focus on the key processes, which are redistribution due to erosion and a change in inputs due to the change in vegetation. This is usually achieved by combining erosion models with carbon cycling models (Lacoste et al., 2015). We discussed erosion models above, and now we discuss carbon cycling models.

There are three main approaches to modelling carbon cycling within a particular land use (Smith, 2006). The first two are substrate-based, treating substrates either as several discrete pools or as one continually changing pool. The third is ecologically or food web based.

Substrate-based models that categorise soil organic carbon into several discrete pools are the most used and developed of soil carbon models. In these models, carbon is lost from each pool according to first-order kinetics, and each pool has a different rate coefficient (or 'constant'). Pools comprise plant residue inputs (one or several pools), microbial biomass (one or several pools) and dead organic matter (several pools). The dead organic matter pools have the lowest rate coefficients and some models have an inert pool that does not decompose at all. Carbon lost from each pool is routed into another pool (usually biomass pool or a pool with lower slower rate constant) or lost as CO_2, with the proportions specified by 'carbon use efficiency' factors.

Microbial carbon use efficiency factors are a fundamental component of carbon models but are simply treated in most models. Geyer et al. (2016) have recently defined the concepts involved in a way that will be useful for future advances in carbon modelling.

The most well-known and used multi-pool models are CENTURY (Parton, 1996) and its daily time-step version DAYCENT (Grosso et al., 2001), and Roth-C (Coleman and Jenkinson, 2014). Other examples include APSIM (Probert et al., 1998), Biome-BGC (Thornton and Rosenbloom, 2005), Candy (Franko, 1996), CN-SIM (Petersen et al., 2005), DAISY (Mueller et al., 1996), DNDC (Li et al., 1994; Li, 2007), MIMICS (Wieder et al., 2014), NCSOIL (Molina et al., 1997) and ORCHIDEE (Krinner et al., 2005). Some of these models model nitrogen as well as carbon. Their structure and assumptions have been reviewed by Molina and Smith (1998) and Smith (2006).

Multi-pool models can successfully simulate soil organic carbon contents (Smith et al., 1997), but there are limitations to applying them. One problem is how to initialise the model and another is how to measure or validate pool sizes, rate decay coefficients and efficiency factors. Those things have been done by measuring the quantity of carbon in physically separated fractions and then either a) using the measured fractions as values for the pools and calculating rate coefficients or efficiency factors by fitting the model (Skjemstad et al., 2004), or b) using existing coefficients and splitting the measured pools to match the model pools (Zimmermann et al., 2007). However, the parameters derived in those ways for those situations are not necessarily applicable in other situations. Petersen et al. (2005) showed that a seven-pool model could satisfactorily simulate soil organic carbon contents for several sites using the same parameters, but the sites had similar climate, soil properties and management. Alternatively, chemically characterised pools have shown promise for modelling (Corbeels et al., 2005a,b). Application of multi-pool models within earth system models has resulted in widely varying fits with observed values, so there is considerable scope for improvement (Todd-Brown et al., 2013; Wang et al., 2014; Luo et al., 2016).

The second substrate-based approach to modelling soil carbon cycling is to treat each addition of plant residues as a cohort or single pool that changes with time, rather than assigning all soil carbon to several discrete pools (Bosatta and Ågren, 1995; Yang and Janssen, 2000; Manzoni et al., 2012). In these models the decomposition rate coefficient of each cohort decreases continually with time. This approach is attractive compared to multi-pool models because it reflects the continuously variable nature of substrates and because it requires less parameters. Mono-pool models can be used to simulate soil organic carbon dynamics with a similar degree of accuracy to multi-pool models (Manzoni et al., 2012). However, single- and multi-pool approaches both suffer from the difficulty of measuring pools that correspond with the conceptual pools, determining parameters (decay coefficients in particular) that can be transferred from one situation to another, and initialisation.

Food web-based models have been reviewed by Smith (2006) and Luo et al. (2016). They simulate the decomposing activities of soil organisms in various ways, usually focusing on microbial biomass or components of it, or production and activity of enzymes. For example, Kaiser et al. (2015) showed that soil organic matter accumulation is influenced by interactions between microorganisms that produce catabolic enzymes and those that do not. Food web-based models are able to simulate some aspects of carbon cycling that substrate-based approaches cannot, such as the priming effect and responses to wet-drying cycles. However, they require much more information than substrate-based models. Furthermore, they may not provide much better descriptions of reality because decomposition tends to be limited by substrate supply and physico-chemical conditions, and microbial populations adapt accordingly.

All models modify decomposition rate or decomposer activity according to environmental variables. The most important variables are temperature and the availability of water and oxygen. Their effects are modelled using empirically derived factors, such as an Arrhenius function for temperature and an optimal response function for water content (Luo et al., 2016). The function for water content commonly uses a parameter such as water-filled pore space, which to some extent integrates effects of water film thickness and oxygen supply on microbial activity. Responses to pH, availability of other nutrients, litter quality, litter layer thickness and clay content may also be modelled. Mineralogy, structure and structural stability, particularly pore size distribution, are not included in most models, even though they are known to affect stabilisation of organic matter.

Finally, we need to mention depth and management. The most commonly used models initially focused on a topsoil layer, assumed to be uniform. Only recently has the distribution of processes with depth been incorporated (Jenkinson and Coleman, 2008). Carbon inputs and environmental parameters change with depth, and carbon is moved vertically by fauna. In a similar vein, addition or loss of soil through erosion and deposition, a process that occurs in three dimensions, influences soil depth and the distribution of carbon with depth. Management factors may be included in some carbon models, or their effect on carbon cycling may be accounted for in agricultural system models.

Of the soil carbon modelling approaches, the only one that has been adapted to oil palm systems is the multi-pool substrate-based approach. It is used to model carbon and nitrogen cycling in litter and soil in the agricultural system models APSIM and WaNuLCAS (Probert et al., 1998; Noordwijk et al., 2011). APSIM uses a simplified approach, wherein all soil carbon in the deepest soil layer is assumed to be inert, and the size of the inert pool is assumed to be the same in all layers. It has successfully simulated long-term soil organic carbon contents in other crops (O'Leary et al., 2016). When carbon models are incorporated into crop system models they must realistically simulate not only soil organic carbon contents, but also the mineralisation and immobilisation of nitrogen and its supply to the plant.

The ability of carbon models to simulate soil organic carbon contents in oil palm plantations has not yet been tested. This is mostly because there are no long-term data sets of soil carbon content under oil palm. Such model testing would be very useful, given the peculiarities of oil palm systems. For example, none of the current carbon and nitrogen models mechanistically model the spatio-temporal variability in cycling processes in oil palm plantations, which may or may not be important. Temporal variability in carbon and nitrogen cycling processes is mostly related to the large input of organic matter at the end of the cycle and presence of legume cover crop at the start of the cycle. Spatial variability is largely due to placement of pruned fronds into concentrated stacks, root distribution, weeding and placement of mill by-products (Haron et al., 1998; Nelson et al., 2014; Carron et al., 2016; Goodrick et al., 2016).

5.4 Nitrogen models

The nitrogen cycle is fundamentally important for production and environment, but is complex and challenging to model. The main potential and observed environmental impacts are pollution of ground and surface waters and emission of GHGs, especially N_2O (Choo et al., 2011; Comte et al., 2012). For instance, during the cultivation period, 48.7% of the GHGs emitted to produce 1 t of palm oil fruit are due to nitrogen fertilisation (Choo et al., 2011). The main inputs of nitrogen to the field are manufactured fertilisers, palm oil mill by-products and biological fixation by legume cover crops (Fig. 12).

The largest losses of nitrogen from the field are due to leaching of mineral nitrogen (nitrate in particular) and volatilisation of NH_3, but the magnitude of these and other losses is highly uncertain (Pardon et al., 2016a). In their review, Pardon et al. (2016a) also emphasised that internal nitrogen fluxes in oil palm are large and important compared to other agricultural systems. As measurement of nitrogen losses is difficult and expensive, modelling is useful to help identify management practices that could reduce losses.

The most commonly used process-based models of nitrogen cycling are the same models used for carbon cycling, for example, CENTURY (Parton, 1996), DAYCENT (Grosso et al., 2001), DNDC (Li et al., 1994; Li, 2007) and DAISY (Mueller et al., 1996). They generally model mineralisation and immobilisation of nitrogen based on the C:N ratio of the pools, in the one-dimensional soil–plant–atmosphere continuum, at the scale of the soil profile or the field. Modelling nitrogen cycling is generally more complex than modelling carbon cycling, and several dissolved and gaseous nitrogen compounds are important for plant growth and the environment. In addition to the comprehensive nitrogen cycling models, some models focus on specific fluxes, such as the emission of N_2O (Henault et al., 2005). The concepts and challenges for modelling N_2O emissions have been discussed by Farquharson and Baldock (2008) and Butterbach-Bahl et al. (2013).

When the aim is to identify practices to mitigate environmental impacts, process-based nitrogen models suffer the same limitations as all complex models; they require precise soil and climate input data to function, they tend be not very sensitive to management practices and they are generally calibrated for temperate climate and annual crops and not easily adaptable to other conditions. In the case of nitrogen loss modelling, few models are available for tropical crops, and even fewer for perennial tropical crops such as oil palm (Cannavo et al., 2008; Pardon et al., 2016a,b).

In this context, simpler empirical agri-environmental indicators, such as the INDIGO® method (Bockstaller et al., 2008), may be useful for environmental impact modelling. Indicators combine quantitative and qualitative data based on expert knowledge and are hence suitable in situations where data is limited (Girardin et al., 1999). Their structure is adaptable to new cropping systems, allowing for the accounting of context-specific practices. For instance, Thiollet (2003) developed an indicator of nitrogen losses for vineyards, which might be adapted for oil palm systems.

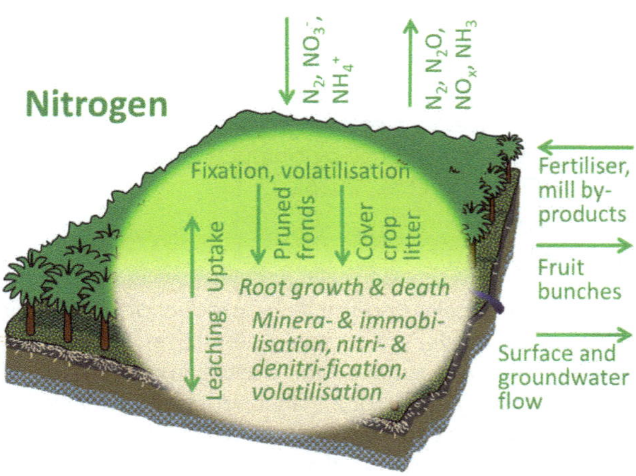

Figure 12 Fluxes of nitrogen into and out of the field, and transformations of nitrogen compounds. Nitrogen-containing compounds are important for many environmental impacts.

In a recent study, Pardon et al. (2016b) compared estimates of nitrogen losses from oil palm plantations made by 11 models and 29 sub-models. Three of the 11 models were process-based cropping system models [WaNuLCAS (Noordwijk et al., 2011); SNOOP (Barros, 2012); APSIM (Huth et al., 2014)], but only APSIM-Oil Palm was validated for production. Two models used a nitrogen budget approach at the scale of the field (Banabas, 2007; Schmidt, 2007). The others were simpler static models that calculate nitrogen losses using empirical relationships (Mosier et al., 1998; Roy et al., 2005; IPCC, 2006; Nemecek and Kägi, 2007; Brockmann et al., 2014; Meier et al., 2014).

The models were compared using the same scenario, a plantation in the Riau region of Sumatra, on a typical Ultisol, over a 25-year growth cycle, using standard management practices. Estimates of total nitrogen losses differed substantially between models, ranging from 21 to 139 kg N ha^{-1} yr^{-1} (Fig. 13). Leaching was the most important pathway, accounting for about 80% of the losses. On average, 31% of the losses occurred during the first three years, which represents 12% of the cycle duration. Losses of nitrogen through leaching and run-off seemed to be overestimated by some models and underestimated by others, with no clear relationship between the complexity or comprehensiveness of models and the magnitude of the losses predicted. All models seemed to underestimate NH_3 volatilisation compared to measured values. Modelled N_2O emissions were similar to field measurements, although the minimum modelled emissions were higher than the minimum losses measured in the field. The most influential variables across the three pathways were related to leaching losses, that is, soil clay content, rooting depth, and nitrogen uptake by the palms. A recent sensitivity analysis using APSIM-Oil Palm identified nitrogen fertiliser rate, drainage and fraction of legume in groundcover vegetation as the factors having most influence on nitrogen losses (Pardon et al., 2017).

Several challenges for modelling nitrogen losses were identified by Pardon et al. (2016b). First, estimation of any one flux depends on accurate estimation of all other fluxes, because they are interdependent. For example, the amount of nitrogen fixed biologically has a large influence on losses, but there is surprisingly little data and only very simple models. Legumes can regulate nitrogen fixation rates depending on the mineral nitrogen content of the soil (Giller and Fairhurst, 2003). Hence, it could be useful to model legume fixation with a rate of nitrogen fixation changing according to the soil mineral N content, such as in the EPIC crop model for soybean (Bouniols et al., 1991). Similarly, the amount of nitrogen released in the soil by residue decomposition is significant, and the way it is modelled has a large influence on losses. Indeed, residue decomposition can be accompanied by temporary immobilisation of nitrogen in the litter and soil organic matter, and the magnitude and timing of losses depend on whether or not the nitrogen immobilisation is considered.

Another challenge is to find parameters that accurately reflect the most influential factors, such as soil water retention and conductivity. The availability of appropriate climate data is also an important limitation.

5.5 Pesticide and contaminant fate models

The use of pesticides is limited in oil palm plantations, but their fate must be considered (Henson, 1994). Herbicides are used to keep the weeded circle and harvest path clear and to control woody weeds. Usually the compounds used rapidly degrade or become inactivated on contact with soil. Insecticides are occasionally used to control outbreaks of pests, in conjunction with biological control methods. The most preferred method of

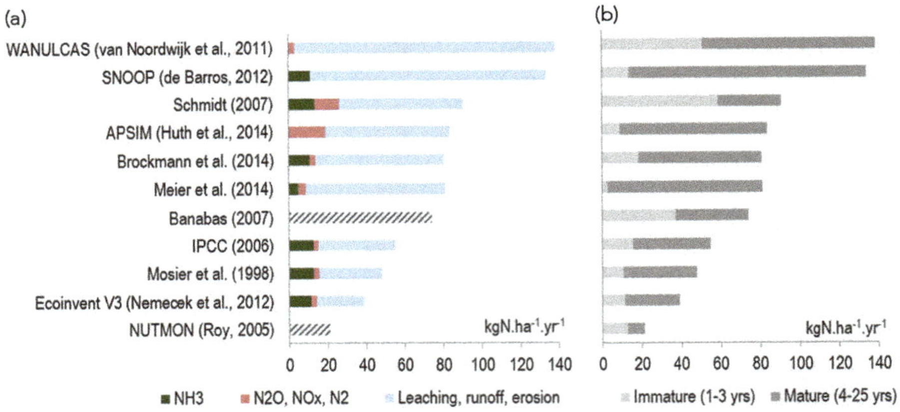

Figure 13 Estimates of N losses from an oil palm field in Sumatra by 11 models. (a) Distribution of the annual average losses between three loss pathways. The hatched bars represent estimates in which separate pathways were not distinguished. (b) Distribution of the annual average losses between the immature and the mature phases (Reproduced from Pardon et al. 2016b, *Biogeosciences* 13, 5433–5452).

application for most insect pests is trunk injection, so opportunities for movement into soil and water are limited. The site of most intense insecticide use, and also fungicides and herbicides, is in nurseries. Rodenticides may also be used in plantations, although alternatives such as encouragement of barn owls through deployment of nesting boxes, is common practice.

To ensure the non-target impact of the pesticides is minimised, several questions must be answered. These include: which pesticide to use, when, where and how? Will these pesticides move offsite (Fig. 14) and if so, what potential impact they may have on the ecosystem and human health?

What best management practices can be adopted/adapted to minimise such impacts? Models and other decision support tools, for example risk indicators, can help answer such questions and assist decision makers (Kookana et al., 2007).

A plethora of models with various degrees of complexity and comprehensiveness have been developed to assess the fate and behaviour of pesticides in the environment. Many have been designed for a specific purpose or with a focus on certain transport pathways (e.g. surface water or groundwater transport, and macropore flow). Therefore, while choosing the model it is important to match the specific question or purpose the user has in mind with the strength of the model.

Pesticide fate models may be classified, on the basis of their intended use and complexity, into categories such as screening, research and management models (e.g. Addiscott and Wagenet, 1985). Screening models are used to compare likely pesticide behaviour under different conditions, or to categorise them into classes differing in potential mobility or toxicity. These are often simple (e.g. analytical solutions) with low data requirement. Risk indicators used in Europe and elsewhere (Reus et al., 2002; Kookana et al., 2007) are this type of model. Research models are designed to quantitatively estimate pesticide transport and transformations. They tend to be comprehensive in terms of fate processes and commensurately hungry for input data. Management models are designed to assess

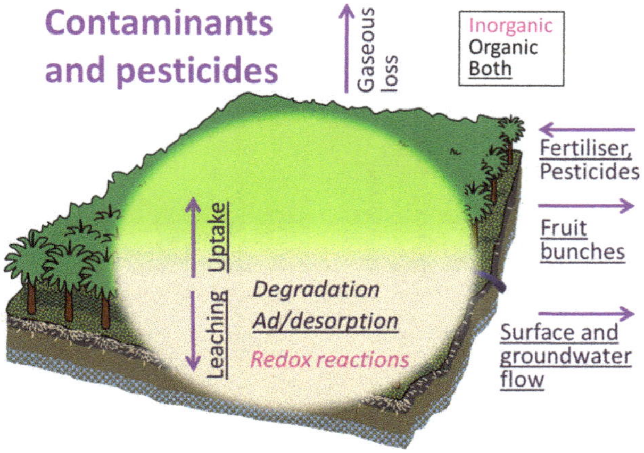

Figure 14 Fluxes and transformations of contaminants and pesticides in the field.

the effect of management practices on pesticide behaviour in field conditions (Bockstaller et al., 2008). The input data requirement of such models is generally somewhere between the research and screening models. A description of some of the models from these categories and their potential uses are described in Table 3.

The fate and behaviour of pesticides in tropical conditions can differ substantially from that in temperate conditions. Currently the database on pesticide behaviour in tropical regions is not well developed and most models rely on pesticide fate data from temperate regions, which may not accurately reflect pesticide behaviour in the tropics. For example, pesticides degrade many times faster in tropical than temperate conditions (Laabs et al., 2000; Kookana et al., 2010). The nature of tropical soils also needs to be considered.

Highly weathered tropical soils, rich in iron and aluminium (hydr)oxides, carry variable charge and may have significant anion exchange capacity at ambient pH (Uehara and Gillman, 1981). Some ionisable pesticides have been found to be substantially sorbed on such soils (Regitano et al., 2000). Soil physical properties are also important. Some tropical soils tend to have substantial macroporosity (biopores), so pesticides may bypass the soil matrix in percolating water. Such preferential flow can result in significant leaching, despite high degradation rates (Laabs et al., 2000). Even if only a very small fraction of applied pesticide (e.g. 0.1%) reaches groundwater, it may breach stringent groundwater standards such as the one used in Europe (0.1 μg/L for any pesticide).

It is therefore important that as much as possible the local data on pesticide fate are used in the models. For degradation, correction factors for persistence based on temperature may need to be applied. For pesticide sorption, techniques such as mid- or near-infrared spectroscopy, together with chemometrics, may allow a rapid and cost-effective estimate of pesticide sorption coefficients (K_d) in a large number of soils (Forouzangohar et al., 2008; Kookana et al., 2014). Ismail and Ooi (2012) found, for a variety of Malaysian soils, adsorption of metsulfuron-methyl was related to soil organic matter and clay content, which are readily estimated.

Some models have been applied and/or tested in tropical conditions (Table 3), but mostly in annual cropping systems. For example, Bannwarth et al. (2014) used a SWAT hydrological model (Table 3) coupled with ANSLEM to simulate the fate of three

Table 3 Some examples of different types of models developed for predicted fate and impact of pesticides from non-point sources

Model name	Class	Processes covered	Intended application	Comments
BAM (Jury et al. 1987)	Screening	Transport, sorption, degradation, volatilisation	To compare behaviour under specific conditions	Useful in grouping pesticides in behavioural classes
PIRI(Kookana et al. 2005)	Screening	Risk indicator integrating mobility, toxicity, application rate and method, topography, soil and environmental conditions	To group pesticides mobility and toxicity to non-target organisms	Useful for identifying relative risk of different pesticides
PRoMPT (Whelan et al. 2007)	Screening	Leaching to groundwater based on a simple water balance model	To identify areas prone to pesticide leaching	Applied in oil palm in Kenya by Unilever
PEARL (Leistra et al. 2000)	Research/ regulatory	Inputs, transport, sorption, degradation, volatilisation, leaching, uptake, transformation	To assess groundwater contamination at regional and local scale	Used for regulatory purposes in Europe
LEACHM (Wagenet and Hutson 1989)	Research	Transport, sorption, transformation, degradation, volatilisation, plant uptake, management	To assess groundwater contamination under cropping systems	Does not cover surface run-off, foliar wash-off, erosion and preferential flow
MACRO (Larsbo and Jarvis 2003)	Research	Specially designed to incorporate the preferential flow paths for pesticide leaching	To assess leaching in soils with macropores	
SWAT (Arnold et al. 2011)	Research/ management	Surface and subsurface hydrological processes. Fate and transport of pesticides is modelled based on GLEAMS model (Leonard et al. 1987)	To predict concentrations in rivers at daily time step	Performed well in tropical northern Thailand (Bannwarth et al. 2014)
PRZM (Suárez et al. 2005)	Regulatory/ management	Surface run-off, foliar wash-off, erosion losses and plant uptake. Not volatilisation	To assess effects of management practices	
PESTFADE (Clemente et al., 1991)	Management	Transport via run-off, erosion and leaching (unsaturated homogenous soil). Includes nutrients	To assess effects of management practices on transport at field, farm and watershed scale	Evaluated in tropics, in Thailand (Shrestha and Datta, 2015)
RZWQM (Ahuja et al. 2000)	Management	Transport via run-off, erosion and leaching (unsaturated homogenous soil). Includes nutrients	To assess effects of management practices on transport at field scale	Evaluated in tropics, in Thailand (Shrestha and Datta, 2015)
I-Phy (Bockstaller et al. 2008)	Management	Leaching, run-off, erosion, drift, volatilisation	To assess effects of management practices on transport at field scale	Being evaluated for glyphosate in oil palm (Marichal et al. 2016)

pesticides in a mountainous catchment of northern Thailand. The predicted daily pesticide concentrations in the river water over two seasons (2008 and 2010) matched the observed concentrations reasonably well, despite the complexity involved at the catchment scale. In terms of sensitivity of different parameters, the percolation parameter was identified as a key parameter.

Reliability of data on pesticide application rate and timing in the catchment is important in such simulations. For pesticide leaching through a soil profile, Shrestha and Datta (2015) evaluated the performance of two models (RZWQM and PESTFADE) for fate and transport of metribuzin herbicide under field conditions in Thailand. While they were generally satisfied with both models, they observed that RZWQM performed better in simulating water content of the soil profile, whereas PESTFADE was better at simulating the herbicide residue. Pesticide transport models are generally not well developed for plantation systems in tropical regions.

Risk indicators are being increasingly used in decision making by pesticide users, natural resource managers and regulators (Kookana et al. 2007). Some of the advantages of risk indicators over comprehensive models are their simplicity, ease of use and usability when input data are scarce. Risk indicators are helpful in assisting (i) growers to choose pesticides that are likely to be more environmentally acceptable, thereby also facilitating adoption of and compliance with integrated pest management or environmental management systems; (ii) growers and regulators to compare risk management options, identify potential hot spots, evaluate management practices and develop appropriate monitoring programmes; (iii) researchers to prioritise pesticides for greater understanding of their environmental fate and toxicology; and (iv) regulators and policy makers to analyse risk trends and develop appropriate policy interventions.

Risk indicators are often classified into two broad types (Reus et al., 2002; Feola et al., 2011). The first type uses simple algorithms such as a scoring table developed based on expert judgement, while the second type employ an exposure-to-toxicity ratio (ETR), that is, comparing the predicted exposure concentration to the toxicity parameter for organisms (e.g. LC_{50}). In the context of potential utility of risk indicators in emerging economies, Feola et al. (2011) used a case study based on small agricultural holdings in Colombia to assess the suitability of a set of indicators drawn from the ETR group (POCER, EPRIP and PIRI) and the non-ETR group (EIQ and PestScreen). They concluded that user-friendly ETR indicators were better suited than non-ETR indicators for reliably estimating environmental risk.

Several risk indicators have been used in tropical tree crops, including oil palm. One recommended risk indicator, PIRI (Kookana et al., 2005) (Table 3), has been promoted over the last decade for use in several tropical countries of South America, Asia and Africa by the joint division of FAO and IAEA on Food and Environment. For a comparative risk assessment of pesticides in papaya plantations, Hernández-Hernández et al. (2007) used SYNOPS risk indicator (an ETR-type risk indicator) and ranked 15 pesticides in terms of their chronological biological risk index. The authors recommended SYNOPS_2 for potential use in other tropical fruit plantations. The INDIGO® risk indicator for pesticides, I-Phy (Bockstaller et al., 2008), is being adapted for oil palm plantations and evaluated for glyphosate (Caliman et al., 2006; Marichal et al., 2016).

In addition to pesticides, trace metals and metalloids such as cadmium, mercury and lead can contaminate soil and water due to inputs as trace contaminants of fertiliser. Although heavy metal contamination has not been detected in oil palm planted soils (Sulaiman et al., 2016), it is a possible threat depending on the source of fertilisers and other amendments applied. The behaviour of trace metals in soils can be modelled to some extent using geochemical models, but soil- and situation-specific parameters are needed (Michel et al., 2007; Jacques et al., 2008; Selim and Zhang, 2013).

6 Conclusions and research directions

There are a variety of approaches and tools available for numerically modelling the environmental impacts of agriculture, including oil palm cultivation. They range from models of a single process or impact to models integrating many processes and impacts. In terms of operation, they range from simple empirical relationships to complex process-based dynamic models, operating at different scales and in different dimensions. For any particular purpose and situation, different models may be useful, but the unknowns and uncertainties must be kept in mind. Agricultural system modelling approaches such as APSIM are useful because they enable comparison of the effects of management changes or environmental factors on productivity and several environmental impacts. Such comparison allows managers to assess possible trade-offs between the two.

Environmental modelling is a rapidly developing field and there is still much to do to make models more accurate and useful for growers, researchers and regulators. Models developed for agricultural systems other than oil palm may be applicable, but peculiarities of the oil palm system need to be taken into account. As for other crops, the most important factors are the climatic, topographic and soil conditions, the nature of the prior and surrounding ecosystems, and the ways in which crop establishment, water, pests, nutrition and harvesting are managed. There is great need for more data on these factors to drive and accompany development of better models.

The type, structure and use of environmental models are changing rapidly in response to pressing needs. Focusing modelling on specific, practical problems of interest can guide optimal selection of measurements, advance our understanding of processes, and improve the integration of science into management and policy decisions. Current and likely future research falls into five main themes.

Representing complexity

As our understanding of biological, chemical and physical processes improves, the attendant complexity must somehow be represented in models. There are many ways this is being or could be tackled, including development of more mechanistic process-based models, integration of existing models, assimilation of data into models, or 'model data fusion' (Keenan et al., 2011; Dietze et al., 2013), hybridisation of statistical and probability-based models with process-based models (Aguilera et al., 2011) and alternative ways of modelling complex self-organising systems, such as cellular automata (Favis-Mortlock, 2013). 'Traditional (and still open) challenges in developing reliable and efficient models are associated with heterogeneity and variability in parameters and state variables; non-linearities and scale effects in process dynamics; and complex or poorly known boundary conditions and initial system states' (Paniconi and Putti, 2015).

Representing the oil palm system

Many models are not well suited to the nature of oil palm systems but could be applied to them, given some evaluation using data from the field, or modifications to the models.

Enhancing usability

Much can be done to incorporate sufficient complexity into models but make them operationally simple, accessible, up-to-date and relevant for managers and the broader community.

Evaluation of models

Evaluation of models, including sensitivity and uncertainty analysis, is becoming increasingly important. There is great scope for evaluation of existing models in oil palm systems and for development of better evaluation techniques.

Data from monitoring and experiments

There is a pressing and growing need for data to inform and evaluate environmental impact models. Essential input data on climate and soil are available globally as estimates from remote sensing and ground observations. However, data scarcity in many oil palm growing regions means their accuracy is often untested and may be poor. There is also much scope for exploiting data from new satellite- and ground-based sensors and other sources such as palm oil mills. Long-term monitoring plots, especially those with experimental treatments, have been valuable for calibrating and validating models and would be very useful for oil palm. Finally, urgently needed data on many key processes could also be obtained by shorter-term experiments.

7 Where to look for further information

The main scientific journals focusing on the topic of this chapter are *Environmental Modelling & Software* and *Ecological Modelling*. Relevant conferences include those of the International Environmental Modelling & Software Society' (http://www.iemss.org/society/). Some references that provide a comprehensive treatment of various aspects of the topic in this chapter follow:

- The oil palm system, including modelling and environmental aspects: Corley and Tinker (2016).
- Environmental modelling: Granell et al. (2013), Wainwright and Mulligan (2013).
- Evaluation of environmental models: Jakeman et al. (2006), Bennett et al. (2013), Kelly et al. (2013).
- Life cycle assessment: ISO 14040 series 2000–2006 and the International Reference Life Cycle Data System Handbook http://eplca.jrc.ec.europa.eu/?page_id=86
- Earth system modelling: Heavens et al. (2013).
- Agricultural system (oil palm) modelling: Huth et al. (2014).
- Catchment process modelling: Mulligan and Wainwright (2013b), Paniconi and Putti (2015), Fatichi et al. (2016).

- Atmospheric impact modelling: IPCC (2003, 2006, 2014).
- Aquatic ecosystem modelling: Robson (2014a).
- Terrestrial biodiversity impacts: Foster et al. (2011), Phalan et al. (2011), Lee et al. (2014).
- Soil quality and erosion: Lal and Stewart (2013), Labrière et al. (2015).
- Erosion modelling: Morgan and Nearing (2011).
- Carbon and nitrogen cycle modelling: Molina and Smith (1998), Smith (2006), Cabrera et al. (2008), Campbell and Paustian (2015).
- Nitrogen environmental impact modelling: Cannavo et al. (2008), Pardon et al. (2016b).
- Nitrous oxide emission modelling: Chen et al. (2008), Butterbach-Bahl et al. (2013).
- Pesticide risk modelling: Kookana et al. (2007).

8 References

Achten, W. M. J., Vandenbempt, P., Almeida, J., Mathijs, E. and Muys, B. (2010). Life cycle assessment of a palm oil system with simultaneous production of biodiesel and cooking oil in Cameroon. *Environ. Sci. Technol.* 44, 4809–15.

Acuña, V., Díez, J. R., Flores, L., Meleason, M. and Elosegi, A. (2013). Does it make economic sense to restore rivers for their ecosystem services? *J. Appl. Ecol.* 50, 988–97.

Addiscott, T. and Wagenet, R. (1985). Concepts of solute leaching in soils: A review of modelling approaches. *J. Soil Sci.* 36, 411–24.

Anderson, J. M. (2008). Eco-friendly approaches to sustainable palm oil production. *J. Oil Palm Res.* 127–42.

Babel, M., Shrestha, B. and Perret, S. (2011). Hydrological impact of biofuel production: A case study of the Khlong Phlo Watershed in Thailand. *Agric. water manag.* 101, 8–26.

Bahr, E., Zaragocin, D. C. and Makeschin, F. (2014). Soil nutrient stock dynamics and land-use management of annuals, perennials and pastures after slash-and-burn in the Southern Ecuadorian Andes. *Agric. Ecosyst. Environ.* 188, 275–88.

Baird, A. J., Low, R., Young, D., Swindles, G. T., Lopez, O. R. and Page, S. (2017). High permeability explains the vulnerability of the carbon store in drained tropical peatlands. *Geophys. Res. Lett.* 44, 1–7. doi:10.1002/2016GL072245.

Banabas, M. (2007). Study of nitrogen loss pathways in oil palm (*Elaeis guineensis* Jacq.) growing agro-ecosystems on volcanic ash soils in Papua New Guinea, PhD thesis, Massey University, Palmerston North, New Zealand.

Bannwarth, M., Sangchan, W., Hugenschmidt, C., Lamers, M., Ingwersen, J., Ziegler, A. and Streck, T. (2014). Pesticide transport simulation in a tropical catchment by SWAT. *Environ. Pollut.* 191, 70–9.

Barnes, A. D., Jochum, M., Mumme, S., Haneda, N. F., Farajallah, A., Widarto, T. H. and Brose, U. (2014). Consequences of tropical land use for multitrophic biodiversity and ecosystem functioning. *Nat. Commun.* 5, 5351.

Barros, I. D. (2012). SNOOP: A simulation model for the dynamics of water and nitrogen in oil palm. In 'V Congresso Brasileiro de Mamona/II Simpósio Internacional de Oleaginosas Energéticas & I Fórum Capixaba de Pinhão Manso, Guarapari (ES) – 2012'.

Basri, M. F., Yacob, S., Hassan, M. A., Shirai, Y., Wakisaka, M., Zakaria, M. R. and Phang, L. Y. (2010). Improved biogas production from palm oil mill effluent by a scaled-down anaerobic treatment process. *World. J. Microbiol. Biotechnol.* 26, 505–14.

Bennett, N. D., Croke, B. F. W., Guariso, G., Guillaume, J. H. A., Hamilton, S. H., Jakeman, A. J., Marsili-Libelli, S., Newham, L. T. H., Norton, J. P., Perrin, C., Pierce, S. A., Robson, B., Seppelt, R., Voinov, A. A., Fath, B. D. and Andreassian, V. (2013). Characterising performance of environmental models. *Environ. Model. Softw.* 40, 1–20.

Bessou, C., Basset-Mens, C., Tran, T. and Benoist, A. (2013). LCA applied to perennial cropping systems: A review focused on the farm stage. *Int. J. Life Cycle Assess.* 18, 340–61.

Bessou, C., Chase, L. D. C., Henson, I. E., Abdul-Manan, A. F. N., Milà i Canals, L., Agus, F., Sharma, M. and Chin, M. (2014). Pilot application of PalmGHG, the Roundtable on Sustainable Palm Oil greenhouse gas calculator for oil palm products. *J. Clean. Prod.* 73, 136–45.

Bockstaller, C., Guichard, L., Makowski, D., Aveline, A., Girardin, P. and Plantureux, S. (2008). Agri-environmental indicators to assess cropping and farming systems. A review. *Agron. Sustain. Dev.* 28, 139–49.

Bosatta, E. and Ågren, G. I. (1995). The power and reactive continuum models as particular cases of the q-theory of organic matter dynamics. *Geochim. Cosmochim. Acta* 59, 3833–5.

Boulton, A. J., Boyero, L., Covich, A. P., Dobson, M., Lake, S.and Pearson, R. (2008). Are tropical streams ecologically different from temperate streams? In Dudgeon, D. (Ed) *Tropical Stream Ecology*. Elsevier: London, UK, pp. 257–84.

Bouniols, A., Cabelguenne, M., Jones, C. A., Chalamet, A., Charpenteau, J. L. and Marty, J. R. (1991). Simulation of soybean nitrogen nutrition for a silty clay soil in southern France. *Field Crops Res.* 26, 19–34.

Bouwman, A. F., Boumans, L. J. M. and Batjes, N. H. (2002a). Emissions of N_2O and NO from fertilized fields: Summary of available measurement data. *Global Biogeochem. Cycles*, 16, 6-1 – 6-13.

Bouwman, A. F., Boumans, L. J. M. and Batjes, N. H. (2002b). Modeling global annual N_2O and NO emissions from fertilized fields. *Global Biogeochem. Cycles* 16, 28-1 – 28-9.

Boyer, K. (2003). Riparian mangement for wood in rivers. *Am. Fish. Soc. Symp.* 37, 407–20.

Brisson, N., Gary, C., Justes, E., Roche, R., Mary, B., Ripoche, D., Zimmer, D., Sierra, J., Bertuzzi, P., Burger, P., Bussière, F., Cabidoche, Y., Cellier, P., Debaeke, P., Gaudillère, J., Hénault, C., Maraux, F., Seguin, B. and Sinoquet, H. (2003). An overview of the crop model STICS. *Eur. J. Agron.* 18, 309–32.

Brockmann, D., Hanhoun, M., Négri, O. and Hélias, A. (2014). Environmental assessment of nutrient recycling from biological pig slurry treatment – Impact of fertilizer substitution and field emissions. *Bioresour. Technol.* 163, 270–9.

Brouwer, F. M., Ittersum, M. V. (Eds) (2010). *Environmental and Agricultural Modelling. Integrated Approaches for Policy Impact Assessment*. Springer: Dordrecht.

Brown, H. E., Huth, N. I., Holzworth, D. P., Teixeira, E. I., Zyskowski, R. F., Hargreaves, J. N. G. and Moot, D. J. (2014). Plant Modelling Framework: Software for building and running crop models on the APSIM platform. *Environ. Model. Softw.* 62, 385–98.

Bruijnzeel, L. A. (2004). Hydrological functions of tropical forests: Not seeing the soil for the trees? *Agric. Ecosyst. Environ.* 104, 185–228.

Bruun, T. B., Egay, K., Mertz, O. and Magid, J. (2013). Improved sampling methods document decline in soil organic carbon stocks and concentrations of permanganate oxidizable carbon after transition from swidden to oil palm cultivation. *Agric. Ecosyst. Environ.* 178, 127–34.

Butterbach-Bahl, K., Baggs, E. M., Dannenmann, M., Kiese, R. and Zechmeister-Boltenstern, S. (2013). Nitrous oxide emissions from soils: How well do we understand the processes and their controls? *Phil. Trans. R. Soc. B.* 368, 20130122

Caddy, J. F. (2008). The importance of 'Cover' in the life histories of demersal and benthic marine resources: A neglected issue in fisheries assessment and management. *Bull. Mar. Sci.* 83, 7–52.

Caliman, J. P., Wohlfahrt, J., Carcasses, R., Girardin, P., Perel, N., Wahyu, A., Pujianto, Dubos, B. and Verwilghen, A. (2006). Agri-environmental indicators for sustainable palm oil production. *In* 'XV International Oil Palm Conference', Cartagena de Indias, Colombia.

Cannavo, P., Recous, S., Parnaudeau, V. and Reau, R. (2008). Modeling N dynamics to assess environmental impacts of cropped soils. *Adv. Agron.* 97, 131–74.

Carlson, K. M., Curran, L. M., Ponette-González, A. G., Ratnasari, D., Lisnawati, N., Purwanto, Y., Brauman, K. A. and Raymond, P. A. (2014). Influence of watershed-climate interactions on stream temperature, sediment yield and metabolism along a land use intensity gradient in Indonesian Borneo. *J. Geophys. Res. Biogeosci.* 119, 1110–28.

Carlson, K. M., Curran, L. M., Ponette-González, A. G., Ratnasari, D., Lisnawati, N., Purwanto, Y., Brauman, K. A. and Raymond, P. A. (2015). Consistent results in stream hydrology across multiple watersheds: A reply to Chew and Goh. *J. Geophys. Res. Biogeosci.* 120, 812–17.

Carlson, K. M., Goodman, L. K. and May-Tobin, C. C. (2015). Modeling relationships between water table depth and peat soil carbon loss in Southeast Asian plantations. *Environ. Res. Lett.* 10. doi:10.1088/1748-9326/10/7/074006.

Carron, M. P., Auriac, Q., Snoeck, D., Villenave, C., Blanchart, E., Ribeyre, F., Marichal, R., Darminto, M. and Caliman, J. P. (2015a). Spatial heterogeneity of soil quality around mature oil palms receiving mineral fertilization. *Eur. J. Soil Biol.* 66, 24–31.

Carron, M. P., Auriac, Q., Snoeck, D., Villenave, C., Blanchart, E., Ribeyre, F., Marichal, R., Darminto, M. and Caliman, J. P. (2016). Do the impact of organic residues on soil quality extend beyond the deposition area under oil palm? *Eur. J. Soil Biol.* 75, 54–61.

Carron, M. P., Pierrat, M., Snoeck, D., Villenave, C., Ribeyre, F., Suhardi, Marichal, R. and Caliman, J. P. (2015b). Temporal variability in soil quality after organic residue application in mature oil palm plantations. *Soil Res.*, 53, 205–15.

Chase, L. D. C., Henson, I. E., Abdul-Manan, A. F. N., Agus, F., Bessou, C., Milà i Canals, L. and Sharma, M. (2012). *The PalmGHG Calculator: The RSPO greenhouse gas calculator for oil palm products, Beta-version.* The Roundtable for Sustainable Palm Oil – RSPO: Kuala Lumpur, Malaysia.

Chen, D. L., Li, Y., Grace, P. and Mosier, A. R. (2008). N_2O emissions from agricultural lands: A synthesis of simulation approaches. *Plant Soil* 309, 169–89.

Chew, P. S. and Goh, K. J. (2015). Comment on 'Influence of watershed-climate interactions on stream temperature, sediment yield and metabolism along a land use intensity gradient in Indonesian Borneo'. *J. Geophys. Res. Biogeosci.* 120, 809–11.

Choo, Y. M., Muhamad, H., Hashim, Z., Subramaniam, V., Puah, C. W. and Tan, Y. (2011). Determination of GHG contributions by subsystems in the oil palm supply chain using the LCA approach. *Int. J. Life Cycle Assess.* 16, 669–81.

Ciampalini, R., Follain, S. and Bissonnais, Y. L. (2012). LandSoil: A model for analysing the impact of erosion on agricultural landscape evolution. *Geomorphology* 175–176, 25–37.

Coleman, K. and Jenkinson, D. S. (2014). *RothC – A Model for the Turnover of Carbon in Soil.* Model description and users guide (Windows version) (updated June 2014). Rothamsted Research: Harpenden, UK.

Combres, J.-C., Pallas, B., Rouan, L., Mialet-Serra, I., Caliman, J.-P., Braconnier, S., Soulie, J.-C. and Dingkuhn, M. (2013). Simulation of inflorescence dynamics in oil palm and estimation of environment-sensitive phenological phases: A model based analysis. *Funct. Plant Biol.* 40, 263–79.

Comte, I., Colin, F., Whalen, J. K., Grünberger, O. and Caliman, J.-P. (2012). Agricultural Practices in Oil Palm Plantations and Their Impact on Hydrological Changes, Nutrient Fluxes and Water Quality in Indonesia: A Review. *Adv. Agron.* 116, 71–124.

Cooper, J., Gilliam, J., Daniels, R. and Robarge, W. (1987). Riparian areas as filters for agricultural sediment. *Soil. Sci. Soc. Am. J.* 51, 416–20.

Corbeels, M., McMurtrie, R. E., Pepper, D. A. and O'Connell, A. M. (2005a). A process-based model of nitrogen cycling in forest plantations: Part I. Structure, calibration and analysis of the decomposition model. *Ecol. Model.* 187, 426–48.

Corbeels, M., McMurtrie, R. E., Pepper, D. A. and O'Connell, A. M. (2005b). A process-based model of nitrogen cycling in forest plantations: Part II. Simulating growth and nitrogen mineralisation of Eucalyptus globulus plantations in south-western Australia. *Ecol. Model.* 187, 449–74.

Corley, R. H. V. (2009). How much palm oil do we need? *Environ. Sci. Policy* 12, 134–9.

Corley, R. H. V. and Tinker, P. B. (2016). *The Oil Palm.* 5th Ed. Wiley-Blackwell: Oxford.

Davis, A. M., Pearson, R. G., Brodie, J. E. and Butler, B. (2017). Review and conceptual models of agricultural impacts and water quality in waterways of the Great Barrier Reef catchment area. *Mar. Freshwater Res.* 68, 1–19.

de Graaf, G. (2003). Dynamics of floodplain fisheries in Bangladesh, results of 8 years fisheries monitoring in the Compartmentalization Pilot Project. *Fish. Manag. Ecol.* 10, 191–9.

de Souza, S. P., Pacca, S., de Ávila, M. T. and Borges, J. L. B. (2010). Greenhouse gas emissions and energy balance of palm oil biofuel. *Renew. Energy* 35, 2552–61.

de Vente, J. d., Poesen, J., Verstraeten, G., Govers, G., Vanmaercke, M., Rompaey, A. V., Arabkhedri, M. and Boix-Fayos, C. (2013). Predicting soil erosion and sediment yield at regional scales: Where do we stand? *Earth-Sci. Rev.* 127, 16–29.

de Vente, J. d., Poesen, J., Verstraeten, G., Govers, G., Vanmaercke, M., Rompaey, A. V., Arabkhedri, M. and Boix-Fayos, C. (2014). Corrigendum to 'Predicting soil erosion and sediment yield at regional scales: Where do we stand?' [*Earth-Sci. Rev.* 127 (2013) 16–29]. *Earth-Sci. Rev.* 133, 94.

Don, A., Schumacher, J. and Freibauer, A. (2011). Impact of tropical land-use change on soil organic carbon stocks – a meta-analysis. *Global Change Biol.* 17, 1658–70.

Donatelli, M., Russell, G., Rizzoli, A. E., Acutis, M., Adam, M., Athanasiadis, I. N., Balderacchi, M., Bechini, L., Belhouchette, H., Bellocchi, G., Bergez, J.-E., Botta, M., Braudeau, E., Bregaglio, S., Carlini, L., Casellas, E., Celette, F., Ceotto, E., Charron-Moirez, M. H., Confalonieri, R., Corbeels, M., Criscuolo, L., Cruz, P., Guardo, A. d., Ditto, D., Dupraz, C., Duru, M., Fiorani, D., Gentile, A., Ewert, F., Gary, C., Habyarimana, E., Jouany, C., Kansou, K., Knapen, R., Filippi, G. L., Leffelaar, P. A., Manici, L., Martin, G., Martin, P., Meuter, E., Mugueta, N., Mulia, R., Noordwijk, M. v., Oomen, R., Rosenmund, A., Rossi, V., Salinari, F., Serrano, A., Sorce, A., Vincent, G., Theau, J.-P., Thérond, O., Trevisan, M., Trevisiol, P., Evert, F. K. v., Wallach, D., Wery, J. and Zerourou, A. (2010). A Component-based framework for simulating agricultural production and externalities. In Brouwer, F. M. and Ittersum, M. K. (Eds), *Environmental and Agricultural Modelling: Integrated Approaches for Policy Impact Assessment*. Springer Science+Business Media B.V., pp. 63–108.

Dubos, B., Snoeck, D. and Flori, A. (2016). Excessive use of fertilizer can increase leaching processes and modify soil reserves in two Ecuadorian oil palm plantations. *Exp. Agric.*, 53, 255–68.

European Commission (2009). Directive 2009/28/EC of the European parliament and of the council of 23 April 2009 on the promotion of the use of energy from renewable sources and amending and subsequently repealing Directives 2001/77/EC and 2003/30/EC. *Official Journal of the European Union*, L140, 16–62.

Ewert, F., Ittersum, M. K. v., Heckelei, T., Therond, O., Bezlepkina, I. and Andersen, E. (2011). Scale changes and model linking methods for integrated assessment of agri-environmental systems. *Agric. Ecosyst. Environ.* 142, 6–17.

Fan, Y., Roupsard, O., Bernoux, M., Maire, G. L., Panferov, O., Kotowska, M. M. and Knohl, A. (2015). A sub-canopy structure for simulating oil palm in the Community Land Model (CLM-Palm): Phenology, allocation and yield. *Geosci. Model Dev.* 8, 3785–800.

FAOSTAT (Accessed 2016). http://www.fao.org/faostat/en/#data.

Fargione, J., Hill, J., Tilman, D., Polasky, S. and Hawthorne, P. (2008). Land Clearing and the Biofuel Carbon Debt. *Science* 319, 1235–8.

Farquharson, R. and Baldock, J. (2008). Concepts in modelling N2O emissions from land use. *Plant Soil* 309, 147–67.

Fatichi, S., Ivanov, V. Y. and Caporali, E. (2012a). A mechanistic ecohydrological model to investigate complex interactions in cold and warm water-controlled environments: 1. Theoretical framework and plot-scale analysis. *J. Adv. Model. Earth Syst.* 4, M05002.

Fatichi, S., Ivanov, V. Y. and Caporali, E. (2012b). A mechanistic ecohydrological model to investigate complex interactions in cold and warm water-controlled environments: 2. Spatiotemporal analyses. *J. Adv. Model. Earth Syst.* 4, M05003.

Fatichi, S., Vivoni, E. R., Ogden, F. L., Ivanov, V. Y., Mirus, B., Gochis, D., Downer, C. W., Camporese, M., Davison, J. H., Ebel, B., Jones, N., Kim, J., Mascaro, G., Niswonger, R., Restrepo, P., Rigon, R., Shen, C., Sulis, M. and Tarboton, D. (2016). An overview of current applications, challenges and future trends in distributed process-based models in hydrology. *J. Hydrol.* 537, 45–60.

Favis-Mortlock, D. (2013). Non-linear dynamics, self-organisation and cellular automata models. In Wainwright, J. and Mulligan, M. (Eds), *Environmental Modelling: Finding Simplicity in Complexity*. John Wiley & Sons, Ltd., pp. 45–67.

Fayle, T. M., Turner, E. C., Snaddon, J. L., Chey, V. K., Chung, A. Y. C., Eggleton, P. and Foster, W. A. (2010). Oil palm expansion into rainforest greatly reduces ant biodiversity in canopy, epiphytes and leaf-litter. *Basic Appl. Ecol.* 11, 337–45.

Fayle, T. M., Turner, E. C. and Foster, W. A. (2013). Ant mosaics occur in SE Asian oil palm plantation but not rain forest and are influenced by the presence of nest-sites and non-native species. *Ecography* 36, 1051–7.

Feola, G., Rahn, E. and Binder, C. (2011). Suitability of pesticide risk indicators for less developed countries: A comparison. *Agric. Ecosyst. Environ* 142, 238–45.

Flato, G., Marotzke, J., Abiodun, B., Braconnot, P., Chou, S. C., Collins, W., Cox, P., Driouech, F., Emori, S., Eyring, V., Forest, C., Gleckler, P., Guilyardi, E., Jakob, C., Kattsov, V., Reason, C. and Rummukainen, M. (2013). Evaluation of Climate Models. In Stocker, T. F., Qin, D., Plattner, G.-K., Tignor, M., Allen, S. K., Boschung, J., Nauels, A., Xia, Y., Bex, V. and Midgley, P. M. (Eds), *Climate Change 2013: The Physical Science Basis. Contribution of Working Group I to the Fifth Assessment Report of the Intergovernmental Panel on Climate Change*. Cambridge University Press: Cambridge, United Kingdom and New York, NY, USA.

Forouzangohar, M., Kookana, R. S., Forrester, S. T., Smernik, R. J. and Chittleborough, D. J. (2008). Midinfrared spectroscopy and chemometrics to predict diuron sorption coefficients in soils. *Environ. Sci. Technol.* 42, 3283–8.

Foster, W. A., Snaddon, J. L., Turner, E. C., Fayle, T. M., Cockerill, T. D., Ellwood, M. D. F., Broad, G. R., Chung, A. Y. C., Eggleton, P., Khen, C. V. and Yusah, K. M. (2011). Establishing the evidence base for maintaining biodiversity and ecosystem function in the oil palm landscapes of South East Asia. *Phil. Trans. R. Soc. B.* 366, 3277–91.

Franko, U. (1996). Modelling approaches of soil organic matter turnover within the CANDY system. In Powlson, D. S., Smith, P. and Smith, J. U. (Eds), *Evaluation of Soil Organic Matter Models Using Existing, Long-Term Datasets*. Springer-Verlag: Berlin, pp. 247–54

Frazão, L. A., Paustian, K., Pellegrino Cerri, C. E. and Cerri, C. C. (2013). Soil carbon stocks and changes after oil palm introduction in the Brazilian Amazon. *Glob. Change Biol. Bioenergy* 5, 384–90.

Geyer, K. M., Kyker-Snowman, E., Grandy, A. S. and Frey, S. D. (2016). Microbial carbon use efficiency: accounting for population, community and ecosystem-scale controls over the fate of metabolized organic matter. *Biogeochemistry* 127, 173–88.

Gibbs, H. K., Johnston, M., Foley, J. A., Holloway, T., Monfreda, C., Ramankutty, N. and Zaks, D. (2008). Carbon payback times for crop-based biofuel expansion in the tropics: The effects of changing yield and technology. *Environ. Res. Lett.* 3, 034001.

Giller, K. E. and Fairhurst, T. H. (2003). Legume cover plants. In Fairhurst, T. H. and Härdter, R. (Eds), *"Oil Palm. Management for Large and Sustainable Yields"*. Potash & Phosphate Institute/Potash & Phosphate Institute of Canada and International Potash Institute: Singapore, pp. 151–61.

Girardin, P., Bockstaller, C. and Werf, H. v. d. (1999). Indicators: Tools to evaluate the environmental impacts of farming systems. *J. Sustainable Agric.* 13, 5–21.

Goodrick, I., Nelson, P. N., Banabas, M., Wurster, C. M. and Bird, M. I. (2015). Soil carbon balance following conversion of grassland to oil palm. *Glob. Change Biol. Bioenergy* 7, 263–72.

Goodrick, I., Nelson, P. N., Nake, S., Webb, M. J., Bird, M. I. and Huth, N. (2016). Tree-scale spatial variability of soil carbon cycling in a mature oil palm plantation. *Soil Res.* 54, 397–406.

Granell, C., Schade, S. and Ostländer, N. (2013). Seeing the forest through the trees: A review of integrated environmental modelling tools. *Comput. Environ. Urban Syst.* 41, 136–50.

Grayson, R. and Blöschl, G. (2001). *Spatial Patterns in Catchment Hydrology*. Cambridge University Press: Cambridge, UK.

Gregory, S. V., Swanson, F. J., McKee, W. A. and Cummins, K. W. (1991). An ecosystem perspective of riparian zones. *BioScience*, 41, 540–51.

Grosso, S. J. D., Parton, W. J., Mosier, A. R., Hartman, M. D., Brenner, J., Ojima, D. S. and Schimel, D. S. (2001). Simulated interaction of carbon dynamics and nitrogen trace gas fluxes using the DAYCENT model. In Shaffer, M. J., Ma, L. and Hansen, S. (Eds), *Modeling Carbon and Nitrogen Dynamics for Soil Management*. CRC Press: Boca Raton, USA, pp. 303–32.

Gupta, A. (2011). *Tropical Geomorphology*. Cambridge University Press: Cambridge, UK.
Gurnell, A. M., Bertoldi, W. and Corenblit, D. (2012). Changing river channels: The roles of hydrological processes, plants and pioneer fluvial landforms in humid temperate, mixed load, gravel bed rivers. *Earth-Sci. Rev.* 111, 129–41.
Haddix, M. L., Paul, E. A. and Cotrufo, M. F. (2016). Dual, differential isotope labeling shows the preferential movement of labile plant constituents into mineral bonded soil organic matter. *Global Change Biol.* 22, 2301–12.
Hamza, M. A. and Anderson, W. K. (2005). Soil compaction in cropping systems: A review of the nature, causes and possible solutions. *Soil Tillage Res.* 82, 121–45.
Harmel, R. D., Smith, P. K., Migliaccio, K. W., Chaubey, I., Douglas-Mankin, K. R., Benham, B., Shukla, S., Muñoz-Carpena, R. and Robson, B. J. (2014). Evaluating, interpreting and communicating performance of hydrologic/water quality models considering intended use: A review and recommendations. *Environ. Model. Softw.* 57, 40–51.
Haron, K., Brookes, P. C. Anderson, J. M. and Zakaria, Z. Z. (1998). Microbial biomass and soil organic matter dynamics in oil palm (*Elaeis guineensis* Jacq.) plantations, West Malaysia. *Soil Biol. Biochem.* 30, 547–52.
Harris, G. P. and Heathwaite, A. (2012). Why is achieving good ecological outcomes in rivers so difficult? *Freshwater Biol.* 57, 91–107.
Harsono, S. S., Grundmann, P. and Soebronto, S. (2014). Anaerobic treatment of palm oil mill effluents: Potential contribution to net energy yield and reduction of greenhouse gas emissions from biodiesel production. *J. Clean. Prod.* 64, 619–27.
Harsono, S. S., Prochnow, A., Grundmann, P., Hansen, A. and Hallmann, C. (2012). Energy balances and greenhouse gas emissions of palm oil biodiesel in Indonesia. *GCB Bioenergy* 4, 213–28.
Hateley, L. R., Ellis, R., Shaw, M., Waters, D. and Carroll, C. (2014). Modelling reductions of pollutant loads due to improved management practices in the Great Barrier Reef catchments – Wet Tropics NRM region, Technical Report, Volume 3. Queensland Department of Natural Resources and Mines: Cairns, Queensland.
Heavens, N. G., Ward, D. S. and Natalie, M. M. (2013). Studying and Projecting Climate Change with Earth System Models. *Nature Education Knowledge* 4, 4.
Henault, C., Bizouard, F., Laville, P., Gabrielle, B., Nicoullaud, B., Germon, J. C. and Cellier, P. (2005). Predicting in situ soil N2O emission using NOE algorithm and soil database. *Global Change Biol.* 11, 115–27.
Henley, W. F., Patterson, M. A., Neves, R. J. and Lemly, A. D. (2000). Effects of Sedimentation and Turbidity on Lotic Food Webs: A Concise Review for Natural Resource Managers. *Rev. Fish. Sci.* 8, 125–39.
Henson, I. E. (1994). *Environmental Impacts of Oil Palm Plantations In Malaysia*. Palm Oil Research Institute of Malaysia: Kuala Lumpur, Malaysia.
Henson, I. E., Yahya, Z., Noor, M. R. M., Harun, M. H. and Mohammed, A. T. (2007). Predicting soil water status, evapotranspiration, growth and yield of young oil palm in a seasonally dry region of Malaysia. *J. Oil Palm Res.* 19, 398–415.
Hernández-Hernández, C. N., Valle-Mora, J., Santiesteban-Hernández, A. and Bello-Mendoza, R. (2007). Comparative ecological risks of pesticides used in plantation production of papaya: Application of the SYNOPS indicator. *Sci. Total* Eenviron. 381, 112–25.
Hewitt, C. N., MacKenzie, A. R., Carlo, P. D., Marco, C. F. D., Dorsey, J. R., Evans, M., Fowler, D., Gallagher, M. W., Hopkins, J. R., Jones, C. E., Langford, B., Lee, J. D., Lewis, A. C., Lim, S. F., McQuaid, J., Misztal, P., Moller, S. J., Monks, P. S., Nemitz, E., Oram, D. E., Owen, S. M., Phillips, G. J., Pugh, T. A. M., Pyle, J. A., Reeves, C. E., Ryder, J., Siong, J., Skiba, U. and Stewart, D. J. (2009). Nitrogen management is essential to prevent tropical oil palm plantations from causing ground-level ozone pollution. *Proc. Natl. Acad. Sci. USA* 106, 18447–51.
Hoffmann, M., Donough, C., Oberthür, T., VERA, A. C., van Wijk, M., Lim, C., Asmono, D., Samosir, Y., Lubis, A. and Moses, D. (2015). Benchmarking yield for sustainable intensification of oil palm production in Indonesia using PALMSIM. *Planter* 91, 81–96.

Hoffmann, M. P., Castaneda Vera, A., van Wijk, M. T., Giller, K. E., Oberthür, T., Donough, C. and Whitbread, A. M. (2014). Simulating potential growth and yield of oil palm (Elaeis guineensis) with PALMSIM: Model description, evaluation and application. *Agric. Sys.* 131, 1–10.

Holzworth, D. P., Huth, N. I. and deVoil, P. G. (2011). Simple software processes and tests improve the reliability and usefulness of a model. *Environ. Model. Softw.* 24, 510–16.

Holzworth, D. P., Huth, N. I., deVoil, P. G., Zurcher, E. J., Herrmann, N. I., McLean, G., Chenu, K., Oosterom, E. J. v., Snow, V., Murphy, C., Moore, A. D., Brown, H., Whish, J. P. M., Verrall, S., Fainges, J., Bell, L. W., Peake, A. S., Poulton, P. L., Hochman, Z., Thorburn, P. J., Gaydon, D. S., Dalgliesh, N. P., Rodriguez, D., Cox, H., Chapman, S., Doherty, A., Teixeira, E., Sharp, J., Cichota, R., Vogeler, I., Li, F. Y., Wang, E., Hammer, G. L., Robertson, M. J., Dimes, J. P., Whitbread, A. M., Hunt, J., Rees, H. v., McClelland, T., Carberry, P. S., Hargreaves, J. N. G., MacLeod, N., McDonald, C., Harsdorf, J., Wedgwood, S. and Keating, B. A. (2014). APSIM – Evolution towards a new generation of agricultural systems simulation. *Environ. Model. Softw.* 62, 327–50.

Holzworth, D. P., Huth, N. I. and Voil, P. G. d. (2010). Simplifying environmental model reuse. *Environ. Model. Softw.* 25, 269–75.

Hooijer, A., Page, S., Canadell, J. G., Silvius, M., Kwadijk, J., Wösten, H. and Jauhiainen, J. (2010). Current and future CO2 emissions from drained peatlands in Southeast Asia. *Biogeosciences* 7, 1505–14.

Hurrell, J. W., Holland, M. M., Gent, P. R., Ghan, S., Kay, J. E., Kushner, P. J., Lamarque, J.-F., Large, W. G., Lawrence, D., Lindsay, K., Lipscomb, W. H., Long, M. C., Mahowald, N., Marsh, D. R., Neale, R. B., Rasch, P., Vavrus, S., Vertenstein, M., Bader, D., Collins, W. D., Hack, J. J., Kiehl, J. and Marshall, S. (2013). The community earth system model. A framework for collaborative research. *Bull. Am. Meteorol. Soc.* 94, 1339–60.

Huth, N. I., Banabas, M., Nelson, P. N. and Webb, M. (2014). Development of an oil palm cropping systems model: Lessons learned and future directions. *Environ. Model. Softw.* 62, 411–19.

IPCC (2003). '*Good Practice Guidance for Land Use, Land-Use Change and Forestry*'. Edited by J. Penman, M. Gytarsky, T. Hiraishi, T. Krug, D. Kruger, R. Pipatti, L. Buendia, K. Miwa, T. Ngara, K. Tanabe and F. Wagner. Intergovernmental Panel of Climate Change (IPCC) National Greenhouse Gas Inventories Programme, Japan.

IPCC (2006). '*2006 IPCC Guidelines for National Greenhouse Gas Inventories*', Prepared by the National Greenhouse Gas Inventories Programme, Eggleston, H.S., Buendia, L., Miwa, K., Ngara, T. and Tanabe, K. (eds). Intergovernmental Panel of Climate Change (IPCC) National Greenhouse Gas Inventories Programme: Japan.

IPCC (2014). '*2013 Supplement to the 2006 IPCC Guidelines for National Greenhouse Gas Inventories: Wetlands*', Hiraishi, T., Krug, T., Tanabe, K., Srivastava, N., Baasansuren, J., Fukuda, M. and Troxler, T.G. (Eds). Intergovernmental Panel on Climate Change: Switzerland.

Ismail, B. S. and Ooi, K. E. (2012). Adsorption, desorption and mobility of metsulfuron-methyl in soils of the oil palm agroecosystem in Malaysia. *J. Env. Biol.* 33, 573–7.

Jacques, D., Šimůnek, J., Mallants, D. and Genuchten, M. T. v. (2008). Modelling coupled water flow, solute transport and geochemical reactions affecting heavy metal migration in a podzol soil. *Geoderma* 145, 449–61.

Jakeman, A. J., Letcher, R. A. and Norton, J. P. (2006). Ten iterative steps in development and evaluation of environmental models. *Environ. Model. Softw.* 21, 602–14.

Janssen, A. B. G., Arhonditsis, G. B., Beusen, A., Bolding, K., Bruce, L., Bruggeman, J., Couture, R.-M., Downing, A. S., Elliott, J. A., Frassl, M. A., Gal, G., Gerla, D. J., Hipsey, M. R., Hu, F., Ives, S. C., Janse, J. H., Jeppesen, E., Jöhnk, K. D., Kneis, D., Kong, X., Kuiper, J. J., Lehmann, M. K., Lemmen, C., Özkundakci, D., Petzoldt, T., Rinke, K., Robson, B. J., Sachse, R., Schep, S. A., Schmid, M., Scholten, H., Teurlincx, S., Trolle, D., Troost, T. A., Dam, A. A. V., Gerven, L. P. A. V., Weijerman, M., Wells, S. A. and Mooij, W. M. (2015). Exploring, exploiting and evolving diversity of aquatic ecosystem models: A community perspective. *Aquatic Ecol.* 49, 513–48.

Jenkinson, D. S. and Coleman, K. (2008). The turnover of organic carbon in subsoils. Part 2. Modelling carbon turnover. *Eur. J. Soil Sci.* 59, 400–13.

Jennings, A. P., Naim, M., Advento, A. D., Aryawan, A. A. K., Ps, S., Caliman, J.-P., Verwilghen, A. and Veron, G. (2015). Diversity and occupancy of small carnivores within oil palm plantations in central Sumatra, Indonesia. *Mammal Res.* 60, 181–8.
Jones, J. W., Hoogenboom, G., Porter, C. H., Boote, K. J., Batchelor, W. D., Hunt, L. A., Wilkens, P. W., Singh, U., Gijsman, A. J. and Ritchie, J. T. (2003). The DSSAT cropping system model. *Eur. J. Agron.* 18, 235–65.
Kaiser, C., Franklin, O., Richter, A. and Dieckmann, U. (2015). Social dynamics within decomposer communities lead to nitrogen retention and organic matter build-up in soils. *Nat. Commun.* 6, 1–10.
Kee, K. K., Goh, K. J. and ChewChew, P. S. (1995). Effects of NK fertiliser on soil pH and exchangeable K status on acid soils in an oil palm plantation in Malaysia. In Date, R., Grundon, N., Rayment, G. and Probert, M. (Eds), *Plant-Soil Interactions at Low pH: Principles and Management*. Kluwer: Amsterdam, pp. 809–15.
Kelly (Letcher), R. A., Jakeman, A. J., Barreteau, O., Borsuk, M. E., ElSawah, S., Hamilton, S. H., Henriksen, H. J., Kuikka, S., Maier, H. R., Rizzoli, A. E., van Delden, H. and Voinov, A. A. (2013). Selecting among five common modelling approaches for integrated environmental assessment and management. *Environ. Model. Softw.* 47, 159–81.
Khasanah, N. M., Noordwijk, M. v., Ningsih, H. and Rahayua, S. (2015a). Carbon neutral? No change in mineral soil carbon stock under oil palm plantations derived from forest or non-forest in Indonesia. *Agric. Ecosyst. Environ.* 211, 195–206.
Koh, L. P. (2008). Can oil palm plantations be made more hospitable for forest butterflies and birds? *J. Appl. Ecol.* 45, 1002–9.
Koh, L. P., Miettinen, J., Liew, S. C. and Ghazoul, J. (2011). Remotely sensed evidence of tropical peatland conversion to oil palm. *Proc. Natl. Acad. Sci.* 108, 5127–32.
Kookana, R., Holz, G., Barnes, C., Bubb, K., Fremlin, R. and Boardman, B. (2010). Impact of climatic and soil conditions on environmental fate of atrazine used under plantation forestry in Australia. *J. Environ. Manag.* 91, 2649–56.
Kookana, R. S., Ahmad, R. and Farenhorst, A. (2014). Sorption of pesticides and its dependence on soil properties: Chemometrics approach for estimating sorption. In Chen, W., Sabljic, A., Cryer, S. A. and et. al. (Eds), *Non-First Order Degradation and Time-Dependent Sorption of Organic Chemicals in Soil*, Volume 1174. American Chemical Society: Washington, DC, pp. 221–40.
Kookana, R. S., Correll, R. L. and Miller, R. B. (2005). Pesticide impact rating index–A pesticide risk indicator for water quality. *Water Air Soil Pollut.* 5, 45–65.
Kookana, R. S., Kumar, A., Oliver, D. P. and Correll, R. L. (2007). Pesticide risk indicators: Their role in minimizing off-site impacts of pesticides on water quality. In *American Chemical Society Symposium Serie*, Vol. 966. Oxford University Press: Oxford, UK. pp. 37–52.
Krinner, G., Viovy, N., de Noblet-Ducoudré, N., Ogée, J., Polcher, J., Friedlingstein, P., Ciais, P., Sitch, S. and Prentice, I. C. (2005). A dynamic global vegetation model for studies of the coupled atmosphere-biosphere system. *Global Biogeochem. Cycles* 19.
Kristensen, P. B., Kristensen, E. A., Riis, T., Baisner, A. J., Larsen, S. E., Verdonschot, P. and Baattrup-Pedersen, A. (2013). Riparian forest as a management tool for moderating future thermal conditions of lowland temperate streams. *Hydrol. Earth Syst. Sci. Discuss.* 10, 6081–106.
Kui, L., Stella, J. C., Lightbody, A. and Wilcox, A. C. (2014). Ecogeomorphic feedbacks and flood loss of riparian tree seedlings in meandering channel experiments. *Water Resources Research* 50, 9366–84.
Laabs, V., Amelung, W., Pinto, A., Altstaedt, A. and Zech, W. (2000). Leaching and degradation of corn and soybean pesticides in an Oxisol of the Brazilian Cerrados. *Chemosphere* 41, 1441–9.
Labrière, N., Locatelli, B., Laumonier, Y., Freycon, V. and Bernoux, M. (2015). Soil erosion in the humid tropics: A systematic quantitative review. *Agric. Ecosyst. Environ.* 203, 127–39.
Lacoste, M., Viaud, V., Michot, D. and Walter, C. (2015). Landscape-scale modelling of erosion processes and soil carbon dynamics under land-use and climate change in agroecosystems. *Eur. J. Soil Sci.* 66, 780–91.
Lal, P. (1990). *Soil Erosion in the Tropics – Principles and Management*. McGraw-Hill: New York.

Lal, R. and Stewart, B. A. (2013). *Principles of Sustainable Soil Management in Agroecosystems*. CRC Press: Boca Raton, USA.

Law, M. C., Balasundram, S. K., Husni, M. H. A., Ahmed, O. H. and Harun, M. H. (2009). Spatial variability of soil organic carbon in oil palm. *Int. J. Soil Sci.* 4, 93–103.

Lee, J. S. H., Garcia-Ulloa, J., Ghazoul, J., Obidzinski, K. and Koh, L. P. (2014). Modelling environmental and socio-economic trade-offs associated with land-sparing and land-sharing approaches to oil palm expansion. *J. Appl. Ecol.* 51, 1366–77.

Levin, P. S. and Stunz, G. W. (2005). Habitat triage for exploited fishes: Can we identify essential 'Essential Fish Habitat?'. *Estuar. Coast. Shelf Sci.* 64, 70–8.

Lewis, S. L., Sonke, B., Sunderland, T., Begne, S. K., Lopez-Gonzalez, G., Heijden, G. M. F. v. d., Phillips, O. L., Affum-Baffoe, K., Baker, T. R., Banin, L., Bastin, J.-F. o., Beeckman, H., Boeckx, P., Bogaert, J., Cannie`re, C. D., Chezeaux, E., Clark, C. J., Collins, M., Djagbletey, G., Noe¨, M., Djuikouo, K., Droissart, V., Doucet, J.-L., Ewango, C. E. N., Fauset, S., Feldpausch, T. R., Foli, E. G., Gillet, J.-F. o., Hamilton, A. C., Harris, D. J., Hart, T. B., Haulleville, T. d., Hladik, A., Hufkens, K., Huygens, D., Jeanmart, P., Jeffery, K. J., Kearsley, E., Leal, M. E., Lloyd, J., Lovett, J. C., Makana, J.-R., Malhi, Y., Marshall, A. R., Ojo, L., Peh, K. S.-H., Pickavance, G., Poulsen, J. R., Reitsma, J. M., Sheil, D., Simo, M., Steppe, K., Taedoumg, H. E., Talbot, J., Taplin, J. R. D., Taylor, D., Thomas, S. C., Toirambe, B., Verbeeck, H., Vleminckx, J., White, L. J. T., Willcock, S., Woell, H. and Zemagho, L. (2013). Above-ground biomass and structure of 260 African tropical forests. *Phil. Trans. R. Soc. B.* 368, 20120295.

Li, C. (2007). Quantifying greenhouse gas emissions from soils: Scientific basis and modeling approach. *Soil Sci. Plant Nutr.* 53, 344–52.

Li, C., Frolking, S. and Harriss, R. (1994). Modeling carbon biogeochemistry in agricultural soils. *Global Biogeochem. Cycles*, 237–54.

Lucey, J., Hill, J., Meer, P. v. d., Reynolds, G. and Agus, F. (2014). Change in carbon stocks arising from land-use conversion to oil palm plantations. A science-for-policy paper for the Oil palm Research-Policy Partnership Network.

Luo, Y., Ahlström, A., Allison, S. D., Batjes, N., Brovkin, V., Carvalhais, N., Chappell, A., Ciais, P., Davidson, E. A., Finzi, A., Georgiou, K., Guenet, B., Hararuk, O., Harden, J. W., He, Y., Hopkins, F., Jiang, L., Koven, C., Jackson, R. B., Jones, C. D., Lara, M. J., Liang, J., McGuire, A. D., Parton, W., Peng, C., Randerson, J. T., Salazar, A., Sierra, C. A., Smith, M. J., Tian, H., Todd-Brown, K. E. O., Torn, M., Groenigen, K. J. v., Wang, Y. P., West, T. O., Wei, Y., Wieder, W. R., Xia, J., Xu, X., Xu, X. and Zhou, T. (2016). Toward more realistic projections of soil carbon dynamics by Earth system models. *Global Biogeochem. Cycles* 30, 40–56.

Majid, M. R. and Rusli, N. (2014). Hydrological impact of large scale conversion of rubber to oil palm plantation. *In* "2014 International SWAT Conference", Pernanbuco, Brazil.

Makowski, D., Tichit, M., Guichard, L., Keulen, H. V. and Beaudoin, N. (2009). Measuring the accuracy of agro-environmental indicators. *J. Environ. Manag.* 90, S139–46.

Maliva, R. G. and Missimer, T. M. (2012). *Arid Lands Water Evaluation and Management*. Springer Science & Business Media.

Manik, Y. and Halog, A. (2013). A meta-analytic review of life cycle assessment and flow analyses studies of palm oil biodiesel. *Integr. Environ. Assess. Manag.* 9, 134–41.

Manzoni, S., Piñeiro, G., Jackson, R. B., Jobbágy, E. G., Kim, J. H. and Porporato, A. (2012). Analytical models of soil and litter decomposition: Solutions for mass loss and time-dependent decay rates. *Soil Biol. Biochem.* 50, 66–76.

Marichal, R., Sionita, R., Wholfahrt, J., Bessou, C. and Caliman, J. P. (2016). Validation of I-Phy palm, an indicator of pesticide pollution risk: method and preliminary results. *In* "5th International Conference on Oil Palm and Environment (ICOPE)", Bali, Indonesia.

McCown, R. L., Hammer, G. L., Hargreaves, J. N. G., Holzworth, D. and Huth, N. I. (1995). APSIM – An agricultural production system simulation model for operational research. *Math. Comput. Simul.* 39, 225–31.

McKergow, L. A., Prosser, I. P., Hughes, A. O. and Brodie, J. (2005). Sources of sediment to the Great Barrier Reef World Heritage Area. *Marine Poll. Bull.* 51, 200–11.

Meier, M. S., Jungbluth, N., Stoessel, F., Schader, C. and Stolze, M. (2014). Higher accuracy in N modeling makes a difference. *In* "Proceedings of the 9th International Conference on Life Cycle Assessment in the Agri-Food Sector" (R. Schenck and D. Huizenga, eds.). ACLCA, Vashon, WA, USA, San Francisco.

Mercer, E. V., Mercer, T. G. and Sayok, A. K. (2014). Effects of forest conversions to oil palm plantations on freshwater macroinvertebrates: A case study from Sarawak, Malaysia. *J. Land Use Sci.* 9, 260–77.

Michel, K., Roose, M. and Ludwig, B. (2007). Comparison of different approaches for modelling heavy metal transport in acidic soils. *Geoderma* 140, 207–14.

Molina, J.-A. E. and Smith, P. (1998). Modeling carbon and nitrogen cycling processes in soils. *Adv. Agron.* 62, 253–98.

Molina, J. A. E., Crocker, G. J., Grace, P. R., Klír, J., Körschens, M., Poulton, P. R. and Richter, D. D. (1997). Simulating trends in soil organic carbon in long-term experiments using NCSOIL and NCSWAP models. *Geoderma* 81, 91–107.

Moore, A. D., Holzworth, D. P., Herrmann, N. I., Brown, H. E., Voil, P. G. d., Snow, V. O., Zurcher, E. J. and Huth, N. I. (2014). Modelling the manager: representing rule-based management in farming systems simulation models. *Environ. Model. Softw.* 62, 399–410.

Moore, R., Spittlehouse, D. and Story, A. (2005). Riparian microclimate and stream temperature response to forest harvesting: A review. *J. Am. Water Resour. Assoc.* 41, 813–34.

Morgan, R. P. C. and Nearing, M. A. (2011). *Handbook of Erosion Modelling*. Wiley-Blackwell.

Mosier, A., Kroeze, C., Nevison, C., Oenema, O., Seitzinger, S. and Cleemput, O. v. (1998). Closing the global N2O budget: Nitrous oxide emissions through the agricultural nitrogen cycle. *Nutr. Cycl. Agroecosy.* 52, 225–48.

Mueller, T., Jensen, L. S., Hansen, S. and Nielsen, N. E. (1996). Simulating soil carbon and nitrogen dynamics with the soil-plant-atmosphere system model DAISY. In Powlson, D. S., Smith, P. and Smith, J. U. (Eds), *Evaluation of Soil Organic Matter Models Using Existing, Long-Term Datasets*. Springer-Verlag: Berlin, pp. 275–81.

Muetzelfeldt, R. and Massheder, J. (2003). The Simile visual modelling environment. *Eur. J. Agron.* 18, 345–58.

Mulligan, M. and Wainwright, J. (2013a). Modelling and Model Building. In Wainwright, J. and Mulligan, M. (Eds), *Environmental Modelling: Finding Simplicity in Complexity*. John Wiley & Sons, Ltd., pp. 7–26.

Mulligan, M. and Wainwright, J. (2013b). Modelling catchment and fluvial processes and their interactions. In Mulligan, M. and Wainwright, J. (Eds), *Environmental Modelling: Finding Simplicity in Complexity*. John Wiley & Sons, Ltd., p. 183.

Naiman, R. J., Decamps, H. and Pollock, M. (1993). The role of riparian corridors in maintaining regional biodiversity. *Ecol. Appl.* 3, 209–12.

Nelson, P. N., Banabas, M., Goodrick, I., Webb, M. J., Huth, N. I. and O'Grady, D. (2015a). Soil sampling in oil palm plantations: a practical design that accounts for lateral variability at the tree scale. *Plant Soil* 394, 421–9.

Nelson, P. N., Banabas, M., Huth, N. I. and Webb, M. J. (2015b). Quantifying trends in soil fertility under oil palm: practical challenges and approaches. In Webb, M. J., Nelson, P. N., Bessou, C., Caliman, J.-P. and Sutarta, E. S. (Eds), *Sustainable Management of Soil in Oil Palm Plantings. ACIAR Proceedings No. 144*. Australian Centre for International Agricultural Research: Canberra, Australia, pp. 60–4.

Nelson, P. N., Berthelsen, S., Webb, M. J. and Banabas, M. (2010a). Acidification of volcanic ash soils under oil palm in Papua New Guinea: effects of fertiliser type and placement. In Gilkes, R. J. and Prakongkep, N. (Eds), *Proceedings of the 19th World Congress of Soil Science; Soil Solutions for a Changing World*. International Union of Soil Science: Brisbane, Australia, pp. 8–11.

Nelson, P. N., Webb, M. J., Banabas, M., Nake, S., Goodrick, I., Gordon, J., O'Grady, D. and Dubos, B. (2014). Methods to account for tree-scale variability in soil- and plant-related parameters in oil palm plantations. *Plant Soil* 374, 459–71.

Nelson, P. N., Webb, M. J., Orrell, I., van Rees, H., Banabas, M., Berthelsen, S., Sheaves, M., Bakani, F., Pukam, O., Hoare, M., Griffiths, W., King, G., Carberry, P., Pipai, R., McNeill, A., Meekers, P.,

Lord, S., Butler, J., Pattison, T., Armour, J. and Dewhurst, C. (2010b). Environmental sustainability of oil palm cultivation in Papua New Guinea. *ACIAR Technical Report No. 75*. The Australian Centre for International Agricultural Research: Canberra, Australia.

Nemecek, T. and Kägi, T. (2007). Life cycle inventories of agricultural production systems. Ecoinvent Report No.15. ecoinvent Centre: Switzerland.

Newbold, T., Hudson, L. N., Phillips, H. R. P., Hill, S. L. L., Contu, S., Lysenko, I., Blandon, A., Butchart, S. H. M., Booth, H. L., Day, J., Palma, A. D., Harrison, M. L. K., Kirkpatrick, L., Pynegar, E., Robinson, A., Simpson, J., Mace, G. M., Scharlemann, J. P. W. and Purvis, A. (2014). A global model of the response of tropical and sub-tropical forest biodiversity to anthropogenic pressures. *Proc. R. Soc. B.* 281: 20141371.

Noordwijk, M. v., Lusiana, B. and Khasanah, N. m. (2011). *WaNuLCAS 3.01. Background on a Model of Water, Nutrient and Light Capture in Agroforestry Systems.*

O'Leary, G. J., Liu, D. L., Ma, Y., Li, F. Y., McCaskill, M., Conyers, M., Dalal, R., Reeves, S., Page, K., Dang, Y. P. and Robertson, F. (2016). Modelling soil organic carbon 1. Performance of APSIM crop and pasture modules against long-term experimental data. *Geoderma* 264, 227–37.

Oleson, K. W., Lawrence, D. M., Bonan, G. B., Drewniak, B., Huang, M., Koven, C. D., Levis, S., Li, F., Riley, W. J., Subin, Z. M., Swenson, S. C., Thornton, P. E., Bozbiyik, A., Fisher, R., Heald, C. L., Kluzek, E., Lamarque, J.-F., Lawrence, P. J., Leung, L. R., Lipscomb, W., Muszala, S., Ricciuto, D. M., Sacks, W., Sun, Y., Tang, J. and Yang, Z.-L. (2013). *Technical Description of version 4.5 of the Community Land Model (CLM)*. National Center for Atmospheric Research: Boulder, Colorado.

Paniconi, C. and Putti, M. (2015). Physically based modeling in catchment hydrology at 50: Survey and outlook. *Water Resour. Res.* 51, 7090–129.

Paramananthan, S. (2011). Soil requirements and land evaluation for oil palms for high yields. In Goh, K. J., Chiu, S. B. and Paramananthan, S. (Eds), *Agronomic Principles and Practices of Oil Palm Cultivation*. Agricultural Crop Trust: Selangor Darul Ehsan, Malaysia.

Pardon, L., Bessou, C., Nelson, P. N., Dubos, B., Ollivier, J., Marichal, R., Caliman, J.-P. and Gabrielle, B. (2016a). Key unknowns in nitrogen budget for oil palm plantations. A review. *Agron. Sustain. Dev.* 36, 1–21.

Pardon, L., Bessou, C., Saint-Geours, N., Gabrielle, B., Khasanah, N. m., Caliman, J.-P. and Nelson, P. N. (2016b). Quantifying nitrogen losses in oil palm plantations: Models and challenges. *Biogeosciences* 13, 5433–52.

Pardon, L., Huth, N., Nelson, P. N., Banabas, M., Gabrielle, B. and Bessou, C. (2017) Yield and nitrogen losses in oil palm plantations: main drivers and management trade-offs determined using simulation. *Field Crops Res.* 210, 20–32.

Parton, W. J. (1996). The CENTURY model. In Powlson, D. S., Smith, P. and Smith, J. U. (Eds), *Evaluation of Soil Organic Matter Models Using Existing Long Term Datasets*. Springer-Verlag: Berlin, Germany.

Petersen, B. M., Berntsen, J., Hansen, S. and Jensen, L. S. (2005). CN-SIM – a model for the turnover of soil organic matter. I. Long-term carbon and radiocarbon development. *Soil Biol. Biochem.* 37, 359–74.

Phalan, B., Onial, M., Balmford, A. and Green, R. E. (2011). Reconciling Food Production and Biodiversity Conservation: Land Sharing and Land Sparing Compared. *Science* 333, 1289–91.

Pleanjai, S., Gheewala, S. H. and Garivait, S. (2009). Greenhouse gas emissions from the production and use of palm methyl ester in Thailand. *Int. J. Global Warm.* 1, 418–31.

Prescott, G. W., Gilroy, J. J., Haugaasen, T., Uribe, C. A. M., Foster, W. A. and Edwards, D. P. (2016). Reducing the impacts of Neotropical oil palm development on functional diversity. *Biol. Conserv.* 197, 139–45.

Probert, M. E., Delve, R. J., Kimani, S. K. and Dimes, J. P. (2005). Modelling nitrogen mineralization from manures: Representing quality aspects by varying C:N ratio of sub-pools. *Soil Biol. Biochem.* 37, 279–87.

Probert, M. E., Dimes, J. P., Keating, B. A., Dalal, R. C. and Strong, W. M. (1998). APSIM's water and nitrogen modules and simulation of the dynamics of water and nitrogen in fallow systems. *Agric. Sys.* 56, 1–28.

Rahman, J. M., Seaton, S. P. and Cuddy, S. M. (2004). Making frameworks more useable: using model introspection and metadata to develop model processing tools. *Environ. Model. Softw.* 19, 275–84.

Rahman, J. M., Seaton, S. P., Perraud, J. M., Hotham, H., Verrelli, D. I. and Coleman, J. R. (2003). It's TIME for a new environmental modelling framework. *Proceedings of MODSIM* 4, 1727–32.

Regitano, J., Alleoni, L., Vidal-Torrado, P., Casagrande, J. and Tornisielo, V. (2000). Imazaquin sorption in highly weathered tropical soils. *J. Environ. Qual.* 29, 894–900.

Reus, J., Leendertse, P., Bockstaller, C., Fomsgaard, I., Gutsche, V., Lewis, K., Nilsson, C., Pussemier, L., Trevisan, M. and Van der Werf, H. (2002). Comparison and evaluation of eight pesticide environmental risk indicators developed in Europe and recommendations for future use. *Agric. Ecosyst. Environ* 90, 177–87.

Robson, B. J. (2014a). When do aquatic systems models provide useful predictions, what is changing and what is next? *Environ. Model. Softw.* 61, 287–96.

Robson, B. J. (2014b). State of the art in modelling of phosphorus in aquatic systems: Review, criticisms and commentary. *Environ. Model. Softw.* 61, 339–59.

Roy, R. N., Misra, R. V., Lesschen, J. P. and Smaling, E. M. (2005). *Evaluation du bilan en elements nutritifs du sol. Approches et methodologies.* FAO: Rome.

Sayer, J., Ghazoul, J., Nelson, P. and Boedhihartono, A. K. (2012). Oil palm expansion transforms tropical landscapes and livelihoods. *Glob. Food Sec.* 1, 114–19.

Schmidt, J. H. (2007). Life assessment of rapeseed oil and palm oil. Ph. D. thesis, Part 3: Life cycle inventory of rapeseed oil, Aalborg University: Denmark.

Schmidt, J. H. (2010). Comparative life cycle assessment of rapeseed oil and palm oil. *Int. J. Life Cycle Assess.* 15, 183–97.

Schmidt, J. H. (2015). Life cycle assessment of five vegetable oils. *J. Clean. Prod.* 87, 130–8.

Selim, H. M. and Zhang, H. (2013). Modeling Approaches of Competitive Sorption and Transport of Trace Metals and Metalloids in Soils: A Review. *J. Environ. Qual.* 42, 640–53.

Senior, M. J., Hamer, K. C., Bottrell, S., Edwards, D. P., Fayle, T. M., Lucey, J. M., Mayhew, P. J., Newton, R., Peh, K. S. H., Sheldon, F. H. and Stewart, C. (2013). Trait-dependent declines of species following conversion of rain forest to oil palm plantations. *Biodivers. Conser.* 22, 253–68.

Sheaves, M. (1996). Habitat-specific distributions of some fishes in a tropical estuary. *Mar. Freshwater Res.* 47, 827–30.

Shen, C., Niu, J. and Phanikumar, M. S. (2013). Evaluating controls on coupled hydrologic and vegetation dynamics in a humid continental climate watershed using a subsurface-land surface processes model. *Water Resour. Res.* 49, 2552–72.

Shrestha, S. and Datta, A. (2015). Field measurements for evaluating the RZWQM and PESTFADE models for the tropical zone of Thailand. *J. Environ. Manag.* 147, 286–96.

Singh, M., Malhi, Y. and Bhagwat, S. (2015). Aboveground biomass and tree diversity of riparian zones in an oil palm-dominated mixed landscape in Borneo. *J. Trop. Forest Sci.* 27, 227–39.

Singh, R., Ibrahim, M., Esa, N. and Iliyana, M. (2010). Composting of waste from palm oil mill: A sustainable waste management practice. *Rev. Environ. Sci. Bio.* 9, 331–44.

Skjemstad, J. O., Spouncer, L. R., Cowie, B. and Swift, R. S. (2004). Calibration of the Rothamsted organic carbon turnover model (RothC ver. 26.3), using measurable soil organic carbon pools. *Austral. J. Soil Res.* 42, 79–88.

Smith, D. R., Townsend, T. J., Choy, A. W. K., Hardy, I. C. W. and Sjögersten, S. (2012). Short-term soil carbon sink potential of oil palm plantations. *Glob. Change Biol. Bioenergy* 4, 588–96

Smith, P. (2006). Organic matter modelling. In Lal, R. (Ed.), *Encyclopedia of Soil Science*. CRC Press: Boca Raton, pp. 1196–202.

Smith, P., Smith, J. U., Powlson, D. S., McGill, W. B., Arah, J. R. M., Chertov, O. G., Coleman, K., Franko, U., Frolking, S., Jenkinson, D. S., Jensen, L. S., Kelly, R. H., Klein-Gunnewiek, H., Komarov, A. S., Li, C., Molina, J. A. E., Mueller, T., Parton, W. J., Thornley, J. H. M. and Whitmore, A. P. (1997). A comparison of the performance of nine soil organic matter models using datasets from seven long-term experiments. *Geoderma* 81, 153–225.

Soltani, A. and Sinclair, T. R. (2015). A comparison of four wheat models with respect to robustness and transparency: Simulation in a temperate, sub-humid environment. *Field Crops Res.* 175, 37–46.

Stehfest, E. and Bouwman, L. (2006). N2O and NO emission from agricultural fields and soils under natural vegetation: summarizing available measurement data and modeling of global annual emissions. *Nutr. Cycl. Agroecosys.* 74, 207–28.

Stichnothe, H. and Schuchardt, F. (2010). Comparison of different treatment options for palm oil production waste on a life cycle basis. *Int. J. Life Cycle Assess.* 15, 907–15.

Stichnothe, H. and Schuchardt, F. (2011). Life cycle assessment of two palm oil production systems. *Biomass Bioenergy* 35, 3976–84.

Stockle, C. O., Donatelli, M. and Nelson, R. (2003). CropSyst, a cropping systems simulation model. *Eur. J. Agron.* 18, 289–307.

Straaten, O. v., Corre, M. D., Wolf, K., Tchienkoua, M., Cuellar, E., Matthews, R. B. and Veldkamp, E. (2015). Conversion of lowland tropical forests to tree cash crop plantations loses up to one-half of stored soil organic carbon. *PNAS* 112, 9956–60.

Sulaiman, F. R., Mustaffa, N. F. S. and Mohd Khazaai, S. N. (2016). Preliminary assessment of selected metals in agricultural soils in Jengka, Pahang, Malaysia. *Environ. Earth Sci.* 75, 1–9.

Thiollet, M. (2003). *Construction des indicateurs viti-environnementaux de la méthode INDIGO.* INRA: Colmar.

Thorburn, P. J., Biggs, J. S., Collins, K. and Probert, M. E. (2010). Using the APSIM model to estimate nitrous oxide emissions from diverse Australian sugarcane production systems. *Agric. Ecosyst. Environ.* 136, 343–50.

Thornton, P. E. and Rosenbloom, N. A. (2005). Ecosystem model spin-up: estimating steady state conditions in a coupled terrestrial carbon and nitrogen cycle model. *Ecol. Model.* 189, 25–48.

Todd-Brown, K. E., Randerson, J. T., Post, W. M., Hoffman, F. M., Tarnocai, C., Schuur, E. A. and Allison, S. D. (2013). Causes of variation in soil carbon simulations from CMIP5 Earth system models and comparison with observations. *Biogeosciences* 10, 1717–36.

Tscharntke, T., Tylianakis, J. M., Rand, T. A., Didham, R. K., Fahrig, L., Batáry, P., Bengtsson, J., Clough, Y., Crist, T. O., Dormann, C. F. and Ewers, R. M. (2012). Landscape moderation of biodiversity patterns and processes-eight hypotheses. *Biolog. Rev.* 87, 661–85.

Uehara, G. and Gillman, G. (1981). *The Mineralogy, Chemistry and Physics of Tropical Soils with Variable Charge Clays.* Westview Press: Boulder, Colorado.

Valente-Neto, F., Koroiva, R., Fonseca-Gessner, A. A. and de Oliveira Roque, F. (2015). The effect of riparian deforestation on macroinvertebrates associated with submerged woody debris. *Aquatic Ecol.* 49, 115–25.

van Ittersum, M. K. and Donatelli, M. (2003). Modelling cropping systems – highlights of the symposium and preface to the special issues. *Eur. J. Agron.* 18, 187–97.

van Kraalingen, D. W. G., Breure, C. J. and Spitters, C. J. T. (1989). Simulation of oil palm growth and yield. *Agric. Forest Meteorol.* 46, 227–44.

van Noordwijk, M., Lusiana, B., Khasanah, N. and Mulia, R. (2011). *WaNuLCAS version 4.0: Background on a Model of Water, Nutrient and Light Capture in Agroforestry Systems.* World Agroforestry Centre (ICRAF): Bogor, Indonesia.

Vanclay, J. K., Sinclair, F. L. and Prabhu, R. (2003). Modelling Interactions amongst People and Forest Resources at the Landscape Scale. *Small-scale Forest Econ. Manag. Policy* 2, 117–20.

Vijaya, S., Ma, A. N. and Choo, Y. M. (2009). A gate to gate assessment of environmental performance for production of crude palm kernel oil using life cycle assessment approach. *Am. J. Environ. Sci.* 5, 267–72.

Wainwright, J. and Mulligan, M. (2013). *Environmental Modelling: Finding Simplicity in Complexity.* John Wiley & Sons, Ltd.

Wakelin, S. A., Gerard, E., van Koten, C., Banabas, M., O'Callaghan, M. and Nelson, P. N. (2016). Soil physicochemical properties impact more strongly on bacteria and fungi than conversion of grassland to oil palm. *Pedobiologia* 59, 83–91.

Wang, X., Piao, S., Ciais, P., Friedlingstein, P., Myneni, R. B., Cox, P., Heimann, M., Miller, M., Peng, S., Wang, T., Yang, H. and Chen, A. (2014). A two-fold increase of carbon cycle sensitivity to tropical temperature variations. *Nature* 506, 212–15.

Wannamaker, C. M. and Rice, J. A. (2000). Effects of hypoxia on movements and behavior of selected estuarine organisms from the southeastern United States. *J. Exp. Mar. Biol. Ecol.* 249, 145–63.

Wantzen, K. M. and Junk, W. J. (2000). The importance of stream-wetland-systems for biodiversity: a tropical perspective. In Gopal, B., Junk, W. J. and Davies, J. A. (Eds), *Biodiversity in Wetlands: Assessment, Function and Conservation*, pp. 11–34. Backhuys: Leiden, The Netherlands.

Waters, D. K., Carroll, C., Ellis, R., Hateley, L., McCloskey, G. L., Packett, R. ... and Fentie, B. (2014). Modelling reductions of pollutant loads due to improved management practices in the Great Barrier Reef catchments – Whole of GBR, Technical Report, Volume 1. Queensland Department of Natural Resources and Mines: Toowoomba, Queensland.

Welcomme, R. (1988). Concluding remarks I: On the nature of large tropical rivers, floodplains and future research directions. *J. N. Am. Benthol. Soc.* 7, 525–6.

Wicke, B., Dornburg, V., Junginger, M. and Faaij, A. (2008). Different palm oil production systems for energy purposes and their greenhouse gas implications. *Biomass Bioenergy* 32, 1322–37.

Wieder, W. R., Grandy, A. S., Kallenbach, C. M. and Bonan, G. B. (2014). Integrating microbial physiology and physio-chemical principles in soils with the MIcrobial-MIneral Carbon Stabilization (MIMICS) model. *Biogeosciences* 11, 3899–917.

Wilkinson, S. N., Dougall, C., Kinsey-Henderson, A. E., Searle, R. D., Ellis, R. J. and Bartley, R. (2014). Development of a time-stepping sediment budget model for assessing land use impacts in large river basins. *Sci. Total Environ.* 468–469, 1210–24.

Yacob, S., Hassan, M. A., Shirai, Y., Wakisaka, M. and Subash, S. (2005). Baseline study of methane emission from open digesting tanks of palm oil mill effluent treatment. *Chemosphere* 59, 1575–81.

Yang, H. S. and Janssen, B. H. (2000). A mono-component model of carbon mineralization with a dynamic rate constant. *Eur. J. Soil Sci.* 51, 517–29.

Yusop, Z., Chan, C. and Katimon, A. (2007). Runoff characteristics and application of HEC-HMS for modelling stormflow hydrograph in an oil palm catchment. *Water Sci. Technol.* 56, 41–8.

Yusop, Z., Hui, C. M., Garusu, G. J. and Katimon, A. (2008). Estimation of evapotranspiration in oil palm catchment by short-time period water-budget method. *Malays. J. Civ. Eng.* 20, 160–74.

Zhang, G.-H., Liu, B.-Y., Liu, G.-B., He, X.-W. and Nearing, M. (2003). Detachment of undisturbed soil by shallow flow. *Soil. Sci. Soc. Am. J.* 67, 713–19.

Zhang, G., Liu, G., Wang, G. and Wang, Y. (2012). Effects of patterned Artemisia capillaris on overland flow velocity under simulated rainfall. *Hydrol. Process.* 26, 3779–87.

Zimmermann, M., Leifeld, J., Schmidt, M. W. I., Smith, P. and Fuhrer, J. (2007). Measured soil organic matter fractions can be related to pools in the RothC model. *Eur. J. Soil Sci.* 58, 658–67.

Chapter 13

Certifying sustainability in oil palm cultivation

Marcel Djama, CIRAD and MOISA, University of Montpellier, France and Universiti Putra Malaysia, Malaysia

1 Introduction

Established in response to the resolutions of the 1992 Rio Summit on Sustainable Development, sustainability standards have become a key instrument in the regulation of agricultural value chains.

In 2015, the Ecolabel Index database listed more than 450 certification programmes or ecolabels (Lemeilleur and Balineau, 2016), compared to fewer than 50 prior to 1990. These mechanisms have skyrocketed over the last 20 years, and the agricultural, food and forestry sectors are by far some of the most active ones in this respect (Marx and Wouters, 2015).

In practical terms, sustainability standards establish the principles and criteria governing the production of goods, according to a given approach to sustainable development. In order to increase their credibility, most schemes are based on third-party monitoring and certification mechanisms, which verify commodity suppliers' compliance with such principles and criteria.

There are various types of sustainability standards. Most of them involve voluntary initiatives led by private actors: these are Voluntary Sustainability Standards (hereinafter referred to as VSSs). These private standards cover not only the commitments adopted unilaterally by companies (for example, in the form of codes of conduct or corporate social and environmental responsibility charters), but also multi-stakeholder initiatives (MSIs) entailing cooperation between several actors both within the sector (producers, processors, buyers) and beyond it (NGOs, banks).

http://dx.doi.org/10.19103/AS.2017.0018.25

More recently, several public bodies have taken inspiration from these private standardisation mechanisms to regulate industrial sectors at the national level, giving sustainability principles and criteria a legal, mandatory status.

The palm oil industry is particularly involved in the development of sustainability standards (Rival and Levang, 2014). Driven by growing global demand, palm oil production has expanded rapidly in the last few years. Palm oil is the most widely consumed vegetable oil in the world, and its popularity has grown even more with the emergence of new market opportunities in the biofuel sector, in addition to its traditional food and oleochemical uses.

This strong growth has unquestionably contributed to the economic development of the main producer countries – Indonesia and Malaysia – which account for 87% of global production. Palm oil cultivation provides income for many smallholders, whose produce accounts for around 40% of world palm oil output (Feintrenie et al., 2010).

However, the expansion of palm oil cultivation has also generated serious environmental concerns. It results in tropical deforestation and thus has a major impact on biodiversity loss, with the decline of emblematic species such as orangutan in Southeast Asia. It contributes to climate change through deforestation, but also through the conversion of peatlands, which are of vital importance in soil carbon sequestration. The huge forest and bush fires in recent years in Indonesia which are associated with clearing lands for agricultural or forestry plantations caused severe air pollution and public health problems across the sub-region. In addition, industrial plantations are sometimes responsible for polluting waterways, into which chemical inputs and processing plant waste are dumped.

Moreover, this expansion has sometimes resulted in social abuses and human rights violations, in the form of land grabbing by plantation companies at the expense of local and indigenous communities or of the exploitation of plantation workers.

Condemnation of these abuses by NGOs and growing consumer awareness of the adverse impacts of the expansion of palm oil plantation have driven the development of sustainability standards. Such standards are aimed at transforming production practices in order to mitigate their adverse environmental and social effects.

In 2001, representatives of the food processing and distribution sector launched a dialogue with WWF and plantation companies, leading to the creation in 2004 of the first voluntary sustainability standard in the sector, the Roundtable on Sustainable Palm Oil (RSPO).

There are now 2.41 million hectares of RSPO-certified plantations, while sustainable palm oil accounted for 20% of world trade in this product (RSPO, 2017). Meanwhile, several other initiatives proposing a vision of palm oil sustainability have emerged, positioning themselves as either a complement or an alternative to RSPO.

The development of these initiatives demonstrates the growing awareness among producers, the industry and the public authorities of the need to transform the sector to enable it to contribute to the Sustainable Development Goals (United Nations, 2016).

But this proliferation of sustainability standards itself poses new challenges, even though the environmental and social problems that motivated their emergence remain unresolved (Rival et al., 2016).

This chapter takes stock of sustainable palm oil certification initiatives. The section following this introduction presents a brief review of the literature on palm oil certification and introduces the main topics addressed.

In Section 3, we present the key issues and challenges facing sustainable palm oil certification today.

In Section 4, we explore several avenues for overcoming the current limitations of sustainability initiatives, before concluding.

2 Key themes in sustainability certification research

Condemnation by NGOs of the adverse impacts of the expansion of oil palm plantation and the development of sustainability standards to mitigate these impacts have resulted in a revival of public and academic interest in this crop.

A recent bibliometric study (Hansen et al., 2015) has shown a substantial increase in publications on palm oil sustainability in recent years, from 11 titles in 2004 to 713 in 2013. However, according to the authors the vast majority of the research identified is dominated by studies addressing a specific technical field of palm oil production or processing (crop management techniques or industrial waste management, for example) at the expense of social, economic and environmental issues.

Despite being under-represented in the body of scientific literature on palm oil sustainability, research on certification has quickly developed. Two major themes dominate this literature: first, issues relating to the role of VSSs in the governance and regulation of the sector; and second, the social and environmental impacts of VSSs, addressing problems associated with both the inclusion/exclusion of smallholders and their environmental performance.

Since VSSs in general – and those concerning oil palm cultivation in particular – have only recently been developed and implemented, there is more literature available on their institutional characteristics or their organisational functioning than on ex-post evaluations of their performance.

The rapid development of VSSs has caught the attention of a number of analysts in the fields of political science, international relations and political economy, who see such instruments as real policy innovations (Cashore, 2002). Indeed, despite being traditionally the realm of the public authorities or professional bodies, the definition of 'good agricultural practice' is increasingly reliant on partnerships between companies and NGOs, especially under the framework of multi-stakeholder consultations (Gereffi et al., 2001). VSSs are considered in this context as private institutions with political authority, in the sense that they produce rules and standards accepted as legitimate, not only by the actors that draft or adopt them, but also by states.

This political dimension of VSSs raises four areas of analysis, which can be found in the literature on RSPO.

A first strand of research focuses on the organisational structure of VSSs and, more specifically, on the way in which they develop legitimacy to regulate economic sectors (Schouten and Glasbergen, 2011; Schouten et al., 2012; Geibler, 2013). A second body of research tests the collaborative functioning of MSIs observing the power relations and political rationalities that shape NGO–industry partnerships (Djama et al., 2011; Pesqueira and Glasbergen, 2013). A third area of research expands on this review of the multi-stakeholder dimension, highlighting the various processes involved in the exclusion of a number of stakeholders (Cheyns, 2011, 2014; Pichler, 2013). Finally, some studies have been published which address the relation between private voluntary standards and public regulations (Djama, 2011). This is an emerging issue, with the implementation of new sustainable standards in Indonesia (ISPO) and Malaysia (MSPO), led by some public authorities (Hospes, 2014; Cramb and McCarthy, 2016).

Ex-post evaluations of the environmental performance of VSSs are still under-represented in the scientific literature, due to a lack of hindsight and to the difficulty of generating robust analysis instruments (Milder et al., 2014). Some research is nevertheless based on an analysis of the principles and criteria adopted (Edwards et al., 2012) or on the procedures involved in their implementation (Ruysschaert and Salles, 2014). These studies evidenced that RSPO-type mechanisms have a low environmental impact.

Research on the social impacts of certification is beginning to grow, albeit very unevenly depending on the issue addressed. Some study areas are practically non-existent, such as the analysis of the role played by VSSs in improving working conditions in plantations. Indeed, few scientific studies (with the notable exception of Pye et al. (2012) and Barral (2015) have examined the contemporary situation of migrant workers in plantations. Multi-site assessment of VSSs' structural impact on the improvement of working conditions in oil palm plantations is yet to be conducted. Nevertheless, it is worth noticing that major palm oil-producing countries have a poor record of adequate monitoring and control of working conditions. As an example, the rate of supervision by labour inspectors in Indonesia is one of the lowest worldwide, with one inspector for more than 11 000 companies (ILO, 2017). In this context, it is not appropriate to rely only on VSSs for some sizeable change.

The costs and benefits of certification for smallholders is an emerging issue. In a comprehensive review of VSSs' impact evaluations in several agricultural value chains, Dankers and Loconto (2014) discussed case studies indicating that certification helps to improve household socio-economic conditions, with an impact on yields and quality. The authors also pointed out that such impacts do vary according to the contexts and commodities, making any general conclusions difficult. Where palm oil smallholder certification is concerned, some studies have found that certified producers benefit beyond the mark-up through efficiencies gained by following best practices (WWF, 2012). However the scant research available was found to focus on the complex nature of the certification process for small, poorly organised producers, as well as on the financial, managerial and agricultural obstacles they must overcome (Brandi et al., 2015).

3 Challenges and limitations of sustainability initiatives in palm oil cultivation

Literature surveys highlight various challenges and limitations encountered today in the certification of palm oil cultivation. These challenges relate to the institutional and practical performance of sustainability initiatives.

At the institutional level, the proliferation of sustainability initiatives since the creation of RSPO reflects a real fragmentation of the regulatory framework (Rival et al., 2016). This proliferation also raises the question of the articulation of these voluntary standards with the public regulations and national sustainability standards that producer countries have adopted.

Finally, measures to ensure the sustainability of palm oil cultivation need to bolster their credibility by guaranteeing better inclusion of the millions of smallholders, and by contributing in an effective, measurable way to mitigating the adverse social and environmental impacts of growth in palm oil cultivation. In this field, the role of collaborative and multidisciplinary research in providing strong evidence-based impact evaluation of standards is crucial.

3.1 A fragmented regulatory framework

At present, the sustainability of palm oil cultivation is covered by three private sector VSSs, supplemented by two national sustainability standards that have a regulatory dimension and several initiatives led by some of the main palm oil consumer goods manufacturers (Table 1).

These mechanisms differ not so much in their organisational structure, but rather in the approach to sustainability they propose and the interest groups which are standing behind them.

RSPO is still by far the most important standard, in terms of both surface areas and quantities of certified oil. Created in 2004, it has regularly fine-tuned its principles, criteria and indicators. These gradual improvements to the standard's principles and criteria occur as part of a five-yearly revision programme, but also under the pressure of criticism regularly levelled at the process by NGOs. In 2014, RSPO developed a complementary standard, RSPO-NEXT, aimed at producers wishing to voluntarily align themselves with the more stringent commitments (both environmental and social) adopted by some agri-business companies. Due to its seniority and its dominant position in the sustainable palm oil market, RSPO remains the reference for the drafting of other initiatives, whether complementary or of different nature (mandatory vs voluntary).

Two other private certification systems have since been established, without creating any real competition, since they focus more on niche or specialist markets.

In 2008, in the context of its Sustainable Agriculture Network, the Rainforest Alliance developed a sustainable palm oil standard. This initiative is aimed at producers wishing to benefit from the 'Rainforest Alliance' label, and positions itself more as a complement to RSPO than as competitor.

With the rapid development of biofuels, a sustainable palm oil certification programme aimed at biodiesel production was created in 2011, stimulated by both manufacturers and the public authorities in Germany: International Sustainability and Carbon Certification (ISCC). In view of the market targeted, ISCC competes more with other biofuel certification programmes, such as the Roundtable on Sustainable Biomaterials (RSB – formerly named Roundtable on Sustainable Biofuel), than with RSPO (Ponte, 2014), even if the latter has been included for several years in the list of standards recognised by the European Commission under the framework of its Renewable Energy Directive (RED).

Table 1 The main sustainability initiatives in the palm oil industry

Initiative	Year	Type
Roundtable on Sustainable Palm Oil (RSPO)	2004	Private standard
Rainforest Alliance (SAN - Sustainable Agriculture Network)	2008	Private standard
International Sustainability and Carbon Certification (ISCC)	2010	Private standard
Indonesian Sustainable Palm Oil (ISPO)	2011	National government standard
Palm Oil Innovative Group (POIG)	2013	Private sector commitment
Malaysian Sustainable Palm Oil (MSPO)	2013	National government standard
Sustainable palm oil manifesto-high carbon stock study	2014	Private sector commitment
RSPO-NEXT	2014	Private standard

The sustainability programmes created as a challenge to RSPO include two types of initiatives: first, the sustainable standards adopted by the public authorities in the main producing countries (Indonesia and Malaysia), which consider RSPO as being too stringent for the majority of their producers; and second, initiatives designed by downstream companies in association with NGOs, which on the contrary consider RSPO to be too permissive on certain social and environmental aspects.

The Indonesian Sustainable Palm Oil (ISPO) national standard was launched in 2009 and consists mainly in the harmonisation of a number of national laws regulating the palm oil industry. Its Malaysian equivalent (MSPO) was launched in 2013. Unlike VSSs, these two national mechanisms are binding and are aimed at ensuring compliance along the whole palm oil value chain in these countries. So far, these national standards have fallen short of RSPO requirements on key environmental aspects: for example, by refusing to adopt methodologies enabling the identification and protection of High Conservation Value (HCV) areas or High Carbon Stock (HCS) forests, as well as criteria forbidding planting on peatland. Nor does the Indonesian national standard recognise the international consultation mechanism for local and indigenous communities' land rights, based on the principle of their free, prior and informed consent (FPIC).

In contrast, it is because criticism of RSPO's environmental and social limitations by several NGOs that several European companies using palm oil have launched initiatives aimed at sourcing oil from producers guaranteeing production methods based on 'no deforestation, no peat, no exploitation' principles by 2020.

These industrial initiatives aimed at 'tightening' the sustainability criteria for palm oil cultivation are supported and adopted by a number of European countries. The 'Amsterdam Declaration', for example, signed in 2015 by Denmark, France, Germany, the Netherlands and the United Kingdom translates the same commitments on tighter sustainability standards by 2020.

To some extent, the proliferation of these private and public initiatives reflects acknowledgement of the concept of sustainable palm oil although it also raises the question of the alignment of these processes.

The fragmentation of the industry's regulatory system could foster market segmentation according to the level of sustainability required, with the concomitant development of uncertified palm oil demand on emerging country markets (Schleifer, 2016) and demand for sustainable oil on the European market. It also contributes to the growing cost of certification for producers, who are obliged to diversify their sustainable standards according to the markets targeted with supplementary costs.

3.2 Finding synergies between private initiatives and public regulation

VSSs have often been hailed as a sign of the emergence of the political role of non-state actors in a transnational framework. However, this interpretation is somewhat misleading, since in practice, these mechanisms have almost always been encouraged or even funded by states (Bartley, 2007), which usually represent the interests of the consumer countries of the North.

The implementation of the EC RED, which imposes sustainability criteria for feedstocks intended for the production of its biofuels, is a good example of the proactive, incentive role played by public authorities. This directive defines those VSSs that are eligible on the EU biofuels market, since they comply with its sustainability criteria.

Where sustainable palm oil is concerned, the commitments made by EU member states in the context of the 'Amsterdam Declaration' are another illustration of this. Outside the European Union, Norway introduced in 2016 a clause into its public procurement policy on mandatory sourcing of 'no deforestation' certified palm oil.

The producing countries, on the other hand, have responded to the rapid development of VSSs by creating national sustainability standards, which are often less stringent but have the advantage of establishing sustainability along the supply chain as a legal matter. Indonesia and Malaysia have also established the Council of Palm Oil Producer Countries (CPOPC) in November 2015. This intergovernmental body aims to control the global crude palm oil (CPO) supply, stabilise palm oil prices, improve the welfare of smallholders and promote sustainable practices in the palm oil industry.

At present, the position of private and government sustainability standards is more one of competition than of cooperation. They thus illustrate the power relations that structure the palm oil global value chain, according to a North/South rationale. They would nevertheless benefit from building synergies.

Palm oil sustainability will not become more effective if it is limited to only producers exporting to European markets. Indeed, VSSs have no impact on the vast majority of producers, who are not exposed to sustainability sensitive markets.

Moreover, the issues underlying the uncontrolled expansion of palm oil are the result of localised political, social and economic processes that are beyond the scope of the transnational scale of VSS action, and sometimes even of the central government (McCarthy et al., 2012).

Dealing with the source of the adverse impacts of oil palm cultivation, therefore, implies active cooperation between VSSs and the local and national authorities.

3.3 Including smallholders in sustainability initiatives

The RSPO standard defines smallholders as producers farming a surface area lower than 50 ha. Globally, oil palm smallholders account for 40% of all land under oil palm cultivation and 30% of world palm oil production.

This smallholder category in fact covers a very wide range of different situations. In the main producing countries in Southeast Asia (Indonesia and Malaysia), there are two types of smallholders: schemed or managed smallholders and independent smallholders. The former are generally bound by contract to one or several oil mills belonging to the same corporate organisation. They benefit from the provision of planting material and inputs, technical support and access to credit. In return, they must comply with the organisation's specifications, especially on cultivation methods and crop delivery to the mill. Independent smallholders are more autonomous in terms of how they manage their farms and market their production. They benefit from technical support from government extension services in countries where this is available.

In early 2016, less than 13% of sustainable palm oil certified by RSPO was produced by smallholders (SPOTT, 2017).

Smallholder certification is becoming a key issue for VSS and food companies, since they contribute to the legitimation of standards and to the reputation of companies wishing to consolidate their Corporate Social Responsibility commitments. Its dissemination comes up against various economic and organisational constraints.

Within RSPO, traders and consumer goods manufacturers managed to adjust the premiums for sustainable oil to the minimum level. Consequently, at present, the price differentiation between sustainable oil and unsustainable oil is not much of an incentive.

The assumption guiding smallholder certification is that the adoption of good agricultural practices will improve producers' income by enabling them to use fewer inputs while increasing yields, thereby reducing the yield gap with the large plantations (Molenaar et al., 2013). Improving the productivity of small farms would also have a positive environmental impact, since it would help to limit cropland expansion by producing more on areas already cultivated through ecological intensification (Rival and Levang, 2014). However, although some studies have documented that good agricultural management practices are shown to increase yields and are predicted to increase income (SENSOR, 2016), the hypothesis of a positive linkage between an increase in yields and a reduction in cropland expansion is nevertheless very controversial, with some studies showing on the contrary that the profitability of cash crops encourages their expansion to the detriment of adjacent forests (Angelsen and Kaimovitz, 2001).

These weak economic incentives are combined with the heavy organisational and managerial constraints imposed by certification. In order to limit the costs of certification and to facilitate internal monitoring of producers, RSPO is promoting group certification for smallholders (Brandi et al., 2015). Groups can be formed based on existing collective dynamics (cooperatives, for example), but in most cases they are formed under the leadership of a third-party organisation or 'group manager' – an oil mill in the case of schemed smallholders or an NGO in the case of independent smallholders. The assistance provided to smallholders by the group manager is often crucial in terms of technical support and financial and managerial assistance although it does not help to empower producers. Several experiments have shown that certification dynamics do not withstand the departure of the group manager and thus remain highly dependent on external support (Dankers and Loconto, 2014; Oosterveer et al., 2014).

3.4 Effectiveness of sustainable initiatives

VSSs devote substantial resources and energy to the implementation of stakeholder participation, consultation and deliberation procedures. In the name of transparency, thousands of documents are published on their websites to report on new initiatives, the outcomes of consultations, research by working groups or even complaints filed against offending companies.

The impact of sustainability standards is evaluated according to the number of members, the certified surface areas and the amount of sustainable oil marketed.

Their performance is also assessed in a more qualitative manner, through the scope of the sustainability sectors covered and the stringency of their requirements.

In benchmarking exercises, the Indonesian and Malaysian government standards are regularly downgraded in relation to private mechanisms. Their mode of governance is less participatory and their fields of action are not sufficiently rigorous in their coverage of key sustainability sectors. As indicated above, contrary to private voluntary standards, they have not adopted the biodiversity management instruments recommended by the conservation community, or some international instruments that safeguard local communities' rights (with ISPO at least, for the latter point).

Benchmarking also operates between private sector initiatives, often on a 'highest bidder wins' basis in terms of sustainability. The ambiguity surrounding RSPO's definition of forest areas resulted in the development of the 'High Carbon Stock Forest' concept, under the impetus of Greenpeace, and new methodologies to identify and map forest areas have been recently designed. These new instruments underpin the 'no deforestation' policies adopted by many companies.

Paradoxically, the plethora of procedures, documents and instruments to manage sustainability provides little information about the impact of the initiatives actually implemented.

Have sustainable standards fundamentally transformed practices within the palm oil industry? Do they make a significant contribution to mitigating environmental and social impacts?

The sustainable standards approach consists less in confronting this type of evaluation than in monitoring producers' compliance with the principles and criteria defined, through auditing of plantations and factories wishing to obtain certification. However, the limitations of auditing when it comes to identifying structural changes and effective social and environmental impacts go far beyond the recurrent accusations of connivance or collusion between the inspectors and the inspected (EIA, 2015). Indeed, auditing reduces the complexity of environmental and social processes. This complexity is translated in terms of the presence/absence of indicators and compliance with established thresholds. For example, working conditions in plantations are determined based on observations made during inspections, which note the presence or absence of protective gear, of equipped medical centres, of accommodation, etc.; in short, of indicators that only very partially reflect the reality of daily working conditions. The social and environmental processes are further reduced by auditing because of the very limited samples and time spent on inspections, due to cost constraints.

After almost 15 years of existence of sustainability standards, an evaluation of the social and environmental impacts, going beyond the measurement of certified cropland and volumes, or of the costs/benefits of certification for producers, is now more necessary than ever to boost their credibility.

4 The way forward for sustainability certification

Despite their potential, in their current form, sustainable palm oil initiatives appear incapable of significantly mitigating the adverse impacts of cropland expansion under the effect of growing demand.

A virtuous dynamic would entail: first, an upward harmonisation of the sustainable initiatives currently mushrooming; second, efforts to ensure better synergies between VSSs and the sustainability standards adopted by producing countries, but also to better articulate the transnational, national and local levels of action; third, support for smallholders' certification by providing economic incentives and technical support focused more on improving their livelihood than on buyers' interests; and fourth, the production of monitoring and evaluation indicators that track the social and environmental performances of sustainability policies.

4.1 Strengthening the role of international cooperation in the upward harmonisation of sustainability initiatives

Two different approaches are generally adopted in order to regulate the proliferation of standards covering a given sector.

The first consists in developing meta-standards governing drafting procedures for standards to guarantee rigour and credibility without undermining their diversity. For sustainability standards, this is the part played by the ISEAL Alliance, an organisation responsible for establishing good practice in standardisation (Loconto and Fouilleux, 2014). ISEAL Alliance contributes to the legitimation of VSSs by evaluating their modes of operation (sustainability criteria, participation, transparency, neutrality, continuous improvement, etc.).

The second approach is more directive and consists in fostering a type of standard according to its compliance with the requirements of a given market. This approach is the one adopted by the European Commission through its RED, or the one recently advocated by the European Parliament during a deliberation on palm oil in April 2017.

Both approaches nevertheless carry the risk of accelerating market segmentation and thereby encouraging palm oil exports to the emerging markets, which do not have such stringent sustainability requirements.

To avoid these adverse effects, consumer countries in the North – especially the European Union – need to provide financial support, in the context of technical cooperation, for national certification programmes in producer countries, with the goal of progressively aligning them with the best practices. These approaches could take inspiration from experience acquired by the EU in terms of cooperation on tropical timber legislation through the FLEGT programme (Forest Law Enforcement, Governance and Trade).

4.2 Articulating instruments and levels of action in sustainability policy

As indicated above, VSSs have an impact only on the producers and companies which are most exposed to international markets with stringent sustainability requirements.

However, it is also important to note that oil palm expansion dynamics are often part of localised political configurations, over which the central governments have very little control. In Malaysia, for example, the autonomy and powers constitutionally conferred by the federal government to the states of Sabah and Sarawak (Eastern Malaysia) after independence, as well as the political compromises reached between the local elites and the federal government, form the political backdrop to the conversion of forests to plantations in these frontier zones (Jomo et al., 2004). In Indonesia, the decentralisation process launched in the 1980s gave the regional and local authorities (in the provinces and districts) greater negotiation opportunities with plantation companies, which circumvent the policies of the central government's regulation agencies (McCarthy et al., 2012).

Influencing these socio-political configurations implies associating the local authorities and different stakeholders in a given territory with sustainable development strategies. New initiatives founded on this principle of giving sustainable standards a territorial basis have recently emerged, around the concepts of 'landscape approaches' or 'jurisdictional approaches' (Reed et al., 2016). These are aimed at encouraging local actors to discuss the multiple, often competing, uses of resources in an administratively defined territory, and to find compromises that reconcile conservation challenges with agricultural intensification and poverty eradication. An increasing number of agri-businesses are attempting to articulate their 'no deforestation, no peat, no exploitation' procurement policies with these territorial dynamics (Wolosin, 2016).

4.3 Rethinking smallholder inclusion in certification

The line of thinking that links increases in smallholder income with higher yields and better farm productivity is very hypothetical in its capacity to limit cropland expansion and to guarantee economic security for producers. This approach, which consists in replicating at smallholder level the intensive, monocropping model of the large plantations, threatens the diversification strategies and the multifunctional nature of agricultural activities, which are fundamental to food security for farmers.

To address the difficulties smallholders have in financing initial certification costs, RSPO has implemented an incentive fund for producers committing to the process. But this *ad hoc* support cannot offset the heavy organisational constraints and the low economic benefits which currently limit smallholder commitment in the long-term.

Two interconnected courses of action should be considered for better smallholder integration in sustainability mechanisms.

The first entails rethinking technical support for smallholders according to principles that reconcile sustainable practices and food security. In practical terms, this would involve integrating oil palm cultivation into agricultural processes based on agroecology and fostering crop diversification.

The second consists in proposing incentive payment models for independent certified smallholders. An 'equitable' business model could, for example, be built on buyers' procurement initiatives, within the framework of the territorialised 'no deforestation' approaches.

4.4 Evaluating the social and environmental performances of sustainability initiatives

The sustainability of palm oil cultivation needs to be evaluated not only according to whether practices comply with a standard, but also to the capacity of these practices, to effectively mitigate the industry's environmental and social problems.

However, this process is not without challenges. First, the causal links between sustainability standards and their impacts are difficult to establish and cannot necessarily be reproduced in all production contexts due to the wide range of different situations (Miteva et al., 2012). Next, the certification programmes do not include a theory of change in their approach. As indicated above, producers are evaluated according to their compliance with a set of principles and criteria. Impact assessments would imply identifying an initial state of reference and monitoring the progress of states over time.

This impact evaluation process requires the mobilisation of the scientific community in order to produce theories, methods and evaluation data. It also calls for active cooperation between standard setters and stakeholders (especially smallholders) to include elements of a theory of change in their approach to sustainability.

5 Future trends and conclusion

Sustainability certification has established itself as a key instrument in the regulation of the palm oil industry.

The main effects of the emergence of VSSs in response to campaigns accusing the industry have been the organisation of a dialogue between stakeholders and the search for solutions to environmental and social problems. Initially driven by the private sector alone, sustainability standards have now also been adopted by governments in producing countries.

Innovative synergies are also beginning to emerge at the territorial level, linking public authorities with multinational companies, producers and civil society. These initiatives look promising, in that they operate as close as possible to the fronts where the unregulated expansion of oil palm plantations is taking place.

However, these sustainability initiatives cannot be consolidated unless producers see the adoption of 'good practices' in palm oil cultivation as being far more profitable than maintaining the status quo. The introduction of a new business model fostering a more equitable inclusion of independent smallholders in value chains is the condition for their long-term commitment to sustainable agriculture. One of the arguments developed in this chapter is that this sustainability process is less a matter of focusing on strategies to improve yields and productivity than of supporting crop diversification and social changes in plantation systems which have poorly evolved since colonial times, when rural labour was docile and abundant.

Research now needs to assist palm oil sustainability by developing agroecological models, agricultural practices and modes of organisation capable of accompanying extension services and growers in their return to a multifunctional approach to agricultural activities. This process needs to reconcile the role of palm oil in the production of food and energy products with food security imperatives as well as landscape and environmental protection.

Finally, encouragement is needed for research aimed at developing monitoring and evaluation tools for the environmental and social performances of sustainability standards.

6 Where to look for further information

Some of the current works on standards and certification in the oil palm value chain can be found in the references section.

However, two recent books are suggested as the best introduction to the main issues and challenges oil palm is currently facing:

Cramb, R. and McCarthy, J. (2016), *The Oil Palm Complex. Smallholders, Agribusiness and the State in Indonesia and Malaysia*, NUS Press, Singapore.

Rival A. and Levang P. (2014), *Palms of Controversies: Oil palm and development challenges,* Centre for International Forestry Research, Bogor.

Below are some well documented and useful websites on sustainability standards and on palm oil sustainability initiatives.

6.1 Some resources on voluntary sustainability standards

ISEAL Alliance is the global association of sustainability standards, in charge of the development of a Code of Good Practices for standard setting organizations. www.isealalliance.org

The State of Sustainability Initiatives is a project led by the International Institute on Sustainable Development (IISD), monitoring the growth, impacts and opportunities associated with voluntary sustainability initiatives. https://www.iisd.org/ssi/

6.2 Some resources on palm oil certification schemes

Sustainability Policy Transparency Toolkit (SPOTT) is an online platform developed by the Zoological Society of London (ZSL) to promote and monitor the implementation of environmental and social best practices for the sustainable production and trade of global commodities. SPOTT delivers an up to date state of the art of sustainability standards and initiatives in the palm oil industry, and also provide links to their official website. https://www.spott.org/palm-oil-resource-archive/standards/.

Forest Peoples Program (FPP) is a non-governmental organisation registered in The Netherlands and the UK, supporting 'the rights of peoples who live in forests and depend on them for their livelihoods'. FPP is active in documenting abuses on forest communities associated to oil palm expansion, especially regarding land grabbing issues. The organisation played a pivotal role in RSPO adoption of principles and criteria consistent with international human rights laws and with indigenous peoples' rights. FPP carries out field studies to monitor standards implementation by oil palm companies and delivers studies on standards and certification. http://www.forestpeoples.org.

WWF is one of the founding member of the Roundtable Sustainable Palm Oil (RSPO), the first sustainability standard and certification scheme in palm oil industry. Among its various initiatives to support best environmental and social practices in the industry, WWF has developed Palm Oil scorecards to monitor buyers' commitments to source RSPO certified sustainable palm oil. http://palmoilscorecard.panda.org/.

7 References

Angelsen, A. and Kaimovitz, D. (2001), *Agricultural Technologies and Tropical Deforestation*, CABI Publishing, Wallingford, UK.

Barral, S. (2015), *Capitalismes agraires. Economie politique de la grande plantation en Asie du Sud-Est*, Presses de Sciences Po., Paris, France, p. 240.

Bartley, T. (2007), 'Institutional emergence in an era of globalization: The rise of transnational private regulation of labor and environmental conditions', *American Journal of Sociology*, 113(2), 297–351.

Brandi, C., Cabani, T., Hosang, C., Schirmbeck, S., Westermann, L. and Wiese, H. (2015), 'Sustainability standards for palm oil: Challenges for smallholder certification under the RSPO', *Journal of Environment and Development*, 24(3), 292–394.

Cashore, B. (2002), 'Legitimacy and the privatization of environmental governance: How non-state market-driven (NSMD) governance systems gain rule-making authority', *Governance: An International Journal of Policy, Administration and Institutions*, 15(4), 503–29.

Cheyns, E. (2011), 'Multi-stakeholder initiatives for sustainable agriculture: Limits of the Inclusiveness paradigm', In: Gibbon, P., Ponte, S. and Vestegaard, J. (Eds), *Governing through Standards*, Palgrave, London, pp. 210–35.

Cheyns, E. (2014), 'Making minority voices heard in transnational roundtables: The role of local NGOs in reintroducing justice and attachments', *Agriculture and Human Values*, 31(3), 439–53.

Cramb, R. and McCarthy, J. (2016), *The Oil Palm Complex: Smallholders, Agribusiness and the State in Indonesia and Malaysia*, NUS Press, Singapore.

Dankers, C. and Loconto, A. (2014), *Impact of International Voluntary Standards on Smallholder Market Participation in Developing Countries: A Review of Literature*, FAO, Rome, Italy.

Djama, M. (2011), 'Regulation the globalized economy: Articulating private voluntary standards and public regulations', Perspectives 11, CIRAD, Montpellier – France.

Djama, M., Fouilleux, E. and Vagneron, I. (2011), 'Standard-setting, certifying and benchmarking: A governmentality approach to sustainability in the Agro-Food sector', In: Gibbon, P., Ponte, S. and Vestegaard, J. (Eds), *Governing through Standards*, Palgrave, London, pp. 184–209.

Edwards, D. P. and Laurance, S. G. (2012), 'Green Labelling, sustainability and the expansion of tropical agriculture: Critical issues for certification schemes', *Biological Conservation*, 151(2012), 60–4.

EIA (2015), *Who Watches the Watchmen? Auditors and the Breakdown of Oversight in the RSPO*, Environmental Investigation Agency, London, UK.

Feintrenie, L., Chong, W. K. and Levang, P. (2010), 'Why do farmers prefer oil palm? Lessons learnt from Bungo district, Indonesia', *Small-Scale Forestry*, 9(3), 379–96.

Geibler, J. (2013), 'Market-based governance for sustainability in value chains: Conditions for successful standard setting in the palm oil sector', *Journal of Cleaner Production*, 56(2013), 39–53.

Gereffi, G., Garcia-Johnson, R. and Sasser, E. (2001), 'The NGO-Industrial complex', *Foreign Policy*, July-August 2001.

Hansen, S. B., Padfield, R., Syayuti, K., Evers, S., Zakariah, Z. and Mastura, S. (2015), 'Trends in global palm oil sustainability research', *Journal of Cleaner Production*, 100(2015), 140–9.

Hospes, O. (2014), 'Marking the success or end of global multi-stakeholder governance? The rise of national sustainability standards in Indonesia and Brazil for palm oil and soy', *Agriculture and Human Values*, 31(3), 425–37.

ILO – The International labor organisation (2017), *Factsheet on Labour Inspection in Indonesia.* http://www.ilo.org/wcmsp5/groups/public/---asia/---ro-bangkok/---ilo-jakarta/documents/publication/wcms_549704.pdf.

Jomo, K. S., Chang, Y. T. and Khoo, K. J. (2004), *Deforesting Malaysia. The Political Economy and Social Ecology of Agricultural Expansion and Commercial Logging*, Zed Books, London, UK.

Lemeilleur, S. and Balineau, G. (2016), 'Tackling the climate change challenge: What roles for certification and ecolabels?', In: Torquebiau, E. (Ed.), *Climate Change and Agriculture Worldwide*, Springer, Dordrecht, pp. 289–99.

Loconto, A. and Fouilleux, E. (2014), 'Politics of private regulation: ISEAL and the shaping of transnational sustainability governance', *Regulation and Governance*, 8(2), 166–85.

Marx, A. and Wouters, J. (2015), 'Competition and cooperation in the market of voluntary sustainability standards', In: Delimatis, P. (Ed.), *The Law, Economics and Politics of International Standardization*, Cambridge University Press, Cambridge, UK, pp. 215–41.

McCarthy, J. F., Gillespie, P. and Zen, Z. (2012), 'Swimming upstream: Local Indonesian production networks in Globalized palm oil production', *World Development*, 40(3), 555–69.

Milder, J., Arbuthnot, M., Blackman, A, Brooks, S., Giovannucci, D., Gross, L., Kennedy, E., Komives, K., Lambin, E., Lee, A., Meyer, D., Newton, P., Phalan, B., Schroth, G., Semroc, B., Rikxoort, H. and Zrust, M., (2014), 'An agenda for assessing and improving conservation impacts of sustainability standards in tropical agriculture', *Conservation Biology*, 29(2), 309–20.

Miteva, D. A., Pattanayak, S. K. and Ferraro, P. J. (2012), 'Evaluation of biodiversity policy instruments: What works and what doesn't?', *Oxford Review of Economic Policy* 28(1), 69–92.

Molenaar, J. W., Persch-Orth, M., Lord, S., Taylor, C. and Harms, J. (2013), *Diagnostic Study on Indonesian Oil Palm Smallholders. Developing a Better Understanding of Their Performance and Potential*, IFC (International Finance Corporation – World Bank Group), Jakarta, Indonesia.

Oosterveer, P., Adjei, B. E., Vellema, S. and Slingerland, M. (2014), 'Global sustainability standards and food security: Exploring unintended effects of voluntary certification in palm oil', *Global Food Security*, 3(3), 220–6.

Pesqueira, L. and Glasbergen, P. 2013, 'Playing the politics of scale: Oxfam's intervention in the Roundtable on Sustainable Palm Oil', *Geoforum*, 45, 296–304.

Pichler, M. (2013), 'People, Planet & Profit: Consumer-oriented hegemony and power relations in palm oil and agrofuel certification', *The Journal of Environment & Development*, 22, 370–90.

Ponte, S. (2014), 'Roundtabling sustainability: Lessons from the biofuel industry', *Geoforum*, 54, 261–71.

Pye, O., Daud, R., Harmono, Y. and Tatat (2012), 'Precarious lives: Transnational biographies of migrant oil palm workers', *Asia Pacific Viewpoint*, 53, 330–42.

Reed, J., Van Vianen, J., Deakin, E. L., Barlow, J. and Sunderland, T., (2016), 'Integrated landscape approaches to managing social and environmental issues in the tropics: Learning from the past to guide the future', *Global Change Biology*, 22, 2540–54

Rival, A. and Levang, P. (2014), *Palm of Controversies: Oil Palm and Development Challenges*, CFOR, Bogor, Indonesia.

Rival, A., Montet, D. and Pioch, D. (2016), 'Certification, labelling and traceability of palm oil: Can we build confidence upon trustworthy standards?', *Oilseeds and Fats, Crops and Lipids (OCL)*, 23(6), D.609, p. 11.

RSPO (2017), Impacts of RSPO, Accessed from: https://www.rspo.org/about/impacts.

Ruysschaert, D. and Salles, D. (2014), 'Towards global voluntary standards: Questioning the effectiveness in attaining conservation goals. The case of the Roundtable on Sustainable Palm Oil (RSPO)', *Ecological Economics*, 107, 438–46.

Schleifer, P. (2016), 'Private governance undermined: India and the Roundtable on Sustainable Palm Oil', *Global Environmental Politics*, 16(1), 38–58.

Schouten, G. and Glasbergen, P. (2011), 'Creating legitimacy in global private governance: The case of The Roundtable on Sustainable Palm Oil', *Ecological Economics*, 70, pp. 1891–9.

Schouten, G., Leroy, P. and Glasbergen, P. (2012), 'On the deliberative capacity of private multi-stakeholder governance: The Roundtables on Responsible Soy and Sustainable Palm Oil', *Ecological Economics*, 83, 42–50.

SENSOR (2016), Costs and benefits of RSPO certification for independent smallholders, http://www.standardsimpacts.org/sites/default/files/Costs-and-benefits-of-RSPO-certification-for-independent-smallholders-FINAL(2).pdf.

SPOTT (2017), Smallholders, Accessed from: https://www.spott.org/palm-oil-resource-archive/smallholders/.

United Nations (2016), Sustainable Development Goals, Accessed from: www.un.org/sustainabledevelopment/sustainable-development-goals.

Wolosin, M., (2016), Jurisdictional approaches to zero deforestation commodities, WWF Discussion Paper, WWF, USA.

WWF (2012), Profitability and sustainability in palm oil production analysis of incremental financial costs and benefits of RSPO compliance, WWF Report, Accessed from: http://www.cdcgroup.com/Documents/ESG%20Publications/profitabilityandsustainabilityinpalmoilproduction.pdf.

Chapter 14

Balancing oil palm cultivation with forest and biodiversity conservation

Carl Traeholt, South East Asia Programme Director, Copenhagen Zoo, Malaysia

1 Introduction

The past decade has seen a significant increase in environmental degradation and loss of ecosystem services in most parts of the world (Bradshaw et al., 2009; Butler and Laurance, 2009; Craft et al., 2009; De Jong et al., 2015; Dobson et al., 2006; FAO, 2009; Ghazali et al., 2016; Greenpeace, 2007; Goodman and Mulik, 2015; Hoeinghaus et al., 2009; Kettunen and ten Brink, 2006; Koh and Wilcove, 2008; Malcolm et al., 2006; Petrenko et al., 2016; Sodhi et al., 2004; UNEP, 2007; UNESCAPS, 2007). In Southeast Asia, the main driver of deforestation is agriculture, specifically for palm oil production (Hansen et al., 2013; Miettinen et al., 2011; Stibig et al., 2014), although pulp and logging in legal commercial concessions in Indonesia is reportedly higher (Abood et al., 2015). The rapid agricultural development resulted in habitat loss and fragmentation and, consequently, in the loss of biodiversity (Canale et al., 2012; Gibson et al., 2013). Indeed, rivers, lakes and peat forests suffered enormous siltation, pollution loading and dehydration. The draining of extensive peat forests of Sumatra and Borneo lead to the occurrence of the most damaging event when more than 45 000 km^2 of forest burnt by wildfires in 1997 (Heil et al., 2001) and resulted in forest degradation and deforestation cost of US$1.62–2.7 billion (Tacconi, 2003). The associated cost of the smoke pollution was estimated at US$799 million (Tacconi, 2003), which caused an estimated 20 million people in Indonesia alone to suffer from respiratory illnesses (Brauer, 1997; Emmanuel and Lim, 1998; Heil and Goldhammer,

http://dx.doi.org/10.19103/AS.2017.0018.26

2001; WHO, 1998). The recent 2013–2015 wildfires have caused extreme carbon emission (Gaveau et al., 2014; Huijnen et al., 2016) and serious health hazards (Koplitz et al., 2016).

The Roundtable of Sustainable Palm Oil (RSPO) was formed in 2004 to put in place mechanisms aimed at promoting effective environmental-friendly practices within the sector. RSPO introduced its principles and criteria (RSPO P&C) in 2007 (RSPO, 2007) that presented a set of due diligence requirements for RSPO members. Whereas RSPO membership is voluntary, it is mandatory for each member to commit to the RSPO P&C; however, many seemed unaware of the real on-the-ground ramifications of committing to these. For the RSPO P&C to have any meaningful positive conservation impact, it needs to be (1) mainstreamed within each member company and (2) operationalized and integrated into a respective company's standard operational procedures (SOP). In practice, this means that the conservation agenda, as stipulated in the RSPO Vision and guided by the RSPO P&C, must be made familiar to all employees in a company, in order to ensure that environmental concerns are as important as traditional agricultural activities such as pruning, harvesting and fertilizing. Embracing the RSPO concept of producing 'sustainable' palm oil means that agricultural, environmental and social aspects are suddenly equally important (Fig. 1). Missing these three main pillars of sustainable production will lead to failing an RSPO certification audit. In many cases, growers misinterpret the 'sustainability' demands as affording an equal financial investment in each component, with the argument that social and environmental concerns do not constitute its core business. Indeed, many companies fail to understand that these components must become part of the core business, if they aspire to become environmental-friendly producers and pass RSPO's 'sustainable certification' audit. It is not akin to each pillar requiring an equal amount of investment and/or operational budget (Fig. 2). The required investment into each pillar depends entirely on the extent of each challenge in the relevant production area. In most cases, the planted production areas dwarf the conservation areas and, consequently, the agricultural activities

Figure 1 The three main pillars of 'sustainable palm oil production': agriculture, community and environment.

Figure 2 The three main pillars of 'sustainable palm oil production' do not necessarily require the same amount of financial input.

will need a higher operational budget than conservation activities. The outcome of the RSPO concept is that the palm oil sector plays an active role in conserving biodiversity as part of safeguarding important ecosystem processes and human welfare (Dobson et al., 2006; Duffy, 2009; European Communities, 2008; Ewel, 2009; Gamfelt et al., 2008).

In 2008, Malaysia-based United Plantations Bhd (UP) became the first RSPO-certified 'sustainable' palm oil producer. Many environmental NGOs perceived the certification as premature and untimely (Greenpeace, 2008a,b) and called for a temporary halt to certifying more. In addition, critical publications and reports focused on issues and recommendations that were often technical in nature (Bradshaw et al., 2009; Butler and Laurance, 2009; Koh et al., 2009; Laurence et al., 2010), rather than providing much needed useful operational support to the industry (Traeholt and Schriver, 2011). This, combined with a slow transformation progress within the industry has resulted in the delivery of positive conservation impact on the ground remain miniscule (Ruysschaert and Salles, 2014). The most serious evidence of the limited on-the-ground effect of RSPO certification was the recurrence of devastating forest fires in 2015, where World Bank reported more than 26 000 km^2 burnt, resulting in an estimate cost of more than US$16 billion to Indonesia alone. Petrenko et al. (2016) reported that 20% of all wildfires were caused by palm oil-related development, and Marlier et al. (2015) stated that Indonesian fire emissions arose primarily from within logging, paper pulp and oil palm concessions. While 88% of the total emissions in Sumatra in 2006 was caused by pulp and paper concessions, 67% of all emissions in Kalimantan in the same year was caused by palm oil concessions (Marlier et al. 2015). With this in mind, sceptics across the world questioned the real value of the RSPO 'sustainability' certificate, if such devastating events could happen again after the 1997 disaster. Recent studies, however, suggest that RSPO has had a positive effect in reducing wildfires in members' concessions in the period 2012–2015, although only when the fire likelihood is relatively low (Cattau et al., 2016).

Most consumers, however, are unaware of RSPO being a voluntary organization and that a major part of the industry still operates outside the RSPO regulation, with only 17% of total production being certified as 'sustainable' (RSPO, 2016). It is not clear to many either that RSPO has had a profound positive effect in terms of member companies undertaking proper due diligence, that is, the RSPO P&C has resulted in a positive preventative effect. In 2015, more than 144 000 ha of high conservation value habitat have been spared (RSPO, 2015). The real issue is that preventative measures only make up half of the challenge vis-à-vis compliance set aside is not enough. The other half constitutes actively managing conservation areas, including active population management of endangered species. While the RSPO sustainability certificate ensures that a certified company has taken the necessary precautionary steps to avoid indiscriminate development detrimental to local societies and environment, it does not make a company 'sustainable' *per se*. The key is for a company to be able to measure environmental impacts from its plantation operations and document that biodiversity and ecological processes remain intact in the post-development phase (Tallis et al., 2009; Tierny et al., 2009). This also includes the application of biological pest management to reduce dependency of toxic chemicals and prevent these from accumulating in the food chain. If key ecological parameters show downward trends, the operations evidently have negative environmental impacts and, hence, are not 'sustainable'. The question remains, is it possible to document the operational impact on the environment?

The present case study presents the results generated through six years of collaboration between Copenhagen Zoo (Zoo) and UP, building environmental and biodiversity capacity in UP, in the aim of measuring the ecological impacts of palm oil operation and make biodiversity concern an integrated part of daily operational procedures.

2 Research methods

This case study took place in United Plantations Bhd subsidiary PT Suria Sawit Sejati estates in Central Kalimantan, Indonesia. Of a total landbank of 15 560 ha, approx. 6000 ha have been set aside as conservation land (UP annual report, 2015). Part of that land bank is made of high conservation value (HCV) forest and riparian zones and tracts of wetland areas in need of rehabilitation.

The basic idea of producing environmentally 'sustainable' palm oil is meaningless without tools to measure any form of ecological footprints of plantation operations. Consequently, we considered the meaning of 'sustainable production' in an environmental context and defined it as follows:

> *Sustainable palm oil production is applying agricultural practices that do not significantly undermine biodiversity values and/or degrade ecological processes in the area of operation and elsewhere.*

3 Measuring operational impacts on biodiversity

To be able to assess the environmental impact of plantation operations, it is necessary to measure various ecological parameters *before* and *after* such plantation operations have begun. To measure biodiversity 'values' before commencing plantation operations, we undertook substantial biodiversity surveys throughout the estate landscape and subsequently monitor the same ecological parameters at various time intervals. We used a range of standard biodiversity surveying methods, such as:

Camera trapping: Camera traps were used to record, primarily, medium-large mammals and other elusive species. For consistency, 60 cameras were set for 30 consecutive days in various study sites; however, after the basic surveys cameras remained active 24 hr/day across a range of monitoring sites. Mammals were identified from Payne and Francis (2007), Francis (2001) and Phillipps and Phillipps (2016) and birds from Myers (2009) and Robson (2005).

Ground traps: Terrestrial small mammals were recorded by deploying 60 small mammal traps, baited with coconut and banana. Trapping cycles lasted for nine consecutive days to increase the probability of trapping saturation (Anan et al., 1998; Lundahl and Olsson, 2000; Wilson et al., 1996). Species were identified from Payne and Francis (2007), Francis (2001) and Phillipps and Phillipps (2016).

Mist netting and harp traps: Three mist nets and three harp traps were set for nine consecutive days on various types of habitats, for example, understory tall dipterocarp forests, peat forest, mangrove, forest edges and grassland. These were used primarily to evaluate the understory bird and bat communities. Captured mammal species were identified using Payne and Francis (2007), Francis (2001) and Phillipps and Phillipps (2016), and birds were identified from Myers (2009) and Robson (2005).

Fish nets and traps: Local made box traps made of bamboo were placed in rivers and streams, and active fishing using scoop nets were carried out in streams, pools and rivers. Fish were collected and identified while still alive, following Inger and Inger (2002) and Atack (2006), and voucher specimens were stored for further identification.

Amphibians and reptiles: Amphibians and reptiles were collected opportunistically during daytime. Amphibians were also collected after darkness using torchlight to identify eye shine and prepared following Heyer et al. (1994), Inger (1966), Inger and Stuebing (1997) and Das (2010).

Biodiversity monitoring: Spatial Monitoring and Reporting Tool (SMART) was setup as the primary platform that integrates all observations, patrolling results and effort with the main GIS database. This allowed the team to detect relative density changes in endangered and vulnerable species, make seasonal predictions and identify specific areas in need of attention.

Tree censuses: Tree censuses were conducted in all study sites. Species were identified from Engel and Phummai (2008) and LaFrankie Jr. (2010).

Rehabilitation processes: Habitat rehabilitation and restoration were designed to reinstall native plant species. In order to ensure habitat similarity and appropriate species composition, seeds were collected from adjacent similar forests and propagated in a purpose-made tree nursery. Out planting was designed to maximize the growth rate for faster natural recovery (Farikhah and Traeholt, 2014). Rehabilitation protocols followed the recommendations of Schumann and Joosten (2008) and Nuyim (2005). In the aim of measuring the effect on biodiversity of rehabilitation, mist netting for understory birds are undertaken regularly for five consecutive days.

4 Measuring operational impacts on abiotic factors

Land clearing for agriculture, infrastructure and urbanization causes extreme type of habitat change. Apart from a physical removal of living species with subsequent loss of biodiversity, many abiotic conditions are altered and the impact thereof is often detected in waterways. Soil erosion, loss of top soil, fertilizer and pesticide washout, changes in microclimate and organic waste efflux are some of the serious negative results of land clearing. To measure the degree and impact of erosion, fertilizer/pesticide washout and organic contamination of catchment areas and downstream river systems, we setup a water monitoring system with strategic sample points, (1) inflow to UP's estates, (2) a few downstream sampling points and (3) at the point where rivers exit the estates. All water samples are tested for the following parameters:

Biological oxygen demand (BOD) is the amount of dissolved oxygen (DO) needed by aerobic organisms to break down organic material present in a given water sample at a certain temperature over a specific period. We measured the BOD as expressed in milligrams of oxygen consumed per litre of water sample during five days of incubation at 20°C. This is used to assess the degree of organic pollution of streams/rivers in the estates and is monitored weekly (Fig. 3).

Nitrate (NO_3), Phosphate (PO_4) and total dissolved solids (TDS). Nitrate and phosphate constitute two of the most important fertilizer components, along with potassium (K) and a number of macro- and micronutrients (e.g. Ca, Mg, S, Mn, MO). Poor application of fertilizer combined with heavy rains can cause undesired run-off of these important chemical compounds that, in turn, will contaminate the waterways and cause excessive plant and algae growth, and decrease yield of palms. To capture a relative level of run-off of other chemical elements, we also measured TDS, primarily at water sources supplying households, to gauge the effect of water filtering systems.

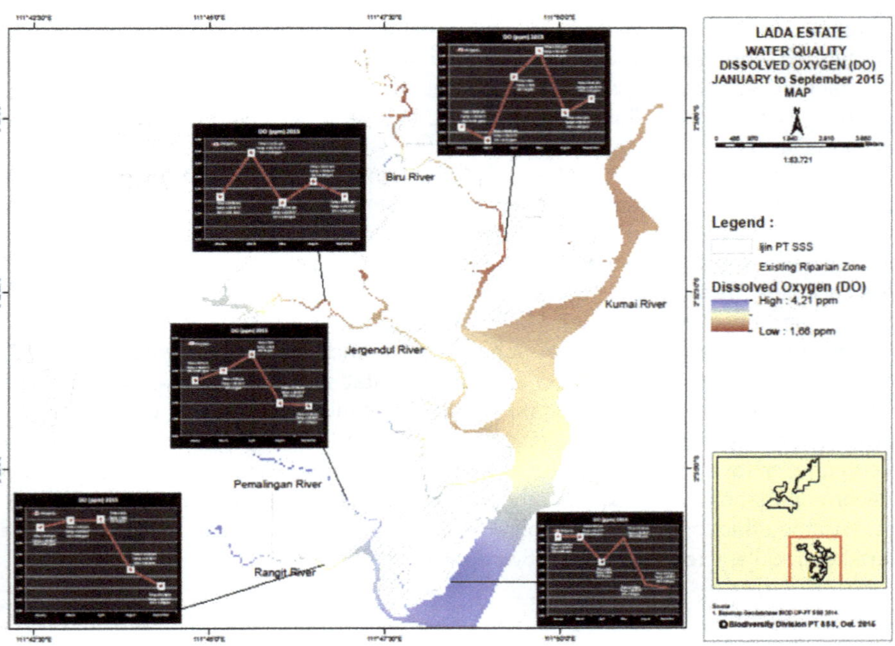

Figure 3 Water quality monitoring in Lada Estate, 2015. From various sampling stations it is possible to assess which streams/rivers are not performing to a certain threshold value. Non-performing waterways will initiate an immediate response from the management, to identify a source of possible contamination. If it emerges from within the estates, remediation actions will be taken.

Total suspended solids (TSS). TSS can include a wide variety of material, such as silt, decaying plant and animal matter, industrial wastes and sewage. High concentrations of suspended solids can cause many problems for stream health and aquatic life. In a plantation, this is primarily manifested as organic contamination (e.g. empty fruit bunches (EFB), palm fronds and overflowing septic tanks).

pH is measured across all waterways to assess possible deviations from the baseline. Such data are particularly important to collect in peat swamp areas, where many species of plants and aquatic life are specialized to survive in very acidic conditions. If a peat swamp turns increasingly basic, this is an indication that the ecological processes are being undermined by activities outside the peat dome itself.

5 Biological pest management

Some of the most common pests in a palm oil estate belong to the animal kingdom. Rats (e.g. *Rattus argentiventer*, *Rattus tiomanicus*), rhinoceros beetles (e.g. *Oryctes rhinoceros*) and Lepidoptera larvae can inflict serious loss of fruits (rats), kill young palms (beetles) and defoliate palms (Lepidoptera). The rhinoceros beetles are controlled by installing pheromone traps, and beneficial plants (e.g. *Turnera subulata* and *Antigonon leptopus*) are planted to attract predatory and parasitic wasps from in the Spheciformes and Vespoidea families that predate or deposit their own larvae on Lepidoptera larvae (Wäckers, 2004; Winkler et al., 2009). Rats can cause more than 5% economic loss in the absence of proper control (Khoo, 1984) and have traditionally been controlled with rat poison, until biological control in the form of barn owls, *Tyto alba*, were introduced in the 1980s (Duckett, 1984). To what degree these birds are effective is still questionable, and they need careful management (Duckett, 1991; Duckett and Karuppuah, 1989; Martin, 2009). The barn owl is not considered 'native' but a vagrant to the southern part of Central Kalimantan, and it would contravene biological sense if these were introduced. Instead, already-present predators in the estate landscape were assessed for their effect as rat predators, namely the leopard cat, *Prionailurus bengalensis*, and black cobra, *Naja sumatrana*. Both are common in the plantation landscape and feed primarily on rats.

A total of 11 leopard cats and 2 black cobras were captured and fitted with radio collars. Dispersal range, territory, mate monopolization, food and foraging were studied along with population dynamics.

6 Results and discussion

6.1 Biodiversity

The biodiversity surveys produced 278 terrestrial vertebrate species (55 mammals, 158 birds, 35 reptiles and 30 amphibians), 79 fish species and 107 tree species (Fig. 4). Five mammals and one bird species are listed as endangered on the IUCN Red List (Table 1), with a total of 25 listed as vulnerable. All the endangered species, except for orangutan (*Pongo pygmaeus*) and the proboscis monkey (*Nasalis larvatus*), are very rarely recorded. A total of 79 fish species were recorded of which many are endemic to Borneo.

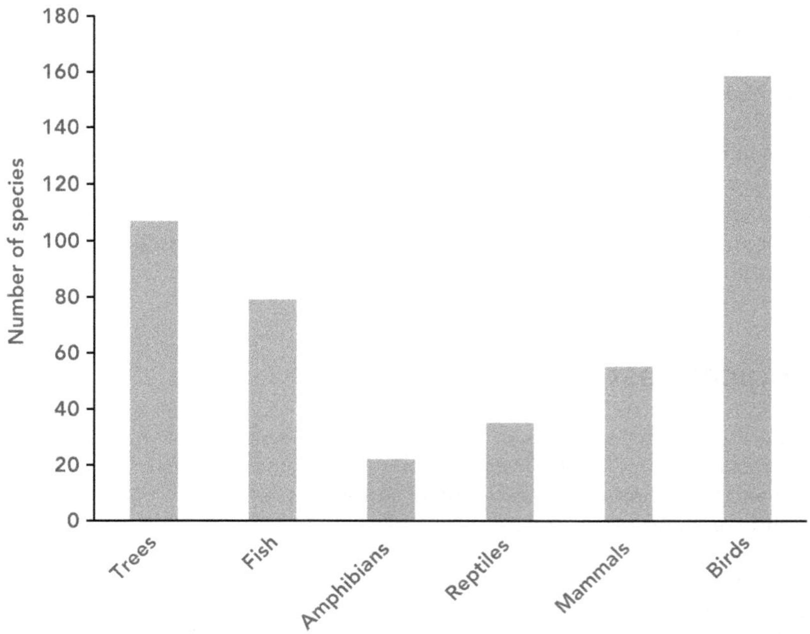

Figure 4 Baseline biodiversity of various taxon recorded in UP/PTSSS plantation landscape.

Table 1 Endangered vertebrates recorded in UP/PTSSS plantations, Central Kalimantan

Common name	Latin name	IUCN Red List status
Flat-headed cat	*Prionailurus planiceps*	Endangered
Pangolin	*Manis javanicus*	Endangered
Bornean gibbon	*Hylobates muelleri*	Endangered
Orangutan	*Pongo pygmaeus*	Endangered
Proboscis monkey	*Nasalis larvatus*	Endangered
Storm's stork	*Ciconia stormi*	Endangered

The significant number of vertebrate species reflects the importance of maintaining HCV areas. Future surveys are likely to uncover more species, in particular reptiles, amphibian and fish species. The study site consists of a complex hydrological landscape affected by both seasonal changes (flooding/drought) and tidal interchanges. There remains little information about the area's aquatic fauna, exemplified by the fact that a majority of the recorded fish species have not yet been evaluated for the IUCN Red List.

Two tree species, *Shorea belangeran* and *Dipterocarpus elongatus*, are listed as critically endangered and six species, *Agathis borneensis, Anisoptera laevis, Anisoptera marginata, Shorea ovali, Shorea parvifolia* and *Shorea cf acuminate*, are listed as endangered on the IUCN Red List (Table 2). Further six species are listed as vulnerable (e.g. Ulin, *Eusideroxylon*

Table 2 Endangered tree species recorded in UP/PTSSS plantations, Central Kalimantan

Common name	Latin name	IUCN Red List status
Belangiran	*Shorea belangeran*	Critically endangered
Keruing tempudau	*Dipterocarpus elongatus*	Critically endangered
Ketimpun	*Anisoptera laevis*	Endangered
Ketimpun	*Anisoptera marginata*	Endangered
Borneo kauri	*Agathis borneensis*	Endangered
Meranti	*Shorea cf acuminata*	Endangered
Meranti luang	*Shorea ovalis*	Endangered
Light-red meranti	*Shorea parvifolia*	Endangered

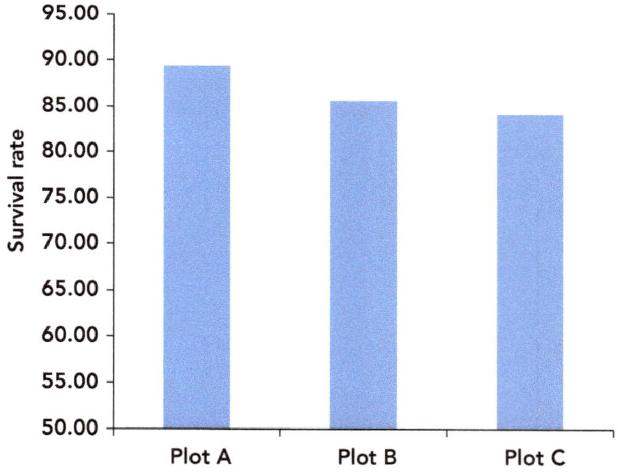

Figure 5 The survival rate of seedlings from five tree species (*Gluta renghas*, *Calophyllum castaneum*, *Cratoxylum arborescens*, *Shorea balangeran* and *Mesua* sp) in a rehabilitation test plot. A = 100% palm stand, B = 50% palm stand and C = cleared (From Farikhah and Traeholt, 2014).

zwageri). Most species turned out to be relatively easy to propagate, although certain species with hard-shelled seeds (e.g. Ulin, *Eusideroxylon zwageri*) required some excision to promote sprouting in 1–2 months (control seeds sprouted only after 9–13 months).

Survival rate of seedlings planted in semi-shaded habitat, that is, with oil palm stands intact or reduced by 50% were higher than survival rates in clear-cut areas (Farikhah and Traeholt, 2014) (Fig. 5). Areas planted with oil palms but in need of restoration, therefore, underwent gradual replacement of oil palms with trees, rather than removing all oil palms before commencing rehabilitation planting. Rehabilitation of non-peat habitat progressed well, and various species of trees planted in 2012 have reached a height of 5–8 m (Fig. 6). Due to the ongoing canopy formation, low undergrowth consisting of grasses (e.g. *Imperata cylindrica*) and ferns (e.g. *Dicranopteris linearis*) is progressively suppressed

Figure 6 Mohd Silmi, manager of the UP/BioD division, standing next to trees planted at the end of 2011. Note the absence of *Dicranopteris linearis* and the relatively short growth of lalang *Imperata cylindrical*.

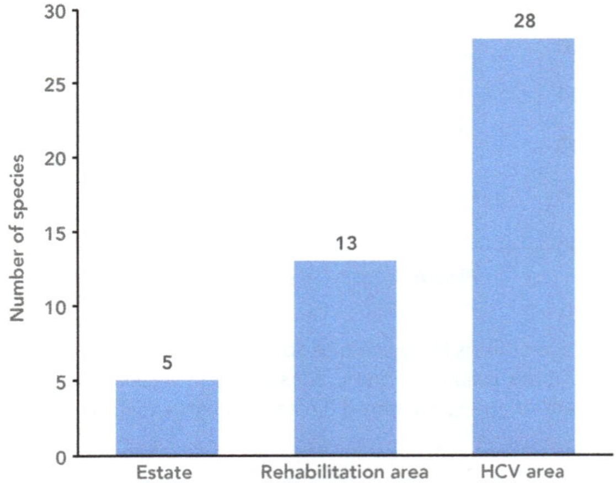

Figure 7 The number of bird species in three different types of habitat. The results reflect species count from three adjacent areas four years after planting the first seedlings in the rehabilitation area.

due to, primarily, photosynthesis competition and changes to the soil surface microclimate (e.g. Germer, 2003). Subsequently, the number of bird species recorded is increasing and significantly higher than in an oil palm plantation (Fig. 7).

Peatland rehabilitation has been less successful. While the collection and propagation of seedlings in nurseries was successful, the mortality rate of outplanted seedlings reached 95%. This may have been aggravated by the El Niño episode, which occurred during the 2014–2015 season, which caused extreme wet conditions in 2012–2014, resulting in extended flood periods that drowned most seedlings. The following outplanting process succeeded a little better, suffering extended drought periods, combined with occasional fires, which decimated 95% of the seedlings. At present, the biodiversity team experiments with planting seedlings on 50–80 cm earth mounds, thereby reducing the risk of mortality due to flooding.

6.2 Water quality

Water quality fluctuated seasonally, according to the location, types of vegetation and size of the waterways. Generally, the concentration values measured for NO_3, PO_4 and BOD fell within the legal limits throughout the entire study period. The TSS dropped sharply in all areas once planting commenced, except for a few streams that were still affected by outside infrastructure development and large abandoned agricultural sites. On a few occasions, severe organic contamination was detected and the source could be identified as EFB. Due to their high content in urea, phosphate and potash, EFBs are recycled and brought back to the field as an organic fertilizer after being milled for oil. In two recorded cases, a truck carrying EFBs to the fields had tilted into a drain and offloaded its cargo of EFBs. In three cases, EFB deposition had been undertaken poorly, with several EFBs ending up in streams, and, in one case, heavy rains washed out effluents from piles of EFBs, before these had been dispersed in the field. Similarly, palm oil mill effluent (POME) is recycled as a liquid organic fertilizer. After passing through decomposition processes in purpose made ponds (a large portion of POME is fed into a biogas plant) POME is pumped into the fields. To avoid serious risk of organic contamination, POME is deposited into 'cascading flatbeds' consisting of an elaborate system of 4–6 m^3 holes. This system is designed to prevent washouts; however, during heavy rains many of the 'flat beds' can fill up and overflow and cause run-off into adjacent waterways. During the rainy season, this is the most common cause of organic contamination. Weather monitoring and prediction, combined with effective management has reduced the number of overflows to a minimum by (1) building mini bunds around the flat beds, (2) digging more flat beds and (3) reducing volume in high-risk areas.

6.3 Pest management

Within three months, the two radio-tracked black cobras were killed and eaten by king cobras (*Ophiophagus hannah*). Therefore, it was not possible to quantify the effect of black cobras as rat predators, despite observing several cobras hunting and killing rats. The preliminary results from radio-tracking leopard cats conclude that the average number of rats killed and eaten by a cat is 2 rats/day, or approx. 730 rats/year/cat. While leopard cat population density depends on the habitat, the population density in some areas was estimated to be approx. 1 cat per 10 hectares (1000 individuals per 100km^2). This is significantly higher than the 12–16 individuals per 100km^2 recorded by Mohamed et al. (2013) in disturbed forests habitats and 89 individuals per 100 km^2 in Pulang Tekong, Singapore (Chua et al., 2016). Chua et al. (2016) observed that the predation on murids were far higher in monocrop plantations than in diverse forest habitats. Based on its energy needs, Duckett and Karuppuah (1989) suggested a pair of barn owls could kill 1300 rats in a year, or 650 rats/year/owl. Considering the optimal number of barn owls per area unit is approx. 8 ha/owl, there are potentially more leopard cats in an oil palm plantation with suitable habitat. In the study site, suitable habitat was created by leaving palm fronds stacks to overgrow with natural vegetation. This vegetation does not compete with palms for nutrients or photosynthesis and it provides ideal cover for leopard cats, even during harvesting rounds. In the study site, no chemical rat baiting has been required since the project commenced in 2012, which suggests that the leopard cats and other natural predators are keeping the rat population at a level where potential crop damage is negligible. Whereas these generalist rat predators thrive in an oil palm landscape, better biological pest management results may be developed by maintaining pockets of forests refuges within the plantation landscape to increase the mammalian diversity foraging for rats and other rodents in the plantations (Sasidhran et al., 2016; Yue et al., 2015).

7 Future trends and conclusion

The formation of an internal Biodiversity Division gathering a wide variety of experts from limnology to forestry and GIS has proven fundamental for UP in the aim of reaching 'sustainable' production of palm oil. The division is responsible for protecting and managing UP's conservation areas, as well as essential in a management process designed to ensure 'sustainable' palm oil production. This means that all sampling results are used in an adaptive management system, that is, results are processed by the Biodiversity Division and, if deemed necessary, corrective and/or preventative actions will be initiated. The present study also demonstrates that measuring the operational impact of palm oil production on the environment is possible, provided the right internal capacity is in place at the estate level, rather than a 'sustainability officer' in a far-away urban headquarter. Indeed, it is well within the obligations of many growers to develop internal capacity to conserve and manage its natural resources, including biodiversity and ecological processes.

Biological pest management forms a key part of sustainable palm oil production. The more natural pest predation that takes place, the less need there is for highly toxic chemicals for controlling pest outbreaks. The decision to use leopard cats as rat controllers instead of importing barn owls is a direct result of having a knowledgeable, dedicated and effective biodiversity team in place. It resulted in saving thousands of barn owls and nest boxes and, from an environmental perspective, more importantly, less rat poison has been deployed if compared to previous methods.

There remains significant future challenges in biodiversity management. Due to the ongoing agricultural and urban development, more natural habitat is being replaced with monoculture cropland or fast-wood plantations and habitats are increasingly fragmented. This effectively isolates many species found in plantation conservation areas, thereby inhibiting the flow of genetic diversity. As a result, plantation companies with isolated tracts of HCV forests harbouring small populations of endangered species risk losing a major part of its biodiversity due to stochastic variables (e.g. inbreeding, catastrophes and diseases). Whereas previous studies have documented that small HCV areas in plantation landscapes are important for biodiversity (Phalan et al., 2009), preventing the loss of isolated populations constitutes one of the biggest future management challenges for the industry.

Removing individuals of an endangered species does not constitute a meaningful management intervention, because it usually does not solve a conservation problem but merely transfers it to another location. Habitat protection, enrichment and extension are required; however, with the ongoing habitat loss and fragmentation, the management of particularly large mammals and birds require dedicated meta-population management (Byers et al., 2013; Steinmetz et al., 2009). This is a tried-and-tested process amongst zoos, livestock producers and safari game reserves worldwide; however, it requires a central management authority that is committed to and capable of coordinating such efforts. For example, if a plantation company harbours a small population of 8–10 of endangered orangutans in an isolated plot of HCV forest, the population is considered genetically non-viable, because there is a high risk that inbreeding depression will occur already in the first generation. In addition, it can significantly undermine endangered species conservation, if such a 'population' is captured and sent to already bulging poorly financed rehabilitation centres. Such isolated populations need to be managed as an ecological closed population that requires regular genetic supplementation and exchange in the form of external unrelated individuals. Such intensive species management is unlikely to happen, unless a responsible company invests in the necessary in-house capacity with a responsibility to engage.

The lack of progress in integrating ecosystem services with plantation activities has often lead to the general perception that the palm oil industry lacks the commitment to incorporate environmental divisions. Traeholt and Schriver (2011) suggested that limited conceptual understanding of the RSPO P&C combined with inadequate guidance explained the slow progress in driving the industry towards sustainable production. However, the future market requirements and present environmental emergencies will impose far greater obligations for RSPO member companies, which want to be certified as 'sustainable palm oil producers'. It is imperative that the necessary investments are made to ensure environmental and biodiversity concerns forms part and parcel of SOP.

This study documents a range of biodiversity conservation and environmental management activities that could never have taken place without the necessary internal capacity put in place by the plantation company itself. It also demonstrates that the process of measuring, monitoring and managing biotic and abiotic parameters as part of an adaptive management process is needed to document ecological sustainability in palm oil production.

8 Where to look for further information

Information about palm oil and sustainability, Global markets and statistics can be found on the following websites:

Food and Agricultural Organization of the United Nations: www.fao.org/
Roundtable for Sustainable Palm Oil: www.rspo.org
Indonesia sustainable palm oil: http://ispo-org.or.id/index.php?lang=ina
Malaysian Palm Oil Board: http://www.mpob.gov.my/
Malaysian Palm Oil Council: http://www.mpoc.org.my/
United Plantations Bhd: http://unitedplantations.com/

More information about climate change, biodiversity conservation, environmental issues at strategic and policy level, ecological sustainability, agricultural and social impact of palm oil and other agricultural products can be found on the following sites:

Convention on Biological Diversity: https://www.cbd.int/
NASA: https://climate.nasa.gov/

International Union for Conservation of Nature: https://www.iucn.org/
United Nations Development Programme: http://www.undp.org/
Center for International Forestry Research: http://www.cifor.org/
Copenhagen Zoo: www.zoo.dk
European Association for Zoos and Aquaria: www.eaza.net
American Association for Zoos and Aquariums: www.aza.org
World Resources Institute: http://www.wri.org/
Alliance for Leading Environmental Researchers & Thinkers (ALERT): http://alert-conservation.org/
Union of Concerned Scientists: http://www.ucsusa.org/
HCV Resource Network: https://www.hcvnetwork.org/
Forest Peoples Programme: http://www.forestpeoples.org/
Sawit Watch: http://sawitwatch.or.id/
International Work Group for Indigenous Affairs: http://www.iwgia.org/

9 References

Abood, S. A., Lee, J. S. H., Burivalova, Z., Garcia-Ulloa, J. and Koh, L. P. (2015). Relative contributions of the logging, fiber, oil palm, and mining industries to forest loss in Indonesia. *Conserv. Lett.*, 8(1), 58–67.

Anan, S., Azmin, M. R., Abdul, J. and Lim, B.-L. (1998). A survey of small mammals at Sungai Chenderoh, Perlok, Krau Wildlife Reserve, Pahang. *J. Wildl. Parks*, 16, 104–10.

Andrade, G. S. M. and Rhodes, J. R. (2012). Protected areas and local communities: An inevitable partnership toward successful conservation strategies? *Ecol. Soc.*, 17(4), 14. http://dx.doi.org/10.5751/ES-05216-170414.

Atack, K. (2006). *A Field Guide to the Fishes of Kuching Rivers, Sarawak, Malaysian Borneo*. Natural History Publications: Kota Kinabalu, Malaysia.

Bay, R. A., Rishnan, U. R. and Hadly, E. A. (2015). A call for tiger management using reserves of genetic diversity. *J. Hered.*, 105(3), 295–302. doi:10.1093/jhered/est086.

Boissière, M., Sheil, D., Basuki, I., Wan, M. and Le, H. (2009). Can engaging local people's interests reduce forest degradation in central Vietnam? *Biodivers. Conserv.*, 18, 2743–57. http://dx.doi.org/10.1007/. https://doi.org/10.1007/s10531-009-9627-1

Borrini-Feyerabend, G., Kothari, A. and Oviedo, G. (2004). *Indigenous and Local Communities and Protected Areas: Towards Equity and Enhanced Conservation*. IUCN: Gland, Switzerland and Cambridge, UK.

Bradshaw, C. J. A., Sodhi, N. S. and Brook, B. (2009). Tropical turmoil: a biodiversity tragedy in Progress. *Front. Ecol. Environ.*, 7(2), 79–87.

Brauer, M. (1997). *Assessment of Health Implications of Haze in Malaysia*. Mission Rep RE/97/0441 Rev.1, November 1997. WHO Regional Office: Manila.

Butler, R. A. and Conway, S. (2007). Can peatland conservation be more profitable than palm oil? *Jakarta Post*. http://old.thejakartapost.com/yesterdaydetail.asp?fileid=20070822 (22 August).

Butler, R. A. and Laurance, W. F. (2009). Is oil palm the next emerging threat to the Amazon?. *Trop. Conserv. Sci.*, 2(1), 1–10. https://doi.org/10.1177/194008290900200102.

Butler, R. A., Koh, L. P. and Ghazoul, J. (2009). REDD in the red: palm oil could undermine carbon payment schemes. *Conserv. Lett.*, 2, 67–73.

Byers, O., Lees, C., Wilcken, J. and Schwitzer, C. (2013). The one plan approach: The philosophy and implementation of CBSG's approach to integrated species conservation planning. *WAZA magazine*, 14, 2–5.

Canale, G. R., Peres, C. A., Guidorizzi, C. E., Gatto, C. A. F. and Kierulff, M. C. M. (2012). Pervasive defaunation of forest remnants in a tropical biodiversity hotspot. *PLoS One*, 7(8), e41671. doi:10.1371/journal.pone.0041671.

Cattau, M. E., Marlier, M. E. and DeFries, R. (2016). Effectiveness of Roundtable on Sustainable Palm Oil (RSPO) for reducing fires on oil palm concessions in Indonesia from 2012 to 2015. *Environ. Res. Lett.*, 11, 105007.

Chua, M. A. H., Sivasothi, N. and Meier, R. (2016). Population density, spatiotemporal use and diet of the leopard cat (*Prionailurus bengalensis*) in a human-modified succession forest landscape of Singapore. *Mammal Res.*, 61, 99–108. doi:10.1007/s13364-015-0259-4.

Craft, C., Clough, J., Ehman, J., Joye, S., Park, R., Penning, S., Guo, H. and Machmuller, M. (2009). Forecasting the effect of accelerated sea-level rise on tidal marsh ecosystem services. *Front. Ecol. Environ.*, 7(2), 72–8.

Das, I. (2010). *A Field Guide to the Reptiles of Southeast Asia*. New Holland Publishers (UK) Ltd.: United Kingdom.

De Jong, E. B. P., Ragas, A. M. J., Nooteboom, G. and Mursidi, M. (2015). Changing water quality in the Middle Mahakam Lakes: Water quality trends in a context of rapid deforestation, mining and palm oil plantation development in Indonesia's Middle Mahakam Wetlands. *Wetlands*, 35(4), 733–44.

Dobson, A., Lodge, D., Alder, J., Cumming, G. S., Keymer, J., McGlade, J., Mooney, H., Rusak, J. R., Sala, O., Wolters, V., Wall, D., Winfree, R. and Xenopoulos, M. A. (2006). Habitat loss, trophic collapse, and the decline of ecosystem services. *Ecology*, 87(8), 1915–24.

Duckett, J. E. (1991). Management of the barn owl *(Tyto alba javanica)* as a predator of rats in oil palm (*Elaeis guineensis*) plantations in Malaysia. *Birds Prey Bull.*, 4, 11–24.

Duckett, J. E. (1984). Barn Owls *(Tyto alba)* and the 'second generation' rat-baits utilised in oil palm in Malaysia. *Planter*, 60, 3–11.

Duckett, J. E. and Karuppuah, S. (1989). A guide to the planter in utilizing barn owl (*Tyto alba*) as an effective biological control of rats in mature oil palm plantations. Proceedings of the PORIM International Oil Palm Development Conference, Malaysia, pp. 357–72. Palm Oil Research Institute of Malaysia: Kuala Lumpur, Malaysia.

Duffy, J. E. (2009). Why biodiversity is important to the functioning of real-world ecosystems. *Front. Ecol. Environ.*, 7(8), 437–44.

Emmanuel, S. C. and Lim, O. P. (1998). Singapore country report on the 1997 smoke-haze. In: Proceedings of Biregional workshop on health impacts of haze-related air pollution, Kuala Lumpur, 1–4 June, World Health Organisation.

Engel, D. H. and Phummai, S. (2008). *A Field Guide to Tropical Plans of Asia*. Marshall Cavendish Editions: Singapore.

European Communities (2008). The Economics of Ecosystems and Biodiversity – An Interim Report.

Ewel, K. C. (2009). Appreciating tropical coastal wetlands from a landscape perspective. *Front. Ecol. Environ.*, 8(1), 20–6. doi:10.1890/080090.

FAO (2009). *The State of the World's Forest*. FAO: Rome, Italy. p. 168.

Farikhah, U. and Traeholt, C. (2014). Growth and survival rates of tree seedlings in a rehabilitation area of Central Kalimantan, Indonesia. *J. Indonesian Nat. Hist.*, 2(2), 32–7.

Francis, C. M. (2001). *A Photographic Guide to Mammals of South-East Asia: Including Thailand, Malaysia, Singapore, Myanmar, Laos, Vietnam*. Ralph Curtis Publishing: Sanibel Island, Florida, USA. p. 127.

Fuller, D. O., Jessup, T. C. and Salim, A. (2004). Deforestation trends in a tropical landscape and implications for endangered large mammals. *Conserv. Biol.*, 18(1), 249–54.

Gamfeldt, L., Hillebrand, H. and Jonsson, P. R. (2008). Multiple functions increase the importance of biodiversity for overall ecosystem functioning. *Ecology*, 89(5), 1223–31.

Gaveau, D. L. A., Salim, M. A., Hergoulc'h, K., Locatelli, B., Sloan, S., Wooster, M., Marlier, M. E., Molidena, E., Yaen, H., DeFries, R., Verchat, L., Murdiyarso, D., Nasi, R., Holmgren, P. and Shiel, D. (2014). Major atmospheric emissions from peat fires in Southeast Asia during non-drought years: Evidence from the 2013 Sumatran fires. *Sci. Rep.*, 4, 6112. doi:10.1038/srep06112.

Germer, J. U. (2003). *Spatial Undergrowth Species Composition in Oil Palm (Elaeis guineensis Jacq.) in West Sumatra*. University of Hohenheim: Germany.

Ghazali, A., Asmah, S., Syafiq, M., Yahya, M., Aziz, N., Peng, T., Norhisham, A. R., Puan, C. L., Turner, E. C. and Azhar, B. (2016). Effects of monoculture and polyculture farming in oil palm smallholdings on terrestrial arthropod diversity. *J. Asia Pac. Entomol.*, 19(2), 415–21.

Gibson, L., Lynam, A. J., Bradshaw, C. J. A., He, F., Bickford, D. P., Woodruff, D. S., Bumrungsri, S. and Laurance, W. F. (2013), Near-complete extinction of native small mammal fauna 25 years after forest fragmentation. *Science*, 341, 1508–10.

Gilpin, M. E. and Soulé, M. E. (1986). Minimum viable populations: processes of species extinction. In Soulé, M. E. (Ed.), *Conservation Biology: The Science of Scarcity and Diversity*. Sinauer Associates: Sunderland, MA.

Goodman, L. K. and Mulik, K. (2015), *Clearing the Air: Palm Oil, Peat Destruction, and Air Pollution*. Union of Concerned Scientists: Cambridge, MA, USA.

Greenpeace (2008a). *United Plantations Certified Despite Gross Violations of RSPO Standards*. Greenpeace Netherlands: Amsterdam, The Netherlands.

Greenpeace (2008b). *The Impact of RSPO Partial Certification Standards: A Case Study in Kotawarining Barat, Central Kalimantan*. Greenpeace International: The Netherlands.

Greenpeace (2008c). *The Hidden Carbon Liability of Indonesian Palm Oil*. Greenpeace International: The Netherlands.

Greenpeace (2007). *How the Oil Palm Industry is Cooking the Climate*. Greenpeace International: The Netherlands.

Hansen, M. C., Potapov, P. V., Moore, R., Hancher, M. and et al. (2013). High-resolution global maps of 21st-century forest cover change. *Science*, 342(6160), 850–3. doi:10.1126/science.1244693.

Heil, A. and Goldhammer, J. G. (2001). Smoke-haze pollution: a review of the 1997 episode in Southeast Asia. *Reg. Environ. Change*, 2, 24–37.

Heyer, W. R., Donnelly, M. A., McDiarmid, R. W., Hayek, L.-A. C. and Foster, M. S. (1994). *Meauring and Monitoring Biological Diversity: Standard Methods for Amphibians*. Smithsonian Institution Press: Washington, DC, USA.

Hoeinghaus, D. J., Agostinho, A. A., Gomes, L. C., Pelicice, F. M., Okada, E. K., Latini, J. D., Kashiwaqui, E. A. L. and Winemiller, K. O. (2009). Effects of River Impoundment on Ecosystem Services of Large Tropical Rivers: Embodied Energy and Market Value of Artisanal Fisheries. *Conservation Biology*, 23(5), 1222–31.

Houghton, R. A. (2003), *Emissions and sinks of carbon from land use change*. DN-LUCF.pdf. Report to the World Resources Institute from the Woods Hole Research Center, Washington DC.

Huijnen, V., Wooster, M. J., Kaiser, J. W., Gaveau, D. L. A., Flemming, J., Parrington, M., Inness, A., Murdiyarso, D., Main, B. and van Weele, M. (2016). Fire carbon emissions over maritime Southeast Asia in 2015 largest since 1997. *Sci. Rep.*, 6, 26886. doi:10.1038/srep26886.

Inger, R. (1966). The systematics and zoogeography of the amphibia of Borneo. *Fieldiana: Zoology Volume* 52. Field Museum of Natural History: Chicago, USA.

Inger, R. F. and Inger, C. P. K. (2002). Fresh-Water Fishes of North Borneo. Natural History Publications: Kota Kinabalu, Malaysia.

Inger, R. and Stuebing, R. B. (1997). *A Field Guide to the Frogs of Borneo*. Natural History Publications: Kota Kinabalu, Malaysia.

Kenney, J., Allendorf, F. W., McDougal, C. and Smith, J. L. D. (2014). How much gene flow is needed to avoid inbreeding depression in wild tiger populations?. *Proc. R. Soc. B.*, 281, 20133337. http://dx.doi.org/10.1098/rspb.2013.3337.

Kettunen, M. and ten Brink, P. (2006). Value of biodiversity- Documenting EU examples where biodiversity loss has led to the loss of ecosystem services. Final report for the European Commission. Institute for European Environmental Policy (IEEP): Brussels, Belgium.

Kinnaird, M. F., Sanderson, E. W., O'Brien, T. G., Wibisono, H. T. and Woolmer, G. (2003). Deforestation trends in a tropical landscape and implications for endangered large mammals. *Conserv. Biol.*, 17(1), 245–57.

Koh, L. P. (2008a). Birds defend oil palms from herbivorous insects. *Ecol. Appl.*, 18, 821–5.
Koh, L. P. (2008b). Can oil palm plantations be made more hospitable for forest butterflies and birds? *J. Appl. Ecol.*, 45, 1002–9.
Koh, L. P., Butler, R. A. and Bradshaw, C. J. A. (2009). Conversion of Indonesia's peatlands. *Front. Ecol. Environ.*, 7(5), 238.
Koh, L. P. and Wilcove, D. S. (2008). Is oil palm agriculture really destroying tropical biodiversity? *Conserv. Lett.*, 2, 1–5.
Koh, L. P. (2007). Impending disaster or sliver of hope for Southeast Asian forests? The devil may lie in the details. *Biodivers. Conserv.*, 16, 3935–8.
Koplitz, S. N., Mickley, L. J., Marlier, M. E., Buonocore, J. J., Liu, T., Sulprizio, J. P., DeFries, R. S., Jacob, D. J., Schwartz, J., Kim, P. S. Pongsiri, M. and Myers, S. S. (2016). Public health impacts of the severe haze in Equatorial Asia in September – October 2015: Demonstration of a new framework for informing fire management strategies to reduce downwind smoke exposure. *Environ. Res. Lett.*, 11, 094023.
Laurance, W. F., Koh, L. P., Butler, R., Sodhi, N. S., Bradshaw, C. J. A., Neidel, J. D., Consunji, H. and Vega, J. M. (2010). Improving the performance of the Round Table on Sustainable Palm Oil for nature conservation. *Conserv. Biol.*, 24(2), 377–81.
LaFrankie, J. (2010). *Trees of Tropical Asia: An Illustrated Guide to Diversity*. Black Tree Publications Inc.: Philippines.
Langton, M., Zane, M. R. and Palmer, L. (2005). Community-oriented protected areas for indigenous peoples and local communities. *J. Pol. Ecol.*, 12, 23–9.
Lundahl, S. L. L. and Olsson, A. (2000). Small mammal composition in Krau Wildlife Reserve in Pensinsular Malaysia -with reference to human disturbance. Unpublished M.Sc. thesis. Zoological Institute, Copenhagen University. Denmark.
Malcolm, J. R., Liu, C., Neilson, R. P., Hansen, L. and Hannah, L. (2006). Global warming and extinctions of endemic species from biodiversity hotspots. *Conserv. Biol.*, 20(2), 538–48.
Marlier, M. E., DeFries, R. S., Kim, P. S., Koplitz, S. N., Mickley, L. J. and Meyers, S. M. (2015). Fire emissions and regional air quality impacts from fires in oil palm, timber, and logging concessions in Indonesia. *Environ. Res. Lett.*, 10, 085005.
Mbile, P., Vabi, M., Meboka, M., Okon, D., Arrey-Mbo, J., Nkongho, F. and Ebong, E. (2005). Linking management and livelihood in environmental conservation: Case of Korup National Park Cameroon. *J. Environ. Manage.*, 76, 1–13. http://dx.doi.org/10.1016/j.
Miettinen, J., Shi, C. and Liew, S. C. (2011). Deforestation rates in insular Southeast Asia between 2000 and 2010. *Global Change Biol.*, 17, 2261.
Mohamed, A., Sollmann, R., Bernard, H., Ambu, L. N., Lagan, P., Mannan, S., Hofer, H. and Wilting, A. (2013). Density and habitat use of the leopard cat (Prionailurus bengalensis) in three commercial forest reserves in Sabah, Malaysian Borneo. *J. Mammal.*, 94(1), 82–9.
Myers, S. (2009). *A Field Guide to the Birds of Borneo*. New Holland Publishers (UK) Ltd.: United Kingdom.
Nuyim, T. (2005). *Guideline on Peat Swamp Forests Rehabilitation and Planting in Thailand*. Global Environment Centre: 2nd Floor, Wisma Hing, 47300 Petaling Jaya, Malaysia.
Payne, J. and Francis, C. M. (2007). *A Field Guide to the Mammals of Borneo*. The Sabah Society: Kota Kinabalu, Malaysia.
Petrenko, C., Paltseva, J. and Searle, S. (2016). *Ecological Impacts of Palm Oil Expansion in Indonesia*. ICCT: Washington DC, USA.
Phalan, B., Fitzherbet, E. B., Rafflegeau, S., Struebig, M. J. and Verwilghen, A. (2009). Conservation in oil-palm landscapes. *Conserv. Biol.*, 23(2), 244–5.
Phillipps, Q. and Phillipps, K. (2016). *Phillipps' Field Guide to the Mammals of Borneo and Their Ecology: Sabah, Sarawak, Brunei, and Kalimantan*. Princeton University Press: USA.
Robson, C. (2005). *Birds of Southeast Asia*. Princeton University Press: USA.
RSPO (2007). *RSPO Principles and Criteria for Sustainable Palm Oil Production Including Indicators and Guidance*. RSPO: Kuala Lumpur, Malaysia.
RSPO (2015). Impact Report 2015. RSPO Secretariat. Unit A-37-1, Menara UOA Bangsar, Kuala Lumpur, Malaysia.

RSPO (2016). http://www.rspo.org/about/impacts. Accessed10 December 2016.

Ruysschaert, D. and Salles, D. (2014). Towards global voluntary standards: Questioning the effectiveness in attaining conservation goals: The case of the Roundtable on Sustainable Palm Oil (RSPO). *Ecol. Econom.*, 107, 438–46.

Sasidhran, S., Adila, N., Hamdan, M. S., Samantha, L. D., Aziz, N., Kamarudin, N., Puan, C. L, Turner, E. and B. Azhar (2016). Habitat occupancy patterns and activity rate of native mammals in tropical fragmented peat swamp reserves in Peninsular Malaysia. *For. Ecol. Manage.*, 363, 140–8.

Schumann, M. and Joosten, J. (2008). *Global Peatland Restoration Manual*. Institute of Botany and Landscape Ecology, Greifswald University: Germany.

Simberloff, D. and Cox, J. (1987). Consequences and costs of conservation corridors. *Conserv. Biol.*, 1(1), 63–71.

Sodhi, N. S., Koh, L. P., Brook, B. W. and Ng, P. K. L. (2004). Southeast Asian biodiversity: an impending disaster. *Trends Ecol. Evol.*, 19, 654–60.

Stibig, H. J., Achard, F., Carboni, S., Rasi, R. and Miettinen, J. (2014). Change in tropical forest cover of Southeast Asia from 1990 to 2010. *Biogeosciences*, 11, 247–58.

Steinmetz, R, Chutipong, W. and Seuaturien, N. (2006). Collaborating to conserve large mammals in Southeast Asia. *Conserv. Biol.*, 20(5), 1391–401.

Tacconi, L. (2003). *Fires in Indonesia: Causes, Costs and Policy Implications*. CIFOR: Bogor, Indonesia.

Tallis, H., Goldman, R., Uhl, M. and Brosi, B. (2009). Integrating conservation and development in the field: Implementing ecosystem service projects. *Front. Ecol. Environ.*, 7(1), 12–20.

Tierney, G. L., Faber-Langendoen, D., Mitchell, B. R., Shriver, W. G. and Gibbs, J. P. (2009). Monitoring and evaluating the ecological integrity of forests ecosystems. *Front. Ecol. Environ.*, 7(6), 308–16.

Traeholt, C. and Schriver, C. (2011). RSPO P&C – The challenge of making the principles and criteria operational. *Oil Palm Ind. Econ. J.*, 11(1), 1–11.

UNEP (2007), *The Last Stand of the Orangutan – State of Emergency: Illegal Logging, Fire and Palm Oil in Indonesia's National Parks*, edited by Nellemann, C., Miles, L., Kaltenborn, B. P., Virtue, M. and Ahlenius, H. United Nations Environment Programme, GRID-Arendal: Norway, www.grida.no.

UNESCAP (2007), *Millennium Development Goals: progress in Asia and the Pacific 2007*. UN Economic and Social Commission for Asia and the Pacific: Bangkok.

United Plantations (2015). *Annual Report*. United Plantations Bhd.: Jendarata, Perak, Malaysia.

Wäckers, F. (2004), Assessing the suitability of flowering herbs as parasitoid food sources: Flower attractiveness and nectar accessibility. *Biol. Control*, 29(3), 307–14.

Wilson, D. E., Cole, F. R., Nichols, J. D., Rudran, R. and Foster, M. S. (1996). *Meauring and Monitoring Biological Diversity: Standard Methods for Mammals*. Smithsonian Institution Press: Washington DC, USA.

Winkler, K., Wäckers, F. L., Kaufman, L. V., Larraz, V. and van Lenteren, J. C. (2009). Nectar exploitation by herbivores and their parasitoids is a function of flower species and relative humidity. *Biolog. Control*, 50(3), 299–306.

Yue, S., Brodie, J. F., Zipkin, E. F. and Bernard, H. (2015). Oil palm plantations fail to support mammal diversity. *Ecol. Appl.*, 25(8), 2285–92.

Chapter 15

Waste management and recycling in oil palm cultivation

Salman Zafar, BioEnergy Consult, India

1 Introduction

The abundance of wastes from the palm oil industry, coupled with the huge environmental impact of palm oil production, has made sustainable management of plantations and mills an urgent necessity. The palm oil industry generates huge quantities of waste, which must be fully utilized in order to avoid enormous disposal problems. The situation is compounded by the fact that tens of millions of oil palms are cut and/or burnt each year to make way for new plantations.

The palm oil industry, besides producing crude palm oil (CPO) and palm kernel oil, produces palm kernel shell (PKS), press fibre, empty fruit bunches (EFBs), palm oil mill effluent (POME), palm trunk (during replanting) and palm fronds (during pruning). These solid by-products can be converted to value-added products in order to generate additional profit for the flourishing palm oil industry in Southeast Asia and other parts of the world, in addition to saving the environment from greenhouse gas (GHG) emissions and ecological degradation.

2 Palm oil mills and their wastes: overview

2.1 Palm oil mills

The palm oil fruits grow in branches and when mature and harvested, are referred to as fresh fruit bunches (FFBs). Palm trees typically begin flowering and producing fruit after

http://dx.doi.org/10.19103/AS.2017.0018.27

Table 1 Fuel characteristics of FFB residues

	Moisture (%)	Ash (%)	Chloride (%)	LHV (MJ/kg)	HHV (MJ/kg)
Empty fruit bunch	58.6	2.25	0.38	7.2	8.7
Fibres	38.5	3.45	–	11.4	12.3
Shells	10.9	2.10	–	18.2	18.5

3–4 years, and profitable yields are obtained by the owner well enough after 6–8 years. Oil palm plantations generally remain profitable for 25 years, after which they need to be replanted. Peak palm oil yields occur anywhere between 10 and 18 years of age, and gradually decline thereafter (Shiel, 2009). After harvesting, the FFBs are transported to the oil mills.

The capacities of the palm oil mills are measured in input tons per hour of FFB, and more than 60% of the oil mills in Southeast Asia have capacities exceeding 30 tons per hour. In the palm oil mill, the FFBs are sterilized (treated with heat under pressures of up to 3.6 bar) after which the oil fruits can be removed from the branches. The EFBs are left as residues, and the fruits are pressed in oil mills. The palm oil fruits are then pressed, and the kernel is separated from the press cake (mesocarp fibres). The palm kernels are then crushed, and the kernels then transported and pressed in separate mills.

EFBs, fibres and kernel shells are the main waste products from the palm oil industry that may be utilized as fuel for energy production due to their high-energy value, as shown in Table 1, and many of the palm oil mills today use their fibres and a part of the shells for energy purposes. Due to steam sterilization, the EFBs have, without separate treatment, a moisture content of 60–70% (Zafar, 2015).

In a typical palm oil plantation, almost 70% of the FFBs are turned into wastes in the form of EFBs, fibres and shells, as well as liquid effluent. Fibres and shells are traditionally used as fuels to generate power and steam. Effluents are sometimes converted into biogas that can be used in gas-fired engines.

Like sugar mills, palm oil mills have traditionally been designed to cover their own energy needs by utilizing low-pressure boilers and backpressure turbogenerators. Thus, palm oil factories have the potential of generating large amounts of electricity, for internal consumption and public grid, using their own wastes.

2.2 Wastes from the palm oil industry

Oil palm is a perennial tree crop, which is cultivated extensively under the humid tropics. For efficient productivity, the average planting of a palm tree spans over a period of about 25 years. Because of the conversion of solar radiation to carbohydrates by photosynthesis, the chemical energy content of the harvested palm fruit and biomass exceeds the energy input through the farming system. Thus, oil palm can act as a net source of useful energy.

For example, palm oil plantations in Malaysia are planted with a density of 148 palms per hectare. One stand of oil palm tree occupies 0.0068 hectare (ha) of land. Each palm yields about 150 kg of FFBs per year. Therefore, the average total production of FFBs per palm tree for 23 productive years is 3.45 tons (Hill, 2013).

The pruning of fronds is carried out at each round of harvesting in order to allow cutting of ripe fruit bunches. The dry weight of fronds from annual pruning is more than 10 tons/ha (von Uexkull and Fairhurst, 1991). For 23 years of productive period for a palm tree that bears fruit, the total dry weight of frond per palm from pruning is 1.8 tons. On the other hand, total biomass that fell during the replanting after 25 years is about 0.71 ton of trunk and fronds per palm. With an estimated calorific value of 10 GJ/ton, the total biomass energy available per palm tree, excluding the FFB, amounts to 25.1 GJ (Yusoff, 2006).

The main residues obtained after processing of FFBs include:

- EFBs
- POME
- Mesocarp fibre
- PKS
- Palm kernel cake

Ng et al. (2013) observed that the processing of 1000 kg of FFBs approximately leads to:

- Production of CPO – 200 kg
- Generation of EFBs – 230 kg
- Generation of POME – 600–700 kg
- Generation of PKS and mesocarp fibres – 190 kg
- Utilization of 20–35 kWe per ton FFB
- Utilization of 730 kg steam per ton FFB

Until recently, most of the wastes from palm oil mills were either burnt in the open or thrown away in waste ponds, despite the fact that palm oil wastes are rich in energy content (as shown in Table 2). The palm oil industry contributes significantly to global climate change by emitting carbon dioxide and methane even if substantial efforts are being made for recycling both plantation and mill wastes. Like sugar and rice mills, palm oil mills have traditionally been designed to cover their own energy needs (process heat and electricity) by utilizing low-pressure boilers and backpressure turbogenerators. Efficient energy conversion technologies that can utilize all palm oil residues, including EFBs, are currently available. Thus, palm oil factories have the potential of generating large amounts of electricity for captive consumption as well as export of surplus power to the public grid.

In conclusion, each ton of FFB gives residue equivalent to 4.2 MJ of heat. The milling process for a ton of FFB requires 20–25 kWh electricity and steam of 0.73 ton (2.5 bar), for which a total primary energy of 2.2 MJ is required. Thus, surplus energy from the mill is 2 MJ for each ton of FFB being processed (Prasertsan and Sajjakulnukit, 2006). Interestingly, the demand for palm shell has increased considerably in countries such as Malaysia, Indonesia and Thailand, resulting in a price close to that of coal. Nowadays, cement industries are using PKS to replace coal mainly because of the benefits of carbon

Table 2 Energy value of different palm oil wastes

Biomass residue	Energy value
Palm oil mill effluent (POME)	434.3 MJ per m^3
Empty fruit bunches (EFBs)	4.6 MJ/kg
Pressed mesocarp fibre	9.6 MJ/kg
Palm kernel shell (PKS)	17.4 MJ/kg

credits (Kanadasan and Razak, 2015). In fact, the demand of PKS for commercial power generation and industrial heating is steadily increasing in Asia-Pacific countries such as Japan and South Korea.

3 Residues after processing: kernel shells, mesocarp fibres and effluent

3.1 Palm kernel shells

PKS are the shell fractions left after the nut has been removed after being crushed in the palm kernel oil mill. Kernel shells are a fibrous material and can easily be handled in bulk directly from the product line to the end use. Large and small shell fractions are mixed with dust-like fractions and small fibres. Moisture content in kernel shells is low compared to other biomass residues with different sources suggesting values between 11 and 13%.

PKS contain residues of palm oil, which accounts for its slightly higher heating value than average lignocellulosic biomass. Compared to other residues from the industry, it is a good-quality biomass fuel with uniform size distribution, easy handling, easy crushing and limited biological activity due to low moisture content.

PKS generated by the mills are traditionally used as solid fuels for steam boilers. The steam generated is used to run turbines for electricity production. These two solid fuels alone are able to generate more than enough energy to meet the energy demands of a palm oil mill. Most palm oil mills in Malaysia and Indonesia are self-sufficient in terms of energy generation as they make use of kernel shells and mesocarp fibres in cogeneration units (Zafar, 2015).

The problems associated with the burning of these solid fuels are the emissions of dark smoke, and the carry-over of partially carbonized fibrous particulates due to incomplete combustion of the fuels can be tackled by commercially proven technologies in the form of high-pressure boilers.

Dual-fired boilers capable of burning either diesel oil or natural gas are the most suitable ones for burning palm oil waste since they could also facilitate the use of POME-derived biogas as a supplementary fuel. However, there is a great scope for the introduction of high-efficiency combined heat and power (CHP) systems in the palm oil industry, which will result in substantial supply of excess power to the public grid.

3.2 Palm oil fibres

Palm oil fibre is produced from EFB, which is considered as waste after the extraction of oil palm fruits. To become the useable fibre, the EFBs go through a process that involves shredding, separation, refining and drying.

Palm oil fibres are non-hazardous, clean, non-carcinogenic and free from pesticides and soft parenchyma cells. Palm fibres are versatile and stable and can be processed into various dimensional grades to suit specific applications such as erosion control, mattress cushion production, soil stabilization, horticulture and landscaping, ceramic and brick manufacturing, paper production, acoustics control, livestock care, compost, fertilizer and animal feed (Abdullah and Sulaiman, 2013). Palm fibres can also be used as fillers in thermoplastics and thermoset composites that have wide applications in furniture and automobile components.

Oil palm fibres and shells can be dried using pyrolysis and be used as fuel for boiler. The ash formed after burning can be utilized as an absorbent for removing pollutant gases such as nitrogen oxide and sulphur oxide. Ash thus formed contains high amounts of calcium, silica, potassium and alumina which can all be utilized to synthesize active compounds to absorb the pollutant gases into absorbent.

Carbon molecular sieve (CMS) is a material containing tiny pores of a precise and uniform size, which is used as an adsorbent for gases and liquids, and used for separating nitrogen from other gases contained in it. CMS is produced from oil palm wastes from lignocellulosic materials. Basically, there are three steps involved in preparing CMS from oil palm wastes – carbonization of the wastes, activation of the chars produced and pore modification of the activated carbons to obtain CMS.

3.3 Palm oil mill effluent (POME)

Palm oil processing also gives rise to huge quantities of highly polluting wastewater, known as POME, which is often treated in clarification ponds where catabolization occurs, resulting in the leaching of contaminants that pollute the groundwater and soil, and in the release of methane gas into the atmosphere. Methane slippage from conventional treatment systems for POME can account for up to 70% of the total GHG emissions in CPO production.

POME contains high concentrations of protein, nitrogenous compounds, carbohydrate, lipids and minerals that could be converted into useful material using microbial process, called anaerobic digestion, to produce biogas and fertilizer. Anaerobic digestion of POME results in energy-rich biogas which consists of 60–70% methane, 30–40% CO_2 and trace amounts of hydrogen sulphide. At many palm oil mills, this process is already in place to meet water quality standards for industrial effluent. The gas is flared off or used in generators to produce electricity.

In a conventional palm oil mill, 600–700 kg of POME is generated for every ton of processed FFB. As per conservative estimates, more than 100 million m^3 of POME are generated each year in Malaysia and Indonesia. Anaerobic digestion is widely adopted in the industry as a primary treatment for POME. Palm oil mills in Malaysia, Indonesia and Thailand use open lagoon systems for anaerobic and aerobic treatments of POME before discharge to a recipient due to their low costs and operational simplicity (Langerak, 2015).

A conventional open lagoon system consists of four types of ponds: a fat pit, cooling pond, anaerobic pond and aerobic pond. The fat pit recovers any remaining CPO in the POME and is generally recovered by the mill operator. The cooling pond decreases the temperature of POME, creating optimal conditions for the decomposition of organic material in the anaerobic and aerobic ponds. After treatment in these four ponds, the effluent is safe to be discharged to waterways or to be applied on land as a fertilizer. Although the open lagoon system is economical, it is land- and time-intensive, and releases large volumes of methane to the atmosphere primarily from the organic decomposition that occurs in the anaerobic pond.

Technologies for the effective treatment and handling of POME have been applied in several palm oil mills in Malaysia and Indonesia. The present systems typically involve the anaerobic decomposition of the organic components of POME and are sufficient to meet the required final effluent biochemical oxygen demand limits imposed by the respective governments. As to the biogas produced during POME treatment, there are no government regulations yet requiring palm oil mills to prevent its release to the

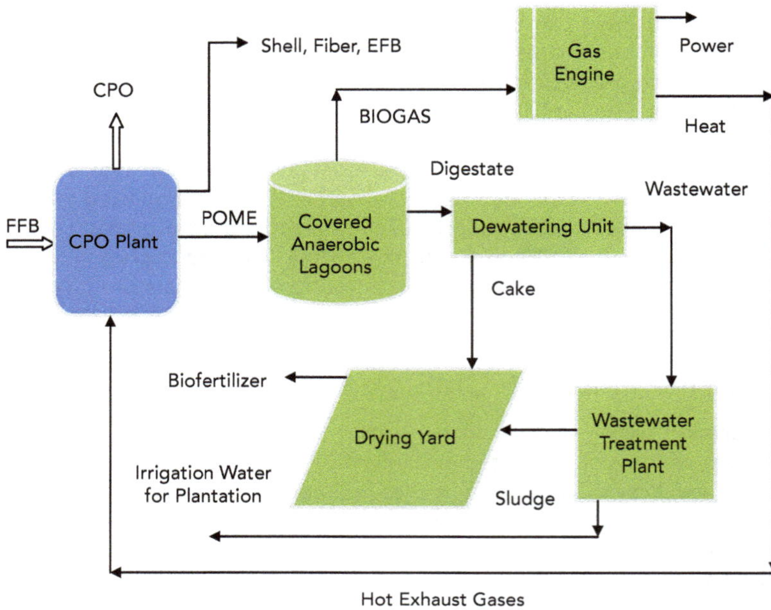

Figure 1 Schematic of biogas production from palm oil mill effluent.

atmosphere. However, with advancements in anaerobic digestion technologies, it is now possible to generate biogas in a profitable and sustainable manner, as shown in Fig. 1.

4 Residues after processing: empty fruit bunches (EFBs)

In a typical palm oil mill, EFBs are abundantly available as fibrous material of purely biological origin. EFB contains neither chemical nor mineral additives, and depending on proper handling operations at the mill, it is free from foreign elements such as gravel, nails, wood residues, waste and so on. However, it is saturated with water due to the biological growth combined with steam sterilization at the mill. Since the moisture content in EFB is around 67%, pre-processing is necessary before EFB can be considered as a good fuel (Carlos et al., 2015).

In contrast to shells and fibres, EFBs are usually burnt causing air pollution or returned to the plantations as mulch. EFBs can be conveniently collected and are available for exploitation in all palm oil mills around the world. Since shells and fibres are easy-to-handle, high-quality fuels compared to EFB, it will be advantageous to utilize EFB for on-site energy demand while making shells and fibres available for off-site utilization which may bring more revenues as compared to burning on-site.

Pyrolysis technology can be used to convert palm oil waste, especially EFB, into biochar which can be further utilized for the improvement of soil, thus minimizing the waste management problem arising from palm oil plantation and mills. The yield and characteristics of biochar from EFBs are dependent on various factors such as operating parameter, heating rate, holding time, as well as the type of reactor. The EFB biochar in soil applications has the potential as a carbon sequestration due to its capability of storing carbon for long periods, thus preventing it from being released to the atmosphere and mitigating GHGs (Ariffin et al., 2014). In fact, oil palm biochar with high yield and higher heating value under low energy requirement is required for improved waste management and utilization in the palm oil industry (Juferi et al., 2015).

4.1 Processing and utilization of EFB

Unprocessed EFB is available as very wet whole EFBs each weighing several kilograms, while processed EFB is a fibrous material with a fibre length of 10–20 cm and a reduced moisture content of 30–50%. Additional processing steps can reduce fibre length to around 5 cm and the material can also be processed into bales, pellets or pulverized form after being dried (Thuy et al., 2016).

There is a large potential for transforming EFB into renewable energy feedstock that could meet the existing energy demand of palm oil mills or other industries. Pre-treatment steps such as shredding/chipping and dewatering (screw pressing or drying) are necessary in order to improve the fuel property of EFB. Pre-processing of EFB will greatly improve its handling properties and reduce the transportation cost to the end user, that is, power plant. Under such scenario, kernel shells and mesocarp fibres which are currently utilized for providing heat for mills can be relieved for other uses off-site with higher economic returns for palm oil millers.

The fuel could be either prepared by the mills before being sold to the power plants, or handled by the end users based on their own requirements. Besides, centralized EFB collection and pre-processing system could be considered as a component in the EFB supply chain. It is evident that the mapping of available EFB resources would be useful for EFB resource supply chain improvement. This is particularly important as there are many different competitive usages. With proper mapping, assessment of better logistics and EFB resource planning can lead to better cost-effectiveness for both supplier and user of the EFB.

A covered yard is necessary to supply a constant amount of this biomass resource to the energy conversion sector. Storage time should, however, be short, for example, five days, as the product, even with 45% moisture, is vulnerable to natural decay through fungi or bacterial processes. This gives handling and health problems due to fungi spores, but it also contributes through a loss of dry matter through biological degradation (Zafar, 2015). Transportation of EFB is recommended in open trucks with high sides which can be capable of carrying an acceptable tonnage of this low-density biomass waste.

For EFB utilization in power stations, the supply chain is characterized by size reduction, drying and pressing into bales. This may result in significantly higher processing costs, but transport costs are reduced. For use in co-firing in power plants, this would be the best solution, as equipment for fuel handling in the power plant could operate with very high reliability having eliminated all problems associated with the handling of a moist, fibrous fuel in bulk.

4.2 Important issues in combustion of EFB

A comparison between EFB and wheat straw suggests that the two fuels have similar combustion characteristics. This implies that the risk of slagging and fouling, common in straw combustion systems, would also be present in EFB systems. These problems can be resolved by using lower temperature in the combustion zone (Nyakumaa et al., 2013). Additional solutions to EFB combustion include the following:

- Fuel handling facilities should include shredder and screw press to pre-treat the EFB into suitable size and moisture. The conveying systems ought to be specially designed.
- Controlled combustion will be useful to minimize the generation of nitrogen oxides and to promote better degradation of carbon in EFB.
- Co-combustion or use of additives could be the option for EFB combustion.
- Higher thermal efficiencies can be achieved by selecting higher boiler pressure and lower condenser pressure thermodynamic.

4.3 EFB-based pyrolysis plant in Malaysia

In cooperation with the Malaysian-based Genting Sanyen Bhd, BTG Biomass Technology Group BV has completed the first pyrolysis plant based on rotating cone pyrolysis technology in which EFBs are converted into pyrolysis oil. Usually, the wet EFBs (moisture content approximately 65 wt%) are combusted on-site, yielding only ash which can be recycled to the oil palm plantations. The palm mill produces about six tons per hour of this wet EFB, and as a new alternative to combustion, the EFB can be converted into fast pyrolysis oil. Prior to feeding it into the pyrolysis plant, the EFB is comminuted and dried to a moisture content of about 5–10%. In this way, all the wet EFBs from the palm are converted into approximately 1.2 ton per hour of pyrolysis oil (Venderbosch et al., 2007).

4.4 Composting of EFB

In recent years, composting of EFB is gaining traction across Southeast Asia due to the numerous benefits associated with the use of EFB as an organic fertilizer for oil palm plantations. Composting of EFB reduces its volume which makes it easier to be transported and spread on the fields. In addition, it helps in protecting the soil and crops by reducing the spread of weeds, parasites and pathogens, a common problem associated with the use of manure as a fertilizer (Siddiquee et al., 2017). EFB can also be co-composted along with POME, thereby further reducing costs and minimizing waste disposal issues at mills (Krishnan et al., 2017).

Composting of EFB not only offers oil palm plantations a cost-effective and environmentally friendly solution to dispose wastes, but also provides additional revenue stream, thus helping the mills in maintaining a competitive edge over their rivals.

5 Cogeneration technologies

The characteristic feature of energy systems installed in palm oil mills is the stress on cost-effectiveness, rather than efficiency, due to plentiful availability of on-site biomass resources. Most of the existing combustion systems in palm oils utilize low-efficiency low-pressure boilers. The average conversion efficiencies of steam and electricity generation processes are 35% and 3%, respectively, while the average overall cogeneration efficiency is 38%.

Commercially proven technologies are available in the international market for efficient production of power and heat from all kinds of palm oil wastes – PKS, EFB and POME. The state-of-the-art modern technologies utilize efficient high-pressure boilers. Some of these boilers are capable of dual-fuel burning, utilizing either liquid (e.g. diesel oil) or gas (e.g. natural gas) fuel as the supplementary energy source. Dual-fired boilers can be used in palm oil waste-fired boilers to facilitate the use of POME-derived biogas as supplementary fuel.

Local manufacturing capacity of efficient high-pressure steam generators in palm oil production is presently low. Most of the equipment for a palm oil waste-based power generation and cogeneration (also known as CHP) has to be imported, making the capital cost of a conventional biomass power plant or CHP facility high (typically around US$ 1500/kW). Moreover, with the market potentials of palm oil waste-based power projects and a suitable government policy on power pricing, the local boiler industry could possibly take up the manufacturing of high-pressure palm oil biomass boilers, when the market and demand for efficient biomass power technology takes off in major palm oil-producing countries.

6 Conclusion

Since the palm oil mills have abundant waste resources, their energy systems have traditionally been designed to be of low cost instead of being efficient. Most of the existing biomass combustion systems in the region utilize low-efficiency low-pressure boilers where the overall cogeneration efficiency is less than 40%. An additional source of energy in palm oil mills is the biogas produced in the anaerobic decomposition (for wastewater treatment purposes) of POME. However, at that time, POME-derived biogas (mostly methane, CH_4) was not being recovered and used but allowed to dissipate into the atmosphere. Retrofitting existing palm oil mills with new cogeneration equipment and biogas systems capturing the methane released from the wastewater ponds are promising commercial opportunity as well as sustainable waste management options.

The current palm oil production system, around the world, is generally seen as unsustainable because of detrimental effects on biodiversity such as loss of virgin forests and GHG emissions associated with current waste disposal methods. The use of by-products for energy recovery and value-added products offers fresh perspectives for designing a sustainable palm oil production chain that is essential for the development of a bio-based economy. The recognition that utilizing by-products for added value is beneficial to the sustainability of palm oil production is essential for certifying the sustainability of the palm oil biomass energy and products.

7 Where to look for further information

Wastes from the palm oil industry can be converted into heat, power and a wide range of bio-based products. Research activities may range from conventional applications such as compost, mulching mats and etc. to latest applications like biomass pellets and biofuels. Future research in palm oil waste management may be directed towards commercializing the following:

1 Biomass fuel for power generation
2 Use of palm oil waste in concrete production
3 Use of palm oil waste in production of biochar and activated carbon
4 Use of palm oil waste in production of polymer and metal matrix composite.

Organizations involved:

1 Malaysian Palm Oil Board (MPOB), Malaysia
http://www.mpob.gov.my/
2 Oil Palm Biomass Center (OPBC), Malaysia
http://www.be-basic.org/international/malaysia/oil-palm-biomass-centre.html
3 Indonesian Oil Palm Research Institute (IOPRI), Indonesia
http://www.iopri.org/
4 Roundtable on Sustainable Oil Palm (RSPO)
http://www.rspo.org/
5 ASEAN Centre for Energy
http://www.aseanenergy.org/
6 Oil Palm Research Institute, Ghana
http://csir.org.gh/index.php/research-inst/135-oil-palm-research-institute
7 ICAR - Indian Institute of Oil Palm Research, India
http://dopr.gov.in/
8 Food and Agriculture Organization, FAO
http://www.fao.org/
9 Japan International Research Center for Agricultural Sciences
https://www.jircas.go.jp/en
10 IEA Bioenergy
http://www.ieabioenergy.com/

8 References

Abdullah, N. and Sulaiman, F. (2013), *The Oil Palm Wastes in Malaysia, Biomass Now: Sustainable Growth and Use*, InTech, Croatia, pp. 75–93.

Ariffin, M. A., Wan, M. W. M. F., Mohammad, N. M. T. and Ramizi, M. (2014), Potential of oil palm Empty Fruit Bunch (EFB) biochar from gasification process, *Australian Journal of Basic and Applied Sciences*, 8(19), pp. 149–52.

Carlos, A. G. V., Valentina, A. M, Carlos, A. and Cardona, A. (2015), Stand-alone and Biorefinery ways to produce bioenergy from solid biodiesel wastes in Colombia, In *Conference Proceedings*, http://uest.ntua.gr/tinos2015/proceedings/pdfs/Carlos_Garcia_et_al.pdf (accessed on 2 January 2017).

Hill, R. D. (2013), *Agriculture in the Malaysian Region*, NUS Press, Singapore, pp. 279–80.

Juferi, I., Shirai, Y., Anduo, Y., Ahmad Amiruddin, M. A., Othman, M. R., Ibrahim, I., Husen, R. and Hassan, M. A. (2015), Improved yield and higher heating value of biochar from oil palm biomass at low retention time under self-sustained carbonization, *Journal of Cleaner Production*, 104, pp. 475–9

Kanadasan, J. and Razak H. A. (2015), Utilization of palm oil clinker as cement replacement material, *Materials*, 8(12), pp. 8817–38.

Krishnan, Y., Bong, C. P. C., Azman, N. F., Zakaria, Z., Othman, N., Abdullah, N., Ho, C. S., Lee, C. T. and Hansen, S. B. (2017), Co-composting of palm empty fruit bunch and palm oil mill effluent: Microbial diversity and potential mitigation of greenhouse gas emission, *Journal of Cleaner Production*, 146, pp. 94–100.

Langerak, J. (2015), POME as a source of biomethane, *BioEnergy Consult*, http://www.bioenergyconsult.com/pome-biogas/ (accessed on 25 November 2016).

Ng, R. T. L., Ng, D. K. S. and Tan, R. R. (2013), Systematic approach for synthesis of integrated palm oil processing complex, Industrial and Engineering Chemistry Research, 52(30), pp. 10221–35.

Nyakumaa, B. B., Joharia, A. and Ahmad, A. (2013), Thermochemical analysis of palm oil wastes as fuel for biomass gasification, *Jurnal Teknologi*, 62(3), pp. 73–6.

Prasertsan, S. and Sajjakulnukit, S. (2006), Biomass and biogas energy in Thailand: Potential, opportunity and barriers, *Renewable Energy*, 31(5), pp. 599–610.

Shiel, D. (2009), *The Impacts and Opportunities of Oil palm in Southeast Asia*, Center for International Forestry Research, Bogor, pp . 1–7.

Siddiquee, S., Shafawati, S. N. and Naher, L. (2017), Effective composting of empty fruit bunches using potential Trichoderma strains, *Biotechnology Reports*, 13, pp. 1–7.

Thuy, M., Junginger, M., Chavez, R. and Visser, L. (2016), Assessment of sustainable lignocellulosic biomass export potentials from Indonesia to the European Union, Project Report, *BioTrade2020plus*, http://www.biotrade2020plus.eu/images/case_studies/D3.1._CS_Indonesia_Final_01.06.2016.pdf (accessed on 10 December 2016).

Venderbosch, R. H., Gansekoele, E., Florijn, J. F., Assink, D. and Ng, H. Y. (2007), Pyrolysis of palm oil residues in Malaysia, *PyNe*, Issue 19, pp. 2–3.

von Uexkull, H. R. and Fairhurst, T. H. (1991), *Fertilizing for High Yield and Quality: The Oil Palm*, International Potash Institute, Horgen, pp. 57–8.

Yusoff, S. (2006), Renewable energy from palm oil – innovation on effective utilization of waste, *Journal of Cleaner Production*, 14, pp. 87–93.

Zafar, S. (2015), Biomass wastes from palm oil mills, *BioEnergy Consult*,http://www.bioenergyconsult.com/palm-biomass/ (accessed 24 December 2016).

Zafar, S. (2015), Palm kernel shells as biomass resource, *BioEnergy Consult*,http://www.bioenergyconsult.com/palm-kernel-shells-as-biomass-resource/ (accessed 28 December 2016).

Zafar, S. (2015), Storage systems for biomass, *Cleantech Solutions*,http://www.cleantechloops.com/biomass-storage-systems/ (accessed 6 December 2016).

Chapter 16

Understanding smallholders in oil palm cultivation: a case study from Sumatra

Pierre-Marie Bosc and Cédric Gaillard, CIRAD, France

1 Introduction

The province of Riau, located on the Indonesian island of Sumatra, has developed through the explotation of natural resources (mainly oil and gas), and also agriculture through the cultivation of cash crops and forestry (timber) supported by waves of transmigration. The development of the palm oil sector started in Riau in the early twentieth century (1920) at about the same time the first state plantations were set up in North Sumatra by the Dutch colonial authorities.

After independence, the Indonesian government promoted the palm oil sector through public and private schemes involving national and international companies with the required labour force being brought from outside the province. Starting in the 1980s, this development policy was supported by large-scale transmigration programmes within Indonesia with more than 500 000 transmigrants arriving from Java in the 40 years leading to 2001 (Potter and Badcock, 2001).

The continued expansion has confirmed Riau's reputation as Indonesia's foremost producer of crude palm oil (CPO). Production reached 5.4 million tons in 2010, an increase

http://dx.doi.org/10.19103/AS.2017.0018.35

of 12.5% from 2008. The trend of fast expansion was maintained with a 20% increase between 2010 and 2013, with up to 6.5 million tons produced in 2013. The Riau province also ranks first in terms of cultivated areas with 1.6 million ha under oil palm cultivation and ranks second in terms of productivity with an average of 4 t/ha of CPO. The proportion of smallholders, including those in partnerships with companies, is particularly high and this population of family farmers contributes to half of the area and more than half of the production (Mansuetus, 2015). 'Independent' smallholders have also become successful through technical spillovers from companies. Policies have long favoured the expansion of oil palm cultivation, and local governments rely on the development of the palm oil sector for poverty alleviation (Baudoin et al., 2017).

The first cultivated areas to be managed as independent plots were developed by workers from the local oil palm plantations with limited capital although their connections with the agro-industry helped them to some extent[1]. The arrival of transmigrants opened a new phase (Levang, 1997): indeed, the number of smallholders holding 'plasma' and independent plots increased and independent smallholders also started planting oil palm on their own. Estate wage workers shifted to family farming (Barral, 2012), many of them being spontaneous migrants from North Sumatra, but also through the inclusion of indigenous people in the industrial schemes (Potter and Badcock, 2004). Timing has played an important role in independent ownership of plots since the opportunity to access plasma plots is now reduced and the dynamism of the sector now rests on the smallholders (Barral, 2015).

The growing importance of smallholders is illustrated by their increasing share in planted areas and production of fresh fruit bunches (FFBs) contributing everyday more to the production of CPO even if they do not own and manage their own mills. Up to now, there is little knowledge on the smallholders engaged oil palm production. The definition provided by Roundtable on Sustainable Palm Oil (RSPO)[2] is wide: '*Smallholders are farmers who grow oil palm alongside with subsistence crops where the family provides the majority of labour and the farm provides the principal source of income and the planted oil palm area are is less than 50 hectares.*' The issue of clearly defining who is a smallholder is far from being specific for oil palm growers. It is a new issue for both agricultural economists and social scientists. Agreeing on an accepted definition is a purely academic issue but it has policy implications when it comes to policy-makers to define who is a smallholder and who is not; they need to use reliable information based on a definition covering actual situations.

In this chapter, we will provide insights about oil palm smallholders in the Riau province of Sumatra, Indonesia, based on two recent sets of data. Our samples were rather limited in size and were by no means comparable, but the two surveys conducted in 2013 and in 2015 provided some elements to characterise smallholders' production structures. To do so, we relied on the methodological framework elaborated in 2012 by the international initiative World Agriculture Watch (FAO, 2012). This initiative was set up in the FAO[3] to facilitate the documentation of the dynamics of agricultural holding transformations,

1. The agro-industrial model is based on a large holding (Estate) directly managed by the industry through hired labour close to a mill. Additional supply for the mill is provided by smallholders (Nucleus) supervised by the industry that are included in the scheme through access to a 2-ha plot which investment they have to refund through selling their production to the mill.
2. http://www.rspo.org/smallholders/rspo-smallholders-definition (access 1 March 2017).
3. The French Government, IFAD and FAO have supported this international initiative since 2011. The Land and Water Division (CBL) of FAO is hosting the Executive Secretariat of WAW.

weighing their respective contributions in order to inform policy debates. The present work opens new directions for further studies on larger samples, which would allow filling knowledge gaps on these dynamics. This knowledge is of foremost importance if we consider that large-scale schemes appear outdated in the light of current challenges facing agriculture and rural areas (Hayami, 2010).

2 Defining smallholders

It is difficult to consider 'smallholder agriculture', 'smallholder farming' and 'small-scale farming' or 'smallholding' as a conceptual or analytical category. The word 'small' refers here to the size of the holding which is generally expressed through the cultivated area, the number of animals raised or the economic dimension of the holding [standard margin in the European Union or the sales value in the United States (Hoppe et al., 2007)]. Nevertheless, it is widely used both in development and academic literature.

The term 'smallholder' is of course used in English-speaking countries where reference to family farming is much less frequent (Brookfield and Parsons, 2007; Hayami, 2002). It was influenced by history when the colonial administrations wanted to make a difference between 'indigenous' farming, mainly dedicated to food crops and the estate-type plantations created by colonial settlers for tropical export crops (Wickizer, 1960; Goldthorpe, 1987; Hayami, 2010). This distinction was maintained by the independent states to develop perennial crops (oil palm, rubber, etc.) with 'smallholder plantations' or 'village plantations' as opposed to industrial plantations (or estates).

The distinction does not retain all its meaning today. While the large plantations (corporate or entrepreneurial farming) use paid labour, the smallholder category often includes family farmers and family business farmers who need to hire permanent workers to complement family labour. In the oil palm sector in Indonesia, it is not uncommon for smallholders to gradually hold fairly large plantations to justify/allow the use of permanent employees to carry out bulk of the work (Barral, 2012). Consequently, the term 'smallholder' includes very different categories.

> The basic distinction between smallholding and plantation operations therefore becomes quite clear once the management/hired labour criterion is applied. Smallholder producers are defined as independent decision-makers who use family labour which works on its own or in conjunction with some hired workers on farms which are termed smallholdings. (...) In this definition the distinction between smallholdings and plantation is not based solely on the scale of operations but on internal structure, management control and the employment of agricultural workers.
>
> (Goldthorpe, 1987)

On such a basis, there would be a correspondence between smallholder agriculture and family farming. However, when referring to the nature of work and depending on both the context and the sector, the term 'smallholder' may cover a wide diversity of production structures, to the extent of including family business holdings when permanent hired workers complement family labour (Bélières et al., 2015). Consequently, the notion of smallholder is quite ambiguous due to its inner relative nature which varies depending on each context (be it national or even regional within the same country): size may depend on several factors such as natural resource endowments, the dynamics of agrarian colonisation, the types of production system and the type of investments favoured by public policies, among others.

But the definition of a smallholder can also vary due to the diversity of qualitative variables which depend on the objectives of the study:

> The most obvious measure is farm size and several sources define small farms as those with less than 2 hectares of cropland. In a similar but less precise vein, others describe small farms as those with 'limited resources', a definition that includes land as well as capital, skills and labour. Other authors emphasize variously: the low technology often used on small farms, dependence on household members for most of the labour and subsistence orientation where the primary aim of the farm is to produce the bulk of the household's consumption of staple foods.
>
> (Hazell et al., 2007)

Several issues can undermine a rigorous use of the concept: (i) Does smallholder farming include indirect tenure situations (lease or share cropping situations)? (ii) (How) Does it include activities in holdings, all or part of which are based on resources under common or shared ownership such as pastoral systems (Manoli et al., 2014) or include livestock raising as part of mixed farming systems? [see Traoré et al. (2013) for Africa, Davendra and Thomas (2002) for Asia or Ryschawy et al. (2014) for a global review]. (iii) More broadly, is pluriactivity part of the activity system or livelihood strategies developed by most of agricultural households? (iv) Focusing on the area does not include the qualitative dimensions (soil quality) or the type of investments among which labour-based land development (terraces, irrigation schemes, etc.) are accessible to less endowed holdings and the other types of capital – human, social, financial and physical – as well as the capacity to access them through organisations, institutions and policies (HLPE, 2013).

As a result of these considerations, what would be the boundary between a 'small' and a 'large' holding (Johnston and Kilby, 1975; Hubbard, 2009)? How would the threshold be defined? On the basis of which parameters? Does it include the other activities or not? The reference to the size of the agricultural area raises more questions than it provides useful answers for the analysis. However, it is now recognised that the vast majority of agricultural smallholdings correspond to family forms defined as follows:

> Labour is a key feature of smallholder agriculture. We consider a smallholding to be an agricultural holding run by a family using mostly (or only) their own labour and deriving from that work a large but variable share of its income in kind or in cash. The family relies on its agricultural activities for at least part of the food consumed – be it through self-provision, non-monetary exchanges or market exchanges. The family members also engage in activities other than farming, locally or through migration. The holding relies on family labour with limited reliance on temporary hired labour but may be engaged in labour exchanges within the neighbourhood or a wider kinship framework. Reciprocal relationships are important here for product or productive factor exchanges.
>
> (HLPE, 2013)

3 Surveys and data collection

We started with the definition of the RSPO as quoted above. We conducted two phases of data collection in 2013 and 2015 under a research project SPOP[4] (Sustainable Palm Oil Production) funded by the French National Agency for Research (ANR). Most holding areas were under the limit of 50 ha, but we also included larger holdings, which are also part of the 'smallholder' landscape. We found some of them during our first round of surveys.

3.1 The 2013 survey location and survey sample

The first study in 2013 was designed to cover the diverse smallholdings in the neighbourhoods of four Sinar Mas estate plantations owned by PT Smart Tbk in the Kampar district. It included smallholders having access to plasma plots as well as 'independent' smallholders having developed their plantations outside industrial schemes and a combination of both with smallholders being at the same time involved in an industrial scheme and on their own 'independent' plots.

Four estates with a Nucleus-Plasma system are located in Kijang, Sungai Tapung and Amartha Jaya (under PIR schemes) and Rama-Rama (under the KKPA scheme).

As KKPA is a scheme where smallholders' plots are fully managed by the company, we had to look for additional smallholders in Kijang, Sungai Tapung and Amartha Jaya and their surroundings in order to cover the diversity of combinations between the different types of plots (from fully managed by the company to plots where decisions are all under smallholders' control). Villages out of plasma scheme, where fully independent growers were found, were identified near the lands belonging to the company: Libo Jaya, Libo Baru and Libo Waduk, close to Kijang and Sungai Tapung; and Padang Tarap and Air Tiris close to Rama-Rama and Amartha Jaya. These groups were sampled according to the following criteria: farming activities, non-farm activities, owned area, social status and origin.

The choices were based on informal discussions with local key informants: plantation managers, heads of cooperatives, heads of farmer groups, hired workers and independent fruit collectors.

3.2 The 2015 survey location and survey sample

In 2015 we decided to focus on and explore the diversity among the independent smallholders through selecting new sites and explicitly asking the surveyor to randomly conduct interviews only with independent smallholders outside any industrial scheme and exploring other locations.

4. The SPOP project's goal is to better understand the impacts of palm oil production with respect to its economic, ecological and social dimensions, combining the expertise of researchers from various backgrounds (agronomy, social sciences and agro-ecology) in order to generate scientific knowledge and to propose tools for the assessment of sustainability under different production systems. The approach has involved stakeholders in the research process via specific tools such as surveys and interviews (face to face or focus groups), multi-agent modelling or workshops. Having identified constraints, it analysed whether these were due to an intrinsic inability of the cultivation system to adapt, or rather to a limited stakeholders' access to knowledge and tools. We mainly adopted the framework developed by the international initiative 'World Agriculture Watch', established in the FAO, while orienting research towards smallholders – a major group of stakeholders that accounts for 40% of palm oil global production (adapted from Bessou et al., 2017).

Four main sites were selected to conduct the interviews with independent smallholders. The sites, located in three regencies of the Riau province, correspond either to previous study sites of the SPOP project in 2013 (Siak regency) or other sites documented in the literature (see References).

3.2.1 Siak regency: two sites

Kecamatan Minas: Libo Jaya village (Desa) is close to PT Smart private plantation, but our 2013 survey indicated the presence of independent smallholders. The proximity and knowledge of PT Smart staff helped identify the respondents.

Kecamatan Sungai Apit (villages/desa: Pebadaran, Sungai Limau, Perincit, Benaya and Pusako): This site was selected since the Siak regency administration promoted a pro-poor policy through settling smallholders in oil palm production with support from state companies (PT), as analysed by McCarthy et al. (2012). In the same area, Ela, a local NGO supported by Greenpeace, has also promoted the intensification of oil palm production practices. A qualitative survey was conducted in 2012 as a field component of a Master's thesis (Van der Enden, 2013).

3.2.2 Kampar regency: one site

Kecamatan Tapung (around Petapahan City): This zone is referred to in a 2001 study (Suyanto et al., 2001) on the underlying causes of fires in Indonesia, which was published later (Suyanto, 2004). The 2001 study describes precisely the chronology of the settlements and provides a basis to choose where to survey independent smallholders. Local communities settled around Petapahan consisted of two hamlets: Kampung Asli and RK Petapahan Pillial, which recently was divided into three sub villages, namely: sub village 1, sub village 2 and sub village 3. There is another settlement of local people named Batu Gajah. This has been a site for logging and oil palm plantations through private companies and has also been a site for transmigrant settlements. The first flow of migrants settled down in the 1980s over 2 ha of land, with one ha being converted into an oil palm plantation under a KKPA scheme in collaboration with PT Sawit 4, later in 1995. The villages' names are SP1, SP2, SP3, SP4 and SP5, respectively. The flow of transmigration continued during the 1990 but under the NES scheme established five new villages again, each family receiving 2.5 ha of land out of which oil palm plantation accounted for 2 ha: UPT 1 (NES 1), UPT II (NES 2), UPT III (NES 3), UPT IV (NES 4) and UPT 5 (NES 5). In this area, random sampling through local informers led to the selection of independent smallholders within Petapahan hamlets (Kampung Asli, Petapahan Pillial and Batu Gajan) and settlements SP1, SP2, SP3, SP4 and SP5 for interview.

3.2.3 Bengkalis regency: one site

Kecamatan Mandau (near towns of Pinggir and Semurai): This zone is close to the extreme north west of the Siak Regency but located in the Bengkalis regency. The SPOTT image presents a zone including a PT rubber estate (PT Adei Crumb Rubber) starting north of Balaipungut and getting rather close to south Duri city. To the south of Duri and west of the rubber PT, there is a loose zone of farming area. This zone is located between the south of the concession owned by PT. Darmali Armali Jaya Lestari and the north west of the concession of PT Murini Samsam.

3.3 Selection of the smallholders

Due to the lack of a list of smallholder farmers (leaving out plasma or independent oil palm growers) that would allow sampling by sorting out randomly from the list, we decided to start with a snowball rationale supported by the knowledge of people working in oil palm plantations in the different locations selected. We also narrowed our selection by asking our respondents to identify smallholders that present marked differences in their farming system (either including annual crops or livestock or other perennial crops) or in their activity system (through off-farm and non-farm activities). Diversity also comes from the more or less recent engagement in oil palm cultivation being part of local communities or more or less recent settlers through either transmigration programmes or self-migration.

The 2015 survey was tough to implement, as our surveyor faced rather strong reluctance to answer from smallholders in the various locations we identified to survey. He felt people feared losing their lands mostly to corporates looking to expand their activities. He had to deploy communication skills and demonstrations on compost production as a sustainable way of fertilising to gain their confidence before getting them to engage in the survey.

4 Results

4.1 Human assets

In the 2013 survey, 50% of the household heads were between 40 and 50 years of age, with 35% over 50 and 15% below 40 years.

The age distribution revealed in the 2015 survey is roughly similar even though the sample was slightly younger, with 43% between 40 and 50 years of age and 21% below 40. The other side – over 50 years – has a similar weight of 36%.

The composition of the households is rather similar, with most family sizes ranging from 4 to 6 people per household (Fig. 1).

The distribution of household heads with regard to their education level is quite similar, with a limited number of household heads lacking any educational background. Education being compulsory until the end of junior high school in Indonesia, we observed that 54% of households' heads have reached this level of education or pursued higher education (Fig. 2).

The labour involved in independent holdings (survey 2015) was found to be mostly family labour essentially provided by the head of household (87%). Two additional external workers were found in only 6.5%. Remaining cases with 3 or 5 family workers accounted for only 6.5% of our sample. Thus, most of these holdings rely on limited family labour which is mostly provided by the household's head. A majority of them rely on agricultural activities (over 76%) with a limited share of off-farm work. When considering permanent hired labour, 40% have at least one permanent hired worker, and 90% of the total do not hire temporary workers formally.

For the studied independent holdings, we obtained a kind of dualistic picture of small-scale family holdings relying quasi exclusively on family labour on the one hand and of more consolidated holdings (starting around 4 ha and over) with a structural tendency to supplement the family labour by permanent external workers, on the other hand (Fig. 3).

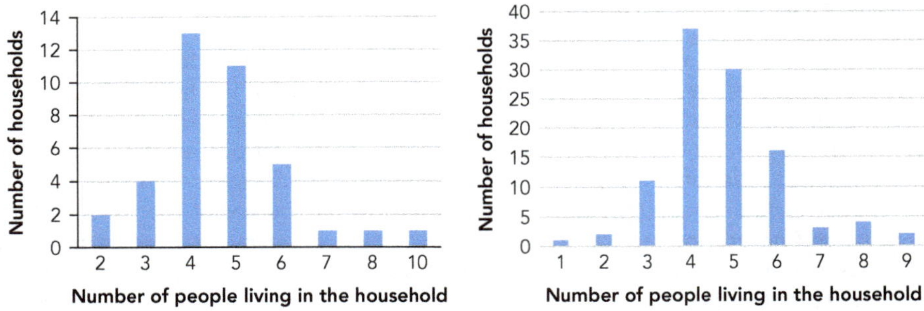

Figure 1 Household populations, 2013 and 2015. Source: Lé (2013) and Survey 2015.

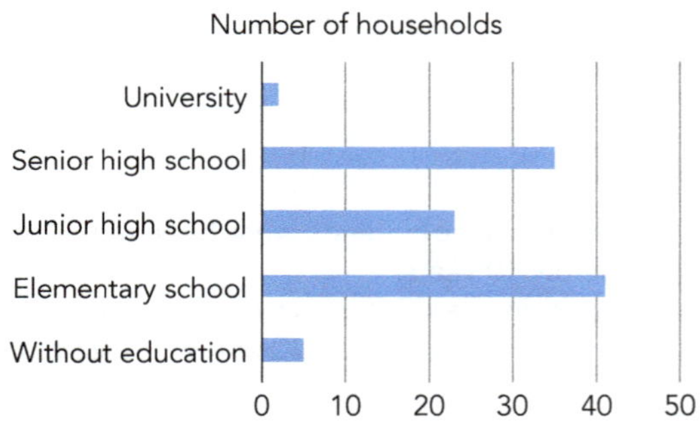

Figure 2 Education level of the head of household. Source: Survey 2015.

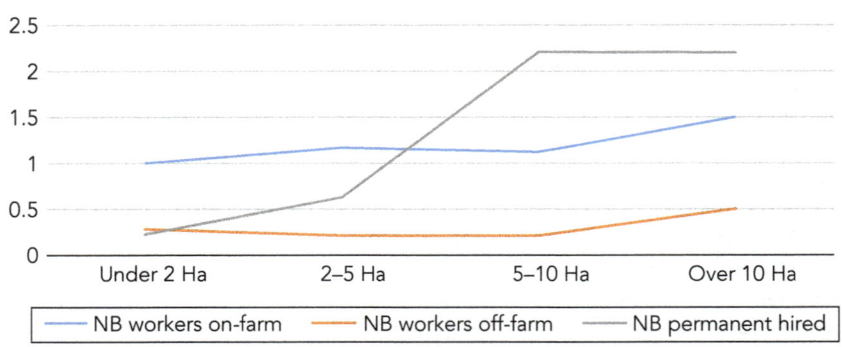

Figure 3 Type of labour and holding size. Source: Survey 2015.

4.2 Social assets

The term 'social assets' includes their origin, involvement in local associations or in elected positions.

In the 2013 sample, most of the households interviewed were either migrants (from north Sumatra) or transmigrants (from Java). However, some migrants coming from North Sumatra were descendants of earlier Javanese transmigrants and could apply to the transmigration scheme using their parents' origin and networks. In this sample, 74% of household heads were already living in the Sumatra Island before settling in Kampar.

From the 2015 sample, we obtained a different distribution with 62% being migrants and 38% natives. The higher share of natives may be linked to the choice of focusing on independent smallholders, but the migrants show a comparable profile with those of Kampar since they are Javanese by origin and migrated from North Sumatra to settle in Riau.

In the 2013 sample, as 76% of holdings have plasma plots (31% having only plasma), they are direct members of KUD[5] through plot ownership. Within this sample, 34% of household heads have a direct responsibility to run the KUD as head or secretary of a farmers' group, while 5 out of 38 were heads of their village or hamlet.

In contrast, the 2015 sample showed a lower rate of involvement in organisations with barely 25% of households belonging to groups, mostly farmers' groups or local development groups.

'Collective' assets appear to be more influential for farmers having at least one plasma plot meaning being integrated in an industrial scheme. The term 'collective' is somewhat misused since membership is through the plot; the plot 'belongs' to or is partially 'supervised' by the co-op (KUD). So to say, that for a farmer owing a plasma plot the sense of being part of a collective undertaking is rather limited, no action is decided and implemented collectively and autonomously unless decided and supervised by the company .

4.3 Physical assets

We chose to include the plantations in the physical assets. Being planted by men, they are not part of 'natural assets' even if the quality of their establishment and development is partially based on the nature of the natural assets (type of soil, slope, etc.). The other physical assets include cultivation means and transportation means.

4.3.1 2013 survey

The share of oil palm in the farming systems is paramount: 75% of the holdings grow only oil palm. The remaining 25% (ten holdings) grow other crops, but their contribution to the household economy is very limited (mostly below 1 ha except for one single case), and except for rubber also in one case out of 38. Food crops are rare: fruit trees, rice, maize and cassava were identified during the survey, but farmers were not in a position to quantify either the area or the consumption of those products.

Figure 4 shows the composition of lands under oil palm cultivation in each of the holdings surveyed. The high influence of the agro-industrial schemes is evidenced through the

5. KUD stands for Koperasi Unit Desa in Indonesian: Village Cooperative System which comprises rural saving and loan cooperatives established within the Nucleus Estate and Smallholders model.

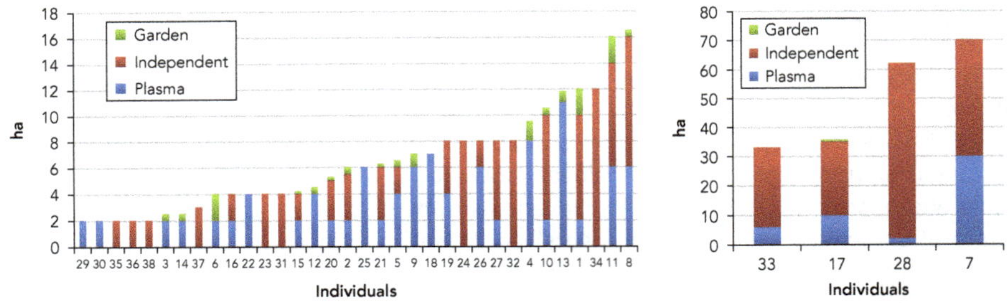

Figure 4 Composition and distribution of the types of plot per holding (large scale holdings are presented separately on the right). Source: Lé (2013).

importance of 'plasma lands' in 76% of the households surveyed (Table 1). Independent smallholders account for 24% in this sample compared to 100% in our 2015 survey. The conversion of gardens (0.5 ha plots attached to the house initially dedicated to food crops) into oil palm cultivation illustrates the strong incentive to plant oil palm (Table 2).

Our sample is not representative of the holding size distribution in Kampar. Since we aimed at representing the diversity of holdings, we have an over-representation of large-scale holdings and we under-represent holdings below 5 ha (Table 3).

Manual tools are commonly owned by the holdings. The differentiation is more on transportation means. All holdings have at least one means of transport: motorbikes are dominant (37/38 with an average of nearly three per household), 15/38 own at least one car and 6/38 own at least one truck to carry bunches. The diversity of the holding sizes is

Table 1 Distribution of oil palm land per category

	Plasma only	Plasma + Independent lands	Independent lands only	Total
NB holdings	13 (31%)	16 (45%)	9 (24%)	38 (100%)

Source: Lé (2013).

Table 2 Sizes of oil palm cumulated plantation areas for surveyed independent and plasma smallholders

Surfaces in production (ha)	% of all smallholders	% of independent smallholders	% of plasma smallholders
Less than 4	28.2	55.5	43.5
Between 4 and 12	56	29.6	52.2
More than 12	15.8	14.9	4.3
		Area across all the independent: Min = 0; Max = 110 ha; Mean = 9.41; Standard deviation = 22	Area across all the plasma: Min = 0; Max = 18; Mean = 3.32; Standard deviation = 3.74

Source: Baudoin et al. (2015).

Table 3 Distribution of owned area for cash crop per household in 2010

Class of area	Number of households	Share (%)	Shares of the category in our sample (%)
Less than 5 ha	2958	71.5	40
5–10 ha	1127	27	34
10–50 ha	50	1	26
Over 50 ha	8	0.2	0

Source: Lé (2013) from BPS and our data.

reflected by the diversity of transport means found in our sample. As all have at least one means, it also reflects the standard of living attached to oil palm cultivation.

4.3.2 Plantations in the 2015 survey

The size of the plantations reflects a majority of small sizes with 89% of the holdings being less than 10 ha – nearly all area being planted with oil palm (the area devoted to other crops is negligible). The majority of holdings are in fact below 8 ha in size as shown in Fig. 5 (Table 4).

This polarisation is reflected in the mean size of 4.9 ha with a median size of 3.75 ha, but the distribution is unbalanced with 12 holdings reporting areas of over 10 ha (Fig. 5 which shows numbers of households with holdings of different sizes, ranging from less than 1 hectare to over 20 hectares).

The relationship between the size of the holding and the age of the holding establishment is statistically significant (chi^2. $p < 0.05$): the size of the holding tends to increase with the duration since starting as an oil palm farm. This observation would illustrate a positive dynamic of accumulation for those holdings outside the agro-industrial schemes (Table 5).

Considering the size of each plot, we can observe the heterogeneity in plot sizes with an important share of small-sized plots below 2 ha (Fig. 6).

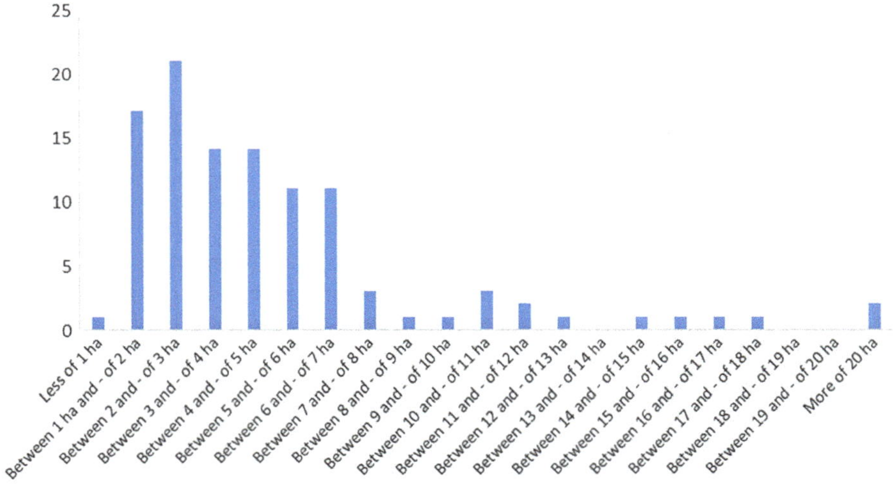

Figure 5 Distribution of the holding sizes.

Table 4 Distribution of holdings according to size class

Size class	Number of ha	%
Under 2 ha	18	16.98
From 2 to 5 ha	49	46.23
From 5 to 10 ha	27	25.47
Over 10 ha	12	11.32
Total	*106*	*100.00*

Table 5 Relationship between the holding size and the duration of establishment as a farm

	Less than 2 ha	2 to less than 5 ha	5 to less than 10 ha	Over 10 ha	Total
Less than 5 years	2	8	3	0	13
5–10 years	9	17	3	0	29
10–15 years	4	15	11	3	33
15–20 years	2	6	3	5	16
Over 20 years	1	3	7	4	15
Total	18	49	27	12	106

Results for other physical assets from the 2015 survey contrast sharply with regard to vehicles' ownership with only 18% of households owning a car and 4% owning two cars. Motorcycles are more widely owned (1.9 in average per household) with only 8% without any, while 92% have at least one motorcycle and 69% more than two per household. Here the studied smallholder population had less opportunity to invest in transport means. The majority of farmers only own motorcycles and the equipment is composed of a variety of hand tools: hand cutting instruments (to harvest the FFB or for small-scale weeding): sprayers (with batteries), wheelbarrows to carry FFBs from the palms to the collecting points. Thus, production techniques remain essentially manual and this applies also for larger-scale plantations.

The majority of holdings own cattle (68%), but there is no association between the size of the holdings and the ownership of animals.

4.4 Financial assets

In our sample, productive loans were found with a high variability and they are present in consolidated farms working mainly with hired workers. The other holdings – the majority of the sampled population – have limited access to banking services either for production or for domestic purposes.

The holdings surveyed in 2015 showed very limited access to financial assets: only five out of 106 have taken out a loan, and out of these, only three were used to fund the purchase of a plantation. Out of the remaining two loans, one is used to fund higher education for a child and the other one to buy a car. As a whole, we observed that independent smallholders do experience a true limitation in terms of access to productive assets.

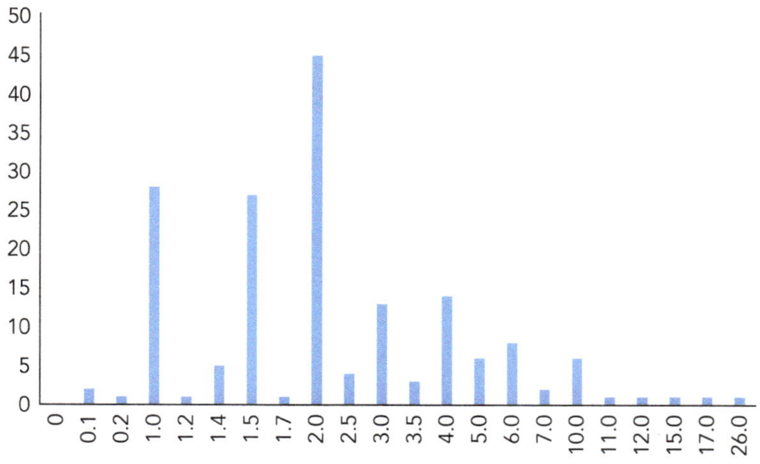

Figure 6 Distribution of plot sizes.

5 Towards a typology of independent smallholders

We realised a factor analysis of mixed data (qualitative and quantitative) based on a selection of variables on 100 observations (2015 survey). The average values for each of the variable used for factor analysis are reported in Table 6.

The first three axes (Fig. 7) will be retained in the analysis as they explain almost 53% of the total variability. The contributions of each variable to the construction of the axes lead us to note a rationale related to our study. The first axis is built around the living condition variables, while the second axis around the variables representing the economic size such as production, number of permanent hired workers and area of the farm. The third axis is built around households' access to roads, hospitals and equipment.

We thus select the first three axes and use an ascending hierarchical classification following Wald's method (Wald, 1939) on these axes. Four groups of farms were evidenced through this analysis.

Some variables had little influence on this classification: use of family labour is the most common way of farming with a limited number of individuals engaged in agriculture (ranging from 1 up to 1.47 across our groups). For these smallholders, FFB production relies mostly on the work performed by the head of household which is consistent with the assets analysis. The same observation applies for livestock, the influence of which on the grouping is also limited.

For income estimates, we could only rely on monthly averages. Therefore, we used it as a basis for estimating high and low levels corresponding to the two production seasons.

Among the four groups on a sample of 100 smallholders, we can consider that the economic results come for the oil palm plot(s). Tables 7 and 8 summarise the results obtained. If we compare our data with those collected by McCarthy and Zen (2016) in 2009 in the neighbouring province of Jambi, the order of magnitude for the income per

Table 6 Variables used for factor analysis

Description	Mean for quantitative variable or frequency % for qualitative variable
Size of household	5.1
Area farmed	4.4
Presence of livestock	0.64
Production of oil palm - monthly (IDR)	3 702 300
Number of permanent hired workers	1.09
Number of items of equipment	4.14
Number of family workers on farm	1.16
Presence of hygiene in the farm	43%
Presence of water in the house	59%
Presence of electricity power in the house	53%
Presence of bathroom in the house	77%
Number of motorcycles per household	2
Distance to hospital (in km)	29.4
Presence of village road	27%

ha is quite similar (around 8 M IDR per ha for unselected seed plots and around 20 M IDR per ha for selected seed plots – detailed data per type of holding). Our estimates for the income (gross sales of declared harvested FFBs) were found to range between a low level (nine months of harvest) and a higher level (12 months of harvest). Even if we adopt this rather optimistic assumption, the economic results for the oil palm lay far behind the results from Jambi. Our groups, respectively, show per ha income (IDR) of 6, 15, 8 and 5 M. Therefore, we may formulate the following hypothesis: (i) the oil palm

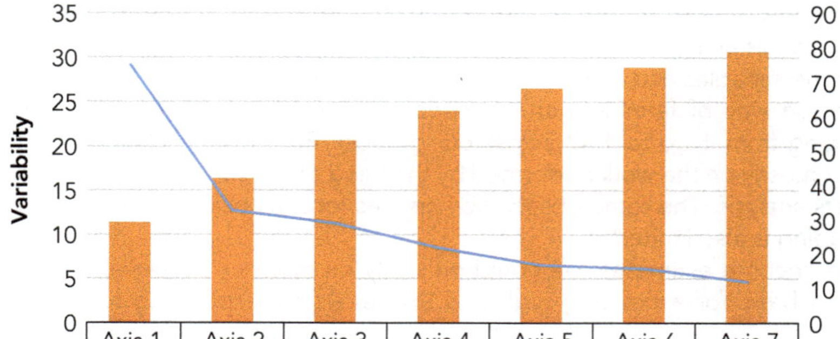

	Axis 1	Axis 2	Axis 3	Axis 4	Axis 5	Axis 6	Axis 7
%cumulated	29.1	41.9	53.1	61.6	68.1	74.2	78.8
Variability %	29.1	12.7	11.2	8.5	6.5	6.1	4.6

Figure 7 Hierarchical classification: variability per axis.

Table 7 Main characteristics of the four groups

	Group 1	Group 2	Group 3	Group 4
Total area (ha)	3.0	6.8	3.8	5.7
Area per family worker (ha/unit)	2.8	5.8	3.8	3.9
Number of family workers (unit)	1.1	1.2	1	1.5
Average hired workers per holding (unit)	0.4	3	1	0.9
Level of fertilisation	221 000	720 000	326 500	177 391
Level of fertilisation (Euros/ha)	4.9	7.0	5.8	2.1
Level of fertilisation per ha	74 161	105 727	86 376	31 231
No. of items of equipment	3.8	5.4	5.1	2.7
Equipment per ha	1.3	0.8	1.3	0.5
Area per equipment of the holding	0.8	1.3	0.7	2.1
	144	656	239	223
Value of production (Euros, monthly average)	121	656	207	184
Average value of production per ha	41	96	55	32
Type of seeds/seedlings	Improved	Balanced	Balanced	Balanced

Table 8 Average incomes per group

	Group 1	Group 2	Group 3	Group 4
Family workers	1.1	1.2	1.0	1.5
Hired workers (av)	0.4	3.0	1.0	0.9
Total workers	1.5	4.2	2.0	2.4
ha	3.0	6.8	3.8	5.7
Income/month (Euro)	121	656	207	184
Income/month (IDR)	1 815 000	9 840 000	3 105 000	2 760 000
Annual income (12)	21 780 000	118 080 000	37 260 000	33 120 000
Annual income (9)	16 335 000	88 560 000	27 945 000	24 840 000
An inc/ha (12) – (RP)	7 260 000	17 364 706	9 805 263	5 810 526
An inc/ha (9) – (RP)	5 445 000	13 023 529	7 353 947	4 357 895
An inc/ha (12) – (US$)	538	1286	726	430
An inc/ha (9) – (US$)	403	965	545	323
Inc/holding (12) (US$)	1613	8747	2760	2453
Inc/holding (9) (US$)	1210	6560	2070	1840
Inc/holding (12)/FW	1467	7289	2760	1636
Inc/holding (9)/FW	1100	5467	2070	1227

seeds or seedlings used are mostly and certainly unselected and uncertified; (ii) Group 2 may have had access to some selected seeds; (iii) average incomes are far from those expected from plasma or independent plots planted with certified seedlings; (iv) the level of fertiliser applied is really low.

5.1 Group 1: family farms with low production but good living conditions

Group 1 holdings rely on family labour with limited external labour (0.35 workers on average (lower than the sample mean of 1.1). The average farm size is the lowest of all groups, and it is lower than the mean average value for the sample and the household size follows the same pattern. Production level appears limited: it is the lowest of the four groups, but the household welfare is higher than for the other groups.

5.2 Group 2: large-scale farms employing permanent labour

Group 2 holdings' size is the highest among the four groups (6.8 ha), and the family is slightly more involved on farm (1.2 work unit in average). To farm nearly 7 ha, households hire permanent workers (three on average, which is about three times higher than the average of the sample) and are better equipped (comparable to the one in Group 3, but with a larger area to farm compared to Group 3). They show the highest level of production far beyond the level of the three other groups, around three times the mean of the sample. The living conditions reach higher standards.

5.3 Group 3: farms using external labour force with low production and poor living conditions

Group 3 holdings farm slightly more land than those of Group 1. The bulk of family work is on the head of household like in Group 1, but unlike this group, households tend to hire permanent external workers (three times more than Group 1, with one work unit). Their dependence on equipment is comparable to that of Group 2 (over the mean of the sample). They raise more livestock than any of the other groups, but this activity remains limited. They appear less efficient in terms of production as a consequence of their relatively poor standard of living.

5.4 Group 4: family farms benefiting from larger areas, underequipped and experiencing very poor living conditions

The average size of Group 4 holdings is just behind Group 2, with 5.7 ha, and relying more strongly on family labour than any of the other groups, but with the help of nearly one unit of permanent worker (like in Group 3). The area farmed is larger than that in Group 1. The households are less efficient in terms of production and their domestic equipment and standard of living are lower than those in any of the other group.

Table 9 summarises the major characteristics of the four groups, combining estimates of production performances and estimates of 'standard of living'.

Table 9 Group distribution according to production performances and standard of living

		Production			
		Low		High	
Living conditions	Low	Group 3	23	Group 4	23
	High	Group 1	37	Group 2	17

6 Conclusion

First, it is worth noting that present results are based on a limited number of samples which were collected under difficult circumstances in 2015. Data could not be compared with other series and some relevant data are missing. We would definitely need to implement surveys on a larger scale in order to obtain an informed picture of smallholder agriculture in several key regions of the Indonesian oil palm sector.

Nevertheless, several interesting results are worth discussing as they stand and will certainly help in designing paths for further research.

Smallholders' population in the oil palm sector is much more diverse than the dualistic presentation 'Estate vs Smallholder' or the homogeneous distribution of plots within the industrial schemes would suggest. If the initial allocation of plasma plots in the industrial system is based on a 2-ha area for each family with a ban on plot sales, reality becomes much more complex. On one side, the industrial plantation keeps the fictional homogeneous distribution alive, considering only the plot as their basic unit for the smallholder sector, ignoring the 'holding level' and the actual market of agricultural lands in these schemes. Our data show that some holdings have more than 2 ha as plasma plots (2013 survey). Interviews with selected employees of the schemes or with smallholders confirm the reality of the land market for planted plots within different industrial schemes. 'Farming' these plots is highly remunerative because most or all (depending on PIR vs KKPA scheme) of the operations are under the control of the plantation company. The high incomes generated by such plots have formed the basis for accumulation of a large number of smallholders that progressively and somewhat rapidly have become strong family business farmers with consolidated properties (Barral, 2012; Baudoin et al., 2015).

These success stories are undeniably credited to the past policies, but one cannot deny that a large number of smallholders could not benefit from the same opportunities and have had or have today more limited facilities with regard to access to land, capacity to invest and finally to catch up with an accumulation process. Our 2015 survey strictly focused on independent smallholders. Through our random selection, we captured a majority of small-scale oil palm producers showing less heterogeneity than in 2013 in terms of scale of operations. Oil palm plantations are smaller and the economic status of these holdings is certainly more precarious. Considering smallholders' oil palm production as defined by RSPO (area under 50 ha) leads to a wide definition that is of little use for policy-making. Even if a smallholder operating 50 ha of oil palm can definitely be considered as a wealthy farmer, it is not useful to use the same criterion for smaller-scale oil palm growers. Our 2015 survey explored a vast majority of smallholders cultivating less than 4 ha and it evidenced a need for a more refined analysis. We agree on the recommendation to deepen the comparative knowledge between smallholders and large-scale plantations[6] (Dislich et al., 2016), although we consider it equally important to scrutinise the heterogeneity within the smallholder population.

For the majority of tropical crops, the twentieth century saw triumph of the family farming model because of a variety of reasons analysed by Daviron (2002). Oil palm in Indonesia could question this statement with the consolidation of large-scale state-owned plantations in the second half of the twentieth century and their privatisation during the last decades of this century. Studying the issue of large-scale plantations versus family farms, Hayami (2002) pointed out three major reasons that may justify:

> the development of plantation agriculture despite the high cost of management: (i) close coordination between farm-level production and large-scale processing/marketing facilities is required for certain crops, (ii) basic infrastructure is absent in the land-opening stage so that the large-scale production unit is needed for internalizing the gain from its supply of infrastructure, and (iii) concessions on the use of large tracts of virgin land are granted for exclusive use by certain power elite, such as Western planters during the colonial period.

In the case of the province of Riau, basic infrastructure has been already established to take advantage of sparsely populated lands with large-scale estate plantations. For the majority of them, large-scale plantations are now under the control of large corporate and financial groups, owned by Asian or international firms. They occupy a dominant position in the sector but face strong international and national dissent because of social and environmental impacts of the industrial palm oil model. As land is becoming scarce with a growing rural population, the necessity to maintain the plantation-based schemes is no longer so obvious for the state, and corporates may have grown reluctant to invest in a context where they face increasing criticism from social movements, when, at the same time, smallholder agriculture is ready to take over and has already started to do so (ref).

We may then assume that we are reaching a turning point in Riau where two of the three conditions that Hayami (2002) framed and could justify large estate plantations are no longer present. Therefore, increasing knowledge on this growing population of oil palm smallholders is a strategic challenge for the present period.

Many argue that being in the vicinity of industrial estates promotes the dissemination of technical improvements and innovations. Indeed, proximity, network effects and more or less informal relationships among farmers can lead to the dissemination of improved techniques. Our data may both validate and question this optimistic view. The 2013 survey points out trajectories of economic accumulation for smallholders who could have access to plasma plots or be able to plant on rather large areas on their own. This situation will become less and less frequent for the establishment of new agro-industrial schemes, which tend to become increasingly difficult in this area when at the same time, arable land is getting scarcer due to population growth. The sites, which we surveyed in 2015, were also in the vicinity of large-scale plantations operated by corporates, but we chose to focus strictly on independent smallholders. So, we found small-scale smallholders around these agro-industrial schemes: for 63% of our sample, holdings under 5 ha were established less than 15 years ago for 52% of them (Table 5) for whom it might be difficult to accumulate at

6. This is the initial core concern of WAW international initiative.

a comparable pace than earlier settlers. The geographical vicinity does not systematically induce a process of economic growth that could be captured by the size of holdings or an estimate of their living conditions. So in our 2015 sample, 'low conditions' affect 40% of the holdings and the 'low production' affects 60% of the holdings. Even if our data limit the scope of our conclusions, these orders of magnitude indicate that we are far from achieving a uniform dissemination of technical improvements and the expected wealth. Within the same locality, smallholders share a geographical proximity, but the date of their installation creates an economic distance with the first settlers. We may guess – in the absence of refined data at provincial level – that these smallholders are the most numerous and therefore would require more attention regarding technical and economic support. The most important efforts relate to investment, both economically and technically. Funding the establishment of a plantation requires standing for five years at least without any FFB yield, which can be compensated by intercropping or off-farm jobs when available. The quality of the seedlings ranks first as a factor determining yield [39% of potential production in average for unselected seeds compared to 100% for selected seeds, production starts later – at least one year later and the production remains lower than 5t/ha/year compared to over 20 t/ha/year after four years in the case of selected seeds and proper management – (Jacquemard, 2011)].

Hence, except for Group 2, the return to family labour is much less attractive than it is for smallholders that could have access either to plasma lands or could afford to invest in selected planting material. The similarities of these smallholdings with those surveyed in 2013, which had access to plasma plots, are striking. For the majority of the smallholders from the three other groups, income from oil palm is barely enough to cover family needs, although too low for making further investments. This result calls for further studies and research on a wider scale with large sample sizes in order to be more representative of the diversity. We should then be in a better position to picture a reality, which is far from the stereotypes and closer to the actual productive systems today. This has a strategic meaning given the importance of the smallholder component in the palm oil sector. Information is scarce, dispersed and lacks a consolidated view.

7 Future trends

The dynamics of smallholder production in the Indonesian oil palm sector is still under documented. This is a growing concern given the share of smallholders in cultivated area as in production.

No doubt that the performances of this sub-sector are less than those of the area under the industrial control and management. In most studies, smallholders are almost left out and would require special attention because they hardly make a living with oil palm given their technical background and lack of financial support to bear the plantation cost, leading to overconsumption of area planted with unselected material and low yields. An undertaking of a large-scale survey of oil palm smallholders with a specific focus on independent growers would be timely. The current period is strategic since most of the 'plasma estates' schemes were established during the 1980s and 1990s, inducing the dynamism of independent smallholders development following the waves of transmigration. The challenges related to smallholders are twofold: (i) How to support established smallholders facing the replantation challenge sooner or later? (ii) How to support smallholders willing to shift to a productive and sustainable oil palm production system under decent labour and income conditions?

Characterising smallholder diversity is of interest for the oil palm sector because of the dual systems and the dynamic independent family farms operating outside the industrial system. The characteristics of these family farms are not known. This is a challenge for the academic community but there is also a need to feed the policy debate with appropriate figures. Without this information, it is not possible to design differentiated policies.

Regarding the specificities of the sector in Indonesia, it is also crucial for smallholders and their representative organisations such as SPKS (http://www.spksnasional.or.id/) for several reasons: (i) the land issue, since not all the smallholders have an official document to prove the ownership of their plantations; (ii) the need for smallholders by 2020 to trace the origin of their FFBs in order to access markets and (iii) the possible reshaping of development models through collective action based on the appropriate knowledge of the smallholder population. SPKS has started developing an ambitious programme to document the diversity of their members in relation to the government agencies.

At the level of FAO, the international initiative WAW aims to document the transformation of the different types of holdings ranging from the diversity of family farms up to corporate farms boosted by the large-scale land acquisition dynamic after 2008. The first challenge is to identify and weigh (in numbers and contributions to Sustainable Development Goals) the family farms sector when the United Nations general assembly has just declared the decade of family farming (see http://undocs.org/A/C.2/72/L.12). The second challenge is to define the type of farming model which would be the most suitable to face the food and related challenges of the rural worlds where family farms dominate, representing the largest employment sector at world level and also concentrating on hunger and poverty, which is not a minor paradox.

8 Acknowledgements

The authors want to thank the French National Research Agency (ANR), which supported the SPOP project (http://spop.cirad.fr/) under the framework of the Agrobiosphere programme. They are also very grateful to PT SMART and CIFOR Indonesia for helping with the fieldwork logistics, and to Karine Lé and Pak Albertus Magnus C.K. for the 2013 field data collection. Wiyono deserves special and warm thanks for having performed data collection in 2015, under difficult circumstances of various kinds, including facing the distrust of smallholders. The 2015 data collection was possible thanks to Cecile Bessou and Alain Rival, as coordinators of the project. Many thanks to Raphael Marichal (CIRAD/PT Smart) who played a key role in organising and supporting the 2015 data collection. The authors want to also thank Dr. Patrice Levang from IRD and CIFOR for his guidance throughout the SPOP project and support to undertake the 2015 survey.

9 References

Barral, S., 2012. Le nouvel esprit du capitalisme agraire. Les formes de l'autonomie ouvrière dans les plantations de palmier à huile en Indonésie Paris. Thèse de doctorat, Paris: EHESS, p. 480.

Barral, S., 2015. Entre firmes et exploitations patronales. De petites plantations familiales éphémères en Indonésie. In: *Diversité des agricultures familiales: Exister, se transformer, devenir* (Bosc Pierre-Marie, Sourisseau Jean-Michel, Bonnal Philippe, Gasselin Pierre, Valette Elodie and Bélières Jean-François (Eds)). Versailles: Ed. Quae, pp. 149–61 (Nature et société).

Baudoin, A., Bosc, P.-M., Moulin, M., Wohlfahrt, J., Marichal, R., Caliman, J.-P. and Bessou, C. 2015. Linking the transformation of production structures to a multidimensional sustainability assessment grid of smallholders' oil palm plantations. *International Journal of Sustainable Development and World Ecology* 22(6): 520–32.

Baudoin, A., Bosc, P.-M., Bessou, C. and Levang, P., 2017. Review of the diversity of palm oil production systems in Indonesia. Case study of two provinces: Riau and Jambi. Woking paper n° 219. Bogor, Indonesia: CIFOR. doi: 10.17528/cifor/00646210/2017. https://www.cifor.org/library/6462/review-of-the-diversity-of-palm-oil-production-systems-in-indonesia-case-study-of-two-provinces-riau-and-jambi/

Bélières, J.-F., Bonnal, P., Bosc, P.-M., Losch, B., Marzin, J. and Sourisseau, J.-M., 2015. *Family Farming Around the World. Definitions, Contributions and Public Policies*. Paris, Montpellier, France: CIRAD and AFD.

Bessou, C., Rival, A., Levang, P., Feintrenie, L., Bosc, P.-M., Cheyns, E., Djama, M., Wohlfahrt, J., Marichal, R., Roda, J.-M., Caliman, J.-P. and Pacheco, P., 2017. Sustainable palm oil production project synthesis. Understanding and anticipating global challenges. Info Brief Cifor, n° 164. Bogor, Indonesia: CIFOR. January 2017. doi: 10.17528/cifor/006361.

Brookfield, H. and Parsons, H. 2007. *Family Farms: Survival and Prospects. A Worldwide Analysis*. Abingdon, UK and New York, NY: Routledge, Taylor and Francis Group, xvii + 272pp.

Daviron, B., 2002. Small farm production and the standardization of tropical products. *Journal of Agrarian Change* 2(2): 162–84.

Devendra, C. and Thomas, D., 2002. Crop–animal interactions in mixed farming systems in Asia. *Agricultural Systems* 71(1–2): 27–40.

Dislich, C., Keyel, A. C., Salecker, J., Kisel, Y., Meyer, K. M., Auliya, M., Barnes, A. D., Corre, M. D., Darras, K., Faust, H., Hess, B., Klasen, S., Knohl, A., Kreft, H., Meijide, A., Nurdiansyah, F., Otten, F., Pe'er, G., Steinebach, S., Tarigan, S., Tölle, M. H., Tscharntke, T. and Wiegand, K., 2017. A review of the ecosystem functions in oil palm plantations, using forests as a reference system. *Biological Reviews* 92: 1539–69.

FAO, 2012. *World Agricultures Watch, Methodological Framework*, version 2.6. Rome: FAO, April 2012, p. 47.

Goldthorpe, C. C., 1987. A definition and typology of plantation agriculture. *Singapore Journal of Tropical Geography* 8(1): 26–43.

Hayami, Y., 2002. Family farms and plantations in tropical development. *Asian Development Review* 19(2): 67–89.

Hayami, Y., 2010. Plantations agriculture. In: *Handbook of Agricultural Economics*, Volume 4 (P. Pingali and R. Evenson (Eds)). Amsterdam: Elsevier, pp. 3305–22.

Hazell, P., 2007. The future of small farms for poverty reduction and growth. 2020 Discussion Paper 52. Washington DC: International Food Policy Research Institute.

HLPE, 2013. Investing in smallholder agriculture for food security. A report to the Committee of World Food Security by The High Level Panel of Experts on Food Security and Nutrition. Rome: CFS, 112pp. http://www.fao.org/fileadmin/user_upload/hlpe/hlpe_documents/HLPE_Reports/HLPE-Report-6_Investing_in_smallholder_agriculture.pdf.

Hoppe, R. A., Korb, P., O'Donoghue, E. J. and Banker, D. E. June 2007. Structure and finances of U.S. Farms: Family farm report, 2007 Edition, EIB-24, U.S. Dept. of Agriculture, Economic Research Service.

Hubbard, C., 2009. Small farms in the EU: How small is small? In: *111th EAAE-IAAE Seminar 'Small Farms: Decline or Persistence'*. Canterbury, UK: University of Kent.

Jacquemard, J.-C. (2011). *Le palmier à huile*. Versailles: Ed. Quæ, CTA, Presses agronomiques de Gembloux, 258pp.

Johnston. B. F. and Kilby, P., 1975. *Agriculture and Structural Transformation: Economic Strategies in Late-Developing Countries*. New York: Oxford University Press.

Lé, K. 2013. Analysis of smallholders oil palm production systems, Kampar district Indonesia. Master dissertation, Montpellier (France): Montpellier Supagro, p. 75.

Levang, P., 1997. *La terre d'en face: la transmigration en Indonésie*. Paris: ORSTOM, 419pp. (A Travers Champs).

Manoli, C., Ancey, V., Corniaux, C., Ickowicz, A., Dedieu, B. and Moulin, C. H., 2014. How do pastoral families combine livestock herds with other livelihood security means to survive? The case of the Ferlo area in Senegal. *Pastoralism: Research, Policy and Practice* 4: 3. http://pastoralismjournal.springeropen.com/articles/10.1186/2041-7136-4-3. doi:10.1186/2041-7136-4-3.

Mansuetus, A. H., 2015. Fair partnership by oil palm smallholders, *SPKS - The Oil palm Smallholder Union*, Indonesia. p. 36.

McCarthy, J. and Zen, Z., 2016. Agribusiness, agrarian change and the fate of oil palm smallholders in Jambi. In: *The Oil Palm Complex. Smallholders and the State in Indonesia and Malaysia* (R. A. Cramb and J. McCarthy (Eds)). Singapore: National University of Singapore, pp. 109–54.

McCarthy, J., Gillespie, P. and Zen, Z., 2012. Swimming upstream: Local Indonesian production networks in 'Globalized' palm oil production. *World Development* 40(3): 555–69.

Potter, L. and Badcock. S., 2001. *The Effects of Indonesia's Decentralisation on Forests and Estate Crops in Riau Province: Case Studies of the Original Districts of Kampar and Indragiri Hulu*. Bogor, Indonesia: CIFOR.

Potter, L. and Badcock, S., 2004. Tree crop smallholders. Capitalism. and *adat*: Studies in Riau Province, Indonesia. *Asia Pacific Viewpoint* 45(3): 341–56.

Ryschawy, J., Joannon, A. and Gibon, A., 2014. L'exploitation de polyculture-élevage: définitions et questions de recherche. Une revue. *Cahiers Agricultures* 23: 346–56. doi: 10.1684/agr.2014.0727.

Suyanto, S., Applegate, G., Permana, R. P., Khususiyah, N. and Kurniawan, I., 2004. The role of fire in changing land use and livelihoods in Riau-Sumatra. *Ecology and Society*, 9(1): 15.

Suyanto, S., Permana, R. P., Khususiyah, N., Kurniawan, I. and Applegate, G., 2001. *The Role of Fire in Changing Land Use and Livelihoods in Petepaha Area*. Riau Province: ICRAF, CIFOR, 48p.

Traoré, M., Nacro, H. B., Ouédraogo, D. and Sanou, M. R., 2013. Dynamique et performance économique des systèmes de production agricole à base de coton dans les villages de Karaborosso et de Kotoura (Ouest du Burkina Faso). *Science et changements planétaires/Sécheresse* 24(2): 115–28. doi:10.1684/sec.2013.0383.

Van der Enden, S., 2013. Smallholders and sustainable palm oil production: A better understanding between policy arrangements and real-life practices. A case study of the Siak smallholders site. Riau province. Sumatra. Master thesis, International Development Studies, Environmental Policy Group, ENP 80433. Wageningen: Wageningen University, 93pp.

Wald, A., 1939. Contributions to the theory of statistical estimation and testing hypotheses. *Annals of Mathematical Statistics* 10(4): 299–326.

Wickizer, V. D. 1960. The smallholder in tropical export crop production. *Food Research Institute Studies* 1(1): 49–99.

Chapter 17

Closing yield gaps for small- and medium-scale oil palm producers: improving cultivation practices

J. I. Sanz, M. Mosquera and J. A. Beltrán, Colombian Oil Palm Research Center – Cenipalma, Colombia

1 Introduction

The structure of the Colombian oil palm sector has changed during the past decade. The proportion of small- and medium-scale growers has increased both in terms of area planted and in number of growers. However, a yield gap has been found among producers, which depends upon their size. Large scale producers tend to reach higher yields in comparison to small- and medium-scale producers. On average, the annual yield gap between the aforementioned groups, when the oil palm crop has reached maturity, has been estimated in 6.6 ton FFB per hectare[1]. Among the reasons that could help explaining this yield gap are the lack of adoption of technologies that could overcome various limiting factors that hamper the proper development of oil palm cultivation.

The Colombian oil palm sector is organized in clusters made up of one main node and several peripheral nodes. The main node produces fresh fruit bunches (FFB) in large scale and it has a mill where palm oil is extracted. Peripheral nodes also produce FFB that are sold to the main node. Such peripheral nodes are made of small- and medium-sized FFB producers who are located near the main node. Fedepalma (The Colombian Federation of Oil Palm Growers) suggested to the main nodes that they should change their view about small- and medium-scale growers; they should consider them not only as raw material providers, but

1. According to the Colombian Federation of Oil Palm Growers Federation, a large producer owns a plantation larger than 500 ha, a medium-scale grower owns a plantation between 50 ha and less than 500 ha and a small-scale producer is defined as a grower whose plantation is smaller than 50 ha.

http://dx.doi.org/10.19103/AS.2017.0018.28

also as strategic partners. Therefore, the main node should be in charge of organizing technical assistance for fruit providers and also, it should be in charge of facilitating the access of FFB providers to credit, inputs and services (such as machinery). In Colombian rural areas, this is important because of the lack of access to formal credit for smallholders and the absence of governmental institutions in charge of technology transfer. The FFB suppliers honour their debts to the main node at the moment they receive the payments from the fruit they sell (periodically). The fruit suppliers' plantations using best agronomic practices will be less prone to attacks by pests and diseases, which improves not only their own sanitary conditions, but also those of the nearby main node plantations, as pests and diseases will not be transferred from one plantation to another. This is therefore a win-win strategy, given the fact that fruit suppliers, having better conditions, will have higher yields and their income will increase, while at the mill, the use of the installed capacity will increase, which in turn, implies a decrease in the fixed costs of the oil extraction process.

Based on the framework described above, the Colombian Oil Palm Research Center (Cenipalma) implemented a strategy known as 'Narrowing the yield gap'. The goal was to enhance technology adoption by small- and medium-scale growers and consequently to improve their actual yields. This strategy required technical and financial support from the main nodes and it was developed from a novel approach for technology transfer, which was named 'producer to producer'.

Briefly, the 'producer-to-producer' strategy begins with the identification of producers who are leaders and are willing to implement the best practices suggested by extension officers (from the main node and Cenipalma). Such leaders must be willing to share with their peers the results obtained by implementing the best practices in their own plantation. Afterwards, technical staff from the main node (and Cenipalma) are in charge of determining the limiting factors that hamper the actual generation of potential yields. Based on this diagnostics, a set of low-cost best practices is proposed for implementation by the identified leader. Once results from best practices are evident, the leading producer is in charge of sharing his experience during field days that will take place in their plantations.

This chapter presents results from the first four years of research undertaken in the Colombian Northern Zone where water deficit is predominant and is worsening with climate change. The practices implemented in plantations owned by producers identified as leaders were as follows: application of organic matter and wide furrow irrigation. The latter was implemented at plantations with access to irrigation. Results indicate that after four years of continuous implementation, the annual yield increased on average by 6 ton FFB/ha and the production costs of a ton of FFB decreased by 6%, on average as well. Production costs are expressed in terms of US$ per ton of FFB (i.e. US$/ton FFB). The maximum increase in annual yield was found to reach 13.5 ton FFB/ha with a consequent decrease in unit costs of 16%. These numbers indicate that small- and medium-sized producers can be part of an economic activity that is not only sustainable, but also profitable and inclusive.

In the introductory paragraphs of this chapter, we present literature reviews on three topics that are crucial for developing this work. We refer to yield gap, organic matter application to oil palm stands and irrigation on oil palm.

1.1 Yield gap in oil palm production

The yield gap is defined as the difference between potential yield and actual yield. The potential yield is obtained when the crop is grown in ideal conditions of climate, soil and management. Meanwhile, the actual yield is obtained by a grower who needs to handle the limiting factors that affect his/her own plantation (Sadras et al., 2015). Yield gap analyses contribute to determining the best practices to implement with the aim of narrowing the yield gap measured in oil palm cultivation.

In practice, determining the potential yield is a very difficult task because there are many interactions among variables such as variety, soil characteristics, soil nutritional content, rainfall, temperature, solar radiation and management practices, among others. Determining the potential yield for a perennial crop brings additional complexities such as considering long-term yield (Affholder et al., 2013). The following information presents a synthesis of the methods that are commonly used to determine the yield potential of a given crop. Please note that it is possible to use more than one method (Lobell et al., 2009).

Crop simulation models: in order to develop these types of models large amounts of data (over long periods of time) are required, which is a time-consuming and expensive process. The main limitation of these models is that they tend to be site specific (Tittonell et al., 2008; Fisher, 2015). Based on the gathered data, scientists develop mathematical models, which are intended to predict the yield response to changes in environmental conditions and crop management (Ittersum et al., 2013).

Yields measured in experimental stations: in research stations, the crop management conditions are ideal, in the sense that any limiting factor is adequately controlled. Such limiting factors include water deficit, lack of nutrients, inappropriate nutrient balance, and pests and diseases, among others. Keeping these ideal conditions along the lifespan of a perennial crop is very costly, so this may be the most expensive way of determining the potential yield. Nonetheless, scientists can generate growth curves that are possibly the closest to potential yield curves (Beddow et al., 2014; Ittersum et al., 2013; Affholder et al., 2013; Sadras et al., 2015).

Highest yields obtained by farmers: this method is based on yield reports provided by growers. Researchers interested in establishing potential yield need to study the appropriate yield records. Commonly, the yield data are organized by regions or by places where environmental conditions are similar (Licker et al., 2010). This is the least expensive manner to determine the yield gap. However, because the yield gap is calculated based on information reported by growers, the main drawback of this method is the reliability of the gathered information (Ittersum et al., 2013).

The yield gap tends to affect the net income of a grower[2], because when the actual yield is less than the potential one, resources are not being used efficiently (Beddow et al., 2014). Once the relationship between yield gap and net income is taken into account, it is evident that the yield gap affects the profit maximization goal of the grower. Academics have established that the lag between actual yield and potential yield is 40% on average at a global level. For instance, small-scale growers from Indonesia tend to produce 10–15 ton FFB/ha, while large-scale producers reach 25 ton/ha (Woittiez et al., 2015).

2. The net income is understood as the result of subtracting the unit production cost from the gross income (per unit).

1.2 Applying organic matter to oil palm stands

Most of the studies on the use of by-products from the oil palm agro-industry, such as empty fruit bunches (EFB), pruned fronds and palm oil mill effluent (POME) at oil palm stands have focused on soil preservation, and on determining the nutrient content of the different types of residues. Several authors have reported the benefits of this practice in oil palm crops:

- The organic matter contributes to keeping moisture in soils and it also provides nutrients to oil palms, and helps to neutralize the soil exchangeable aluminium (Corley and Tinker, 2003; Pulver, 2014).
- The organic matter helps to increase fertilization efficiency, as it stimulates the development of adsorbent roots (Ruiz Alvarez and Molina, 2014).
- The organic matter improves soil structure, thus allowing water to penetrate into the soil (permeability) (Caliman et al., 2001).

1.2.1 Organic matter and soil conservation

With respect to soil preservation, one must refer to the work by Comte et al. (2013), who present results from a long-term research project undertaken in Indonesia that was aimed at evaluating the response of low fertility tropical soils to application of mineral and organic fertilizers. Comte et al. (2013) recommended that in order to preserve low fertility tropical soils, organic matter should be applied as fertilizer. Moradi et al. (2014) came to the conclusion that nutrients are released faster, and in larger amounts, at the initial stages of the decomposition process of the organic matter.

Another interesting result is presented by Moradi et al. (2012). They evaluated soil conservation methods for a three-year period, in some plantations in Malaysia. The soil preservation methods considered were application of EFB, application of compressed EFB mat and application of pruned oil palm fronds. Their results showed that the application of EFB provides the highest amount of dry matter and the highest concentration of nutrients, when compared to the application of other mulching materials (Moradi et al., 2012).

1.2.2 Nutrient content of organic matter

The literature on nutrient contents from the different types of by-products of the oil palm agro-industry may be grouped as follows:

Leaves: studies on organic fertilization have arrived at the conclusion that from a highly productive oil palm plantation, one may obtain about 10 ton/ha of dry matter from the pruned fronds. A ton of fronds contain 12.5 kg N, 0.5–1 kg P, 0.71–1.5 kg Mg, 15 kg K and 6.38 kg Ca (Comte et al., 2012; Moradi et al., 2014).

Empty fruit bunches: according to Taillez (1998), the application of EFB may reduce the need of chemical fertilizers by 50% in young oil palm stands and 5% in mature oil palm stands. Results published by academics on nutrient content from EFB relate that dispensing one ton/ha of FFB, is equivalent to applying: 2.7–8.7 kg N, 0.31–0.51 kg P, 8.1–18.9 kg K, 0.5–1.22 kg Mg and 0.5–2.04 kg Ca (Comte et al., 2013; Moradi et al., 2014).

Compressed EFB mat: one ton of compressed EFB mat contains 6 kg N, 0.3 kg P, 11.26 kg K, 1.72 kg Ca and 0.5 kg Mg (Moradi et al., 2014). Note, in the process of obtaining the EFB mat there is a loss of nutrients. However, this loss is compensated by the lower transportation costs for the EFB mat.

Palm oil mill effluent: dispensing one ton of POME is equivalent to applying 1 Kg N, 0.3 Kg P, 0.3 kg K and 1.3 kg Mg (Comte et al., 2013).

1.2.3 Composting

Yuliansyah et al. (2009) have analysed alternatives to deal with solid waste produced in oil palm plantations and palm oil mills. They mention that, in the past, it was common to incinerate EFB and the resulting ash was used as fertilizer. However, this procedure had its main drawbacks in the large amount of energy required, and in the emissions from the incineration process. As an alternative, the direct application of EFB as mulch was suggested. This had a positive effect on yields with increases ranging from 8 to 34% (Loong et al., 1987; Singh et al., 1989; Yussof, 2006; Beltrán et al., 2015).

As indicated above, EFB direct application imposes high transportation costs and high costs of distribution in the field, and in addition there are traces of palm oil residues in EFB (Schuchardt et al., 2002). In order to overcome these limitations, efforts were made to compost EFB combined with POME. The time taken by the process of composting varies between 2 and 12 weeks, depending upon the presence of nitrogen, which accelerates the decomposition process, and the POME is used as a source of nitrogen (Schuchardt et al., 2002; Silalahi and Foster, 2006). During the process of composting, compounds containing carbon and nitrogen are transformed into more stable organic matter (Singh et al., 2010). Furthermore, using POME for composting brought the extra benefit of dealing with CPO milling residues.

With respect to the content of nutrients in compost, it was found that the application of 15 tons of compost (EFB+POME) supplied the same amount of mineral fertilizers (N, P and Mg) that are commonly applied in Sumatran plantations, with the exception of potassium (Tohiruddin and Foster, 2013). Comte et al. (2012) found that when EFB and POME are composted, the resulting compost contributes to maintaining soil fertility and decreasing the required investment in mineral fertilizers.

1.3 Irrigation

Water management at an oil palm plantation must deal with the stress to oil palm trees caused by water deficit and water excess; both stresses are detrimental to yields. In this chapter, we focused our attention on water deficit literature because this is the condition found in the plantations where this research was undertaken. There are no absolute figures quantifying the yield responses to drought, but according to academics that have studied this topic, it is safe to say that there is a yield loss of about 10% for every 100 mm increase in the potential soil water deficit (Carr, 2011).

In order to avoid water deficit stress, it is necessary to irrigate oil palm trees. This has proven to be effective in increasing yields because watering increases the number of bunches (female inflorescences) and reduces losses caused by abortions (Carr, 2011). For instance, experiments undertaken in Southern Thailand showed that yields increased between 14 and 23% for oil palms under irrigation, when compared to oil palm trees that were not irrigated. They also found that there is a positive yield response to irrigation, if there are enough nutriments available for oil palms (Palat et al., 2008). The experiments undertaken in Southern Thailand were aimed at determining the best method to irrigate oil palms. The irrigation systems considered were dripper irrigation, sprinklers, microsprayers and furrow irrigation. The most cost-effective results were obtained by drip irrigation (Palat et al., 2008).

In Colombia, Alvarez et al. (2007) conducted an experiment in which they implemented two strategies oriented towards improving the water use efficiency of a flood irrigation system. The first strategy consisted of using wide furrow irrigation in order to cut the amount of water required. The second strategy consisted of covering the water conduction channels with geomembrane, in order to avoid the water loss by infiltration. Their results indicated that water use efficiency improved from 5.8 to 20%. The authors concluded that it would be necessary to carry out major improvements in water conduction. The latter is due to the fact that the implemented system of water conduction channels, had ignored relevant aspects of the topography of the plantation (Alvarez et al., 2007).

Among recommendations to avoid drought impact on yields one may highlight:

- Selected planting sites where there are deep water retentive soils (Carr, 2011).
- Subsoiling before planting, as it allows roots to grow and to distribute all over the field (Carr, 2011).
- Mulching with organic materials, since it helps to conserve water (Carr, 2011; Comte et al., 2012).
- Implementation of the most efficient irrigation system, in terms of water use (Palat et al., 2008).

This chapter is organized into four sections. In the first one, we present a brief contextualization of the Colombian oil palm agro-industry. In the second section, we present the institutional arrangement that is being promoted by the Colombian Federation of Oil Palm Growers (Fedepalma). Additionally, we describe the technology transfer strategy implemented by Cenipalma, which is oriented towards increasing the actual yields from Colombian small- and medium-scale producers. In the third section, we present the results in terms of yield, of four years of implementation of practices under the *producer-to-producer* technology transfer system. These results correspond to plantations of small- and medium-scale producers in the Colombian Northern Zone where the main limiting factor is the water scarcity. The practices implemented were oriented towards water deficit mitigation and towards increasing the efficiency of the irrigation system (where available). Finally, in the fourth section, we present some concluding remarks.

2 Analysing yield gaps

2.1 Structural change in the Colombian oil palm agro-industry

During 2000–10, many investors entered the Colombian oil palm sector by means of a public policy, known as *Productive Alliances*. In order to illustrate the structural change, one must refer to the figures on total area planted, according to the plantation size. In 1998, large-scale plantations accounted for 52% of the total area planted with oil palm in Colombia (288.500 ha), while the remaining 138.457 ha were owned by small- and medium-scale oil palm growers. In 2011 large-scale growers accounted for 30% of the total area planted with oil palm (404.101 ha), while medium- and small-sized plantations reached 282.871 ha. In summary, there was a decrease in the participation of large scale plantations in the total area planted with oil palm trees in Colombia, mainly due to bud rot outbreaks, while the area owned by medium- and small-scale growers doubled its size (Girón and Mahecha, 2015). Unfortunately, most newcomers lack, or have, very limited access to financial and technological resources that allow them to implement technologies that help them overcome limiting factors to the proper development of their crops. The latter has caused poor adoption of available technologies, with subsequent low yield (Fontanilla et al., 2015).

2.2 Yield gap among oil palm growers in Colombia

The Colombian large-scale companies usually own a mill for the extraction of palm oil. The mill is fed with oil palm fruit from their own fields and from fields owned by small- and medium-scale growers located in its area of influence. It should be remembered that the large-scale growers who own a mill are referred to as main nodes. Their relationship with small- and medium-scale growers is governed by the biology of the oil palm because once the oil palm fruit is cut, its oxidation process has to be stopped. This is carried out at the mill by means of sterilization (high pressure at high temperatures). If sterilization is delayed, the quality of the oil obtained at the mill will decrease. In Colombia, price penalization is imposed by oil buyers when oil quality is decreased by the presence of free fatty acids ocurring in more than 3% of the total volume of oil.

Figure 1 depicts the yield gap measured in 19 main nodes and their fruit suppliers (i.e. crops from medium- and small-scale growers). An average yield gap of 6.6 tons of FFB per hectare is measured. These data correspond to 34% of the total area planted with oil palm in Colombia and represent all areas where oil palm is planted in the aforementioned country (Fedepalma, 2015). These data do not allow the interpretation of what part of the yield gap is due to lack of productivity and what part is due to the age of plantations (Fedepalma, 2015). However, it is possible to foresee the existence of a trend that links the obtained yield to the scale of the grower (Beltrán et al., 2015). Finally, it must be

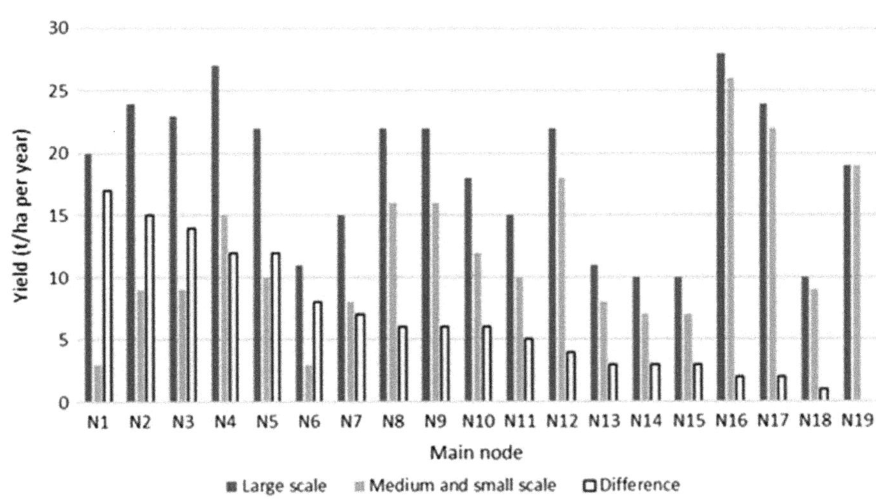

Figure 1 Yields at plantations of large-scale oil palm growers (main nodes) vs yields at plantations of medium- and small-scale oil palm growers, at their areas of influence.

highlighted that yields in main nodes are far from reaching oil palm potential yield; however, they serve as a goal for small- and medium-sized growers.

2.3 The impact of climate change

Climate change is evidenced in Colombia through extreme weather conditions that oscillate between intense rains causing floods and long drought periods (Jarvis and Escobar Carbonari, 2014). Palm oil yields are affected by both water deficit and water excess (Cornaire et al., 1994). In the case of the Colombian Northern Zone, the main obstacle faced by oil palm cultivation is water deficit (Pulver, 2014).

Figure 2 illustrates records of monthly rainfall during the 2010–15 period. The records were provided by Sevilla (Magdalena) weather station, which is the area where this research was undertaken. The horizontal line at 150 mm of precipitation in Fig. 2 indicates the monthly water requirements of the oil palm. It is evident that over the period of study, the number of months for which the oil palm water requirement is met has decreased gradually. In fact, during 8 months of 2010, the rainfall was enough to cover the water requirement of the crop whereas in 2015, the water requirement was only covered during one month of the year.

In summary, the low yields recorded for oil palm in Colombia are the result of several factors. A public policy was put in place which provided incentive to small- and medium-sized growers in order to enter the oil palm business, but then this policy left planters alone once they had entered. As a consequence, they lacked both financial resources and access to technologies to adequately face the obstacles of oil palm cultivation (Bardhan and Udry, 1999). In this chapter, we present the results of four years of research (2010–14) and extension, which was aimed at helping medium- and small-scale growers increase their yields.

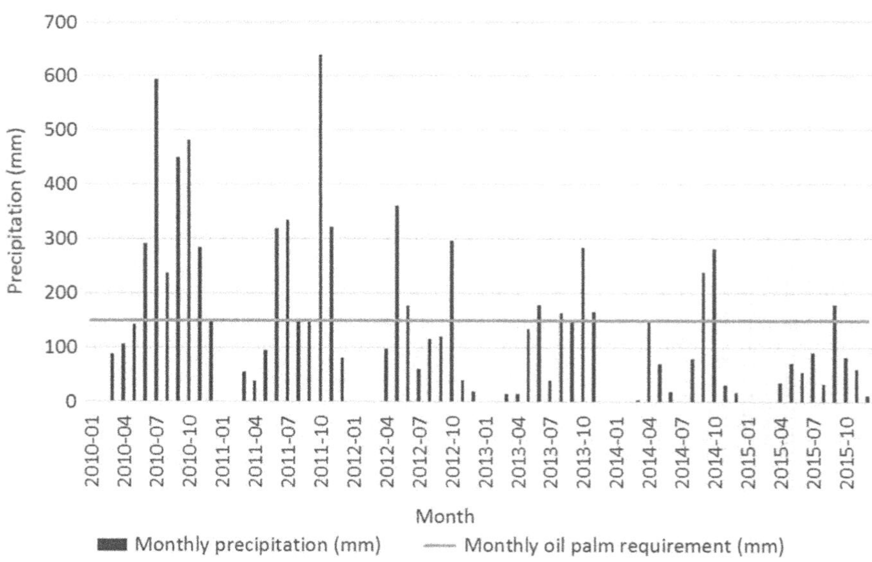

Figure 2 Monthly precipitation during the period 2010–15. Records from the weather station at the Municipality of Sevilla, Magdalena (Colombia).

3 Strategies to improve yields

3.1 Institutional arrangements

In the 1980s, after major Latin American countries had failed to pay their debt, agencies such as the World Bank and the International Monetary Fund requested that Latin American nations adopted orthodox fiscal and monetary policies (Bulmer-Thomas, 2003). Among other consequences, the national budgets for agricultural research and development (R&D) in Latin America were reduced. As a consequence, R&D has been progressively undertaken by the private sector (Franko, 2007). This trend favoured sectors that were already organized and profitable, such as agro-industries. However, this trend left behind the small farming communities and the result was a lack of success of inclusive rural development (Sili et al., 2008).

Fedepalma, as the organization representing the interests of the Colombian oil palm growers, has been promoting a strategy that allows medium- and small-sized growers to have an access to inputs and services through the company buying their fruits: the main node. This means that the main node should go further than only providing technical assistance to their fruit providers. It should also help them to overcome their main obstacle, precisely, their limited access to formal credit. For instance, the main node may lend resources to their fruit providers for buying fertilizers or pesticides. Additionally, the main node will provide services that require economies of scale such as implementing the use of machinery (for instance, drainage or irrigation systems). Small- and medium-sized growers pay their debts to the main node with the fruit they sell. This mechanism is formalized by means of contracts between the main nodes and their fruit suppliers.

As mentioned before, this is a win-win strategy for both the fruit suppliers and the main nodes. The fruit suppliers' plantations which use the best agricultural practices are less likely to be attacked by pests and diseases, increasing their productivity and therefore the fruit suppliers' income. In addition to the fruit suppliers improving their own sanitary conditions they will also improve those of the nearby main node plantations, as they will not be transferring pests and diseases to these plantations (Beltrán et al., 2015). Given the fact that these fruit suppliers will have higher yields the mill figures for use of the installed capacity will improve. Nowadays, the Colombian indicator on mill capacity use averages 60%. This low percentage indicates that there is plenty of room to reduce fixed costs by increasing mill use. Under this strategy set by Fedepalma, the main nodes become the target of the technology transfer research programme and extension service of Cenipalma (Colombian Oil Palm Research Center). Using this approach a win-win strategy is created for both the fruit suppliers and the main nodes, because everyone in the chain gains by the improved conditions, which lead to increased yields, and so greater mill usage, and increased income for everybody.

3.2 Cenipalma's technology transfer strategy

Cenipalma was in charge of coordinating the research and extension strategy aimed at increasing the yields obtained by medium- and small-sized oil palm growers in Colombia. The participation from main nodes that were capable of switching from the traditional view of fruit suppliers to strategic allies was a determinant factor in the success of this initiative.

A technology transfer strategy known as 'producer to producer' was implemented (Pulver and Jennings, 2010). This strategy may be divided into four stages, which will be described in this section:

1 Selection of leader producers
2 Determination of the main limiting factors to attaining potential yields
3 Implementation of demonstrative plots at plantations of leader producers
4 Field trips to demonstrative plots

Selection of leader producers: this stage was carried out based on information collected by the main nodes regarding their fruit suppliers. The focus was on yield records, use of available technology and leadership ability. The selected producers, as well as being willing to implement the suggested practices, had to be respected by their peers, and they also needed to be good at communicating their experiences and results. Once such producers were identified, they were asked to participate in the study on a voluntary basis.

Determination of the main limiting factors to attaining potential yields: a methodology proposed by Cenipalma was used that groups oil palm productive activities (Franco et al., 2012), from the establishment stage (crop planning, soil preparation, selection of planting material, etc.) to crop management (nutrition, pest management, harvesting, weed control, among others). Each set of activities was evaluated in the field by technical staff (from Cenipalma and the main node) and a score was given. The result is a diagnostic of the use of available crop management technologies. Additionally, these results are used to prioritize the low-cost practices to be implemented at the demonstration plot level.

Implementation of demonstrative plots in plantations of leader producers: leader producers participated in implementing the best practices at the demonstrative plots in their own fields. In case there were several factors limiting oil palm development, the most critical ones were tackled first. This is important because it is not wise to overwhelm growers by implementing too many practices at once. The follow-up process to the practices implemented and their results was set on a monthly basis and it was carried out by technical assistants and extension officers. At each visit, technical staff were asked to point out and highlight the improvements observed in the field so they could help leader producers understand their achievements. In addition, this helped leader producers to internalize and share their experiences with peers. Finally, the main advantage from establishing demonstrative plots was the fact that neighbouring producers could witness on a daily basis the improvements at the demonstrative plot of the leader producer. This strategy fostered the technology adoption process.

Field trips to demonstrative plots: it was recommended that transfer technology events took place after results in the field were evidenced. Neighbouring producers were invited to these field trips and the message on practices implemented and the results obtained was transmitted by the leader producer. Technical staff from Cenipalma and the main node were there in a supporting role and helped in the logistics of the field trip, but the leading voice was that of the leader producer. As well as bringing neighbours to the events, field trips were organized for medium- and small-scale growers from distant regions so that they could have a chance to witness the results from implementing low-cost practices.

4 Results of Cenipalma's strategy

4.1 Identification of yield-limiting factors

The oil palm plantations studied in this research were cultivated under dry and warm climates on floodplains and terraces (sub-recent and ancient). With respect to floodplains, the main groups of soils were namely Haplusterts (30%), Fluvaquents (15%), Endoaquepts (10%), Endoaquerts (30%), Haplustepts (10%) and Ustifluvents (5%). In terraces, the most representative groups of soils were Haplustepts (50%), Ustifluvents (20%) and the remaining 30% are soils such as Plintustults, Natrustalfs, Haplustolls and Haplusterts (Instituto Geográfico Agustín Codazzi, 2009).

The physical properties of the soils from the area of study were found to have good drainage and effective depth, which implies that these soils were suitable for the development of oil palm plantations. From a soil physics perspective, the main limiting factors to oil palm cultivation are compaction and the presence of sandy textures with low moisture-holding capacity. With respect to the chemical characteristics of the soils, calcium and magnesium are abundant and available. However, the availability of potassium tends to be compromised by imbalance of bases. In addition, the high content of calcium may limit the availability of iron and manganese (Paramananthan, 2003).

The measurement of the level of technology was conducted in the plantations of fourteen small- and medium-scale producers, following the methodology of Franco et al. (2012). It was established that the score for technology adoption was on average 36 points out of a possible total of 100 points. The scores ranged from 25 to 47 points. These results reflect a very low level of technology adoption from medium- and small-scale growers in the Colombian Northern Zone. The most limiting factors that were found at plantations owned by fruit suppliers in this zone were water deficit, inefficiency of flood irrigation systems and lack of nutrients in the soil.

The yield attained by growers depended upon their access to water. Thus, growers with access to irrigation obtained an average annual FFB yield of 24 ton/ha, despite their delay in implementing proper crop management practices and the deficiency of the irrigation system. Meanwhile, growers that had no access to irrigation obtained, on average, annual FFB yields of 15 ton/ha, as a result of low adoption of technology and their exposure to water deficit. Based on the results just mentioned, the technical staff from the project decided to implement several practices in each demonstration plot. These practices consisted of reaching a nutritional balance and the application of organic matter (mulch) around palm circles. Additionally, for those plots that had access to flood irrigation, a wide furrow irrigation system was implemented, which increased the overall efficiency of the irrigation system.

In addition, from a sample of the oil palm fruit produced, the cost of production per ton FFB was estimated for producers, using the method proposed by Mosquera et al. (2014). The estimation of production costs was of crucial importance in order to determine whether the extra resources required to implement the practices mentioned above represented a profitable investment. The costs' estimation results from the baseline showed that producers who had no access to irrigation had a production cost of US$49.06/t FFB while those who had access to irrigation had a production cost of US$53.75/t FFB. The difference is explained by the costs of installing the irrigation system, its maintenance costs and the payments for the use of water (Fontanilla et al., 2015). However, the net income per hectare was 50% higher in irrigated plots due to higher yield. In fact, it was estimated that the annual net income per hectare without irrigation was US$500 per hectare, while the annual net income in plots with irrigation was US$750 per hectare (Fontanilla et al., 2015).

4.2 Implemented practices

4.2.1 Applying organic matter around oil palm circles

The mulch was placed around the circles of oil palm trees, forming a ring-shaped cushion of organic matter. Two types of mulch were used in this study: EFB and oil palm leaves. The first were brought from the mill, where EFB constitute a by-product. The leaves came from harvest and pruning (Fig. 3).

4.2.2 Costs of applying organic matter

The cost of applying EFB and leaves were considered separately. Regarding the application of EFB, the amount placed around the oil palm circles was 400 kg per palm, every three years. A worker may be able to apply EFB to 20–50 oil palm circles in a working day. It

Figure 3 Organic matter around oil palm circles. Above: Empty fruit bunches. Below: Leaves that were cut at pruning. Photo: Yasmin Penagos. Extensionist at Cenipalma.

Table 1 Costs related to applying organic matter around oil palm circles

Organic matter	Description	Quantity
	Amount of EFB applied (kg/palm)	400
	Labour (US$/ton of EFB applied)	0.47–1.25
	EFB transportation costs (US$/ton)	3.13–6.25
Empty fruit bunch	Average EFB application costs (US$/ha)*	299.41
Leaves	Prunes per year	2
	Extra costs of placing the leaves around the circle (US$/palm)	0.03
	Annual total extra costs of placing leaves around oil palm circles (US$/ha)	8.94
Organic matter application annual costs (US$/ha)		108.55

*The EFB application is done every three years.

was determined that it is easier to apply EFB during the dry season. With respect to EFB transportation costs from the mill to plantations oscillated between US$3.13 and US$6.25 per ton (Table 1).

With respect to the application of leaves while pruning, it was decided to increase the payment per pruned palm by US$0.03, in order to compensate workers for the extra effort of placing the leaves around oil palm circles (Fig. 3). Regarding the application of leaves while harvesting, it was found that placing leaves around the oil palm circle did not mean an extra effort for the worker, but just a modification in where to allocate the leaves. Therefore, it was not necessary for this to incur additional costs. In short, implementing these practices (i.e. to place EFB and leaves, around oil palm circles), increased the annual crop management costs by US$108.55 per hectare (Table 1).

Given that organic matter helps in increasing the number of adsorbent roots and keeping moisture in the soil, fertilization efficiency consequently increases. In other words, the amount of chemical fertilizer that is lost by lixiviation and volatilization decreases. Additionally, the potassium content from EFB (approximately 1.2% wet basis) was considered, and given that organic matter decomposes rapidly under the tropical conditions and nutrient liberation occurs over a short period of time (Khalid et al., 1999), we assumed that the potassium content of EFB was enough to cover the oil palm requirement for this nutrient for one year. However, the potassium applications in the field were not cut.

Table 2 presents two scenarios of increase in the fertilization efficiency (15% and 30%). Our starting point is the cost of the fertilizers prescribed at the demonstrative plots, which averaged US$469.22 per hectare. Beginning from there, we estimated an increase in fertilization efficiency of 15%, which means the oil palms will actually absorb 15% extra from the applied nutrients. The result for this scenario indicates that the grower will not waste 24% of the total value spent on fertilizers (Table 2). Furthermore, if the increase in fertilization efficiency is 30% the grower will not waste 37% of the total value spent on fertilizers (Table 2).

These estimations imply that dispensing organic matter, specifically EFB and cut leaves, around the circles of oil palms is cost efficient. In other words, incurred costs are overcome

Table 2 Economic benefit from increasing fertilization efficiency and value of the potassium content in EFB

Estimated value of increasing the fertilization efficiency		
Fertilization efficiency increase	15%	30%
Value of the total fertilizer prescription per hectare (US$/ha per year)	469.22	
Value of the fertilization efficiency increase (US$/ha per year)	70.38	140.77
Value of the potassium content in EFB		
Fertilization efficiency increase	15%	30%
Value of the potassium content (US$/ha per year)	39.88	32.85
Total economic benefit (US$/ha per year)	110.27	173.61

by their economic benefits. This is a very strong argument that motivated producers to adopt these practices (Fontanilla et al., 2015).

4.2.3 Wide furrow irrigation

As we referred to previously, there is low rainfall in the region where the study took place. These conditions impose the need for watering oil palms. In the region, the most commonly used system is flood irrigation. However, in most plantations, these systems were not designed based on soil studies and topographical characterization. In consequence, the irrigation systems implemented in the region tend to display very low values of water use efficiency. A study was carried out at a plantation in the region and it was determined that the water use efficiency value was 5.8%. It resulted from a conveyance efficiency of 35.6% and 16.1% of application efficiency. In short, for every 100 litres of water that left the water source, just 5.8 litres actually reached the oil palm roots (Alvarez et al., 2007).

Figure 4 Wide furrow irrigation. Photo: Gabriel Enriquez. Extensionist at Cenipalma.

The wide furrow irrigation consists of building two walls that help to guide and control the water path. A width of 1.5 m is recommended in sandy soils and 2.5 m in clayey soils. This system allows for more adequate irrigation and also allows for directing water to where it is needed (Fig. 4). As mentioned above, this practice was implemented in those plantations where small- and medium-scale producers had access to irrigation and used flood irrigation.

4.2.4 Costs of wide furrow irrigation

Implementing wide furrows required two passages of a disc furrower and installing tubes with lids that allowed switching the flow of water from one furrow to the next. The wide furrow implementation had a cost of US$49.21 per hectare. Additionally, the costs related to the maintenance of the irrigation system had to be taken into account. It was determined that in the region where the study took place, the furrows had to be repaired using the disc furrower every two years and, the tubes that allowed the water to pass from one furrow to the next, also had to be replaced with the same frequency. All in all, the annual maintenance costs of the system reached US$8.23 per hectare (Table 3).

From gathered information through the project, we studied the impact of switching from flood irrigation to wide furrow irrigation, in terms of water consumption and labour requirement. A potential evapotranspiration of 1800 mm/year was assumed. Our results indicate that water consumption decreased by 42% and that the area served by an irrigation operator increased by 62% per working day. Since the water runs along the

Table 3 Wide furrow irrigation system. Implementation and maintenance costs

Item	Implementation	Maintenance (Every two years)
Cost of a disc furrower pass (US$/ha)	22.04	11.02
Cost of the tubes for directing the flow of water (US$/ha)	27.16	5.43
Total costs	49.21	16.45

Table 4 Economic benefit from the wide furrow irrigation system

	Irrigation system	
Item	Inundation	Wide furrow
Frequency (episodes of watering per year)	10	10
Labour costs of watering per episode (US$/ha)	9.94	3.75
Water consumption per episode of watering (m^3/ha)	1462	736
Price of water (US$/$m^3$)	0.01	0.01
Irrigation district yearly fee (US$/ha)	35.21	35.21
Total irrigation costs per year (US$/ha)	234.58	118.71
Water savings per year (m^3/ha)		7258
Yearly savings on irrigation costs (US$/ha)		107.2

furrows, the irrigation process is less time-consuming and uses less water, compared to the flood irrigation system that was used before. The irrigation system used previously demanded an operator to unblock the tubes that allowed the water to enter the field. Additionally, the operator was in charge of ensuring that the whole area was irrigated. Summarizing, the savings in annual costs were US$107.2 per hectare, meaning that this technology is also cost efficient (Table 4).

4.2.5 Effect of implemented practices on yields (tons of FFB per hectare)

After four years of implementation of the practices discussed above, the annual FFB yield increased, on average, by 6 ton/ha. This happened, on average, on irrigated plots (from 24 to 30 ton/ha) and on plots without irrigation (from 14 to 20 ton/ha). However, the maximum increase in annual yield observed was of 16 ton/ha in plots with irrigation (reaching 40 ton/ha at the end of the fourth year), and 11 ton/ha in plots without irrigation (reaching 25 ton/ha at the end of the fourth year (Table 5).

In order to assess the economic viability of the implementation of the practices, a variation of the discrete simulation model proposed by Mosquera et al. (2015) was used. This economic model considers cost variations and gross income variations over the long term. We simulated two scenarios of practice implementation (with and without irrigation), and two scenarios of yield increase (average and maximum) for when practices were implemented. In total, we studied six scenarios: three scenarios for plots that have no access to irrigation: (1) without practices, (2) with practices and average increase in yields, and (3) with practices and maximum increase in yields (Fig. 5); and the same three scenarios for plots with irrigation (Fig. 6). The unit area of the model is one hectare and the time unit is one year. The lifecycle length for an oil palm plantation was set as 25 years. From the estimated cash flows, we estimated, for each scenario, the income per hectare, the cost per ton of FFB and the internal rate of return (IRR).

The cash flow from year zero is negative because it represents the financial resources invested in establishing the oil palm plantation. Year one and two are also represented by negative cash flows, because in those periods, growers must invest resources in crop maintenance, but do not receive payments from selling oil palm fruits. At some point, between years three and four, the net cash flows become positive, indicating that gross income from selling fruit is higher than production costs (Fig. 5 and 6). Figures 5 and 6 show the ups and downs of net income curves. This is due to the investment in dispensing organic matter around oil palm circles and to the maintenance costs of the wide furrow irrigation system.

Table 5 Yields at plantations of medium- and small-sized growers, according to adoption of best practices and access to irrigation system

Scenario	Without irrigation	With irrigation
Before implementing the best practices (baseline)	14	24
With adoption of best practices (average yield increase)	20	30
With adoption of best practices (maximum yield increase observed)	25	40

*Tons of oil palm fruit (FFB) per hectare per year.

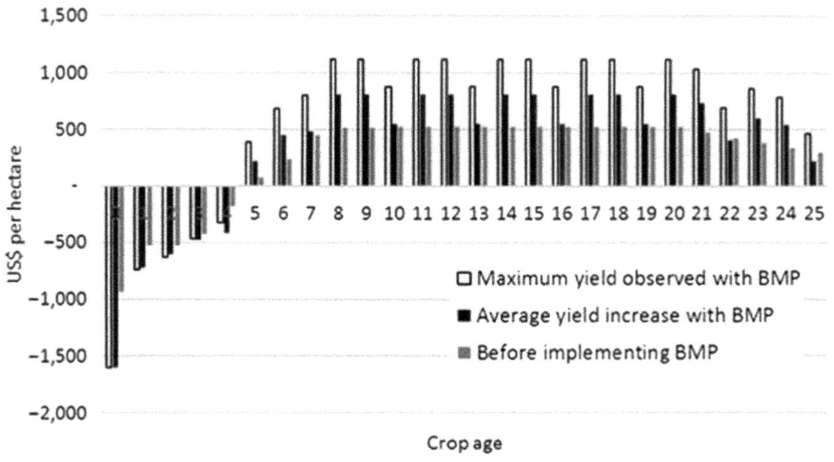

Figure 5 Annual net income attained by small- and medium-scale oil palm growers, according to adoption of best practices and yield. Scenarios without irrigation.

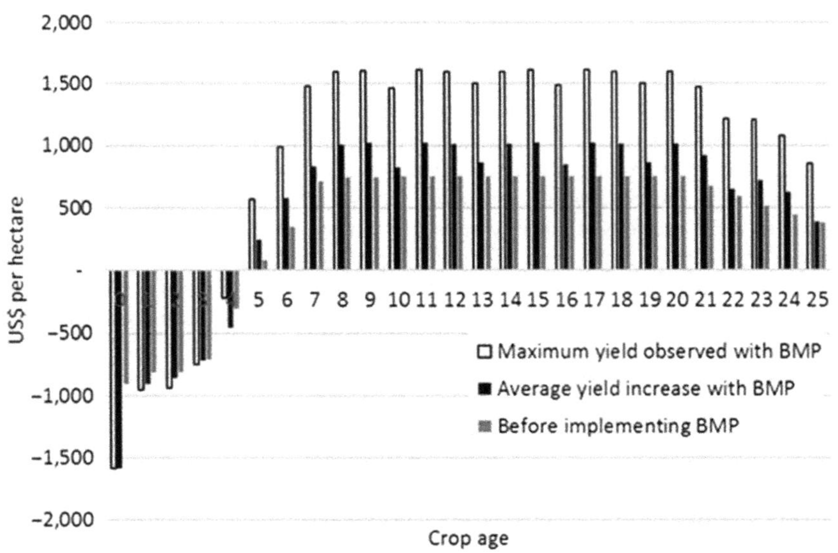

Figure 6 Annual net income attained by small- and medium-scale oil palm growers, according to adoption of best practices and yield. Scenarios with irrigation.

Finally, it can be concluded that the unit cost decreases according to the extent in which yields increase (Table 6). Another very important result is that the internal rate of return (IRR) is positive for all scenarios, meaning that the investment on implementing the practices discussed in this document is viable from an economic standpoint (Table 6).

Table 6 Profitability indicators for medium- and small-scale growers who participated in this project

Access to irrigation	Scenario	Yield[1]*	Unit cost[2]	Annual net income[3]*	IRR[4]
Without irrigation	Before implementing the best practices (baseline)	14	49.06	500.00	11
	With adoption of best practices (average yield increase)	20	45.63	781.25	11
	With adoption of best practices (maximum yield increase observed)	25	41.56	1093.75	15
With irrigation	Before implementing the best practices (baseline)	24	53.75	750.00	12
	With adoption of best practices (average yield increase)	30	51.56	1000.00	12
	With adoption of best practices (maximum yield increase observed)	40	45.00	1593.75	18

[1] Tons per hectare per year; [2] US$ per ton of oil palm fruit; [3] US$ per hectare per year; [4] Internal rate of return expressed in percentage. *Measured in adult palms.

5 Future trends and conclusion

The results of the present research work show that it is possible for small- and medium-scale producers to adapt to climate change with low-cost technologies. Specifically, in the region under study, the limiting condition is water deficit, which is worsening with time as shown by the rainfall records. Water deficit was tackled by applying organic matter around oil palm circles (EFB and leaves) and by installing a more efficient irrigation system.

The impact assessment of these technologies showed that they complied with the objective of increasing productivity and were profitable. As a result, they helped to increase the net income, and thus the welfare of small- and medium-sized growers. All these features indicate that such practices are sustainable.

One of the pillars of success of the work undertaken by Cenipalma and the main nodes was the 'producer-to-producer' strategy implemented for technology transfer: This strategy encouraged leader producers to familiarize themselves with the implementation of best practices and it also assigned them with the role of extensionists. In addition, the neighbouring growers to the leader producer, and those who attended the field trips, witnessed the results at plantations owned by their peers. Having seen the results, it was more likely that they would decide to adopt the best practices.

Another pillar of success for the research undertaken was the participation of technical staff from the main nodes and the willingness of the main nodes to facilitate the access of their strategic partners (i.e. FFB suppliers) to inputs and services required for the implementation of best practices. Those producers that benefited from accessing inputs and services have been paying their debts to the main node, with the money that they receive from selling their fruit. The joint work between the links in the chain allowed medium- and small-sized producers to overcome their most limiting productive obstacles (i.e. credit, access to technology and access to inputs). This fosters integration along the value chain between large-, medium- and small-scale producers.

It should be noted that the joint work between the main nodes and their fruit suppliers is a win-win strategy. On the one hand, the fruit providers improve their yields, their net income and their welfare. On the other hand, the main node improves its indicators of installed capacity use, because of the increase in the amount of fruit processed at their mills.

The oil palm agro-industry is facing market constraints derived from the successful campaign by associations of vegetable fat producers from temperate countries. These associations promoted the idea that palm oil is an unhealthy fat source. This idea is hard to change, even with the recent findings showing that the process of hydrogenating soya oil causes the formation of trans fatty acids that are related to cardiovascular-associated diseases. Naturally, these findings have not caused as many waves as may have been expected, but big corporations from temperate countries are including oil palm stearin into their preparations. Nowadays, the arguments against oil palm have switched from health reasons to sustainability concerns. In fact, oil palm cropping is seen as a threat to the environment. There are many voices in high-income countries saying that in order to establish oil palm plantations, tropical countries are devastating tropical forests and destroying biodiversity hot spots. Another negative aspect that further adds to the poor reputation of the oil palm agro-industry worldwide is that it poorly remunerates its workers. Unfortunately, there are many examples that serve to justify these statements. Finally, the initial capital outlay for oil palm cropping is seen as a barrier for smallholders to enter into the business, and the consequence is that the oil palm agro-industry is not considered to allow inclusive development. The present work in Colombia and the successful Malaysian experience indicate that this is not actually the case, and that when proper access to technology and resources are in place oil palm is a crop that may contribute to rural development.

The market constraints described above have fostered the initiative of certifying oil palm with a sustainability label: roundtable for sustainability of oil palm – RSPO. It allows the trader to show that the palm oil (or fractions/products made out of it) has been properly obtained. This means, environmentally friendly practices and fair trade practices have been used.

From the cropping standpoint, research should be aimed at

- *The development of nutrient recycling practices.* These should be at the core of research since they will lessen the use of mineral fertilizers in oil palm plantations. Additionally, these practices will allow smallholders to have access to low-cost fertilizing inputs.
- *Water management.* In a climate-changing scenario, the water management practices in oil palm plantations under water deficit should be oriented towards improving the water use efficiency, and to exploring alternatives to flood irrigation methods for smallholders.
- *Pest management.* In order to reduce the application of pesticides in oil palm plantations, there is a need to carry out research on pest management based, as much as possible, on biological control, and to provide resistance to pests via improved genetic materials.
- *Mill practices.* At the mill, there is need for research on waste disposal, and on avoiding toxic discharges into water sources. We suggest that composting with combinations of EFB and POME should be favoured.
- *Labour compensation.* Regarding labour, it is necessary to enforce fair labour practices, and from the research standpoint we suggest carrying out research on worker's fair compensation. For instance, in Colombia the oil palm agro-industry

is well known for providing higher payments to their workers when compared to other agro-industries, and for providing social security (retirement fund and health services). The latter are quite rare in rural Colombia.

- *Organizational strategies.* Finally, research on organizational strategies that allow smallholders to access technology and resources is required. Researchers in the field of social sciences are needed to help the agro-industry in designing policies oriented towards inclusion. The model developed by Fedepalma and described in this chapter provides just one successful example on how to target smallholders in order to be part of an inclusive rural development strategy.

Appendix 1 Determination of the level of technology adoption by growers

	Crop management activity	Maximum score
Crop Establishment (Max. 20 points)	Soil characterization studies	2
	Topographic studies	2
	Design of irrigation and drainage channels	6
	Design of agronomical management units	3
	Soil preparation	4
	Establishment of legume covers	3
Crop maintenance (Max. 10 points)	Oil palm circles cleanness	3
	Intercropping lanes cleanness	1
	Pruning	2
	Proper allocation of pruned leaves	2
	Irrigation and drainage channels maintenance	2
Crop nutrition (Max. 30 points)	Foliar sampling for fertilizer prescription	4
	Soil sampling for fertilizer prescription	5
	Future production estimation (counting)	5
	Fertilization efficiency	6
	Fertilization fractioning	4
	Opportune application of fertilizers	4
	Measurement of vegetative growth	2
Pest control (Max. 25 points)	Pest scouting and record keeping	10
	Opportune pest control	10
	Leaf quality	2.5
	Leaf area	2.5
Harvest and yields (Max. 15 points)	Harvest cycle and ripeness criteria	3
	Collection of loose fruits	3
	Quality of fruits that have been harvested	3
	Yields	6

Source: Franco et al. (2012).

6 Where to look for further information

For further information on this topic, readers may refer to the webpage of the Common Fund for Commodities where reports on this research project are kept (CFC-FLIPA Project. Ecuador/Colombia). Additionally, at the *CID-Palmero* from Fedepalma (at Fedepalma's webpage) there are a few written communications that present the details of the rural development initiative named after *'Closing productivity gaps'* (in Spanish: *'Cerrando brechas de productividad'*).

7 Acknowledgements

The authors thank Dr. Edward Pulver and Dr. Jorge Torres for their valuable advice. They also thank Juan Manuel Guerrero for his tireless work, and Álvaro Rincón, Miller Ruiz and Elizabeth Ruiz for their positive contributions. They also thank the researchers and extensionists from Cenipalma. Finally, the authors are grateful to the Fondo de Fomento Palmero (FFP) and to the Common Fund for Commodities (FAO), which co-financed the research presented in this chapter.

8 References

Affholder, F., Poeydebat, C., Corbeels, M., Scopel, E. and Tittonell, P. (2013). The yield gap of mayor food crops in family agriculture in the tropics: assessment and analysis through field surveys and modeling. *Field Crops Research*, 143, 106–18.

Alvarez, O. M., Posada, J. F., Mosquera, M., García, T. and Ardila, G. (2007). Eficiencia en el uso del agua en una plantación de palma de aceite. Importancia económica de su optimización. *Revista Palmas*, 28(2), 11–22.

Bardhan, P. and Udry, C. (1999). Risk and insurance in an agricultural economy. In P. Bardhan and C. Udry (eds), *Development Economics*, pp. 94–109. New York: Oxford University Press.

Beddow, J., Hurley, T. and Pardey, P. (2014). Food security: yield gap. *Encyclopedia of Agriculture and Food Systems*, 3, 352–65.

Beltrán, J., Pulver, E., Guerrero, J. and Mosquera, M. (2015). Cerrando brechas de productividad con la estrategia de transferencia de tecnología de productor a productor. *Revista Palmas*, 36(2), 39–54.

Bulmer-Thomas, V. (2003). *The Economic History of Latin America Since Independence*, Second Edition. Cambridge: Cambridge University Press.

Caliman, J. P., Martha, B. and Saletés, S. (2001). Dynamics of nutrients release from empty fruit bunches in field conditions and soil characteristics changes. *Proceedings of the 2001 PIPOC*. Kuala Lumpur: MPOB.

Carr, M. (2011). The water relations and irrigation requirements of oil palm (*Elaeis guineensis*): a review. *Experimental Agriculture*, 47(4), 629–52. doi:10.1017/S0014479711000494.

Comte, I., Colin, F., Grünberger, O., Follain, S., Whalen, J. K. and Caliman, J. P. (2013). Landscape-scale assessment of soil response to long term organic and mineral fertilizer application in an industrial oil palm plantation, Indonesia. *Agricultura, Ecosystems and Environment*, 169, 58–68. doi:http://dx.doi.org/10.1016/j.agee.2013.02.010.

Comte, I., Colin, F., Whalen, J., Grünberger, O. and Caliman, J. P. (2012). Agricultural practices in oil palm plantations and their impact on hydrological changes, nutrient fluxes and water quality in Indonesia: a review. In D. L. Sparks (eds), *Advances in Agronomy*, vol. 116, pp. 71–124. Burlington: Academic Press.

Corley, R. and Tinker, P. (2003). *The Oil Palm Fourth Edition*. Oxford: Blackwell Science.

Cornaire, B., Daniel, C., Zuily-Fodil, Y. and Lamade, E. (1994). Oil palm performance under water stress. Background to the problem, first results and research approaches. In CIRAD (ed.), *Oléagineux*, 49, 1–11.

Fedepalma. (2015). Objetivo estratégico de incrementar la productividad. In *Informe de Gestión de Fedepalma 2014*. Bogotá: Fedepalma, pp. 87–99.

Fisher, R. (2015). *Definition and Determination of Crop Yield, Yield Gaps, and of Rates of Change*. Field Crops Research.

Fontanilla, C. A., Mosquera, M., Ruiz, E., Beltrán, J. A. and Guerrero, J. M. (2015). Beneficio económico de la implementación de buenas prácticas en cultivos de palma de aceite de productores de pequeña escala en Colombia. *Revista Palmas*, 36(2), 27–38.

Franco, P. N., Arias, N. A. and Beltrán, J. A. (2012). *Calificación del nivel de tecnológico de plantaciones de palma de aceite*. Bogotá: Cenipalma-SENA.

Franko, P. (2007). *The Puzzle of Latinamerican Economic Development*. Lanham: Rowman and Littlefield Publishers.

Girón, E. and Mahecha, X. (2015). Análisis descriptivo de la evolución de la agroindustria de la palma de aceite en Colombia a partir de los censos palmeros de 1997 y 2011. In Fedepalma (ed.), *Revista Palmas*, 36(2), 13–25.

Gómez Cuervo, P. L., Mosquera Montoya, M. and Castilla, C. E. (2005). The oil palm: a sustainable agroindustry in Colombia. In CIRAD (ed.), *Oleaginéaux corps gras lipides*, 2, 121–4.

Instituto Geográfico Agustín Codazzi. (2009). *Estudio General de suelos y zonificación de tierras*. Departamento del Magdalena Escala 1:100.000. Bogotá: Imprenta Nacional de Colombia.

Ittersum, M., Cassman, K., Grassini, J., Tittonell, P. and Hochman, Z. (2013). Yield gap analysis with local to global relevance: a review. *Field Crops Research*, 143, 4–17.

Jarvis, A. and Escobar Carbonari, D. (2014). La adaptación al cambio climático y la palma de aceite. *Revista Palmas*, 35(4), 54–65.

Khalid, H., Zakaria, Z. and Anderson, J. M. (1999). Quantification of oil palm biomass and nutrient value in a mature plantation. Above-ground biomass. *Journal of Oil Palm Research*, 11(1), 23–32.

Licker, R., Johnston, M., Foley, J., Barford, C., Kucharik, C., Monfreda, C. and Ramankutty, N. (2010). Mind the gap: how do climate and agricultural management explain the 'yield gap' of croplands around the world? *Global Ecology and Biogeography*, 1–14.

Lobell, D., Cassman, K. and Field, C. (2009). Crop yield gaps: their importance, magnitudes, and causes. *Annual Reviews Environmental Resources*, 34, 179–204.

Loong, S. G., Nazeeb, M. and Letchumanan, A. (1987). Optimising the use of EFB mulch on oil palms on two different soils. PORIM International Palm Oil development Conference, p. 24. Kuala Lumpur: MPOB.

Moradi, A., Teh Boon Sung, C., Goh, K., Ahmad, H. and Che, F. (2012). Evaluation of four soil conservation practices in a non terraced oil palm plantation. *Soil and Tillage Research*, 145, 62–71.

Moradi, A., Teh, C., Goh, K. J., Husni, M. and Ishak, C. F. (2014). Decomposition and nutrient release temporal pattern of oil palm residues. *Annals of Applied Biology*, 164, 208–19.

Mosquera, M., Evans, E. A. and Ploetz, R. (2015). Assessing the profitability of avocado production in South florida in the presence of laurel wilt. *Theoretical Economics Letters*, 5, 343–56. doi:http://dx.doi.org/10.4236/tel.2015.52040.

Mosquera, M., Ruiz, E., Fontanilla Díaz, C. A., Beltrán, J. A. and Arias, N. A. (2014). *Manual para el registro de costos en plantaciones de palma de aceite de productores de pequeña y mediana escala*. Bogotá: Cenipalma.

Palat, T., Chayawat, N., Clendon, J. H. and Corley, R. H. (2008). A review of 15 years of oil palm irrigation research in Southern Thailand. *The Planter*, 84, 537–46.

Paramananthan, S. (2003). Land selection for oil palm. In T. Y. Fairhurst and R. Härdter (eds), *Oil Palm Management for Large And Sustainable Yields*, pp. 27–57. Norcross: Potash and Phosphate Institute.

Pulver, E. (2014). Ajustándose al cambio climático, ¿qué camino seguimos: estancamiento, paralización o mitigación? *Revista Palmas*, 35(4), 78–85.

Pulver, E. and Jennings, P. (2010). Cerrando la brecha en cultivos alimenticios y sus implicaciones. *XI Conferencia internacional de arroz para América Latina y el Caribe*. Palmira: CIAT Fondo Latinoamericano para el arroz de riego.

Redshaw, M. (2012). Utilization of field residues and mill by-products. In T. H. Fairhurst and R. Härdter (eds), *Oil Palm Management For Large And Sustainable Yields*, pp. 307–19. Kuala Lumpur: MPOB.

Ruiz, E. and Molina, D. (2014). Revisión de literatura sobre beneficios asociados al uso de coberturas leguminosas en palma de aceite y otros cultivos permanentes. *Revista Palmas*, 35(1), 53–64.

Sadras, V., Cassman, K., Grassini, P., Hall, J., Bastiaanssen, W., Laborte, A. and Steduto, P. (2015). *Yield Gap Analysis: Of Field Crops: Methods and Case Studies.* Roma: Food and Agriculture Organization of the United Nations (FAO).

Schuchardt, F., Darnoko, D. and Guritno, P. (2002). *Composting of Empty Oil Palm Fruit Bunch (EFB) with Simultaneous Evaporation of Oil Mill Waste Water (POME).* Bali: International Oil Palm Conference, pp. 8–12.

Silalahi, A. J. and Foster, H. L. (2006). Efficient production of compost from oil palm waste. *Proceedings of the Chemistry and Technology Conference 2006, International Oil Palm Conference (IOPC).* Bali: IOPRI, pp. 427–31.

Sili, M., Ekanayake, I. and Janssen, W. (2008). *Los sistemas de extensión y transferencia de tecnología agropecuaria en América Latina.* Asunción: Banco Mundial.

Singh, G., Manoharan, S. and Toh, T. S. (1989). *United Plantations' Approach to Palm Oil Mill By-Product Management and Utilization.* PORIM International Palm Oil Development Conference. Kuala Lumpur: MPOB, pp. 225–34.

Singh, R. P., Ibrahim, M. H., Esa, N. and Illyana, M. S. (2010). Composting of waste from palm oil mill: a sustainable waste management practice. *Evironmental Science and Bio/Technology*, 9(4), 331–44.

Taillez, B. (1998). Oil palm: a new crop . . . for what future? *Oleaginéaux Corps gras Lipides*, 5, 106–9.

Tittonell, B., Vanlauwe, M., Corbeels, K. and Giller, E. (2008). Yield gaps, nutrient use efficiencies and responseto to fertilisers by maize across heterogeneous smallholder. *Plant Soil* (313), 19–37.

Tohiruddin, L. and Foster, H. L. (2013). Superior effect of compost derived from oil palm mill by products as a replacement for inorganic fertilisers applied to oil palm. *Journal of Oil Palm Research*, 25(1), 123–37.

Woittiez, L., Slingerland, M. and Giller, K. (2015). Yield gaps in Indonesian smallholder plantations: causes and solutions. *Proceddings PIPOC 2015*, 793.

Yuliansyah, A. T., Hirajima, T. and Rochmadi. (2009). Development of the Indonesian palm oil Industry and utilization of solid waste. *Journal of the Mining and Materials Processing Institute of Japan*, 125, 583–9.

Yussof, S. (2006). Renewable energy from palm oil - innovation on effective utilization of waste. *Journal of Cleaner Production*, 14, 87–93.

Chapter 18

Artisanal mills and local production of palm oil by smallholders

Sylvain Rafflegeau, CIRAD, UPR Systèmes de Pérennes, France; Doris Nanda, Université de Yaoundé I, Cameroon; and Claude Genot, INRA-UR BIA, France

1 Introduction

Elaeis guineensis, the most commonly grown oil palm in the world, is native to the Gulf of Guinea. Before colonisation of the African continent, the African people were traditionally red palm oil producers and consumers. The existence of wild palm groves, regionally called 'natural' palm groves, substantiates the origin of such traditional production and consumption (Hartley, 1988). During the colonial period, European colons traded with African custom authorities to export artisanal red palm oil to Europe mainly for industrial uses (Rouzière, 1995). For centuries Africa remained as the main producer of traditionally processed red palm oil. Slaves exported the knowledge about production, culinary traditions and seeds to Brazil (Bahia region) and to some West Indies islands. During the twentieth century, the development of massive processing units occurred until the present-stage industrial mills and refinery with high processing capacities (i.e. 40 t/h).

Owing to the introduction of industrial mills and refineries, the oil palm development took a world scale from the 1960s, particularly in Southeast Asia, and also in Latin America. Nowadays, two types of consumers constitute the world consumption of palm oil on the basis of food uses:

- Artisanal and industrial red palm oil consumers, who are still mainly African people and the descendants of the African slaves; they know and appreciate the taste of red oil palm and its culinary properties, especially its use in the preparation of traditional

http://dx.doi.org/10.19103/AS.2017.0018.36

African and Bahianese meals, although they are also consuming refined oil (Cheyns and Rafflegeau, 2005);
- Refined palm oil consumers are the people in the rest of the world who consume only refined, deodorised and bleached palm oil (after elimination of carotenes, vitamins and taste), because they are not used to the taste of the red palm oil in their culinary recipes.

From this brief historical context, we conclude that Africa is the continent of production and consumption of artisanal red palm oil. This chapter will focus on the artisanal red palm oil production in Africa, where more than 80% of the oil palm production area is controlled by smallholders, including family farms (Rafflegeau et al., 2015). Referring to the history of oil palm development in Africa, the first section will describe the emergence of red palm oil from pre-colonial times up to the recent actual supply chain of artisanal red palm oil, from production to markets.

The existing diversity between smallholders, from family farms to agricultural enterprises, has previously been explained. The different types of smallholders previously described induce a huge range of planting conditions, of technical management of smallholdings, during both immature and mature phases, and of replanting conditions. There are different reasons for smallholders to process by themselves their oil palm fresh fruit bunches (FFB), and a diversity of processing conditions. The second section will show which types of smallholders have embarked on the artisanal processing of red palm oil and will discuss why smallholders decide to do it according to the local context of oil palm development. The third section will describe the artisanal processes and tools associated with artisanal extraction units which are presently at work in Africa.

Due to smallholders' diversity, the processing diversity between smallholders induces a quality diversity of artisanal red palm oil. As a matter of fact, artisanal red palm oil is not a standardised product for the exporting market, but rather a heterogeneous artisanal product sold on the local market. The fourth section will present the quality, properties and uses of artisanal red palm oil. The range found in the proportions of the main chemical components will govern the heterogeneity of quality, which will be assessed in terms of its benefits and risks for human health.

2 Emergence of artisanal extraction of red palm oil in Africa

2.1 Production and consumption zones in Africa

Comparing three African oil crops, namely oil palm, cotton and peanuts, Jannot (2013) established a map showing where colonial agricultural services advised to grow each crop in the 1950s; thus before the independence of African nations, and after this period international institutions started funding the huge national development projects (see Fig. 1). However, the oil palm agricultural area was actually wider than the green line on the map, which just shows the recommended zone, but not the zone classified by colonial agronomists as 'marginal' , where family farms were producing artisanal red palm oil.

Important projects for oil palm development were located in the recommended zones, rather than in the so-called marginal zones, where the oil palm production is less due to climatic

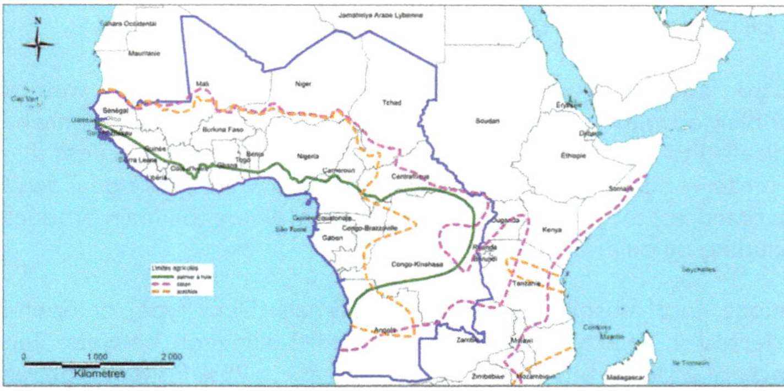

Figure 1 The green line on the map represents the agronomists' recommended zones for oil palm in the 1950s for colonial settlers. Post-independence, oil palm development projects were localised in these recommended zones. Traditional production zones of red palm oil by family farms are wider (from Jannot, 2013).

conditions. In the 1950s, the zone of artisanal red palm oil production by family farms was wider than what the green line on the map represents, and it is still the case. As local traders are selling artisanal red palm oil to south Sahelian population, its consumption zone is even wider.

2.2 Impact of oil palm development history

Ndjogui et al. (2014) described in detail the history of oil palm development in Cameroon, starting from the pre-colonial period. This historical review is not specific to Cameroon as it describes what happened in the palm-oil-producing countries located near the African equatorial zone already described by other authors (Hartley, 1988; Corley and Tinker, 2003). Regarding what we call now 'smallholdings' and 'smallholders', the interest of Ndjogui's review (2014) lies on the description of two parallel transitions due to oil palm development: an agricultural one and a technological one. These two transitions have impacted artisanal red palm oil production.

An agricultural transition

For ages, African men were climbing the palm trees to harvest ripe FFB in 'wild groves' (the community-owned groves of palms described by the European colons) and in agroforestry food crop systems (privately owned plots producing mixed food crops in rotation with fallows). In the 'marginal' zones where there was no planned action for oil palm development, African rural farmers were still harvesting local 'wild' palm in those agroforestry food systems and wild groves. In 'recommended' zones, farmers benefited from oil palm development actions after independence. Most of the smallholders first adopted harvesting tools, thus halting the very dangerous harvesting practice of climbing the palm trees for collecting bunches. They secondly adopted the use of selected planting materials because of its much higher red oil yield, even with a bit lower kernel oil yield. Following agricultural recommendations, they created monospecific oil palm plantation with palms planted in line with a spacing design such as 9 by 9 m in triangle, thus corresponding to 143 palms/ha (see Fig. 2).

Nowadays, palm groves exhibit two different situations:

- oil palm tree areas of high density (from 200 oil palm trees/ha), with trees being sowed or planted with no spacing design, no maintenance and are only harvested;
- area dedicated to food crop production, where few tens of trees (less than 50/ha in order to preserve sunshine for food crops) are saved while clearing fallows with slash-and-burn practices; these are agroforestry systems where the main products are food crops including oil palm bunches.

In both situations, local unselected planting materials (Dura type) or open-pollinated progenies collected in selected plantations (natural pollination between Tenera types giving progenies of 25% Dura type + 50% Tenera type + 25% Pisifera type) are used by family farms. Open-pollinated progenies of unselected materials still produce more oil, even with the 25% sterile Pisifera-type palm trees (Cochard et al., 2001). Figure 2 shows the impact of the type of planting materials on expected oil yields.

A technological transition

For ages, in the oil palm growing area, African people used two traditional methods of oil extraction:

i treading with feet and washing with water in order to extract the floating red oil from the pulp, and
ii heating of the kernel nuts in a saucepan in order to extract kernel oil.

After independence in the 1960s, the huge national development projects funded by international institutions provided access to industrial extraction for smallholders benefiting from the project with plantation credit and technical support. After the funding banks had become bankrupt, the smallholders located inside industrial mill supply areas decided to create their own oil palm plantations in order to benefit from existing industrial mills to

	Types of planting material more generally used			
Smallholders' main cropping systems	Selected (100% T)	Unselected (25%D + 50%T + 25%P)	Local (99% D local)	
Oil palm plantation with design ex: 143 palms/ha	X	X		Agricultural transition ↑ Traditional
Wild grove + isolated palms in food crops plots		X	X	
Palm oil yields % of selected palms yield in the same conditions	100%	40%	10%	
	Agricultural transition ← Traditional			

Figure 2 Oil palm development induced an agricultural transition. The type of planting materials used impacts the oil extraction rate for both artisanal and industrial extraction processes (from Cochard, 2001; Rafflegeau, 2008; Ndjogui, 2014).

process their FFB without having to invest in an extracting tool. In French-speaking African countries, oil palm development was based on the extraction of oil by high-capacity industrial mills which belonged to state-owned development companies. These mills were built to process FFB production from industrial plantations managed by the development company and also from smallholdings developed inside the mill supply area. In English-speaking African countries, oil palm development was based on both industrial mills in large palm oil production area and small-scale processing units for smaller production area (Rouzière, 1995). Artisanal red palm oil production was thus supported by post-independence projects mainly in Nigeria and Ghana and, to a lesser extent, in Sierra Leone and Liberia. Those projects involved national research institutes such as Nigerian Institute for Oil Palm Research (Poku, 2002; Rouzière, 1995).

From the late 1980s, NGOs mostly took over the post-independence projects to promote the artisanal extraction of red palm oil using small-scale mills replicated from English-speaking countries. NGOs – such as APICA and SOWEDA in Cameroon and CFTS in Benin – worked in French- and English-speaking countries, within the recommended zone but also in the marginal zones where many people were involved in traditional oil extraction (Poku, 2002; Rouzière, 1995). Firstly, those NGOs trained blacksmiths to locally produce artisanal extraction tools and secondly they trained smallholders' group to establish and use the small-scale mills. Artisanal extraction immediately boomed because blacksmiths and smallholders copied each other's innovations. On the one hand, turning from traditional extraction of red oil palm into artisanal extraction using a small-scale mill does not affect the quality of the oil and significantly saves time, allowing the same quantity of cooked oil palm fruit to be processed (Fournier et al., 2001). On the other hand, traditional extraction by treading with water facilitated the collection of kernel nuts. As households generally used artisanal red palm oil for food preparation and kernel oil for traditional medicinal purposes, the artisanal extraction process too was adopted, paralleling agricultural innovations. Traditional and artisanal extraction processes coexist, but the artisanal process emerged where and when development occurred.

Figure 3 shows the present mosaic of technological situations for palm oil extraction according to the location – either recommended or marginal – and to the presence – or absence – of oil palm development plans. Trends are thus described: since the

		Types of extraction		
Oil palm agronomic suitability	Oil palm development action	Industrial	Artisanal	Traditional
Recommended zone	With	Frequent	Frequent	Uncommon
	Without	Possible	Common	Uncommon
Marginal zone	With	Never	Frequent	Disappearing
	Without	Never	Possible	Frequent
Oil extraction rate with selected planting material		**21-26%**	**15-18%**	**15-18%**

Figure 3 Oil palm development induced a technological transition as traditional extraction has slowly been disappearing from the 1980s with the boom of artisanal extraction. Artisanal extraction allows some sizeable saving of time to process the same quantity of cooked oil palm fruit.

1980s, traditional extraction has slowly been disappearing with the boom of artisanal extraction.

These two parallel transitions – agronomical and technological – occurred where oil palm development plans were undertaken. Oil palm production can still be traditional elsewhere, if technical innovations were not introduced and implemented. The result is that, nowadays, in a given African country, all types of agricultural and technological situations coexist. Within such a diversity of technical situations, the most important factors for artisanal palm oil production are i) the type of planting materials (see Fig. 2) and ii) the type of extraction method, tools and process, as described in Section 3.

2.3 The present supply chain for artisanal red palm oil

Like in other palm oil producing countries in Asia and Latin America, agro-industries are extracting crude palm oil (CPO) and palm kernel oil from their own industrial FFB and from FFB bought from smallholders. Many African smallholders and palm oil producers without plantation are producing artisanal red palm oil through both artisanal and traditional processes, and figures of such an informal production remain uncovered by international statistics (see Fig. 4). Nevertheless, it represents a huge part of national production in Nigeria. Indeed, Nigeria has not really increased its palm oil production in the last 50 years, dropping from the position of world's leading producer down to fifth rank now – trailing Indonesia, Malaysia, Thailand and Colombia. While it is clear that wild groves cannot satisfy the world's booming demand, artisanal red oil production does meet local demand and foot habits (Cheyns and Rafflegeau, 2005).

Palm groves produce exclusively artisanal red palm oil, but their actual acreage is totally off record and they cannot be considered as plantations because of their specific densities

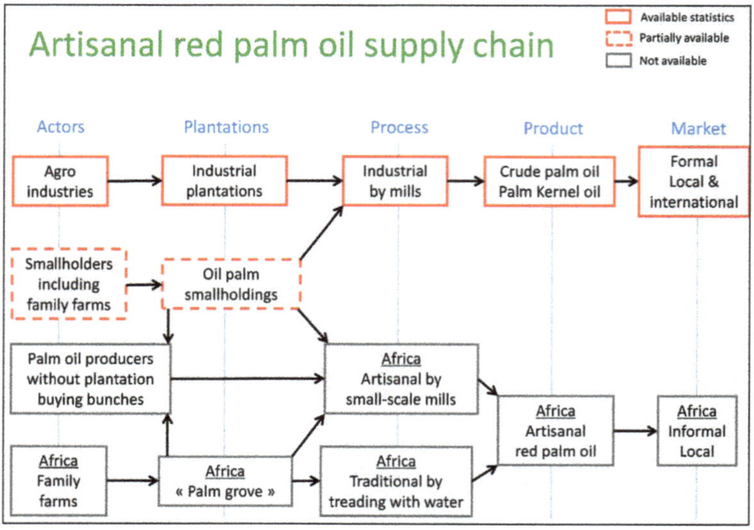

Figure 4 Smallholders are producing artisanal red palm oil by artisanal and traditional extraction, but there is no available production statistics because it is sold on informal market from Lacan et al. (2015).

and structures. They are mainly located out of official palm development zones and thus out of the supply areas of industrial mills.

Figure 4 shows first that artisanal red palm oil is exclusively produced from FFB harvested in smallholdings (except in case of theft from industrial plantations) by local actors: both smallholders and palm oil producers without plantation. Figure 4 also shows that inside a given mill supply area, smallholders can decide to sell their FFB either to the industrial mill or to an artisanal red palm oil producer who does not own any oil palm plantation. They also have the opportunity to process FFB through their own artisanal processing unit or their neighbours' one, the milling cost being traded against a part of the extracted oil (20% in general). In the next section we will describe the reasons governing smallholders' choice for one or the other of these options.

As artisanal red palm oil being sold on the informal market, there is no opportunity to collect production data from custom services or industrial mills similar to those collected for industrial red palm oil and refined palm oil. Because of this informal market, Jannot (2013) explains the difficulties to obtain accurate data of production and uses in Africa which cover both industrial and artisanal red palm oils. When palm oil data from Africa are presented, we must remember that they are only roughly estimated, both for areas (acreage of unselected plantations and palms out of plantation areas are not taken into account) and for production (artisanal palm oil is only roughly estimated or even not taken into account at all).

Artisanal red palm oil can be used for both alimentation and soap production (artisanal and industrial). Industrial refineries do prefer to buy industrial red palm oil rather than the artisanal one in order to reduce the risk of high free fatty acid (FFA) content linked to process conditions and poor preservation.

3 Who is producing artisanal red palm oil and why

We show in Fig. 4 that artisanal red palm oil is produced either by smallholders extracting oil from FFB harvested in their own smallholdings or by oil producers buying smallholders' FFB.

3.1 Diversity of actors producing artisanal red palm oil

Marzin et al. (2015) have characterised the diversity of farm types between entrepreneurial agriculture and family agriculture. Applying their typology to oil palm agricultural actors proved useful in order to show the diversity of farming situations hidden behind the general 'smallholders' name. This typology is built using labour as the main factor, with capital, management, home consumption, legal status and land status being considered as secondary factors. Family farms have no permanent labour, while managerial enterprises employ exclusively paid staff. Family business farms have at least one permanent labour. They could be a 'mature' family farm with enough oil palm area to employ one permanent labour, or an oil palm farm created by a small investor (retail trader, salaried employee, retired person, etc.) involving his family labour force (see Fig. 5).

Only agro-industries are able to invest millions of dollars necessary for setting up an industrial oil mill with high extraction performance and a high level of compliance with environmental requirements, especially regarding the treatments of effluents. Besides, the Caltech model of small-scale hand mill, which costs a few hundred dollars in Cameroon, is within the reach of family business farms and even of family farms. Outside of industrial

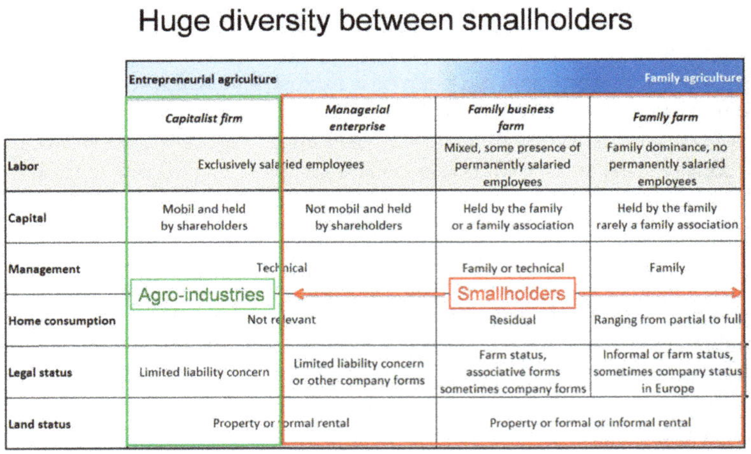

Figure 5 The name 'smallholder' hides a huge diversity of farm types, from family farms with no permanent labour to managerial enterprises with only salaried employees. In contrast, it is easy to imagine what an oil palm agro-industry is with its huge capital invested by multinational capitalist firms and its technical organisation of tasks by salaried labour from Marzin et al. (2015).

mill supply area, when a family business farm invests in some processing equipment which cannot be used at full capacity, it initiates momentum for local development by opening the facility to neighbours. This service will be paid by trading a portion of the produced oil. This is often the case in Africa: a small farmer buys a small-scale mill which he uses only a few days every fortnight. Either inside or outside a mill supply area, clear land tenure situations then allow the synergistic coexistence of family and entrepreneurial forms of production based on market needs for smallholders and on supply needs for all types of mills.

From one ton of FFB produced by selected palms, a small-scale mill usually produces 150–180 kg of artisanal red palm oil, whereas an industrial oil palm mill produces about 250 kg of CPO and palm kernel oil. Despite its low oil extraction efficiency, the small-scale process is considered as economically viable and socially sustainable outside the mill supply basin (Plédran et al., 2016).

Oil producers without oil palm plantations show the same diversity as found for smallholders' farms, with the same characteristics about the factors chosen to build the typology presented in Fig. 5, except for land tenure because such stakeholders do not really need agricultural land. Comparing the levels of investment channelled into extraction units really helps in representing the differences existing among the types of producers' exploitation. Some managerial enterprises of producers can invest to create a semi-industrial extraction unit, in between artisanal extraction units and industrial mills (see Fig. 6 and 7).

Where there is land conflict between state-owned development company and land owners, or between a private agro-industry and people from the neighbouring villages, some oil producers generally steal FFB from inside the industrial plantations at night. Sometimes, theft of FFB can be at the origin of the conflict. Conflicts can also flare up because of FFB stealing, land conflict or jealousy, outside of a mill supply area between a managerial oil palm farm and producers whose farms are adjacent to these managerial farms.

Figure 6 (S. Rafflegeau): Examples of investment in the mini-mill, a semi-industrial extraction unit by managerial enterprises of producers: Global view of a container extraction unit processing stemmed fruit in Côte d'Ivoire.

Figure 7 (S. Rafflegeau): Examples of investment in the mini-mill, a semi-industrial extraction unit, by managerial enterprises of producers: view of a self-made steam FFB cooker in Cameroon.

3.2 Reasons for artisanal extraction by smallholders

The way FFB is processed by smallholders or producers depends on market opportunities for smallholders and on labour available for processing operations. Smallholders have a maximum of four market opportunities for FFB:

1 to sell FFB to the industrial mill or to artisanal producers: this opportunity requires neither investment in processing tools nor labour for processing;
2 to process FFB using their own artisanal extraction unit: this opportunity requires some investment in processing tools and labour for processing;

3 to process FFB using the neighbour's artisanal extraction unit: this opportunity does not require any investment in processing tools, but a proportion of the produced oil must be given to the owner of the extraction unit (20% in general) and for the payment of labour time for processing;
4 to process FFB by traditional extraction: this opportunity requires almost no investment in processing tools, but much more labour for processing than through the artisanal process.

Figure 8 presents a decision chart of the reasons for smallholders and producers to choose between artisanal and traditional processes for the production of artisanal red palm oil.

Outside of industrial mill supply area, the first reason for smallholders to produce artisanal red palm oil is that there is neither industrial mill nor an oil producer for buying FFB. As a consequence, ripe FFB which is not harvested in due time will get rotten on the palm trees. That is the reason why during the peak season of production, extension services do recommend to harvest bunches every 10 days. Ripe FFB which are already harvested will rot much faster than the ones left on the palm tree.

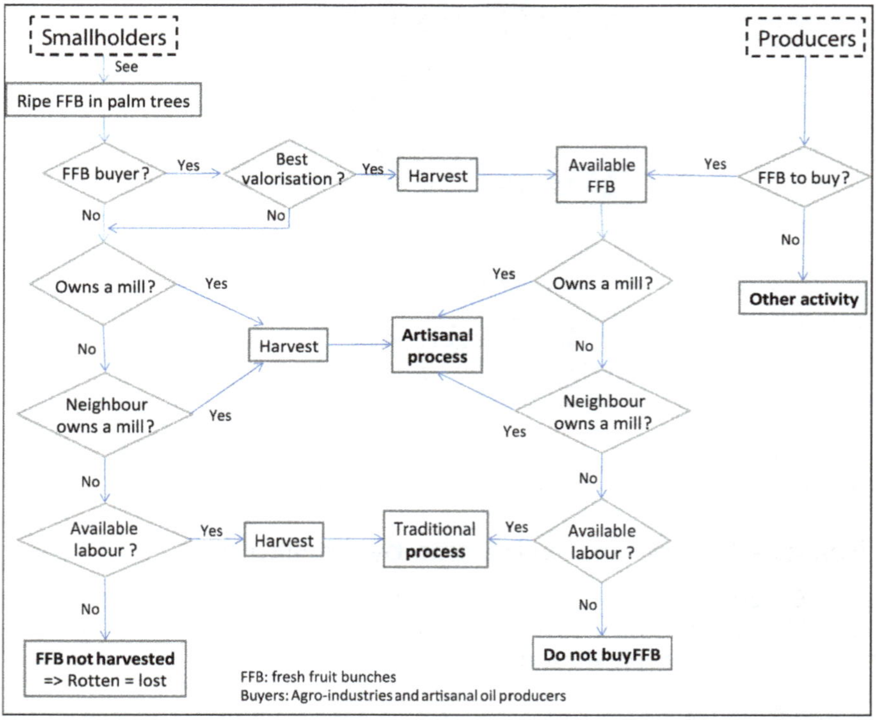

Figure 8 The decision chart shows how smallholders and producers select an extraction process in order to produce artisanal red palm oil. Selling FFB requires no labour for processing, and the traditional process requires the most labour.

Inside an industrial mill supply area, there are three reasons which bring smallholders to invest in an artisanal extraction equipment:

- if the industrial mill is structurally saturated during high production period, collecting from smallholders is temporarily stopped and smallholders lose their FFB;
- if the industrial mill experiences regular breakdowns due to lack of maintenance, smallholders regularly lose FFB;
- if the collecting is consistent during the high production period but voluntarily interrupted during the low production period. This can be due to long travel to farms for a too low quantity of FFB, or high risk for the trucks transporting FFB to get stuck in the mud during heavy rains.

Smallholders located inside an industrial mill supply area can also choose to invest in some processing equipment without any risk of losing FFB, in the following cases:

- if they want to self-process the lost fruit once a fortnight in between two harvests in order to get cash to pay harvesters (day labour) while the mill pay them on a monthly basis;
- if they get more profit from selling the artisanal red palm oil they have processed than from selling FFB to a mill, especially during the low production period when price of artisanal red palm oil is high in the local market, and when labour is not too busy in the food crops plots;
- if there is a disagreement between the industrial mill and the smallholder, like for reimbursing plantation credit.

Artisanal oil extraction is presently booming in Africa, due to investments made by all types of smallholders and producers presented in this section, for all the reasons cited in this section.

4 Major operations and equipment for artisanal processing

Palm oil extraction from FFB is usually done according to the following process (Jacquemard 1995; Nchanji et al., 2013): after harvesting, FFB are transported to the extraction site where bunches are stored. The risk of FFB being stolen increases if FFB stay in the field. The major operations are to stem the fruit, sift and cook them. Subsequently, the cooked fruits are crushed and the palm oil can be extracted by

- washing the mass with water: the lipid top phase obtained is clarified to give artisanal red palm oil;
- pressing the mass: the liquid obtained is clarified to give artisanal red palm oil.

For each operation of the process, we present the corresponding equipment.

4.1 Harvesting FFB

Even if it is not a technological one, this artisanal operation can directly impact oil quality. Harvesting with knife (Fig. 9) can cause cuts on the fruit which promotes the activation of

Figure 9 (S. Rafflegeau): Harvesting palm fruit with a specific tool.

the lipolysis reactions which are responsible for the increase of FFAs content in oil. It is therefore important to handle FFB carefully to limit these reactions (Ngando et al., 2011). When FFB are falling down from the top of a 10-m high old palm tree, some fruits may be damaged.

4.2 Storage

FFB are generally stored for several days (from two to more than ten days) in order to facilitate the destemming of fruit from the bunch. In other cases, bunches are stripped and the spikelets are then stored for a shorter time than bunches. This step induces a fermentation process leading to an increase in FFA content in the fruit due to lipase activity, thus affecting the quality of the oil (Ngando et al., 2011; Owolarafe et al., 2008).

4.3 Threshing (removal of fruit from bunches)

Fruit are removed from bunches and those which remain attached are removed using various objects such as machete, axe, stick and so on (Fig. 10).

Various NGOs have tried to disseminate the use of mechanical threshers but smallholders and producers are not investing in such equipment so they are rarely found.

4.4 Steam sterilising or cooking fruit in boilers

The simplest way of cooking the stemmed fruit is in boiling water in steel barrels (Fig. 11). Managerial smallholders or producers may have invested in the mini-version of industrial mills with the capacity to generate steam to sterilise FFB (Fig. 7). The dried oil cake or dried fibres and kernel, if the kernel oil is not extracted, are often used as cooking fuel. Kernel nuts can be burned as fuel in dried oil cakes, or given to cattle or hand-collected to make traditional kernel oil.

Figure 10 (D. Nanda): Removal of fruit from FFB after few days of storage.

Figure 11 (D. Nanda): Cooking stemmed fruit using firewood.

In most artisanal mills, firewood (with poor energy efficiency) is used, thus causing some environmental impact on surrounding forests (Rouzière, 1995). In some small-scale Ghanaian mills, they even burn old tyres for fuel causing further environmental and food security problems. Industrial mills, however, are self-sufficient in energy, by burning the fibres and nutshells and even produce extra electricity for villages surrounding those mills.

The cooking time very often depends on the amount of fruit and the intensity of the fire. Cooking softens fruit, thus facilitating the crushing process and it also inactivates mesocarp lipase (Babatundé et al., 2003).

4.5 Crushing and mixing

Home-made wood mortar and pestle are the cheapest tools for crushing cooked fruit (Fig. 12), although crushing can also be done mechanically (Fig. 13).

Figure 12 (S. Rafflegeau): Hand crushing.

Figure 13 (S. Rafflegeau): Mechanical crushing and mixing.

4.6 Oil extraction

Oil extracting tools are more often bought, adapted or created by smallholders and producers, whereas other innovative equipment such as clarifiers or mechanical threshers have poor success (Rouzière, 1995; Jannot, 2000). For this reason, there are much more tools for oil extraction than for other processing operations.

Extraction by washing with water

The simplest traditional oil extraction process, which does not require any tool or equipment, was historically the first way to extract oil from the mass of crushed cooked fruits. This method consists in washing it with water and foot treading. This practice still persists in marginal zones despite the harshness and low remuneration of labour (Fournier et al., 2001). A recent adaptation of the traditional extraction consists in building a concrete basin (see Fig. 14) rather than to dig a basin in the ground.

Oil extraction by washing with water can also be done using a motorised water extractor (Fig. 15). It consists of a metal cage inside which there is a rod, around which opposite blades are arranged, connected to a motor. During the oil extraction process water is added. The fruits are mixed and then water is poured into the extractor, which brings

Figure 14 (D. Nanda): Traditional extraction by foot treading and washing with water inside a concrete basin.

Figure 15 (D. Nanda): Workers emptying a motorised water extractor.

out the crude oil to the exhaust of the extractor. This equipment is the highest water-consuming one (Rouzière, 1995).

Extraction by pressing

In marginal zones, smallholders and producers are adapting their existing cassava press (see Fig. 16) or are using a stick and wood structure to press the mass inside a polyethylene bag, such as a rice bag (see Fig. 17). The oil palm cake remains really oily even after two following extraction cycles (pressing, mixing, pressing), showing poor efficiency in oil extraction of such home-made equipment.

There are two major types of oil extraction equipment made by local blacksmiths for smallholders (Jannot, 2000):

- Discontinuous press, with three steps per extraction cycle: filling, pressing and emptying (Fig. 18 and 19). They can be either manual or using hydraulic force to press (Fig. 20) with a manual or motorised pump. They were the first artisanal presses to be disseminated in Nigeria and Ghana. Manual discontinuous press similar to the small Ivorian model shown in Fig. 19 is considered as a valuable tool: it is cheap and provides good performance in terms of labour and oil extraction rate.

Figure 16 (S. Rafflegeau): Adaptation of a traditional cassava press for palm oil extraction.

Figure 17 (S. Rafflegeau): The cooked fruits are manually mashed in the traditional mortar, and then are put in a bag and oil is extracted by pressure using a stick.

Figure 18 (S. Rafflegeau): Pressing with manual and robust discontinuous press imported from Nigeria to Southwest Cameroon for a few hundreds of dollars.

Figure 19 (S. Rafflegeau): Pressing with manual discontinuous press, a smaller model appreciated by women producers in Côte d'Ivoire because of its cheap price (few tens of dollars) and facility of handling.

Figure 20 (D. Nanda): Hydraulic discontinuous press: the cooked fruits are mashed with a digester. The mass is placed in a heavy metal cage and a metal plunger is used to press the materials. The pressure should be increased gradually to allow time for the oil to escape. The plunger can be moved manually or by a motor. If the depth of mass is too great, oil will be trapped in the centre of the mass.

- Continuous equipment: the press is filled from one side, while it is emptied by itself. This is the case of the Caltech-type screw press which also has the advantage of combining two functions into the same equipment: i) crushing the fruit by opposite blades at the beginning of the screw and ii) pressing. They can be either manual (Fig. 21) or motorised (Fig. 22). Today, such artisanal presses are the most appreciated tools because they need less labour, and they can rapidly process fruits (for the motorised ones). Screw presses, due to the turbulence and kneading action exerted on the fruit mass in the press cage, can effectively break open the oil-rich cells and thus release more oil. These presses act as an additional digester and are very efficient for oil extraction. Motorised Caltech-type screw presses are becoming the most popular press in Africa for artisanal palm oil extraction.

The efficiency in oil extraction of the various designs of fruit presses ranges from 60% to 70% for discontinuous presses, 80–87% for hydraulic discontinuous presses and

Figure 21 (S. Rafflegeau): Manual continuous screw press (Caltech type). This vertical model is sold for a few hundreds of dollars in Cameroon.

Figure 22 (D. Nanda): Continuous motorised screw press (Caltech type). This model has a mechanical speed reducer between the engine and the press. Motorised models are sold from about one thousand dollars for the cheapest to a few thousands for the best ones.

75–80% for the continuous Caltech-type screw presses (Poku, 2002). In another study, the press efficiencies ranged from 67% to 88%, with the lowest for discontinuous manual mills and the highest for Caltech-type screw press (Rouzière, 1995).

4.7 Clarification

This is the last major step in the production of artisanal red palm oil. The CPO is cooked with water. At the end of cooking, the palm oil (supernatant) is skimmed (Fig. 23). This operation is either manual or mechanised depending on the production site. The oil collected is then poured into containers.

Figure 23 (D. Nanda): Skimming clarified red palm oil before filling containers.

4.8 Oil drying

Drying the oil consists in heating the oil similar to the clarifying process, but without any water in order to evaporate the humidity that may cause oil alteration by oxidation. Generally, the oil drying operation is not often done after extracting oil by Caltech-type screw presses. However, oil drying is a common operation if extraction has been done with a motorised water extractor (Rouzière, 1995) or traditionally because of too high humidity content.

5 Artisanal extraction units

Artisanal extraction units are combinations of extraction equipment and many authors present the diversity of possibilities in grey literature rather than in scientific papers (Rouzière, 1995; Jannot, 2000; Poku, 2002). For a given extraction unit, the less performing equipment will behave as the bottleneck which will define the potential theoretical extraction capacity, assessed in tons of FFB processed by hour. If the succession of operations in the extraction unit is well organised with the right amount of dedicated labour force, its real extraction capacity equals its theoretical capacity. Traditional extraction units have no proper capacity because each operation is handmade. As most of artisanal extraction units are processing fruits but not FFB, the threshing operation is not taken into account, while it can easily become a bottleneck during the high production period because it requires a lot of labour force. Comparing the press capacity will then give a magnitude of what can be the capacity of the whole extraction unit. In order to rank the various types of artisanal extraction units available, we modified the typology presented by Rouzière (1995) in order to include the traditional extraction process by treading and washing with water (Table 1).

Traditional types of units correspond to marginal zones of palm oil production and also to women producers and young smallholders who are newly established. Micro-mills are the most common situations, as they can offer the best value for a quick investment return. Motorised continuous screw presses are the best systems in terms of oil extraction rate and labour need. Mini-mills are 'businessmen extraction units' showing the lowest labour need and the best extraction rates, but with the longest lag for investment return (Rouzière, 1995).

Table 1 Typology of extraction units

Type of extraction unit	Threshing	Oil extraction tool	Main characteristics	Oil extraction rate	Investment	Labour need
1-A Traditional by washing	Manual	Foot treading, washing with water	Hand process without extraction tool	Medium	None	Highest
1-B Traditional by pressing	Manual	Pressing mass in a bag with a manioc press	Hand process with a home-made tool	Lowest	None	Very high
2-A Micro-mills manual from 200kg to 500kg FFB/h	Manual	Manual press continuous or discontinuous	Hand process with a manual press made by a blacksmith	Medium (discontinuous) high (continuous)	Tens or hundreds of $	Very high (discontinuous) high (continuous)
2-B Micro-mills motorised by washing	Manual	Motorised water extractor	Hand process with an extractor made by a blacksmith	Medium	Tens or hundreds of $	High
2-C Micro-mills motorised from 500kg to 1t FFB/h	Manual or mechanic	Hydraulic continuous press with motorised pump or continuous screw press	Half motorised unit locally made	High	Thousands of $	Medium
3 Mini-mills mainly around 1 or 2 t FFB/h	Mechanic	Continuous screw press	Small-scale reproduction of industrial units extraction, often imported in container	Highest	Tens of thousands of $	Lowest

None of the artisanal extraction units is equipped for the treatment of mill effluents, even with the simplest lagoon for decantation. They quickly cause water pollution in the water courses and rivers surrounding the units and in the worst cases some contaminate the water table. Extraction by washing the mass is probably the process which produces a lot of effluents. Sometimes the quality of the water for processing itself becomes a problem because of the pollution of the rivers by effluents.

6 Artisanal red palm oil composition, quality and uses

6.1 Red palm oil composition and human health

Red palm oil is a lipid extract from the mesocarp of the fruits from oil palm tree (*E. guineensis*). Throughout the world, 90% of red palm oil is used for edible purposes, while the remaining 10% is used for soap and oleochemical manufacturing (Edem, 2002).

Acyl glycerol composition

Palm oil, like all other vegetable oils, consists mainly of acyl esters of glycerol (94–98%) and a minority of other compounds (O'Brien, 1998) (Table 2). TAGs are composed of: 7–10% saturated TAG (mostly tripalmitates) and 6–12% completely unsaturated TAG (Karleskind and Wolff, 1996).

The functional properties of palm oil and its components as ingredient in prepared foods are directly related to the type of triacylglycerols present in these oils (Williams and Hron, 1996). However, the types of triacylglycerols are determined by their fatty

Table 2 Fatty acid ester composition of palm oil (Che Man et al., 1999)

Triglycerides	Content (%)
OOO	3.90
OOL	1.22
PLO	10.02
POO	21.39
OOS	2.78
MPL	3.03
PPL	9.37
PPO	27.39
POS	5.29
SOS	1.39
MMM	0.76
MMP	2.38
PPP	4.81

L: linoleic acid; M: myristic acid; O: oleic acid; P: palmitic acid; S: stearic acid.

acid composition and by the distribution of fatty acids in the individual triacylglycerol molecules. The quantities of each type of triacylglycerols depend on the proportions of the individual fatty acids, the fat or oil source and the processing history of a product (Reske et al., 1997).

Fatty acid composition

The fatty acid composition of palm oil has been widely reported by many researchers (Tan et al., 1981; Che Man et al., 1999) (see Table 3).

Palmitic and oleic acids are the most abundant fatty acids in palm oil. The proportions of saturated, monounsatured and polyunsatured fatty acids are approximately 50%, 40% and 10%, respectively.

Over-consumption of any vegetable oil could have detrimental effects on health. There are two types of fatty acids in oils: the saturated ones and also the unsaturated ones that contain the essential fatty acids. Palm oil is made of about 50% naturally saturated fatty acids (SFA) (palmitic acid), but that does not make it unfit for human consumption. In fact since the 1990s, it has been considered by the food industry as a good alternative to hydrogenated vegetable oil, which could contain trans-fatty acids, a known cancer risk. Nestel et al. (1992) reported a more favourable LDL/HDL ratio with a diet rich in palmitic acid than with a diet rich in trans-fatty acid.

Minor components

They are represented by vitamin E, carotenoids, partial glycerides, sterols and polar lipids.

- Vitamin E (600–1000 ppm): fat-soluble vitamin is predominantly represented by tocotrienols (78–82%) and tocopherols (18–22%) (Tan and Oh, 1981). Tocopherols are structurally characterised by a saturated side chain on the

Table 3 Fatty acid composition of palm oil

Fatty acids (%)	Results[1]	Results[2]
Saturated fatty acids		
Lauric acid C12:0	Trace	0–0.4
Myristic acid C14:0	1–2	0.5–2
Palmitic acid C16:0	43–46	40–48
Stearic acid C18:0	4–6	3.5–6.5
Arachidonic acid C20:0	–	0–1
Unsaturated fatty acids		
Oleic acid C18 :1Δ^{9}	37–41	36–44
Linoleic acid C18: 1$\Delta^{9,6}$	9–12	6.5–12
Linolenic acid C18: 1$\Delta^{9,6,3}$	Trace	0–0.5
Eicosenoic acid C20 :1Δ^{9}	–	0–0.2

Source : Karleskind, 1992[1]; Firestone, 2006[2].

chroman nucleus (Munne-Bosch and Alegre, 2002), whereas the tocotrienols possess an unsaturated side chain of phytyl. Tocotrienols are mainly in the form of alpha (22%), gamma (46%) and delta (12%), and tocopherols in the alpha form. During the refined, bleached and deodorised (RBD) process, the vitamin E content of palm oil is partially lost. Palm stearin, palm olein and palm oil RBD retain approximately 76, 72 and 69%, respectively, of the vitamin E content of red palm oil (Sambanthamurthi et al., 2000).

Palm oil is particularly rich in tocotrienols, unlike other vegetable oils (soya, sunflower) which contain only tocopherols. The α-tocotrienol is known for its protective effect of red blood cells against haemolysis (Sambanthamurthi et al., 2000). Qureshi et al. (1991), on evaluating the cholesterol-lowering effect in humans of tocotrienol-rich palm oil fractions, noted a significant reduction in total cholesterol and LDL in 20 hypercholesteremic subjects.

- Carotenoids, highly unsaturated tetraterpenes biosynthesised from 8 isoprene units (Sambanthamurthi et al., 2000), are responsible for the pronounced red colour of palm oil. They are subdivided into two classes: carotenes (700–800 ppm), which make up the majority of carotenoids (90%) and xanthophylls. Since plants are able to synthesise carotenoids *de novo*, the carotenoid composition of foods and plant origin is variable (Delia *et al.*, 2008), so it is not strange for it to occur in various vegetable oils, for example, yellow maize oil, groundnut oil and soybean oil. However, the concentrations of carotenoids in these vegetable oils are very low, less than 100 ppm (Ong and Tee, 1992). Carotenes are present in the alpha, beta and gamma forms, the first two forms being the most abundant (Goh et al., 1985). They are removed during the refining and deodorisation of the crude oil to produce a light yellow palm oil preferred by consumers (Cottrell, 1991).

β-carotene is the precursor of vitamin A, which is essential for cell growth, differentiation and vision. Because of its rich β-carotene, red palm oil is recommended for the control of vitamin A deficiency (Seshadri, 1996; Gopalan et al., 1992), as it is available year-round, is cheaper and is an accessible resource for most developing countries. In addition, vitamin E, essential for the assimilation of vitamin A '*in vivo*', is present in palm oil (Ames, 1969). Murakoshi et al. (1992) showed the ability of α-carotene and carotenes isolated from palm oil to inhibit skin and liver cancers in mice. These authors concluded that carotenes naturally present in red palm oil possess chemopreventive activities against cancer. The same effects cannot be attributed to synthetic β-carotene.

Carotenoids and vitamin E found in red palm oil are particularly important from a nutritional point of view because of their antioxidant and anti-carcinogenic properties.

- Partial glycerides, generally present in very small quantities, except for oils produced from damaged fruits. Their content will also be high in oils extracted from fruits that were in a state of advanced maturity (Siew and Ng, 1997) or that were not treated immediately after harvest.
- Phytosterols, tetracyclic compounds with 27, 28 and 29 carbon atoms, are present in very small amounts (0.03% of the constituents of palm oil) and have anti-inflammatory properties.

6.2 Artisanal red palm oil uses for human consumption

Home consumption

Palm oil is the most widely consumed edible vegetable oil in the world. In African countries such as Côte d'Ivoire and Cameroon, red palm oil is highly valued and used in the preparation of many traditional dishes (Cheyns et al., 2000; Cheyns and Rafflegeau, 2005). The domestic consumption of red palm oil relies on the local demand for oil, the quality of which is intimately linked to the terroir, the type of planting materials and local processes (Cheyns et al., 2004). Artisanal palm oil is highly appreciated in the local market because of its organoleptic properties (smell, taste and colour) which make this oil an irreplaceable ingredient in several local culinary recipes. With its different characteristics from CPO, artisanal red oil meets local demand for use in cooking specific dishes of the region.

African artisanal palm oil shows the highest values with the FFA level reaching 9% (Ohimain, et al., 2012) and even exceeding 15% (Osei-Amponsah et al., 2012). According to the *Codex Alimentarius* standards, any oil containing more than 5% FFA content is considered as improper for human consumption.

The problem of acidification in artisanal oil is more acute for smallholders and producers who are used to storing FFB for long periods between harvesting and cooking. Indeed, the problem extends from the whole industry down to the consumer in those African countries where artisanal red palm oil is a major commodity oil. There is no guarantee of quality for the artisanal red palm oil found on local markets nor when it is sold to neighbouring countries. Assessing the artisanal red palm oil's quality is also a problem for housewives when they buy oil from the local market. For that reason, market retailers of artisanal palm oil have usually tried to develop loyalty relationship with their customers.

Uses in the food industry

Due to its richness in SFA, palm oil remains the best ingredient used to replace partially hydrogenated fat which may contain *trans*-fatty acids. Intake of high amounts of *trans*-fatty acids is positively correlated with increased risks of coronary heart disease (Hu et al., 1997), inflammation (Mozaffarian et al., 2004) and cancer (Astorg, 2005). The natural solid-fat fraction of palm oil, palm stearin, became a suitable main ingredient in margarines, shortenings and vanaspati.

Low level of polyunsaturated fatty acids and high proportion of vitamin E make palm oil and its products resistant to oxidation (Sambanthamurthi et al., 2000). This makes palm oil a good alternative as frying oil. In addition, its solid and semi-solid state at room temperature due to the presence of SFA is a property which is exploited in food products such as margarines. The combined effect of carotenoids, SFA (about 50% of total fatty acids), tocopherols and tocotrienols provides outstanding oxidation stability over time compared to other vegetable oils (Arora et al., 2006). The choice of the frying oil should be based on economic, nutritional and physico-chemical quality factors, so palm oil is obviously an interesting vegetable oil for food industries.

7 Sustainable development issues for artisanal red palm oil production

Outside the mill supply areas, the artisanal processing of FFB through small-scale extraction units can be considered as economically viable and socially sustainable, while offering job opportunities in rural areas. However, using tyres as fuel for cooking the fruit and clarify the oil has to be banished. It is of general interest to reduce the number of extraction units in order to promote fewer but more performing units – such as the mini-mills described in Table 1 – owned and managed by local businessmen creating managerial enterprises. We recommend to support smallholders and producers for creating cooperatives or better kind of inclusive business by Alliances, for example, the South American 'Alianzas', to really build a win–win partnership between the managerial oil producer and the smallholders who will become shareholders of the Alliance (Rafflegeau and Feintrenie, 2013). In such partnerships, shareholders are taking part to decide the rules, such as fixing the price to pay smallholders.

There are three reasons for supporting such a development of mini-mills outside the supply area of industrial mills, rather than promoting the installation of micro-mills:

- with a higher oil extraction rate, oil palm production can increase without changing anything else;
- as mini-mills are processing FFB rather than stemmed fruit, the delay between harvest and process can be drastically reduced if compared with other technologies, thus having a direct impact on the quality of the artisanal red palm oil; new development with less extraction units of a higher capacity will allow to implement an efficient treatment of mill effluents, thus reducing the environmental impact on water quality in surrounding water courses and rivers; mini-mills are also energetically self-sufficient as they use heaters that are not burning firewood, reducing again the environmental pressure around the mills.

Inside the supply areas of industrial mills, if the mill is saturated during the high production period, the situation becomes exactly the same as outside of mill supply area. If the mill is not saturated during the high production period, it is then a pity to process FFBs through artisanal extraction units, then losing one-third of the oil and polluting the rivers. As we described the reasons for smallholders to process by themselves, industrial mills have to redesign their relationship with smallholders. Win–win partnerships between smallholders and agro-industries can lead to improvements in smallholders' yields, and to increases in the part of FFB originating from smallholdings supplied to industrial mills rather than to small-scale mills (Nkongho et al., 2015; Rafflegeau et al., 2010). Smallholders are not only interested in defining the price of FFB, but also in technical advice; extension services; and supply of certified selected planting materials and of fertiliser, micro-credit and so on.

The present situation of water pollution of the rivers in tropical Africa by a multitude of artisanal extraction units should lead researchers, governments, civil society and actors of development to urgently find adapted solutions for existing extraction units. Secondly, smallholders and producers must be informed and trained for the preservation of the quality of the waters surrounding their artisanal extraction units. This is a major issue hampering the sustainable development of artisanal red palm oil production.

African consumers are expecting more uniformity in quality among artisanal red palm oils which they buy in the local markets. They also need more information about the health properties of red palm oil, which is naturally rich in E vitamins and β-carotene. There is a need for halting the bizarre trend of African people consuming less red palm oil and more refined palm oil. The worst African trend consists in 'bleaching' artisanal red palm oil by over-heating it at home, in order to change the colour of the oil before its incorporation in non-traditional meals. In other parts of the world, the same health reasons led to the promotion of red palm oil consumption rather than refined oil which is poorer in vitamins.

8 Where to look for further information

Biochemical information about palm oil composition and information concerning palm oil consumption's impact on health can easily be found in many scientific papers from specialised journals on chemistry and nutrition. Information about extraction units, artisanal palm oil production and economic analyses is mostly available as grey literature. As the technological development of extraction equipment has already been undertaken, technical information describing artisanal extraction units is available in expert assessment reports which are unfortunately not always available online.

9 References

Ames, S. R. 1969. Factors affecting absorption, transport, and storage of vitamin A. *Am. J. Clin. Nutr.*, 22, 934.

Astorg, P. 2005. Dietary fatty acids and colorectal and prostate cancer: epidemiological studies. *Bull. Cancer*, 92, 670–84.

Babatunde, O. O., Ige, M. T. and Makanjuola, G. A. 2003. Effect of sterilization on fruit recovery in oil palm fruit processing. *J. Agr. Eng. Res.*, 41(2), 75–9.

Che Man, Y. B., Hayati, T., Ghazali, H. M. and Asbi, B. A. 1999. Composition and thermal profile of crude palm oil and its products. *J. Am. Oil Chem. Soc.*, 76, 237–42.

Cheyns, E., Bricas, N. and Aka, A. 2004. Attentes de qualité et structuration des filières alimentaires: la segmentation du marché urbain des huiles de palme rouges en Côte d'Ivoire. *Cah. Agr.*, 13(1), 135–41.

Cheyns, E. and Rafflegeau, S. 2005. Family agriculture and the sustainable development issue: possible approaches from the African oil palm sector. The example of Ivory Coast and Cameroon. *Ol. Corps Gras, Lipides*, 12(2), 111–20.

Cheyns, E., Akindes, F. and Aka Adie, F. 2000. La filière palmier à huile en Côte d'Ivoire 3 ans après la privatisation: état des lieux d'un procès de recomposition institutionnelle. *Ol. Corps Gras, Lipides*, 7(2), 166–71.

Choo YM. 1994. Palm oil carotenoids. *Food Nutr. Bull.*, 15(2) 130–7

Cochard, B., Adon, B., Kouamé Kouamé, R., Durand-Gasselin, T. and Amblard, P. 2001. Intérêts des semences commerciales améliorées de palmier à huile (*Elaeis guineensis* Jacq.). *OCL. Ol. Corps Gras, Lipides*, 8(6), 654–8.

Codex Alimentarius Commission/FAO/WHO food standards: Standard for named vegetable oils CODEX-STAN 210, FAO/WHO, 2005.

Corley, R. H. V. and Tinker, P. B. 2003. *The Oil Palm*, 4th edition. Blackwell Science, Oxford. doi:10.1002/9780470750971.

Cottrell, R. C. 1991. Nutritionnal aspects of palm oil. *Am. J. Clin. Nutr.*, 53, 989S–1009S.

Delia, B., Rodriguez-Amaya, M. K., Godoy, H. T. and Amaya-Farfan, J. 2008. Update Brazilian database on food carotenoids: factors affecting carotenoid composition. *J. Food Comp. Anal.*, 21, 445–63.

Edem, D. O. 2002. Palm oil: biochemical, physiological, nutritional, hematological and toxicological aspects: A review. *Plant Foods Hum. Nutr.*, 57, 319–41.

Firestone, D. (Ed.). 2006. *Physical and Chemical Characteristics of Oils, Fats and Waxes*, 2nd edition. AOCS Press: Champaign, IL, pp. 110–15.

Fournier, S., Ay, P., Jannot, C., Okounlola-Biaou, A. and Pédé, E. 2001. *La transformation artisanale de l'huile de palme au Bénin et au Nigéria*. La Librairie du Cirad: Montpellier, France, p. 139.

Goh, E. M. and Timms, R. E. 1985. Determination of mono- and diglycerides in palm oil, olein and stearin. *J. Am. Oil Chem. Soc.*, 62, 730–4.

Hartley, C. W. S. (1988). *The Oil Palm (Elaeis guinensis* Jacq.*)*. Longman Scientific and Technical Ed.: Essex. (Tropical Agriculture Series).

Hu, F. B., Stampfer, M. J., Manson, J. E., Rimm, E., Colditz, G. A., Rosner, B. A., Hennekens, C. H. and Willett, W. C., 1997. Dietary fat intake and the risk of coronary heart diseases in women. *N. Engl. J. Med.*, 337, 1491–9.

Jacquemard, J. C. 1995. *Le palmier à huile*. Maisonneuve et Larose: Paris, p. 207. (Le Technicien d'agriculture tropicale, 33).

Jannot, C. 2000. Mémo sur le petit matériel de transformation des régimes de palme. CERNA, CIRAD, Freie Universität Berlin, SRPH.

Jannot, C. 2013. Les filières oléagineuses en Afrique de l'ouest et du centre: le grand retour du palmier à huile. In*Déméter 2014: économie et stratégies agricoles*. Club Déméter: Paris, pp. 247–312, http://dx.doi.org/10.4000/vertigo.17757.

Karleskind, A. 1992. Manuel des Corps Gras, LAVOISIER, 1 et 2. 65–78, 115–241, 1072–1089, 1433–1459.

Karleskind, A. and Wolff, J. P. (Eds). 1996. *Oils and Fats Manual: A Comprehensive Treatise-Properties, Production, Application*, Volume 1. Intercept Limited: Andover, UK.

Lacan, X., Rafflegeau, S. and Feintrenie, L. 2015. Agroindustria del aceite de palma y el sector tradicional en África. *Palmas*, 10 (n.spéc. Memorias XVIII Conferencia Internacional sobre Palma de Aceite): p. 171–82. Conferencia Internacional sobre Palma de Aceite. 18, 2015-09-22/2015-09-25, Bogota (Colombie).

Marzin, J., Daviron, B. and Rafflegeau, S. 2015. Family farming and other forms of agriculture. In Sourisseau J.-M. (Ed.), *Family Farming and the Worlds to Come*. Springer [Pays-Bas]: Dordrecht, pp. 71–89. doi:(10.1007/978-94-017-9358-2_5).

Mozzaffarian, D., Pischon, T., Hankinson, S. E., Rifai, N., Joshipura, K., Willett, W. C. and Rimm, E. B. 2004. Dietary intake of trans fatty acids and systemic inflammation in women. *Am. J. Clin. Nutr.*, 79, 606–12.

Munne-Bosch, S. and Alegre L. 2002. The function of tocopherols and tocotrienols in plants. *Crit. Rev. Plant Sci.*, 21, 31–57.

Murakoshi, M., Nishino, H. Satomi, Y. et al. 1992. Potent preventive action of alpha-carotene against carcinogenesis: spontaneous liver carcinogenesis and promoting stage of lung and skin carcinogenesis in mice are suppressed more effectively by alpha-carotene than by beta-carotene. *Cancer Res.*, 52, 6583–7.

Nchanji, Y. K., Tataw, O., Nkongho, R. N. and Levang, P. 2013. *Artisanal Milling of Palm Oil in Cameroon*. IRD, CIFOR.

Ndjogui, T. E., Nkongho, R. N., Rafflegeau, S., Feintrenie, L. and Levang P. 2014. Historique du secteur palmier à huile au Cameroun. Jakarta: CIFOR, p. 56. (Document occasionnel du CIFOR, 109). http://www.cifor.org/library/4789/historique-du-secteur-palmier-a-huile-au- cameroun/

Nestel, P., Noakes, M., Belling, B., Mcarthur, R., Clifton, P., James, E. and Abbey, M. 1992. Plasma lipoprotein lipid and Lp[a] changes with substitution of elaidic acid for oleic acid in the diet. *J. Lipid Res.*, 33, 1029–36.

Ngando, E. G. F., Mpondo, M. E. A., Dikotto, E. E. L. and Koona, P. 2011. Assessment of the quality of crude palm oil from smallholders in Cameroon. *J. Stored Prod. Postharvest Res.*, 2(3), 52–8.

O'Brien, R. O. 1998. *Raw Materials, in Fats and Oils: Formulating and Processing for Applications*. Technomic Publishing Company: Lancaster, PA, pp. 1–46.

Ohimain, E. I., Daokoru-Okukole, C., Isah, S. C. and Alaka, E. E. 2012. Assessment of the quality of crude palm oil produced by smallholder processors in rivers state, Nigeria. *Niger. J. Agri. Food Environ.* 8, 28–34.

Ong, A. S. H. and Tee, E. S. 1992. Natural sources of carotenoids from plants and oils. In John Abelson and Melvin Simon (eds), *Methods in Enzymology*. Academic Press:, San Diego CA, pp. 142–67.

Osei-Amponsah, C., Visser, L. E., Adjei-Nsiah, S. and et al. 2012. Processing practices of small-scale palm oil producers in the Kwaebibirem district, Ghana: A diagnostic study. *NJAS-Wagen J. Life. Sci.* 60, –3 49–5610.1016/j.njas.2012.06.006.

Owolarafe, O. K., Taiwo, E. A. and Oko, O. O. 2008. Effect of processing conditions on yield and quality of hydraulically expressed palm oil. *Int. AgroPhy.*, 22(4), 349–52.

Pledran, O., Rafflegeau, S. and Levang, P. 2016. L'adaptation du contexte institutionnel: condition *sine qua non* du développement durable des palmeraies camerounaises. *VertigO*, 16(2), 24.

Poku, K. 2002. Small-scale palm oil processing in Africa. FAO Agricultural rural services bulletin 148. Rome. http://www.fao.org/docrep/005/y4355e/y4355e00.htm.

Qureshi, A. A., Qureshi, N., Hasler-Rapacz, J. O., Weber, F. E., Chaudhary, V., Crenshaw, T. D., Gapor, A., Ong, A. S. H., Chong, Y. H., Peterson, D. and et al. 1991. Dietary tocotrienols reduce concentrations of plasma cholesterol, apolipoprotein B, thromboxane B_2, and platelet factor 4 in pigs with inherited hyperlipidemias *Am. J. Clin. Nutr.*, 53, 1042s–6s.

Rafflegeau, S. 2008. Dynamique d'implantation et conduite technique des plantations villageoises de palmier à huile au Cameroun: facteurs limitants et raisonss des pratiques. AgroParisTech: Paris, p. 148. Thèse Etat-Agronomie.

Rafflegeau, S., Dubos, B., Ollivier, J. and Michel-Dounias, I. 2010. Building fertilization support for oil palm smallholdings requires partnerships. A Cameroon case study. In Wery J., Shili-Touzi, I., Perrin, A. (Eds), *Proceedings of Agro 2010: the XIth ESA Congress, August 29th - September 3rd, 2010, Montpellier, France*. Montpellier: Agropolis international, pp. 1015–16. ESA Congress. 11, 2010-08-29/2010-09-03, Montpellier (France).

Rafflegeau, S. and Feintrenie, L. 2013. Desarrollo de la palma de aceite: riesgos y oportunidades con base en las lecciones aprendidas de Camerún e Indonesia. *Palmas*, 34(2)(spec), 351–70. Conferencia Internacional sobre Palma de Aceite. 17, 2012-09-25/2012-09-28, Carthagène (Colombie). http://publicaciones.fedepalma.org/index.php/palmas/article/view/10734/10720.

Rafflegeau S., Losch B., Daviron B., Bastide P., Charmetant P., Lescot T., Prades A. and Sainte-Beuve J. 2015. Contributing to production and to international markets. In Sourisseau J.-M. (Ed.), *Family farming and the Worlds to come*. Springer [Pays-Bas]: Dordrecht, pp. 129–44. doi:(10.1007/978-94-017-9358-2_8).

Reske J., Siebrecht J. and Hazebroek J. 1997. Triacylglycerol composition and structure in genetically modified sunflower and soybeans oils. *J. Am. Oil Chem. Soc.*, 74, 989–98.

Rouzière, A. 1995. Etude mini et micro-huileries de palme en Afrique de l'Ouest et du centre (Cameroun, Nigéria, Ghana, mars 1995). Volume 1: rapport principal. Volume 2: annexes. CIRAD-CP: Montpellier, p. 253.

Sambanthamurthi, R., Sundram, K. and Tan, Y.-A. 2000. Chemistry and biochemistry of palm oil. *Prog. Lipid Res.*, 39, 507–58.

Seshadri, S. 1996. In: Use of carotene rich foods to combat vitamin A deficiency in India. Special Report No. 12. New Delhi. Nutrition Foundation of India.

Siew, W. L. and Ng, W. L. 1997. In: Proceedings of the National Seminar on Palm Oil Milling, Refining Technology and Quality, 7–8 August 1997, Kota Kinabalu, Sabah, Malaysia, 124–40.

Tan, B. K. and Oh, FC. H. 1981. PORIM Technol. Palm Oil Research Institute of Malaysia, No.3.

Williams, M. A. and Hron Sr., R. J. 1996. Obtaining oils and fats from source materials. In Y. H. Hui (ed.), *Bailey's Industrial Oil and Fat Products*, 5th edn., Vol. 4, John Wiley & Sons: New-York, pp. 61–155.

Index

Lightning Source UK Ltd.
Milton Keynes UK
UKHW02n0105200318
319686UK00001B/5/P

9 781786 761088